Carotenoids and Retinoids

Molecular Aspects and Health Issues

Editors

Lester Packer
Department of Molecular
Pharmacology and Toxicology
University of Southern California
Los Angeles, California

Klaus Kraemer
BASF Aktiengesellschaft
Ludwigshafen, Germany

Ute Obermüller-Jevic
BASF Aktiengesellschaft
Ludwigshafen, Germany

Helmut Sies
Institute of Biochemistry and Molecular Biology
Heinrich-Heine-University
Düsseldorf, Germany

AOCS PRESS

Champaign, Illinois

AOCS Mission Statement

To be a global forum to promote the exchange of ideas, information, and experience, to enhance personal excellence, and to provide high standards of quality among those with a professional interest in the science and technology of fats, oils, surfactants, and related materials.

AOCS Press, Urbana, IL 61802

The paper used in this book is acid-free and falls within the guidelines established to ensure permanence and durability.

ISBN-10: 1-893997-83-9
ISBN-13: 978-1-893997-83-7

Library of Congress Cataloging-in-Publication Data

Carotenoids and retinoids : molecular aspects and health issues / editors, Lester Packer ... [et al.].
p. cm.
Includes bibliographical references and index.
ISBN 1-893997-83-9
1. Carotenoids--Health aspects. 2. Carotenoids--Molecular aspects. 3. Retinoids--Health aspects. 4. Retinoids--Molecular aspects. I. Packer, Lester.

QP671.C3C373 2004
612'.01528--dc22

2004024198

CIP

Printed in the United States of America.
11 10 09 08 07 6 5 4 3 2

Preface

Carotenoids synthesized in plants are essential for the assembly, function, and stability of photosynthetic pigment–protein complexes. A light-harvesting function of carotenoids allows blue and green sunlight to be used for energy conservation, a process that involves energy transfer from carotenoid excited states to nearby chlorophylls at the active center of oxygenic photosynthetic pigment complexes. Carotenoids also protect against oxidative and photooxidative damage by quenching free radicals that are produced during photosynthesis; this function gains further significance when considering that free radicals or reactive oxygen species are byproducts of metabolism in humans.

Indeed, the presence of carotenoids in the diet and their role in human health has become a subject of unprecedented interest. Some carotenoids are called provitamin A compounds because they are precursors of retinol and retinoic acid. The type of carotenoids found in human plasma depends on the extent to which people consume diets rich in green, yellow/red, or yellow/orange vegetables. Fifty to sixty different carotenoid compounds are typically present in the human diet, including the most abundant forms in plasma: β-carotene, lycopene, lutein, cryptoxanthin, α-carotene, and zeaxanthin. Carotenoids are potent antioxidants known to affect different cellular pathways. For example, lutein and zeaxanthin accumulate in the fovea (macular region) of the human eye and are thought to prevent blue light damage to the eye. A low dietary supply of these carotenoids (xanthophylls) is thought to be associated with age-related macular degeneration, one of the most common causes of irreversible blindness in the Western world. Numerous epidemiological, interventional, and prospective human studies, as well as an incredible array of fundamental research, are currently underway to elucidate the role of carotenoids, vitamin A (retinol), retinoids, and their stereoisomers and metabolites in biological processes and health and disease prevention.

A roundtable discussion on the Safety of β-Carotene and a workshop on Carotenoids and Retinoids: Molecular Aspects and Health Issues were held at the annual Oxygen Club of California (OCC) conference in Santa Barbara, California, on March 10–13, 2004. These events were co-organized by the editors of this volume and sponsored by the Scientific Affairs, Strategic Marketing Human Nutrition unit of BASF, Ludwigshafen, Germany. The chapters in this book represent an account of the information presented at the workshop together with several additional invited contributions to cover topics more completely that are currently at the cutting edge of research. The editors have sought the timely publication of this book in cooperation with AOCS Press.

Some of the highlights of this book on *Carotenoids and Retinoids: Molecular Aspects and Health Issues* are summarized below.

The book commences with comprehensive overview chapters on vitamin A, retinoids, and carotenoids, including different aspects of their uptake, molecular structure, transport, storage, metabolism, transcriptional activity, and roles in human health. There is also a thorough review of the special role that vitamin A intake plays in the health status of developing countries.

Another chapter addresses the essential role of vitamin A in cell signaling and covers historical aspects followed by receptor action, molecular aspects of action with kinases, redox regulation, and the significance of vitamin A in oxygen biology. A chapter on the role of carotenoids and retinoids in cellular/tissue gap junctional communication focuses on structure/activity relationships and interactions with other micronutrients. The molecular structure of carotenoid metabolites and their intracellular distribution are thoroughly reviewed, with an emphasis on human data. Owing to the unique molecular structure of the hydrocarbon chain in carotenoids, resonance Raman spectroscopy is a useful tool for investigating carotenoids in photosynthetic processes, noninvasively measuring carotenoid content in the macular region of the human eye and skin. With this technology, the age-related loss of carotenoids in the human macula has been demonstrated. Recent studies have also provided evidence of a close correlation between serum (high-performance liquid chromatography) and stratum corneum (laser Raman scattering intensity) levels of carotenoids in the skin. This discovery broadens the application of this technology to human studies.

The oxidation of carotenoids and their cleavage reactions result in the formation of metabolites whose biological function requires elucidation. Two chapters describe studies on the actions of these metabolites in cell and mitochondrial systems and the formation of oxidative metabolites in inflamed lung tissue.

Epidemiological studies in two human trials revealed that the presence of β-carotene increased the incidence of lung cancer in individuals exposed to cigarette smoke and asbestos, thus stimulating interest in the scientific community to elucidate the relation to cancer. The molecular targets involved in carotenoid action in smoke-induced lung pathology are described. Another chapter deals with the up-regulation of gap junctional proteins by carotenoids and retinoids and, hence, their cancer-preventive actions. One of the best-studied roles of carotenoids in cancer prevention is the inverse association between consumption of tomato-based foods and lycopene and the incidence of prostate cancer.

A role of phytochemicals in cancer prevention is their induction of Phase II enzymes, which are important in detoxification reactions and antioxidant defense. Recent studies demonstrate that carotenoids and their oxidation products regulate transcription factors that control the induction of key Phase II enzymes in cell culture models. These findings may also serve to explain the antiproliferative effects of carotenoids.

Carotenoids also have been reported to be beneficial in cardiovascular health from studies involving the consumption of fruits and vegetables rich in carotenoids. One of the chapters reports on the evidence of a reduced risk associated with dietary lycopene. In another study, the relationship between consumption of carotenoid-rich fruits and vegetables, their uptake, and the oxidizability of serum carotenoids is reported. Relevant to this topic is the metabolic mechanism of carotenoid oxidizability: The discovery and elucidation of carotenoid oxygenases and dioxygenases has led to remarkable new insights into the role of these cleavage enzymes in vitamin A and retinoid metabolism, as well as their action in developmental processes.

In view of the enormous interest in the health aspects of carotenoids, it is important to summarize and critically evaluate the human epidemiological evidence for the association of carotenoids and human health. A synopsis of the most important and ongoing investigations is given in one of the final chapters. Also, the final chapter presents a brief synopsis of the Round Table Discussion on Safety of β-Carotene attended by many of the leading investigators in the field. Topics discussed were mainly related to the human safety of β-carotene, especially with regard to smokers and individuals exposed to hazardous environments. The author of this insightful account presents the reader with future perspectives and research directions. A feature on Future Horizons of Research on Carotenoids and Retinoids is presented by one of the volume co-editors.

Lester Packer
Ute Obermüller-Jevic
Klaus Kraemer
Helmut Sies

September 10, 2004

Contents

Preface iii

Chapter 1 **Introduction to Retinoids**
Sheila M. O'Byrne and William S. Blaner 1

Chapter 2 **Introduction to Vitamin A: A Nutritional and Life Cycle Perspective**
A. Catharine Ross 23

Chapter 3 **The Essential Role of Vitamin A in Signal Transduction**
Beatrice Hoyos and Ulrich Hammerling 42

Chapter 4 **Chemical and Metabolic Oxidation of Carotenoids**
Frederick Khachik 61

Chapter 5 **Gap Junctional Intercellular Communication: Carotenoids and Retinoids**
Wilhelm Stahl and Helmut Sies 76

Chapter 6 **Raman Detection of Carotenoids in Human Tissue**
Werner Gellermann, Jeff A. Zidichouski, Carsten R. Smidt, and Paul S. Bernstein 86

Chapter 7 **Macular Carotenoids in Eye Health**
Richard A. Bone and John T. Landrum 115

Chapter 8 **β-Carotene Cleavage Products Impair Cellular and Mitochondrial Functions and May Lead to Genotoxicity**
Werner Siems, Ingrid Wiswedel, Avdulla Alija, Nikolaus Bresgen, Peter Eckl, Claus-Dieter Langhans, and Olaf Sommerburg 130

Chapter 9 **Formation of β-Carotene Cleavage Products in View of the Particular Conditions in Inflamed Lung Tissue**
Olaf Sommerburg, Claus-Dieter Langhans, Costantino Salerno, Carlo Crifo, and Werner Siems 149

Chapter 10 **Biological Activity of Lycopene Against Smoke-Induced Lung Lesions: Targeting the IGF-1/IGFBP-3 Signal Transduction Pathway**
Xiang-Dong Wang 168

Chapter 11 **Retinoids and Carotenoids as Cancer Chemopreventive Agents: Role of Upregulated Gap Junctional Communication**
Laura M. Hix, Alex L. Vine, Samuel F. Lockwood, and John S. Bertram 182

Chapter 12 **Lycopene and Risk of Cardiovascular Disease**
Lauren Petr and John W. Erdman, Jr. 204

Chapter 13 **Effect of Feeding and Then Depleting a High Fruit and Vegetable Diet on Oxidizability in Human Serum**
Kyung-Jin Yeum, Giancarlo Aldini, Elizabeth J. Johnson, Robert M. Russell, and Norman I. Krinsky 218

Chapter 14 **Mitochondria as Novel Targets for Proapoptotic Synthetic Retinoids**
Numsen Hail, Jr., and Reuben Lotan 229

Chapter 15 **Molecular Analysis of the Vitamin A Biosynthetic Pathway**
Johannes von Lintig 244

Chapter 16 **Regulation of Transcription by Antioxidant Carotenoids**
Yoav Sharoni, Riad Agbaria, Hadar Amir, Anat Ben-Dor, Noga Dubi, Yudit Giat, Keren Hirsh, Gaby Izumchenko, Marina Khanin, Elena Kirilov, Amit Nahum, Michael Steiner, Yossi Walfisch, Shlomo Walfisch, Michael Danilenko, and Joseph Levy 261

Chapter 17 **Vitamin A in Health and Disease in Developing Countries**
Machteld van Lieshout and Clive E. West 275

Chapter 18 **Lycopene and Prostate Cancer**
Ute C. Obermüller-Jevic and Lester Packer 295

Chapter 19 **Blood Response to β-Carotene Supplementation in Humans: An Evaluation Across Published Studies**
Klaus Kraemer, Gerhard Krennrich, Ute Obermüller-Jevic, and Peter P. Hoppe 303

Chapter 20 **New Horizons in Carotenoid Research**
Helmut Sies and Wilhelm Stahl 315

Chapter 21 **Carotenoids and Cardiovascular Disease**
J. Michael Gaziano and Howard D. Sesso 321

Chapter 22 **Safety of β-Carotene**
Norman I. Krinsky 338

Index 345

Chapter 1

Introduction to Retinoids

Sheila M. O'Byrne[a] and William S. Blaner[a,b]

[a]Institute of Human Nutrition and [b]Department of Medicine, Columbia University, New York, NY 10032

Introduction

What are retinoids? The term retinoid refers to both naturally occurring and synthetic compounds that bear a structural resemblance to vitamin A (all-*trans*-retinol) with or without the biological activity of vitamin A (1). Figure 1.1 shows the chemical structures of some naturally occurring retinoids. The term vitamin A is often used as a general term for all compounds that exhibit the biological activity of retinol. There are many natural and synthetic retinoids. All of the synthetic retinoids were developed as potential pharmacologic agents for use in treating aliments ranging from cancer to acne. For the remainder of this chapter, we will focus only on the metabolism, storage, and transport of natural retinoid forms (vitamin A and its metabolites) that are found in the diet and in the body.

Many of the chapters in this book focus on carotenoids. Some carotenoids may be converted by higher animals to retinoids. These carotenoids are collectively known as provitamin A carotenoids. The best-studied of the provitamin A carotenoids is β-carotene. Within the body, β-carotene can (but does not necessarily have to) undergo cleavage to retinal, which is then reduced to give rise to two molecules of retinol. We will briefly describe the carotene cleavage process as it occurs in the intestine, but we will not deal directly with the metabolism or transport of other members of the large carotenoid family in this chapter.

To understand retinoid metabolism, storage, and transport, we believe that it is necessary to understand the proteins that bind these hydrophobic molecules within the aqueous environment of cells and the extracellular fluids, the enzymes that act upon them to render them biologically active or inactive, and the proteins that are required to facilitate their actions within the body. Many of these proteins will be discussed throughout the course of this chapter to paint a picture of the complexity of the metabolic trafficking that retinoids undergo. The different retinoid forms present in the body relate to and are a result of the actions of these proteins. Most of the enzymatic reactions central to retinoid metabolism, with the notable exception of retinal oxidation to retinoic acid, are reversible. This allows the retinoid metabolism to be finely regulated in response to the body's needs. A listing of binding proteins that shuttle and sequester retinoids, of enzymes that maintain retinoids as inactive forms, and enzymes that activate retinoids, and of nuclear pro-

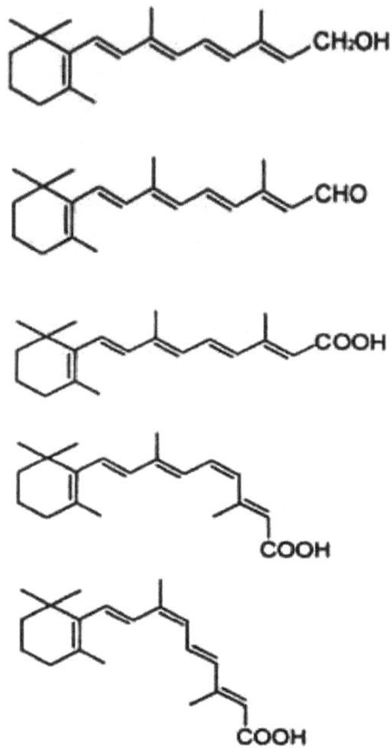

A. All-*trans*-retinol

B. All-*trans*-retinal

C. All-*trans*-retinoic acid

D. 13-*cis*-retinoic-acid

E. 9-*cis*-retinoic-acid

Fig. 1.1. Chemical structures of different natural retinoids. Vitamin A is defined as all-*trans*-retinol (A). All-*trans*-retinol can be oxidized to all-*trans*-retinal (B) which can in turn be further oxidized to all-*trans*-retinoic acid (C). All-*trans*-retinoic acid can undergo isomerization to either 13-*cis*-retinoic acid (D) or 9-*cis*-retinoic acid (E). As mentioned in the text, some 9-*cis*-retinoic acid may be formed through oxidation of 9-*cis*-retinol and 9-*cis*-retinal. Of the retinoic acid isomers, all-*trans*-retinoic acid (C) and 9-*cis*-retinoic acid (E) represent very transcriptionally active retinoid forms.

teins that bring about the transcriptional activator properties of retinoids will be discussed in the text and summarized in tables. Many of the proteins will be discussed in the text and the reader is referred to these tables to facilitate ease of understanding.

Finally, this chapter will not be exhaustive. It is aimed primarily at giving the reader a brief but detailed overview of our current scientific knowledge of mammalian storage and metabolism of natural retinoids.

Dietary Sources of Retinoid

Retinoids are essential micronutrients (i.e., our bodies cannot synthesize them *de novo* and they are required in microamounts) and must therefore be obtained from the diet. There are two sources of retinoid in the diet; the first is as provitamin A carotenoids obtained from dark green and colorful vegetables and the second is as preformed vitamin A from animal products such as meat and dairy (where the retinoid has been preformed from fruits and vegetables consumed by animals).

Provitamin A carotenoids such as β-carotene can be cleaved to form two molecules of retinal, which are then reduced to form two molecules of retinol. The retinoid forms found in animal food products are mainly retinol and retinyl esters. Higher animals are capable of storing retinoid in the liver and to a lesser extent in other tissues, thus alleviating the obligate need for daily dietary intake because the stored retinoid can be mobilized in times of insufficient dietary retinoid intake (2,3).

Major Retinoid Forms in the Body

The different retinoid forms present within the body are generated primarily through modifications to the terminal polar end of the molecule (see Fig. 1.1). Retinol and retinyl esters are the most abundant retinoid forms found in the body. All-*trans*-retinol is by definition vitamin A. When a fatty acyl group is esterified to the hydroxyl terminus of all-*trans*-retinol, a storage form of retinol, the retinyl ester, is formed. The most abundant retinyl esters are those of palmitic, oleic, steric, and linoleic acids (2,3). Retinyl acetate is often used as a dietary supplement, but this short-chain retinyl ester does not occur naturally (2,3). In times of retinoid need or in the intestine upon retinyl ester intake from the diet, the ester bond undergoes hydrolysis to retinol. The formation of retinyl esters makes the retinoid less toxic and allows for its storage within intracellular lipid droplets.

Retinol itself has no known biological activity. However, it can be reversibly oxidized to retinal, which as the 11-*cis* isomer is essential for the visual cycle (4,5). Rhodopsin, the visual pigment responsible for photoperception, consists of 11-*cis*-retinal covalently bound to a lysine residue present in the protein opsin (4,5). As the primal event in vision, a photon of light strikes the rhodopsin molecule, resulting in the photoisomerization of 11-*cis*-retinal to all-*trans*-retinal (4,5). In all tissues aside from the eye, retinal has no other known function beyond serving as an intermediate in the synthesis of retinoic acid (6,7).

The all-*trans*- and 9-*cis*-isomers of retinoic acid are transcriptionally active retinoids and are thought to account for the gene regulatory properties of retinoids within cells and tissues (6,7). The concentration of retinoic acid within tissues is generally very low and is usually 100–1000 times less than that of retinol (2,3). All-*trans*-retinoic acid is formed through the irreversible oxidation of all-*trans*-retinal. This is one of the few irreversible steps in retinoid metabolism and must therefore be finely regulated. All-*trans*-retinoic acid can be isomerized through a nonenzymatic process to form the 9-*cis*- or 13-*cis*-isomers (3). It is possible that some 9-*cis*-retinoic acid may be formed from 9-cis-retinol through a two-step oxidation process similar to that described above for all-*trans*-retinoic acid (8,9). Retinoic acid formed outside of the nucleus can move to the nucleus where it binds and activates one of its nuclear hormone receptors (6,7). This binding leads to the transcription of retinoid-responsive target genes that give rise to the biological activities of vitamin A (6,7). This transcriptional activity is described in more detail in other chapters. Although 13-*cis*-retinoic acid is a naturally occurring retinoid that is present in blood and tissues (2,3), it possesses much less transcriptional activity than either the all-*trans*- or the 9-*cis*-isomer (6).

Various oxo- and hydroxy-forms of retinol and retinoic acid as well as glucuronides of retinol and retinoic acid are present in the body, albeit at very low concentrations relative to retinol and retinyl esters (2,3). Although some of these oxidized and conjugated retinoid forms may have biological/transcriptional activity, it appears likely that most of these forms are catabolic in nature and destined for elimination from the body. Because there are no known enzymes that can reduce retinoic acid to retinal, excessive or unneeded retinoic acid is not recycled and must be catabolized. As described below in more detail, this catabolism is thought to be catalyzed by one of several cytochromes (CYP) (2,3,10–14), giving rise to the more water-soluble retinoid forms that can be easily excreted.

Finally, retro- and anhydro-retinoids are also naturally occurring retinoid forms that can be synthesized by cells and tissues and that are present within the body (15). Enzymes able to catalyze the formation of retro- and anhydro-retinoids were identified (15). It was proposed that the retro- and anhydro-retinoids may have actions in regulating immune function, but the mechanisms responsible for these actions have not yet been elucidated (15,16).

Retinoid-Binding Proteins

To solubilize, protect, and detoxify retinoids in the intracellular and extracellular environment, retinol, retinal, and retinoic acid are usually found bound to specific retinoid-binding proteins. The known retinoid-binding proteins are summarized in Table 1.1. These can be classified using several different criteria. Some of these proteins, specifically retinol-binding protein (RBP), interphotoreceptor matrix retinoid-binding protein (IRBP), epididymal retinoic acid-binding protein (ERABP), and β-trace are found only in extracellular fluids, whereas the remainder are found only intracellularly. Of the intracellular binding proteins, some bind only retinoic acid [cellular retinoic acid-binding protein, type I (CRABP I) and cellular retinoic acid-binding protein, type II (CRABP II)]; some preferentially bind both retinol and retinal [cellular retinol-binding protein, type I (CRBP I) and cellular retinol-binding protein, type II (CRBP II)]; some preferentially bind retinol [cellular retinol-binding protein, type III (CRBP III) and cellular retinol-binding protein, type IV (CRBP IV)]; and one preferentially binds retinal [cellular retinal-binding protein (CRALBP)]. These proteins can also be grouped by the protein families to which they belong. RBP, ERABP, and β-trace are all members of the lipocalin protein family (17–19). CRBP I, II, III, and IV as well as CRABP I and CRABP II are members of the fatty acid-binding protein family of proteins (20–24). CRALBP is a member of the CRAL-TRIO protein family, which also contains a vitamin E–binding protein (25,26).

Each of the known retinoid-binding proteins was proposed to have a role in facilitating retinoid transport and/or metabolism (see Table 1.1). However, none of

TABLE 1.1
Retinoid-Binding Proteins

Protein	Ligand(s)	Proposed functions	Tissue site(s) of synthesis	Refs.
Extracellular				
Retinol-binding protein (RBP)	Retinol	Mobilization of liver stores and transport in blood	Liver, adipose, others	17, 21
Epididymal retinoic acid-binding protein (ERABP)	Retinoic acid	Epididymal lumen transport of retinoic acid	Epididymidis, others	18
β-Trace (Lipocalin-type prostaglandin D synthase)	Retinoic acid	Binds retinoic acid in the cerebral spinal fluid (CSF)	Brain (CSF)	19
Interphotoreceptor matrix retinoid-binding protein (IRBP)	Retinol and 11-*cis*-retinal	Specialized shuttling protein involved in the visual cycle	Eye (retina)	5
Intracellular				
Cellular retinal-binding protein (CRALBP)	Retinol and 11-*cis*-retinal	Specialized binding protein involved in the visual cycle	Eye (retina)	5, 21, 25, 26
Cellular retinol-binding protein, type I (CRBP I)	Retinol and retinal	Cellular retinoid uptake, transport and metabolism	Liver, testis, kidney, lung, most tissues	20–22, 25
Cellular retinol-binding protein, type II (CRBP II)	Retinol and retinal	Dietary retinol uptake and metabolism in the intestinal mucosa	Small intestine	20–22
Cellular retinol-binding protein, type III (CRBP III)	Retinol	Cellular retinol uptake, transport and metabolism	Heart, mammary, fat	23
Cellular retinol-binding protein, type IV (CRBP IV)	Retinol	Cellular retinol uptake, transport and metabolism	Liver, kidney, heart, others	24
Cellular retinoic acid-binding protein, type I (CRABP I)	Retinoic acid and metabolites	Transport and metabolism of retinoic acid within the cell	Most tissues	20, 21
Cellular retinoic acid-binding protein, type II (CRABP II)	Retinoic acid and metabolites	Retinoic acid metabolism shuttling of retinoic acid to the nucleus and involvement in transcription	Skin, uterus, ovaries	20, 21

these proteins can have an essential role in facilitating these processes. The genes for nearly all of these retinoid-binding proteins were ablated in mouse models, and none of the gene disruptions were lethal or even associated with severe phenotypes (27–33). It seems likely that these proteins are used to facilitate optimal retinoid retention, transport, and metabolism. When dietary retinoid availability is not impaired, the actions of the binding proteins are likely not essential or possibly even critical for maintaining retinoid status of the body or the health of the organism. However, in times of dietary retinoid deficiency, the binding proteins and the enhanced metabolic efficiency and retention that they afford convey an advantage to the organism. We note that throughout evolution, dietary retinoid deficiency was the norm rather than the exception it is today.

Transcriptional Activity of Retinoids

Retinoids are required for maintaining reproduction including spermatogenesis, conception, placenta formation, and embryogenesis and processes that are dependent on cell differentiation such as bone remodeling, epithelial and skin differentiation, and immune system function (34). These retinoid actions are mediated by all-*trans*-retinoic acid and 9-*cis*-retinoic acid through effects on retinoid-responsive gene expression. Retinoic acid can bind three retinoic acid receptors (RARα, RARβ, and RARγ) and three retinoid X receptors (RXRα, RXRβ, and RXRγ), which are then activated and can regulate gene expression in the nucleus of cells (6,7,35,36). All-*trans*-retinoic acid binds well and readily transactivates the three RAR and the three RXR, whereas 9-*cis*-retinoic acid binds and transactivates well only the RXR. Thus, all-*trans*-retinoic acid is usually thought of as the natural ligand for the RAR, and 9-*cis*-retinoic acid is considered to be the natural ligand for the RXR (6,7,35,36). Because the RXR can interact with nuclear vitamin D receptors (VDR), thyroid hormone receptors (TR), and peroxisomal proliferator activator receptors (PPAR), retinoic acid helps to regulate a broad spectrum of hormonally responsive genes (6,7,35–37). Well over 500 genes may be regulated by retinoic acid (37). Because these receptors will be discussed in detail in other chapters, we will not discuss them further in this chapter aside from summarizing them in Table 1.2.

Intestinal Absorption and Processing of Retinoids

As mentioned earlier, in most of the Western and developed world, much of the retinoid obtained in the diet arises from animal food sources and therefore consists of retinol and retinyl ester. Because retinol but not retinyl esters can enter the intestinal mucosa, dietary retinyl ester must first be acted upon by a hydrolase to yield free retinol. Retinyl esters can be hydrolyzed within the intestinal lumen by nonspecific pancreatic enzymes such as pancreatic triglyceride lipase and cholesteryl ester hydrolase or at the mucosal cell surface where a retinyl ester hydrolase is associated with the intestinal brush border (2,3). The free retinol formed upon hydrolysis of the retinyl ester or retinol arriving as such from the diet is taken up

TABLE 1.2
Nuclear Retinoic Acid Receptors

Nuclear receptor	Retinoic acid ligand(s)	Function(s)
Retinoic acid receptor α (RARα)	All-*trans* and 9-*cis*	Transcriptional mediator; heterodimerizes with RXR
Retinoic acid receptor β (RARβ)	All-*trans* and 9-*cis*	Transcriptional mediator; heterodimerizes with RXR
Retinoic acid receptor γ (RARγ)	All-*trans* and 9-*cis*	Transcriptional mediator; heterodimerizes with RXR
Retinoid X receptor α (RXRα)	9-*cis*	Transcriptional mediator; homodimerizes; heterodimerizes with RAR, TR, VDR, PPAR and others
Retinoid X receptor β (RXRβ)	9-*cis*	Transcriptional mediator; homodimerizes; heterodimerizes with RAR, TR, VDR, PPAR and others
Retinoid X receptor γ (RXRγ)	9-*cis*	Transcriptional mediator; homodimerizes; heterodimerizes with RAR, TR, VDR, PPAR and others

[a]Abbreviations: TR, thyroid hormone receptors; VDR, vitamin D receptors; PPAR, peroxisome proliferator activator receptors.

into the intestinal cells (2,3). In contrast to dietary preformed vitamin A, dietary provitamin A carotenoids are absorbed unmodified by the same intestinal cells. These carotenoids can be cleaved within the intestinal cells by carotene-15,15′-monooxygenase (also called carotene cleavage enzyme)[1] to retinal (38–42), which is then reduced to retinol by the enzyme retinal reductase (2,3). After carotene cleavage and retinal reduction, the retinol arriving from the diet as preformed vitamin A and as provitamin A carotenoids cannot be distinguished metabolically. CRBP II is present in the small intestine and binds both retinal and retinol (20–22). Retinal bound to CRBP II is the preferred substrate for reduction to retinol by the intestinal retinal reductase. Retinol bound to CRBP II is then reesterified with long-chain fatty acids through the action of the enzyme lecithin:retinol acyltransferase (LRAT), which utilizes preferentially retinol bound to CRBP II as a substrate for esterification (20,21,43). The resulting retinyl esters are then packaged along with the rest of the dietary lipids into nascent chylomicrons and secreted into the lymphatic system for uptake into the general circulation (2,3). Figure 1.2 gives a schematic representation of this process. A partial listing of enzymes and enzyme families that are thought to have roles in catalyzing intestinal metabolism and the metabolic processes discussed later in the chapter is provided in Table 1.3.

[1]The older literature refers to carotene-15,15′-monooxygenase as carotene-15,15′-dioxygenase, but recent mechanistic studies indicate that this enzyme acts through a monooxygenase mechanism rather than that of a dioxygenase (Leuenberger, M.G., Engeloch-Jarret, C., and Woggon, W.-D. (2001) The Reaction Mechanism of the Enzyme-Catalyzed Central Cleavage of β-Carotene to Retinal, *Angew. Chem. Int. Ed. 40*, 2613–2617). As indicated in the text, this enzyme is also referred to in the literature as carotene cleavage enzyme.

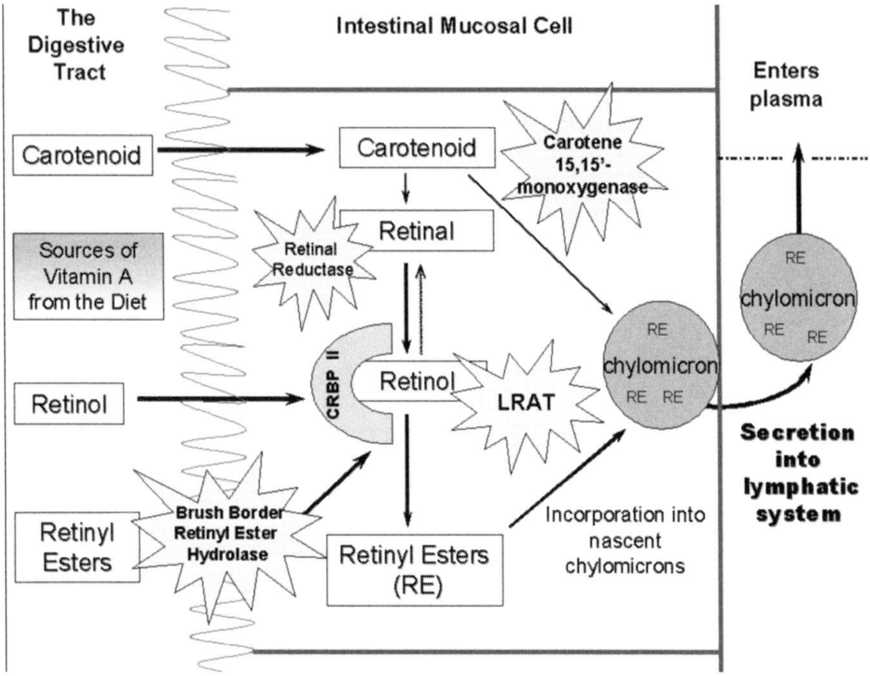

Fig. 1.2. Schematic representation of intestinal absorption and processing of dietary preformed vitamin A and provitamin A carotenoids. Retinoids are present in the diet as preformed vitamin A (primarily as retinol and retinyl esters) or as provitamin A carotenoids (e.g., β-carotene, α-carotene, or β-cryptoxanthin). The dietary carotenoids can enter the intestinal mucosal cell and either be packaged directly into the nascent chylomicrons or be cleaved by carotene 15,15′-monooxygenase to yield two molecules of retinal. The retinal can then be transformed enzymatically to retinol by the action of an intestinal retinal reductase(s). Dietary retinyl esters are unable to pass directly through the intestinal brush border and must first be hydrolyzed to free retinol by either pancreatic retinyl ester hydrolases or a brush border retinyl ester hydrolase. This retinol, like free retinol arriving as such in the diet, readily traverses the intestinal brush border and binds within the enterocyte cellular retinol-binding protein, type II (CRBP II). This protein can also bind the free retinal from the previous step (not shown in this schematic representation). Retinol bound to CRBP II is then esterified to form retinyl esters through the actions of lecithin:retinol acyltransferase (LRAT). These retinyl esters (RE) are incorporated through some undefined mechanism into the nascent chylomicrons along with other dietary lipids; these are secreted into the lymphatic system and subsequently enter the general circulation. Most chylomicron retinyl ester is taken up by the liver, but a substantial percentage is also taken up by peripheral tissues.

TABLE 1.3
Some Retinoid Metabolizing Enzymes and Enzyme Families

Name(s)	Substrate(s)	Proposed physiologic function(s)	Primary location	Refs.
Lecithin:retinol acyltransferase (LRAT)	Retinol	Transfer fatty acid moiety from ecithin to retinol to form retinyl ester	Most tissues particularly eye, intestine, liver	3, 5, 20, 21, 43, 51–53
Acyl CoA:retinol acyltransferase (ARAT)	Retinol	Retinyl ester formation from fatty acyl CoA	Mammary tissue, possibly others	2, 3, 55, 56
Carotene-15,15′-monooxygenase or carotene cleavage enzyme (CCE)	β-Carotene and other provitamin A carotenoids	Central cleavage of carotenoids to form retinal	Intestine, liver, testis, others	38–42
Retinyl ester hydrolases (REH)	Retinyl esters	Mobilization of retinol from stores	Intestinal cells, liver, many others	2, 3, 57–59
Bile salt sensitive-retinyl ester hydrolase (BS-REH) or carboxyl ester lipase (CEL)	Retinyl esters, cholesteryl esters	Hydrolysis of dietary retinyl in intestinal lumen and hepatic retinyl ester	Lumen of the intestine, liver	2, 3, 60
Medium-chain alcohol dehydrogenases (ADH)	Retinol, other alcohols	Oxidation of retinol to retinal; for retinoic acid formation	Most tissues, liver, eye, others	61–65, 69–71
Short-chain dehydrogenase/reductases (SDR)	Retinol, retinal, other alcohols and aldehydes	Oxidation of retinol to retinal or retinal reduction to retinol involved in retinoic acid formation	Most tissues, liver, skin, eye, testis, kidney and lung	3, 61–65
Aldehyde dehydrogenases (ALDH)	Retinal, other aldehydes	Irreversible oxidation of retinal to retinoic acid	Most tissues	73–77
Cytochrome P_{450}, CYP26 isoforms	Retinoic acid	Oxidative metabolism of retinoic acid to control cellular retinoic acid levels	Most tissues, well studied in embryonic tissues	2, 3, 10–14, 78–81
UDP-glucuronosyl transferases	Retinoic acid and retinol	Conjugation to form water soluble glucuronides for elimination from the body	Liver, other tissues	2, 3, 82, 83

Postprandial Retinoid Transport and Hepatic Storage

Retinyl esters in chylomicrons enter the circulation and are taken up by tissues; ~70% of chylomicron retinyl ester is taken up by the liver, and the remainder is cleared by peripheral tissues (44). Before uptake by tissues, chylomicron retinyl ester must undergo hydrolysis. In the liver, the process of retinyl ester hydrolysis occurs as the chylomicron remnant particle is being cleared by hepatocytes (liver parenchymal cells) (2,3) but it is not established what enzyme(s) is responsible for hydrolysis. It was proposed that the enzyme lipoprotein lipase (LPL) performs this function in peripheral tissues, facilitating retinol uptake (45,46). Free postprandial retinol taken up by cells is thought to bind immediately to CRBP I (see Table 1.1) that are present in tissues (20,21,47). It was suggested that CRBP I facilitate/optimize the retinol uptake process (21,47). It is well established that hepatocytes are responsible for the uptake of postprandial retinoid into the liver (3,48).

Because retinoid action is very important for maintaining good health, higher animals have developed the capacity to store retinoid as retinyl esters in the liver (2,3). After postprandial retinoid is taken up by the liver, this retinoid is either secreted back into the circulation bound to RBP (see below) or transferred to the hepatic stellate cells for storage (2,3,48). Within the stellate cells, retinoid is stored as retinyl esters in the large lipid droplets that are characteristic of these cells. Thus, after the postprandial retinoid is taken up by hepatocytes and hydrolyzed to retinol, this retinol must again undergo esterification *via* the actions of LRAT before storage in stellate cell lipid droplets.

The major tissue storage site for retinoid is the liver, although other tissues including the eye, lung, adipose tissue, and skin have the ability to store retinoid, albeit to a much lesser degree than liver. It was estimated that for healthy well-nourished individuals, ~60–80% of the retinoid present in the body will be stored in the liver and ~70% of that is present in the hepatic stellate cells (2,3,48,49). When the body senses a need for retinoid, these esters are hydrolyzed by retinyl ester hydrolase (REH; see below for more detail) to free retinol, which through some poorly characterized process is mobilized from the liver bound to its plasma transport protein, RBP. It remains to be established how a signal is conveyed by the peripheral tissues to the liver in terms of retinoid having to allow for retinoid mobilization from the liver.

Retinol-RBP is secreted from the liver into the circulation as a means of delivering retinol to peripheral tissues (1,17). In the fasted state, for a well-nourished healthy person, >95% of the retinoid in the circulation will be present as retinol-RBP. The liver is the major site of synthesis of RBP in the body, and within the liver, the hepatocyte is the sole cellular site of RBP synthesis (17). Other tissues including adipose tissue, kidney, lung, heart, skeletal muscle, spleen, eye, and testis also express RBP, and this may be important for recycling retinoid from peripheral tissues back to the liver (17). The retinol-RBP complex binds another plasma protein, transthyretin (TTR), and this stabilizes the complex and reduces renal filtering of the retinoid (1,17,50). Once retinol is delivered as retinol-RBP-TTR, it is taken up by cells and either stored within the cell as retinyl ester or oxidized to retinoic acid for use in regulating gene expression. These processes are summarized in Figure 1.3 and discussed in more detail in the text below.

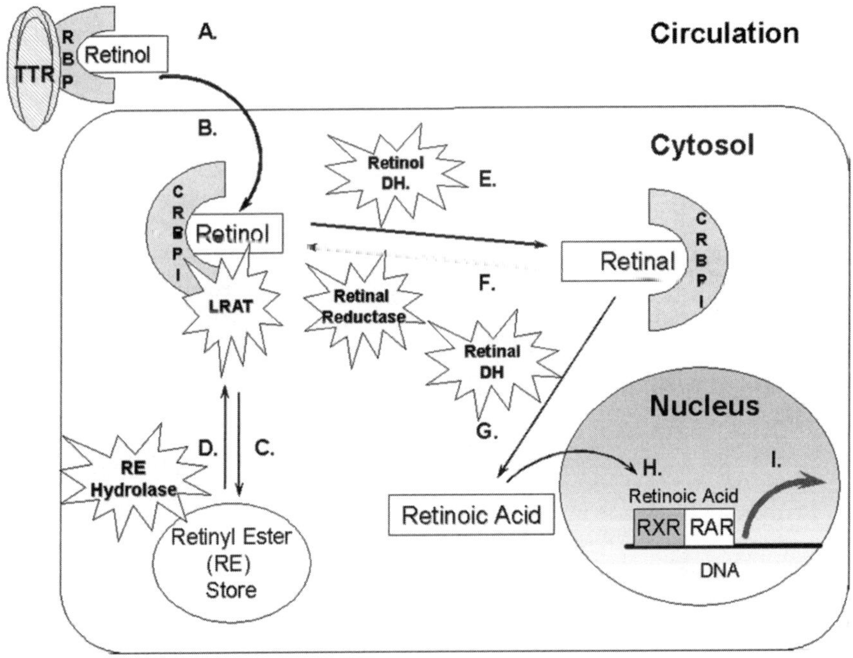

Fig. 1.3. Cellular uptake and processing of retinoids. Retinol is delivered to cells by the circulation after its secretion from the liver of retinol bound to retinol-binding protein (RBP). In the circulation, the retinol-RBP complex undergoes protein-protein interaction with transthyretin (TTR) and circulates as the ternary retinol-RBP-TTR complex (A). Retinol arriving at the cell is then taken into the cell where it is immediately bound by one of the cellular retinol-binding protein family members. In most tissues, this will involve the actions of cellular retinol-binding protein, type I (CRBP I) because CRBP I is widely expressed throughout the body. However, the other three CRBP forms may have roles in this process in tissues where they are expressed (B). Upon cell uptake, there are two possible fates for the retinol. It may be esterified by lecithin:retinol acyltransferase (LRAT) (C) and stored as retinyl esters (RE) in lipid droplets present in the cell. In times of cellular need, a retinyl ester hydrolase(s) can liberate the retinol from the retinyl ester stores (D). Alternatively, retinol can be acted on by one of a number of the retinol dehydrogenases (Retinol DH), which are able to oxidize it to retinal (E). This step is reversible, and retinal can be converted back to retinol by cellular retinal reductases (F). Upon its formation, retinal is usually quickly acted upon by one of several retinal dehydrogenases (Retinal DH) and is irreversibly converted to retinoic acid (G). Retinoic acid then enters the nucleus, binds, and activates one of the retinoid nuclear receptors (H) that regulate transcription of a wide range of target genes (I).

Retinol Esterification and Retinyl Ester Hydrolysis

Although the liver and intestine are the major tissue sites of retinol esterification in the body, many tissues are able to esterify retinol and to accumulate some retinyl ester stores. It appears that the major, if not sole, enzyme responsible for catalyzing retinyl ester formation is LRAT. This enzyme catalyzes retinol esterification with long-chain fatty acids (primarily palmitic, stearic, oleic, and linoleic acids) present in membrane phospholipids (2,3,43,51,52). LRAT in the liver is thought to be identical to intestinal LRAT, which synthesizes retinyl esters from dietary retinol for incorporation into nascent chylomicrons. LRAT is also found in the eye and has an important role in the visual cycle (52,53). This role in vision was verified recently by studies in mice that lack the LRAT protein and whose vision is severely attenuated starting at a young age (53). Interestingly, hepatic but not intestinal LRAT activity is regulated by retinoid nutritional status. This regulation seems to involve the presence of a retinoic acid response element that is present in the LRAT gene and probably the actions of RAR and/or RXR (54). This regulation is suggested to give rise to a positive feedback loop when cellular retinoic acid levels are high.

There is also one other candidate enzyme that may be physiologically relevant for catalyzing retinyl ester formation, i.e., acyl-CoA:retinol acyltransferase (ARAT). ARAT esterifies retinol using fatty acids present in the acyl-CoA pool (55). ARAT also differs in the form of retinol it utilizes as a substrate compared with LRAT because ARAT is incapable of esterifying retinol when it is bound to CRBP I or CRBP II (56).

Because retinyl esters represent a storage form of retinoids, they must first be hydrolyzed to retinol before activation to retinoic acid. Unlike LRAT, which is accepted to be the major enzyme responsible for retinyl ester formation, there are many REH that may be responsible for the generation of free retinol from retinyl ester stores (57–59). One is a bile salt–dependent REH (BS-REH). Most or all of the BS-REH activity in liver probably arises from the actions of bile salt–activated carboxylester lipase (CEL) (60). However, because mice lacking CEL display no alterations in retinoid storage, metabolism, or actions, this enzyme cannot be the sole physiologically relevant REHs (60). Another group of enzymes, collectively known as bile salt–independent REHs were described. There are two groups of REHs based on their pH optima, neutral REHs and acidic REHs. It was reported that the activities of the neutral and acidic REHs are unaffected by retinoid nutritional status (58). There also is evidence demonstrating that three known hepatic carboxylesterases (also known in the literature as ES-2, ES-4, and ES-10) act as REHs *in vitro* (57,59). However, it is not yet established whether any or all of these REHs are physiologically important in retinoid metabolism.

Retinol Oxidation

Members of two enzyme families are proposed to have important roles in catalyzing retinol oxidation to retinal, the first of two oxidative steps required for retinoic acid formation (61–65). The first of these enzyme families is the short-chain dehydrogenase/reductase (SDR) family. The second family is the medium-chain alcohol dehydrogenase (ADH) family. These two enzyme families will be considered separately below.

The SDR Family. At the time this chapter was written, at least a dozen distinct SDRs able to catalyze retinol oxidation were proposed as being important for catalyzing retinol oxidation. The members of the SDR family all range in size between 26 and 34 kDa and are usually associated with cell membrane fractions (61–65). SDRs able to catalyze the oxidation of only all-*trans*-retinol as well as others that can also catalyze oxidation of 9/13-*cis*-retinol have been identified (8,9,61–65). It was proposed that this latter group may play an important role in the synthesis of 9-*cis*-retinoic acid (8,9). These are distributed throughout the tissues of the body, and one or more SDR with retinol dehydrogenase activity are present in tissues such as liver, skin, eye, testis, kidney, and lung, which are known to have a high capacity for retinoid metabolism and/or a great need for retinoic acid. Most SDRs proposed as being physiologically relevant for catalyzing retinol oxidation also possess steroid dehydrogenase activity toward androgens and/or estrogens (62,64–66). It was suggested that the retinoid and steroid metabolisms are connected through these enzymes, which have dual responsibility for the metabolism of each member of each family of very bioactive lipids. However, no data derived from *in vivo* studies are currently available concerning how the metabolism of either retinoids or steroids influences the metabolism of the other. Nevertheless, the possibility of such interactions is very intriguing and one worthy of much future research attention.

Some SDRs are able to utilize retinol bound to CRBP I as a substrate, but usually these enzymes also will catalyze *in vitro* oxidation of retinol that was dispersed in a detergent or solubilized in some other manner. Because most of the retinol present in CRBP I-expressing cells is bound to CRBP I, the ability of SDR to recognize retinol bound to CRBP I as substrate was taken as an indication that SDRs with retinol dehydrogenases are physiologically important for retinoic acid formation. Although the biochemical properties of SDRs that have retinol dehydrogenase activity were thoroughly investigated and characterized, the actions of these enzymes are less well studied in living cells or animals. Thus, it remains difficult to assign definitive physiologic roles for these enzymes as mediators of retinol oxidation in the body.

Interestingly, two members of the SDR family, Ret/SDR and RalR1, were identified as being retinal reductases and proposed as physiologically relevant enzymes for reducing retinal formed upon the cleavage of provitamin A carotenoids (67,68). Although both of these enzymes are expressed in the intestine as well as other tissues that express carotene-15,15′-monooxygenase (38–42), the proposed roles of these SDRs in the formation of retinoid from provitamin A carotenoids have not been investigated in a physiologically meaningful context.

However, no studies convincingly confirmed this physiological role for either Ret/SDR or RalR1.

The ADH Family. Retinol oxidation also can be catalyzed by several members of the ADH family. This enzyme family is composed of cytosolic proteins of ~40 kDa and comprises the "classic" alcohol dehydrogenases that were first identified >70 years ago. There are three major ADH isotypes, ADH1, ADH3, and ADH4, proposed to be involved in retinal formation (62–65,69–71). Each of these ADH isotypes is able to use free or detergent-solubilized retinol as a substrate but not retinol bound to CRBP I.

Both the biochemistry and physiology of the ADH enzymes are well studied. Although the ADH have a very broad substrate specificity for alcohols, comparisons of catalytic efficiencies (k_{cat}/K_m) for these enzymes indicate that retinol is one of the best naturally occurring substrates for the enzymes (63,69–71). One of the important arguments in favor of the physiologic relevance of the ADH in retinoic acid synthesis is that the different isotypes are strongly expressed in embryonic tissues that synthesize and require high amounts of retinoic acid. This provides circumstantial evidence for an involvement of the ADH in retinoic acid synthesis. Targeted disruptions of ADH1, ADH3, and ADH4 were reported and mice lacking one or more of these ADH isotypes are viable (72). However, when ADH4-deficient mice are stressed through administration of a retinoid-deficient diet, they more quickly succumb to retinoid deficiency than do wild-type mice (72). ADH1-deficient mice are no more susceptible to retinoid deficiency than wild-type mice. Interestingly, however, ADH1/4-double knockout mice are more resistant to retinoid deficiency than the single ADH4-knockout mice. This suggests that ADH1 absence in some undefined manner partially rescues ADH4-deficient mice from their increased sensitivity to retinoid deficiency.

Retinal Oxidation

The second and final step needed for retinoic acid synthesis is the oxidation of retinal to retinoic acid. Several distinct cytosolic aldehyde dehydrogenases (ALDH), which are referred to as retinal dehydrogenases (RALDH) when discussed in the context of retinoid physiology, catalyze the irreversible oxidation of retinal to retinoic acid. These RALDH, RALDH1 (ALDH1a1), RALDH2 (ALDH1a2), RALDH3 (ALDH1a3), and RALDH4 (ALDH8a1) are all distinct members of the ALDH protein family (63–65,73,74). The members of the ALDH family catalyze the irreversible oxidation of aldehydes to acids (63–65,73). This irreversible reaction, as the final step of retinoic acid synthesis, must be tightly regulated to control retinoic acid–induced signaling (63–65,73–75). RALDH1, RALDH2, and RALDH3 were studied in the context of all-*trans*-retinoic acid formation (63–65,71,75). RALDH4 was suggested to be involved in catalyzing 9-*cis*-retinal oxidation and consequently might have an important role in the biosynthesis of 9-*cis*-retinoic acid (74). Unlike the situation for retinol oxidation, the physiologic roles of these RALDH are generally well defined.

Of the RALDH, RALDH2 is the best studied. The targeted disruption of the gene for RALDH2 gives rise to embryonic lethality at E11.5 due to severe trunk, hindbrain, and heart defects (75). Growth defects were first observed from E8.5 to E10.5, but these could be reversed upon all-*trans*-retinoic acid administration (75). Using a *LacZ* reporter construct driven by a retinoic acid sensitive promoter, it was possible to demonstrate that the absence of RALDH2 diminished retinoic acid synthesis; consequently, RALDH2 must be viewed as an essential enzyme responsible for retinoic acid synthesis (75). Disruption of the gene encoding RALDH1 does not result in embryonic lethality but does give rise to lessened synthesis of retinoic acid in the developing eye and to microphthalmia in the adult (76). Thus, RALDH1 appears to have an important role in the synthesis of retinoic acid in the developing eye. Like RALDH2, RALDH3 appears to have an essential function during embryogenesis and/or early in the postnatal period because its absence results in lethality shortly after birth due to choanal atresia giving rise to respiratory distress and death (77). Lethality is reversible upon maternal retinoic acid treatment (77). Clearly RALDH1, RALDH2, and RALDH3 each have an important role in catalyzing retinoic acid formation from retinal. Why distinct RALDH forms are needed by the body to catalyze this reaction is not well understood at present.

Oxidative and Conjugative Metabolism of Retinoids

Once retinoic acid has activated its receptor, it is important for the cell's health that the signal be terminated by the removal of the retinoic acid. This is accomplished through the generation of hydroxyl- and oxo-retinoic acid species and glucuronides of retinoic acid (2,3). In the late 1990s, several groups reported the identification of a specific retinoic acid–inducible cytochrome P_{450}-related retinoic acid hydroxylase, CYP26 (10–14). This enzyme shares structural motifs with other cytochrome P_{450} species. Expression of CYP26 in cultured naïve cells conferred on these cells the ability to oxidize all-*trans*-retinoic acid to products identified as its 4-hydroxy- and 4-oxo-metabolites (10–14). Several different isotypes of CYP26 were identified, and each of these seems to have an important role in catalyzing the oxidative breakdown of retinoic acid. Inactivation of CYP26 isotypes through targeted gene disruptions results in impaired embryologic development (10–14,78–81). The lack of CYP26 expression was shown to result in excessive retinoic acid accumulation in embryonic structures (79,81). It is clear from these elegant studies that the oxidative metabolism of retinoic acid catalyzed by CYP26 is an essential mechanism through which the body prevents excessive or unneeded retinoic acid accumulation.

Cytochrome P_{450} species, other than the CYP26 isotypes, are also reported to catalyze *in vitro* oxidation of retinoic acid (2,3). Most of these are present in adult tissues and each may have a role in mediating some aspect of retinoid physiology in adults. However, the precise role(s) of these enzymes in normal retinoid physiology remains to be established.

Both retinol and retinoic acid undergo conjugation with glucuronic acid, giving rise to retinol- and retinoic acid-β-glucuronides (2,3). The retinoid-β-glucuronides are very water soluble, and their formation likely reflects a second pathway through which excess or unneeded retinoids are eliminated from the body. Several UDP-glucuronal transferases were identified; they can catalyze the formation of retinoid-β-glucuronides and have broad substrate specificity for compounds other than retinoids (82,83). There is no information available concerning whether any of these enzymes is specifically responsive to retinoic acid.

Concluding Remarks

In the last 20 years, there has been tremendous growth in our understanding of retinoid metabolism and actions. With the identification and characterization of the retinoic acid nuclear receptors, it became clear how retinoids act physiologically to mediate an apparently diverse array of physiologically essential processes. Many of the enzymes and binding proteins responsible for maintaining normal retinoid physiology also were identified and characterized in this period. We now have a very clear understanding of the enzymatic processes responsible for the storage of retinoids and for the catabolism of excess or unneeded retinoid. We also now have a good understanding of how retinoids are delivered to tissues and how tissues and cells acquire and maintain retinoid pools. However, there are still many areas of retinoid physiology that require elucidation. The true physiologic relevance of many enzymes that are proposed to catalyze retinol oxidation remains to be established. Possible metabolic linkages between steroid and retinoid metabolism require examination. Understanding of how different tissues communicate with each other to facilitate retinoid economy within the body is still lacking. Ultimately, we have a better understanding of the actions of retinoids for maintaining optimal health and preventing disease, but much has yet to be learned if we are to appreciate fully the actions of retinoids within the body.

Acknowledgments

The authors gratefully acknowledge support from the National Institutes of Health through grants R01 DK061310 and R01 DK068437 and a grant (BC031116) from the Department of Defense.

References

1. Goodman, D.S. (1980) Vitamin A Metabolism, *Fed. Proc. 39:* 2716–2722.
2. Blaner, W.S., and Olson, J.A. (1994) Retinol and Retinoic Acid Metabolism, in *The Retinoids: Biology, Chemistry, and Medicine*, 2nd edn., Sporn, M.B., Roberts, A.B., and Goodman, D.S., eds., Raven Press, New York, pp. 229–256.

3. Vogel, S., Gamble, M.V., and Blaner, W.S. (1999) Retinoid Uptake, Metabolism and Transport, in *The Handbook of Experimental Pharmacology, The Retinoids*, Nau, H., and Blaner, W.S., eds., Springer Verlag, Heidelberg, pp. 31–96.
4. Wald, G. (1968) Molecular Basis of Visual Excitation, *Nature 219:* 800–807.
5. Saari, J.C. (1999) Retinoids in Mammalian Vision, in *The Handbook of Experimental Pharmacology, The Retinoids, Volume 139*, Nau, H., and Blaner, W.S., eds., Springer Verlag, Heidelberg, pp. 563–588.
6. Mangelsdorf, D.J., Umesono, K., and Evans, R.M. (1994) The Retinoid Receptors, in *The Retinoids, Biology, Chemistry and Medicine*, 2nd edn., Sporn, M.B., Roberts, A.B., and Goodman, D.S., eds., Raven Press, New York, pp. 319–350.
7. Chambon, P. (1994) The Retinoid Signaling Pathway: Molecular and Genetic Analysis, *Cell Biology 5:* 115–125.
8. Mertz, J.R., Shang, E., Piantedosi, R., Wei, S., Wolgemuth, D.J., and Blaner, W.S. (1997) Identification and Characterization of a Stereospecific Human Enzyme That Catalyzes 9-*cis*-Retinol Oxidation, *J. Biol. Chem. 272:* 11744–11749.
9. Chai, X., Zhai, Y., and Napoli, J.L. (1997) cDNA Cloning and Characterization of a *cis*-Retinol/3α-Hydroxysterol Short-Chain Dehydrogenase, *J. Biol. Chem. 272:* 3125–3131.
10. White, J.A., Beckett-Jones, B., Guo, Y.D., Dilworth, F.J., Bonasoro, J., Jones, G., and Petkovich, M. (1997) cDNA Cloning of Human Retinoic Acid-Metabolizing Enzyme ($hP_{450}RAI$) Identifies a Novel Family of Cytochromes P_{450} (CYP26), *J. Biol. Chem. 272:* 18538–18541.
11. White, J.A., Guo, Y.D., Baetz, K., Beckett-Jones, B., Bonasoro, J., Hsu, K.E., Dilworth, F.J., Jones, G., and Petkovich, M. (1996) Identification of the Retinoic Acid-Inducible all-*trans*-Retinoic Acid 4-Hydroxylase, *J. Biol. Chem. 271:* 29922–29927.
12. Fujii, H., Sato, T., Kaneko, S., Gotoh, O., Fujii-Kuriyama, Y., Osawa, K., Kato, S., and Hamada, H. (1997) Metabolic Inactivation of Retinoic Acid by a Novel P_{450} Differentially Expressed in Developing Mouse Embryos, *EMBO J. 16:* 4163–4173.
13. Ray, W.J., Bain, G., Yao, M., and Gottlieb, D.I. (1997) CYP26, a Novel Mammalian Cytochrome P_{450}, Is Induced by Retinoic Acid and Defines a New Family, *J. Biol. Chem. 272:* 18702–18708.
14. Ross, A.C. (2003) Retinoid Production and Catabolism: Role of Diet in Regulating Retinol Esterification and Retinoic Acid Oxidation, *J. Nutr. 133:* 291S–296S.
15. Vakiani, E., and Buck, J. (1999) *Retro*-Retinoids: Metabolism and Action, in *The Handbook of Experimental Pharmacology, The Retinoids*, Nau, H., and Blaner, W.S., eds., Springer Verlag, Heidelberg, pp. 31–96.
16. Ross, C.A., and Hammerling, U.G. (1994) Retinoids and the Immune System, in *The Retinoids, Biology, Chemistry and Medicine*, Sporn, M.B., Roberts, A.B., and Goodman, D.S., eds., Raven Press, New York, pp. 521–544.
17. Soprano, D.R., and Blaner, W.S. (1994) Plasma Retinol-Binding Protein, in *The Retinoids: Biology, Chemistry, and Medicine*, 2nd edn., Sporn, M.B., Roberts, A.B., and Goodman, D.S., Raven Press, New York, pp. 229–256.
18. Ong, D.E., Newcomer, M.E., Lareyre, J.J., and Orgebin-Crist, M.C. (2000) Epididymal Retinoic Acid-Binding Protein, *Biochim. Biophys. Acta 1482:* 209–217.
19. Tanaka, T., Urade, Y., Kimura, H., Eguchi, N., Nishikawa, A., and Hayaishi, O. (1997) Lipocalin-Type Prostaglandin D Synthase (β-Trace) Is a Newly Recognized Type of Retinoid Transporter, *J. Biol. Chem. 272:* 15789–15795.

20. Ong, D.E., Newcomer, M.E., and Chytil, F. (1994) Cellular Retinoid-Binding Proteins, in *The Retinoids, Biology, Chemistry, and Medicine*, 2nd edn., Sporn, M.B., Roberts, A.B., and Goodman, D.S., eds., Raven Press, New York, pp. 283–318.
21. Noy, N. (2000) Retinoid-Binding Proteins: Mediators of Retinoid Action, *Biochem. J. 348:* 481–495.
22. Levin, M.S., Locke, B., Yang, N.C., Li, E., and Gordon, J.I. (1988) Comparison of the Ligand Binding Properties of Two Homologous Rat Apocellular Retinol-Binding Proteins Expressed in *Escherichia coli, J. Biol. Chem. 263:* 17715–17723.
23. Vogel, S., Mendelsohn, C.L., Mertz, J.R., Piantedosi, R., Waldburger, C., Gottesman, M.E., and Blaner, W.S. (2001) Characterization of a New Member of the Fatty Acid-Binding Protein Family That Binds all-*trans*-Retinol, *J. Biol. Chem. 276:* 1353–1360.
24. Folli, C., Calderone, V., Ramazzina, I., Zanotti, G., and Berni, R. (2002) Ligand Binding and Structural Analysis of a Human Putative Cellular Retinol-Binding Protein, *J. Biol. Chem. 277:* 41970–41977.
25. Saari, J.C., Bredberg, L., and Garwin, G.G. (1982) Identification of the Endogenous Retinoids Associated with Three Cellular Retinoid-Binding Proteins from Bovine Retina and Retinal Pigment Epithelium, *J. Biol. Chem. 257:* 13329–13333.
26. Panagabko, C., Morley, S., Hernandez, M., Cassolato, P., Gordon, H., Parsons, R., Manor, D., and Atkinson, J. (2003) Ligand Specificity in the CRAL-TRIO Protein Family, *Biochemistry 42:* 6467–6474.
27. Gorry, P., Lufkin, T., Dierich, A., Rochette-Egly, C., Decimo, D., Dollé, P., Mark, M., Durand, B., and Chambon, P. (1994) The Cellular Retinoic Acid Binding Protein I Is Dispensable, *Proc. Natl. Acad. Sci. USA 91:* 9032–9036.
28. Lampron, C., Rochette-Egly, C., Gorry, P., Dollé, P., Mark, M., Lufkin, T., LeMeur, M., and Chambon, P. (1995) Mice Deficient in Cellular Retinoic Acid Binding Protein II (CRABPII) or in Both CRABPI and CRABPII Are Essentially Normal, *Development 121:* 539–548.
29. Liou, G.I., Fei, Y., Peachey, N.S., Matragoon, S., Wei, S., Blaner, W.S., Wang, Y., Liu, C., Gottesman, M.E., and Ripps, H. (1998) Early Onset Photoreceptor Abnormalities Induced by Targeted Disruption of the Interphotoreceptor Retinoid-Binding Protein Gene, *J. Neurosci. 18:* 4511–4520.
30. Quadro, L., Blaner, W.S., Salchow, D.J., Vogel, S., Piantedosi, R., Gouras, P., Freeman, S., Cosma, M.P., Colantuoni, V., and Gottesman, M.E. (1999) Visual Defect and Impaired Retinoid Availability in Mice Lacking Retinol-Binding Protein, *EMBO J. 18:* 4633–4644.
31. Ghyselinck, N.B., Bavik, C., Sapin, V., Mark, M., Bonnier, D., Hindelang, C., Dierich, A., Nilsson, C.B., Hakansson, H., Sauvant, P., Azais-Braesco, V., Frasson, M., Picaud, S., and Chambon, P. (1999) Cellular Retinol-Binding Protein I Is Essential for Vitamin A Homeostasis, *EMBO J. 18:* 4903–4914.
32. Saari, J.C., Nawrot, M., Kennedy, B.N., Garwin, G.G., Hurley, J.B., Huang, J., Possin, D.E., and Crabb, J.W. (2001) Visual Cycle Impairment in Cellular Retinaldehyde Binding Protein (CRALBP) Knockout Mice Results in Delayed Dark Adaptation, *Neuron. 29:* 739–748.
33. Zhang, L., Lu, J., Tso, P., Blaner, W.S., Levin, M.S., and Li, E. (2002) Increased Neonatal Mortality in Mice Lacking Cellular Retinol-Binding Protein II, *J. Biol. Chem. 277:* 36617–36623.

34. Gudas, L.J., Sporn, M.B., and Roberts, A.B. (1994) Cellular Biology and Biochemistry of the Retinoids, in *The Retinoids, Biology, Chemistry and Medicine*, 2nd edn., Sporn, M.B., Roberts, A.B., and Goodman, D.S., eds., Raven Press, New York, pp. 443–520.
35. Bastien, J., and Rochette-Egly, C. (2004) Nuclear Retinoid Receptors and the Transcription of Retinoid-Target Genes, *Gene 328:*1–16.
36. Shulman, A.I., Larson, C., Mangelsdorf, D.J., and Ranganathan, R. (2004) Structural Determinants of Allosteric Ligand Activation in RXR Heterodimers, *Cell 116:* 417–429.
37. Balmer, J.E., and Blomhoff, R. (2002) Gene Expression Regulation by Retinoic Acid, *J. Lipid Res. 43:* 1773–1808.
38. von Lintig, J., and Vogt, K. (2004) Vitamin A Formation in Animals: Molecular Identification and Functional Characterization of Carotene Cleaving Enzymes, *J. Nutr. 134:* 251S–256S.
39. Wyss, A. (2004) Carotene Oxygenases: A New Family of Double Bond Cleavage Enzymes, *J. Nutr. 134:* 246S–250S.
40. Redmond, T.M., Gentleman, S., Duncan, T., Yu, S., Wiggert, B., Gantt, E., and Cunningham, Jr., F.X. (2001) Identification, Expression, and Substrate Specificity of a Mammalian Beta-Carotene 15,15′-Dioxygenase, *J. Biol. Chem. 276:* 6560–6565.
41. Paik, J., During, A., Harrison, E.H., Mendelsohn, C.L., Lai, K., and Blaner, W.S. (2001) Expression and Characterization of a Murine Enzyme Able to Cleave Beta-Carotene. The Formation of Retinoids, *J. Biol. Chem. 276:* 32160–32168.
42. Lindqvist, A., and Andersson, S. (2002) Biochemical Properties of Purified Recombinant Human Beta-Carotene 15,15′-Monooxygenase, *J. Biol. Chem. 277:* 23942–23948.
43. Herr, F.M., and Ong, D.E. (1992) Differential Interaction of Lecithin-Retinol Acyltransferase with Cellular Retinol Binding Proteins, *Biochemistry 31:* 6748–6755.
44. Goodman, D.S., Huang, H.S., and Shiratori, T. (1965) Tissue Distribution and Metabolism of Newly Absorbed Vitamin A in the Rat, *J. Lipid Res. 6:* 390–396.
45. Blaner, W.S., Obunike, J.C., Kurlandsky, S.B., al-Haideri, M., Piantedosi, R., Deckelbaum, R.J., and Goldberg, I.J. (1994) Lipoprotein Lipase Hydrolysis of Retinyl Ester. Possible Implications for Retinoid Uptake by Cells, *J. Biol. Chem. 269:* 16559–16565.
46. van Bennekum, A.M., Kako, Y., Weinstock, P.H., Harrison, E.H., Deckelbaum, R.J., Goldberg, I.J., and Blaner, W.S. (1999) Lipoprotein Lipase Expression Level Influences Tissue Clearance of Chylomicron Retinyl Ester, *J. Lipid Res. 40:* 565–574.
47. Noy, N., and Blaner, W.S. (1991) Interactions of Retinol with Binding Proteins: Studies with Rat Cellular Retinol-Binding Protein and with Rat Retinol-Binding Protein, *Biochemistry 30:* 6380–6386.
48. Blaner, W.S. (1994) Retinoid (Vitamin A) Metabolism and the Liver, in *The Liver, Biology and Pathobiology*, 3rd edn., Arias, I.M., Jakoby, W.B., Popper, H., Schacter, D., and Shafritz, D.S., eds., Raven Press, New York, pp. 529–542.
49. Geerts, A., Bleser, P.D., Hautekeete, M.L., Niki, T., and Wisse, E. (1994) Fat-Storing (Ito) Cell Biology, in *The Liver: Biology and Pathobiology*, 3rd edn., Arias, I.M., Boyer, J.L., Fausto, N., Jakoby, W.B., Schachter, D., and Shafritz, D.A., eds., Raven Press, New York, pp. 819–837.
50. van Bennekum, A.M., Wei, S., Gamble, M.V., Vogel, S., Piantedosi, R., Gottesman, M., Episkopou, V., and Blaner, W.S. (2001) Biochemical Basis for Depressed Serum Retinol Levels in Transthyretin-Deficient Mice, *J. Biol. Chem. 276:* 1107–1113.

51. Yost, R.W., Harrison, E.H., and Ross, A.C. (1988) Esterification by Rat Liver Microsomes of Retinol Bound to Cellular Retinol-Binding Protein, *J. Biol. Chem. 263:* 18693–18701.
52. Saari, J.C., and Bredberg, D.L. (1989) Lecithin:Retinol Acyltransferase in Retinal Pigment Epithelial Microsomes, *J. Biol. Chem. 264:* 8636–8640.
53. Batten, M.L., Imanishi, Y., Maeda, T., Tu, D.C., Moise, A.R., Bronson, D., Possin, D., Van Gelder, R.N., Baehr, W., and Palczewski, K. (2004) Lecithin-Retinol Acyltransferase Is Essential for Accumulation of all-*trans*-Retinyl Esters in the Eye and in the Liver, *J. Biol. Chem. 279:* 10422–10432.
54. Matsuura, T., and Ross, A.C. (1993) Regulation of Hepatic Lecithin: Retinol Acyltransferase Activity by Retinoic Acid, *Arch. Biochem. Biophys. 301:* 221–227.
55. Ross, A.C. (1982) Retinol Esterification by Rat Liver Microsomes. Evidence for a Fatty Acyl Coenzyme A:Retinol Acyltransferase, *J. Biol. Chem. 257:* 2453–2459.
56. Randolph, R.K., Winkler, K.E., and Ross, A.C. (1991) Fatty Acyl CoA-Dependent and -Independent Retinol Esterification by Rat Liver and Lactating Mammary Gland Microsomes, *Arch. Biochem. Biophys. 288:* 500–508.
57. Harrison, E.H. (2000) Lipases and Carboxylesterases: Possible Roles in the Hepatic Utilization of Vitamin A, *J. Nutr. 130:* 340S–344S.
58. Matsuura, T., Gad, M.Z., Harrison, E.H., and Ross, A.C. (1997) Lecithin:Retinol Acyltransferase and Retinyl Ester Hydrolase Activities Are Differentially Regulated by Retinoids and Have Distinct Distributions Between Hepatocytes and Nonparenchymal Cell Fractions of Rat Liver, *J. Nutr. 127:* 218–224.
59. Sanghani, S.P., Davis, W.I., Dumaual, N.G., Mahrenholz, A., and Bosron, W.F. (2002) Identification of Microsomal Rat Liver Carboxylesterases and Their Activity with Retinyl Palmitate, *Eur. J. Biochem. 269:* 4387–4398.
60. van Bennekum, A.M., Li, L., Piantedosi, R., Shamir, R., Vogel, S., Fisher, E.A., Blaner, W.S., and Harrison, E.H. (1999) Carboxyl Ester Lipase Overexpression in Rat Hepatoma Cells and CEL Deficiency in Mice Have No Impact on Hepatic Uptake of Metabolism of Chylomicron-Retinyl Ester, *Biochemistry 38:* 4150–4156.
61. Jörnvall, H., Persson, B., Krook, M., Atrian, S., Gonzalez-Duarte, R., Jeffery, J., and Ghosh, D. (1995) Short-Chain Dehydrogenases/Reductases (SDR), *Biochemistry 34:* 6003–6013.
62. Duester, G. (2001) Genetic Dissection of Retinoid Dehydrogenases, *Chem.-Biol. Interact.:* 130–132, 469–489.
63. Duester, G. (2000) Families of Retinoid Dehydrogenases Regulating Vitamin A Function: Production of Visual Pigment and Retinoic Acid, *Eur. J. Biochem. 267:* 4315–4324.
64. Napoli, J.L. (1999) Retinoic Acid: Its Biosynthesis and Metabolism, *Prog. Nucleic Acid Res. Mol. Biol. 63:* 139–188.
65. Napoli, J.L. (1999) Interactions of Retinoid Binding Proteins and Enzymes in Retinoid Metabolism, *Biochim. Biophys. Acta 1440:* 139–162.
66. Biswas, M.G., and Russell, D.W. (1997) Expression Cloning and Characterization of Oxidative 17β and 3α-Hydroxysteroid Dehydrogenases from Rat and Human Prostate, *J. Biol. Chem. 272:* 15959–15966.
67. Haeseleer, F., Huang, J., Lebioda, L., Saari, J.C., and Palczewski, K. (1998) Molecular Characterization of a Novel Short-Chain Dehydrogenase/Reductase That Reduces all-*trans*-Retinal, *J. Biol. Chem. 273:* 21790–21799.

68. Kedishvili, N.Y., Chumakova, O.V., Chetyrkin, S.V., Belyaeva, O.V., Lapshina, E.A., Lin, D.W., Matsumura, M., and Nelson, P.S. (2002) Evidence That the Human Gene for Prostate Short-Chain Dehydrogenase/Reductase (PSDR1) Encodes a Novel Retinal Reductase (RalR1), *J. Biol. Chem. 277:* 28909–28915.
69. Jörnvall, H., Danielsson, O., Hjelmqvist, L., Persson, B., and Shafqat, J. (1995) The Alcohol Dehydrogenase System, *Adv. Exp. Med. Biol. 372:* 281–294.
70. Duester, G., Farrés, J., Felder, M.R., Holmes, R.S., Höög, J.-O, Parés, X., Plapp, B.V., Yin, S.-J., and Jörnvall, H. (1999) Recommended Nomenclature for the Vertebrate Alcohol Dehydrogenase Gene Family, *Biochem. Pharmacol. 58:* 389–395.
71. Duester, G., Mic, F.A., and Molotkov, A. (2003) Cytosolic Retinoid Dehydrogenases Govern Ubiquitous Metabolism of Retinol to Retinaldehyde Followed by Tissue-Specific Metabolism to Retinoic Acid, *Chem.-Biol. Interact. 143/144:* 201–210.
72. Molotkov, A., Deltour, L., Foglio, M.H., Cuenca, A.E., and Duester, G. (2002) Distinct Retinoic Metabolic Functions for Alcohol Dehydrogenase Genes *Adh1* and *Adh4* in Protection Against Vitamin A Toxicity or Deficiency Revealed in Double Null Mutant Mice, *J. Biol. Chem. 277:* 12811–13804.
73. Perozich, J., Nicholas, H., Wang, B.C., Lindahl, R., and Hempel, J. (1999) Relationships Within the Aldehyde Dehydrogenase Extended Family, *Protein Sci. 8:* 137–146.
74. Lin, M., Zhang, M., Abraham, M., Smith, S.M., and Napoli, J.L. (2003) Mouse Retinal Dehydrogenase 4 (RALDH4), Molecular Cloning, Cellular Expression, and Activity in 9-*cis*-Retinoic Acid Biosynthesis in Intact Cells, *J. Biol. Chem. 278:* 9856–9861.
75. Niederreither, K., McCaffery, P., Dräger, U.C., Chambon, P., and Dölle, P. (1997) Restricted Expression and Retinoic Acid-Induced Downregulation of the Retinaldehyde Dehydrogenase Type 2 (RALDH-2) Gene During Mouse Development, *Mech. Dev. 62:* 67–78.
76. Fan, X., Molotkov, A., Manabe, S-I., Donmoyer, C.M., Deltour, L., Fonglio, M.H., Curnca, A.E., Blaner, W.S., Lipton, S.A., and Duester, G. (2003) Targeted Disruption of *Aldh1a1* (*Raldh1*) Provides Evidence for a Complex Mechanism of Retinoic Acid Synthesis in the Developing Retina, *Mol. Cell. Biol. 23:* 4637–4648.
77. Dupe, V., Matt, N., Garnier, J.-M., Chambon, P., Mark, M., and Ghyselinck, N.B. (2003) A Newborn Lethal Defect Due to Inactivation of Retinaldehyde Dehydrogenase Type 3 Is Prevented by Maternal Retinoic Acid Treatment, *Proc. Natl. Acad. Sci. USA 100:* 14036–14041.
78. Taimi, M., Helvig, C., Wisniewski, J., Ramshaw, H., White, J., Amad, M., Korczak, B., and Petkovich, M. (2004) A Novel Human Cytochrome P_{450}, CYP26C1, Involved in Metabolism of 9-*cis* and all-*trans* Isomers of Retinoic Acid, *J. Biol. Chem. 279:* 77–85.
79. Abu-Abed, S., MacLean, G., Fraulob, V., Chambon, P., Petkovich, M., and Dollé, P. (2002) Differential Expression of the Retinoic Acid-Metabolizing Enzymes CYP26A1 and CYP26B1 During Murine Organogenesis, *Mech. Dev. 110:* 173–177.
80. Abu-Abed, S.S., Beckett, B.R., Chiba, H., Chithalen, J.V., Jones, G., Metzger, D., Chambon, P., and Petkovich, M. (1998) Mouse P_{450}RAI (CYP26) Expression and Retinoic Acid-Inducible Retinoic Acid Metabolism in F9 Cells Are Regulated by Retinoic Acid Receptor Gamma and Retinoid X Receptor Alpha, *J. Biol. Chem. 273:* 2409–2415.
81. Abu-Abed, S., Dollé, P., Metzger, D., Beckett, B., Chambon, P., and Petkovich, M. (2001) The Retinoic Acid-Metabolizing Enzyme, CYP26A1, Is Essential for Normal

Hindbrain Patterning, Vertebral Identity, and Development of Posterior Structures, *Genes Dev. 15*, 226–240.

82. Genchi, G., Wang, W., Barua, A., Bidlack, W.R., and Olson, J.A. (1996) Formation of Beta-Glucuronides and of Beta-Galacturonides of Various Retinoids Catalyzed by Induced and Noninduced Microsomal UDP-Glucuronosyltransferases of Rat Liver, *Biochim. Biophys. Acta 1289:* 284–290.
83. Little, J.M, Lehman, P.A, Nowell, S., Samokyszyn, V., and Radominska, A. (1995) Glucuronidation of all-*trans*-Retinoic Acid and 5,6-epoxy-all-*trans*-Retinoic Acid. Activation of Rat Liver Microsomal UDP-Glucuronosyltransferase Activity by Alamethicin, *Drug Metab. Dispos. 25:* 5–11.

Chapter 2

Introduction to Vitamin A: A Nutritional and Life Cycle Perspective

A. Catharine Ross

Department of Nutritional Sciences, The Pennsylvania State University, University Park, PA 16802

Introduction

Vitamin A is an essential, life-maintaining component in the diet of all vertebrates. The forms of vitamin A present in the diet, retinol from foods of animal origin and provitamin A carotenoids from plants, must be metabolized within the animal's tissues to produce the bioactive compounds that actually mediate the actions of vitamin A. Vitamin A has long been recognized as essential for normal vision, embryonic development, growth, reproduction in the male and female, and for a host of other processes. The first molecular functions of vitamin A to be elucidated were those of 11-*cis*-retinal as the chromophore of rhodopsin in rods and opsin in cones of the vertebrate retina. The ability of retinoic acid (RA) to act as a potent hormone that mediates the nonvisual functions of vitamin A was recognized soon after all-*trans*-RA was first isolated in 1946. Since then, and especially after the characterization of the nuclear retinoid receptor system of retinoic acid receptor (RAR) and retinoid X receptor (RXR) genes beginning in the late 1980s, a growing body of research has demonstrated a large number of molecular targets of RA in normal and malignant cell models, the developing embryo, and in normal and genetically manipulated intact organisms. Important recent advances in many of these areas are addressed in detail in this book.

In Chapter 1, O'Byrne and Blaner discussed the proteins, enzymes, and receptors that determine the metabolism and functions of vitamin A. In this chapter, I will explore the functions and physiologic requirements for vitamin A across the life cycle, asking "What aspects of the metabolism or functions of vitamin A are, or may be, crucial at this stage of the life cycle?" First, the nutritional requirement for vitamin A and recent recommended dietary allowances by life stage, as established by the Institute of Medicine (IOM) in 2002 (1) are summarized. The chapter then discusses some of the features and issues regarding vitamin A that are germane to conception, pregnancy, and the perinatal period, and to the postnatal continuum from childhood and adolescence to adulthood and aging.

The Nutritional Requirement for Vitamin A and Recommended Dietary Allowances by Life Stage

Nutritional Equivalency

The dietary requirement for vitamin A is expressed in "equivalents" of retinol, a concept needed because there are two sources of vitamin A, i.e., preformed vitamin A from milk, eggs, and other animal tissues, and provitamin A carotenoids from yellow and green leafy vegetables and some colored fruits. Although strict carnivores and herbivores consume only preformed vitamin A (retinol) or only provitamin A carotenoids (mainly β-carotene, α-carotene and β-cryptoxanthin), respectively, omnivores, a category that includes most human beings, consume a mixture of preformed vitamin A and provitamin A. Consumption patterns from a U.S. survey from 1994 to 1996 showed that the major contributors of vitamin A are vegetables and fruits (~55%) followed by dairy products and meats (~30%) (1). The median adult intake in the U.S. National Health and Nutrition Examination Survey (NHANES III) was equivalent to ~687 μg RAE/d (see below regarding the RAE unit) (1). However, the proportions and absolute amounts of these two forms of vitamin A vary considerably. Factors affecting the intake of vitamin A by individuals and groups include the age of the individual, seasonal availability of foods, affordability of vitamin A-containing foods, and other physiologic and social factors that influence food consumption patterns, and hence the nutrient intake patterns, of individuals and groups.

For humans, retinol and β-carotene are considered to have the same qualitative activity because either form of the vitamin can be converted to retinal for vision and RA for the hormonal activities attributed to vitamin A. However, on a weight basis, retinol and β-carotene do not have the same bioactivity due to differences inherent in their structures and differences in the efficiency of their metabolism. Some carotenoids discussed later in this book are relatively abundant in foods (e.g., lycopene, lutein, and zeaxanthin) and can be absorbed to some extent, but they do not have the activity of vitamin A. Considerable research has gone into establishing the nutritional equivalency of retinol and β-carotene, when they are ingested as purified compounds and as they are present in the complex matrices of foods. The conversion factors thereby obtained have been used to define the vitamin A nutritional value of retinol- and carotenoid-containing foods in "retinol equivalents," a term that describes an amount of vitamin A, whether from retinol or its carotenoid precursors, with the biological activity of 1 μg of all-*trans*-retinol. The units of activity and conversion factors themselves are revised periodically as new data are obtained on the efficiency of bioconversion of various carotenoids to retinol. The Recommended Dietary Allowances (RDA) recently published by the IOM and Health Canada in 2002 (1) are expressed in a new unit of bioactivity, the Retinol Activity Equivalent (RAE). One μg RAE is equal to 1 μg of all-*trans*-retinol, 2 μg of all-*trans*-β-carotene in oil, 12 μg of β-carotene in foods, and 24 μg of other provitamin A carotenoids in foods. These conversion factors are necessarily average values, even though it is understood that the bioactivity of carotenoids differs with the type of foods in which they are present; for example, it is lower for β-carotene present in fibrous vegetables such as spinach than for β-carotene in fruits such as mangoes. Moreover, bioavailability can change as foods are processed, and the efficiency of bioconversion in humans is known to have a high coefficient of variation (2,3).

Dietary Reference Intakes (DRI)

Table 2.1 lists the dietary references intakes for vitamin A by life stage (1). For infants, an Adequate Intake (AI) has been established; it is based on the mean concentration of vitamin A in breast milk, 1.70 µmol/L, as determined in studies of well-nourished mothers, and an average milk intake of 0.78 L/d. For adults, data were available to calculate an Estimated Average Requirement (EAR) using a factorial method. A Recommended Dietary Allowance (RDA) was calculated from the EAR by adding twice the coefficient of variation, 20%, as a safety factor. The RDA is meant to cover the needs of essentially all healthy persons within the age-sex group for which it is set. For children and adolescents, there were insufficient

TABLE 2.1
Recommended Dietary Allowances (RDA) and Tolerable Upper Intake Levels (UL) for Vitamin A by Life Stage[a]

	RDA (µg RAE/d)		UL (µg retinol/d)	
Infants (AI[b])				
0–6 mo	400		600	
7–12 mo	00		600	
Children	Boys	Girls	Boys	Girls
1–3 y	300	300	600	600
4–8 y	400	400	900	900
9–13 y	600	600	1700	1700
14–18 y	900	700	2800	2800
Adults	Men	Women	Men	Women
$\geq$19 y	900	700	3000	3000
Pregnancy				
14–18 y		750		2800
19–50 y		770		3000
Lactation				
14–18 y		1200		2800
19–50 y		1300		3000

[a] *Source:* Reference 1. RAE, retinol activity equivalent.
[b] AI, adequate intake, is the oberved average or experimentally determined intake by a defined population or subgroup. The AI is used if sufficient scientific evidence is not available to derive an estimated average requirement (EAR).

age-specific data to calculate an EAR. It therefore was assumed that the EAR is the same, after being scaled for body weight, for children and adolescents as for adults. The RDA for children and adolescents was then set as the EAR plus 40%. The RDA for pregnancy and lactation were based on the RDA for women of reproductive age, plus an amount to cover the accrual of vitamin A in fetal tissues, or an amount of vitamin A equal to that transferred from the lactating woman into her breast milk, respectively.

Many questions remain regarding optimal vitamin A intakes. It would be desirable to have age-specific information on the absorption and utilization of vitamin A and its metabolites in neonates, children, adolescents, adult men and women, and the elderly, to estimate more precisely the dietary requirement for vitamin A across the life cycle.

Life Cycle

Fertility, Pregnancy, and Fetal Development

Male fertility requires adequate vitamin A. Failure of gametogenesis, reduced testosterone production, and loss of fertility are hallmarks of vitamin A deficiency in men [see (4) for review]. Vitamin A deficiency produces abnormalities in all three main types of testicular cells (Sertoli, germinal and Leydig cells). Mice mutant for the nuclear retinoid receptors, RARα and RXRβ are sterile (4). After ~10 wk of vitamin A deprivation, male rats develop a "Sertoli cell-only situation" (5), in which the proliferation and differentiation of type A spermatogonia and spermiogenesis are halted. After testosterone production falls, the male accessory sex organs atrophy.

In females, vitamin A is necessary for maintenance of a normal estrus cycle, and for the implantation of the blastocyst after conception. Retinol binding protein (RBP) is a major secretory product of the conceptus and the uterus (6). Pig blastocysts, collected on d 15 of pregnancy by lavage, secreted RBP, whereas both d 15 blastocysts and d 15 pregnant endometrium expressed mRNAs transcripts for the cellular retinol-binding protein (CRBP), and for RARα and RARγ (6). It is suggested that retinol transport within developing conceptus and adjacent uterine tissues, as inferred from the patterns of protein and gene expression noted above, may be necessary in embryonic development, and for the differentiation of extraembryonic membranes, and successful implantation and uterine growth (6).

Many studies have addressed the role of RA in regulating gene expression and morphogenesis in the developing embryo (7). By the 1940s, it had been shown that both a deficiency and an excess of vitamin A produced developmental abnormalities in rat fetuses. Subsequent research demonstrated the inductive activity of RA on chick limb structure and established RA as a putative, and possibly critical, morphogen (8). Several molecular targets of RA were identified in early gestational embryos, including homeobox genes, growth factors, and transcription factors that are crucial for proper morphogenesis [see (7,8)]. The fetuses of pregnant rats deprived of RA from embryonic day (ED)11.5 to ED13.5 had neural crest, ocular, and nervous system defects (9), similar to those seen in mutant mice with deletions of the genes for RXRα, RXRα and RARα, and RARα and RARγ. Even moderate vitamin A deficiency in pregnant rats reduced the number of live births and altered the growth trajectory of fetal organs (10). Retinoid biosynthetic and catabolic enzymes are expressed early in mouse embryogenesis. Retinal dehydrogenase (RALDH)2, the enzyme most likely to be rate limiting for the production of RA (see Chapter 1), is expressed as early as ED7.5 (11), whereas cytochrome P_{450} (RA inducible) 26 (CYP26), potentially a counterregulatory enzyme capable of reducing the activity of RA, is expressed nearby during the same period, but usually not in the same cells or identical region of the embryo (12). The expression patterns of these putative biosynthetic and degradative enzymes suggest the induction and suppression of RA biogenesis in waves during morphogenesis. A general model developed from numerous studies is that the RA required for regulation of embryonic gene expression is produced by the embryo using, as substrate, retinol obtained by uptake from the maternal blood supply. Maternal vitamin A deficiency would obviously compromise RA production.

Numerous natural and synthetic retinoids, particularly compounds with a free carboxylate group, are highly teratogenic when administered to pregnant rats, mice, and primates at critical periods of gestation that correspond to times when the embryonic body pattern is established and neural, facial, cardiovascular, and visceral organs are formed (7). The abnormalities produced experimentally in animals closely resemble the birth defects that occurred in infants whose mothers were exposed early in pregnancy to prescription retinoids, before the danger of these treatments was recognized (7). The exposure of the pregnant mother to a teratogenic retinoid during the period of fetal organogenesis is hypothesized to override the closely orchestrated pattern of endogenous fetal RA production and gene expression that is necessary for normal embryonic development.

Despite much evidence that RA mediates the biological effects of retinol in the early embryo, there may also be an essential function for retinol itself, or a metabolite different from RA, for successful pregnancy. When pregnant rats were maintained in a marginally vitamin A–deficient state and supplemented with small amounts of RA sufficient for normal early fetal development, the pregnant rats still required retinol in midgestation to carry their fetuses to term, and for the survival of the newborns (13). By *in situ* hybridization, the genes for RBP and its cotransport protein transthyretin are transcribed in the developing rat embryo by ED7 in the visceral extraembryonic endoderm. They are transcribed in yolk sac and fetal liver by ED10–13, with a gradual increase in the expression in liver in later stages of fetal development (14,15). These patterns of gene expression suggest that the capacity for retinoid transport from mother to fetus and the metabolism of retinoids within fetal tissues is established early in gestation.

Although studies in animal models demonstrated the necessity of vitamin A for gestation, only recently was compelling evidence presented that improving the vitamin A status of pregnant women can actually improve pregnancy outcome. To assess whether routine supplementation with vitamin A or β-carotene could reduce mortality in mother and infants, West and coworkers (16) conducted a series of randomized, placebo-controlled intervention trials in a rural region (Sarlahi district) of Nepal, a region known to have a high rate of low serum retinol values and night blindness (indicators of a high prevalence of vitamin A deficiency in this population). Women were randomly assigned to groups that received a placebo, or either vitamin A or β-carotene, at the same dosage of 7000 μg retinol equivalents/wk, a level that is comparable to the U.S. RDA expressed on a weekly basis. Maternal mortality during pregnancy and 12 wk postpartum was reduced by 40% in the vitamin A group ($P < 0.04$) and by 49% in the β-carotene group ($P < 0.01$). The overall effect of either vitamin A or β-carotene was highly significant ($P < 0.001$). Follow-up studies showed a reduction in the prevalence of night blindness during pregnancy and in some pregnancy-related and perinatal illnesses (17). Together these data support the importance of adequate vitamin A as a means to improve the reproductive health and pregnancy outcome of women in vulnerable populations such as those that still exist in parts of the developing world (18).

Perinatal and Neonatal Periods

By the 3rd trimester of pregnancy, organ systems are formed, but they are still immature. Fetal tissues therefore must undergo both growth and maturation before a normal full-term birth. For many organ systems, maturation continues postnatally. The body's ability to store and metabolize retinoids undergoes significant ontological development, as has been studied in animal models and inferred from studies of human fetal and postnatal tissues.

As discussed in Chapter 1, CRBP solubilize tissue retinoids and facilitate several aspects of retinoid metabolism. The pattern of expression of several of these proteins changes with developmental stage during the perinatal period. The genes for CRBP and CRBP II exhibit different developmental patterns of expression in liver, intestine, lung, kidney, testes, and placenta (19). In rats, CRBP mRNA is detectable in the small intestine by ED16, before the development of a well-differentiated absorptive epithelium; it then remains nearly constant throughout the peri- and postpartum periods. In contrast, intestinal CRBP II mRNA accumulates in parallel with the times of appearance of the intestinal absorptive epithelium, and stays high postnatally and in adulthood (19). Both CRBP and CRBP II genes were expressed in fetal liver but the concentration of CRBP mRNA increased markedly during the suckling and early weaning periods, whereas that of CRBP II mRNA levels fell abruptly after birth. In pregnant and lactating female rats, CRBP and CRBP II were markedly elevated in the liver and small intestine, respectively, before delivery, and declined rapidly after parturition. Given these different patterns of CRBP/CRBP II gene expression, Levin *et al.* (19) suggested that these proteins could serve different physiologic functions. Both of these proteins function as chaperones capable of directing retinol to lecithin:retinol acyltransferase

(LRAT), a microsomal enzyme that catalyzes the formation of retinyl esters (RE), but CRBP II also facilitates the formation of retinol from retinal, and therefore is implicated in the intestinal metabolism of β-carotene (see Chapter 1). Hepatic CRBP is important for RE storage and for controlling the concentration of unesterified retinol, whereas CRBP II is involved in vitamin A absorption. Neither of these retinoid-binding proteins is truly essential, as shown by the viability and growth of mice lacking CRBP or CRBP II, but both proteins appear to confer an adaptive advantage under conditions in which vitamin A is limited. Indeed, a general lesson learned from studies of mice lacking CRBP and/or CRBP II, RBP, or LRAT is that the health of these animals may appear to be normal as long as they are provided with a steady, adequate supply of vitamin A in milk or their post-weaning diet (20–23). However, mice lacking genes for CRBP, CRBP II, RBP, or LRAT are markedly compromised in their ability to conserve and adapt to periods of low vitamin A intake. Thus, these binding proteins, and enzymes, appear to function as "efficiency factors," which provide an important adaptive advantage to species whose access to vitamin A may be infrequent or irregular, whether due to seasonal availability of vitamin A in the food supply, or to other factors that affect nutrient availability.

Vitamin A Status in the Perinatal Period. Birth comprises a major physiologic transition (24). The immediate transition to breathing air is abrupt, followed rapidly by the need to autonomously regulate body temperature, absorb lipid- and protein-rich colostrum from the gut, adapt to the commensal microorganisms that begin to colonize the colon, and resist environmental pathogens through the expression of appropriate immune responses (25). Vitamin A is important for some, if not all, of these processes.

What is the infant's vitamin A status at birth? Dann (26) first addressed this question in the early 1930s, showing that the amount of vitamin A in the liver of newborn rats and rabbits was very low irrespective of the level of carotene that was fed to the pregnant mother. The liver of fetal rats from adequately nourished pregnant mothers contained 1–2 μg retinol/g tissue (27). When rats were fed diets during pregnancy and lactation that contained either a lower level of retinol (0.6 μg/g) or a higher level (15 μg/g) than that in a standard diet (4 μg/g), the retinol concentration of the dam's liver averaged ~130 and 640 μg of retinol/g liver, respectively (28), values that are well above the level of ~20 μg/g that was suggested as an indicator of borderline vitamin A status (29). Yet the concentrations of retinol in the liver of their newborns were only 9 and 15 μg retinol/g, respectively. Dharo *et al.* (30) reported mean liver total retinol values <20 μg/g for a sample of 15 Midwestern infants dying of sudden infant death syndrome for which autopsy specimens were available. Although maternal vitamin A status was not assessed, it was likely to have been adequate in the Midwestern population studied. Thus, newborn animals and humans begin life with low reserves of retinol even when the mother has adequate vitamin A stores.

The quantity and form of vitamin A available for intestinal absorption by neonatal animals and humans depends on the vitamin A intake of the mother and the concentration of vitamin A in her milk, or the type of replacement formula fed to the infant in place of breast milk. In newborn rats, the vitamin A content averaged 7.92 μg in unsuckled pups, and 12.04 μg in pups suckled by normally nourished dams (27). However, vitamin A continues to accrue during the suckling period only if the mother's intake of vitamin A is adequate. The milk of lactating rats fed a high vitamin A diet (15 μg/g) contained two- to threefold more vitamin A/g of fat than did the milk of rats fed 0.6 μg of retinol/g diet, even though both groups had comparable plasma retinol levels (28). In lactating Indonesian women and their infants studied in a community trial of vitamin A supplementation, fewer mothers in the vitamin A–supplemented group had low breast milk retinol (<1.05 μmol/L), and significantly fewer of their infants had plasma retinol values <0.52 μmol/L (31). Humphrey and Rice (32) calculated that in the first 6 mo of life, healthy, well-nourished infants more than double the size of the liver and increase its vitamin A concentration fivefold, increasing total vitamin A by ~70 μmol during this period.

The type of vitamin A in human milk also differs with the form in which the mother consumes vitamin A. In a multicountry survey of lactating women consuming their usual diets, carotenoid intake was documented by a 24-h dietary recall, and milk carotenoid concentration was determined. The carotenoid composition of breast milk varied greatly across countries, with the greatest differences in β-cryptoxanthin (~ninefold) and the least in α-carotene and lycopene (~threefold), whereas breast milk retinol concentrations varied ~twofold (33). From these data it may be concluded that neonates are exposed to different types as well as amounts of vitamin A in breast milk. But the neonate's ability to utilize these different forms is not well understood. With a better understanding of the efficiency of neonatal retinol and carotenoid utilization, better recommendations could be made regarding an optimal intake (amount and form) of dietary vitamin A during lactation.

The recent intake of dietary vitamin A, in addition to maternal plasma retinol, is a determinant of breast milk vitamin A. As observed in a physiologic study in rats, the uptake of RE from chylomicron into the lactating mammary gland increases in direct proportion to chylomicron RE contents (34). Uptake was dependent on active lipolysis in the lactating mammary gland (34). These data suggest two pathways for the uptake of vitamin A into the breast, one in which retinol is obtained from circulating holo-RBP and another from chylomicron RE. The RE component of chylomicrons varies in direct proportion to vitamin A in the diet. Differences in the delivery of chylomicron RE to the lactating mammary glands may explain the changes in milk vitamin A concentration in response to differences in maternal vitamin A consumption.

Physiologic Processes Affected by Vitamin A in the Perinatal Period. The birth transition also requires the development of autonomous body temperature regulation, and deposition of fat reserves for adequate insulation and as an energy reserve. Uncoupling proteins, UCP, are key regulators of thermogenesis in brown adipose tissue. Retinoic acid is a positive regulator of UCP genes. The brown adipose tissue of growing mice fed a vitamin A–deficient diet contained lower levels of UCP1 and UCP2 mRNAs than were present in the brown adipose tissue of mice fed a standard diet. Conversely, brown adipose tissue UCP mRNA levels were elevated by RA (35). These results suggest a possible role and a mechanism for vitamin A in the regulation of thermogenesis. However, these studies did not include newborn mice. Whether vitamin A or RA might improve the health of animal or human neonates by increasing their capacity for thermogenesis warrants exploration.

The lungs of humans, rodents, and numerous other mammals are immature at birth, and undergo extensive capillary remodeling and expansion of the alveolar surface area (alveolarization) in the postnatal period (36). Studies of lung cell cultures and explants showed that retinoids induce the development of the lung stroma, extracellular matrix, and epithelium, and retinoids are produced and metabolized in lung interstitial cells (37). RA is capable of increasing the formation of alveoli in neonatal rodents and improving tissue repair responses in models of emphysema or surgical lung resection (38,39). Because the lungs of premature infants are even less developed than those of term infants, the risk of death, bronchopulmonary dysplasia, and chronic lung disease is elevated in premature infants. It was therefore of interest to determine whether providing vitamin A could improve health outcomes in these very-low-birth-weight premature infants. A small clinical study by Shenai *et al.* (40) showed that administering vitamin A intramuscularly three times weekly could increase plasma retinol and RBP in very-low-birth-weight infants. A subsequent multicenter trial conducted by participating investigators in the National Institute of Child Health and Human Development (NICHD) Neonatal Research Network (41) and a meta-analysis conducted according to the standards of the Cochrane Neonatal Review Group (42) provided evidence that supplementing very-low-birth-weight infants with vitamin A is associated with a reduction in death or oxygen requirement at 1 mo of age, and oxygen requirement among survivors at 36 wk postmenstrual age. Despite these encouraging results, not all neonatal intensive care units use vitamin A as part of standard care (42). Further studies to identify the route, frequency, dose, and form of vitamin A that is most effective in improving neonatal survival and long-term outcome are needed (43).

Interest in vitamin A requirements in the neonatal period was heightened by reports that vitamin A supplementation to newborns may improve outcomes in the neonatal period. Rahmathullah *et al.* (44) recently reported the results of a large randomized, placebo-controlled, community-based trial of vitamin A supplementation on d 1 and 2 of life. All-cause mortality at 6 mo of age was the primary end point. The participants were 11,691 liveborn infants of mothers in the Tamil Nadu province of southern India, a region in which a previous trial of vitamin A supplementation in 6-mo- to 6-y-old children had significantly reduced mortality. In the present study, newborns were treated twice by a healthcare worker within a 24-h interval after birth with an amount of vitamin A (24,000 IU) equal to 8000 μg retinol. Supplementation resulted in a 23% reduction in mortality at 6 mo. Interestingly, the survival curves began to diverge at ~2 wk of age and continued to separate until 3 mo of age, indicating that vitamin A reduced mortality early in this period, after which the curves were parallel, indicating no further effect after 3 mo. The effect of vitamin A was greatest in low-birth-weight infants, although it could not be determined whether these were premature infants because their postmenstrual age was not known. The reduction in infant mortality in this study is consistent with an earlier smaller hospital-based study in Indonesia in which the provision of 50,000 IU of vitamin A at birth reduced mortality at 1 y by 64% (32,45). In that study, the greatest effect of vitamin A was seen in infants with birth weights ≥2500 g. Although it is not known by what mechanisms vitamin A supplementation provides protection to newborns (or to older children, see below), several hypotheses are plausible, including beneficial effects on the lungs, intestinal epithelium, and immune system. Because mortality is a "hard" end point, a leftward shift in the distribution of illness severity, as depicted in the hypothetical illustration in Figure 2.1, in infants treated with vitamin A could result in fewer neonates developing illnesses severe enough to reach the critical end point of mortality, and thereby could reduce the mortality rate in the population at large.

It is now of great interest to determine whether perinatal treatment with vitamin A has any long-term benefit, or at least no adverse effect, in those children who received vitamin A at birth. No adverse effect was observed on neurological functions (mental, psychomotor, and behavioral rating scales) at 3 y of age in Indonesian children who had received 50,000 IU (15 mg) of retinol as newborns (46). Consistent with this, a follow-up study of neurodevelopmental outcomes in 2-y-old children who had received vitamin A or placebo as extremely-low-birth-weight infants in the NICHD-sponsored multicenter vitamin A trial mentioned above, had no adverse effects associated with vitamin A treatment as neonates

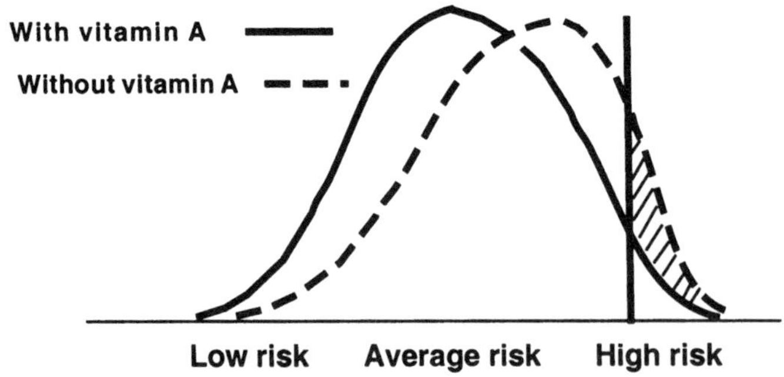

Fig. 2.1. Hypothetical distributions of risk in young children with and without treatment with vitamin A. With mortality as a "hard" end point (noted by vertical line) in trials of vitamin A supplementation, a leftward shift in the population distribution of risk in the vitamin A-treated group may explain the reduction in mortality observed in vitamin A supplementation trials. The difference in mortality with and without vitamin A is illustrated by the section of the distribution between the two lines, to the right of the vertical line.

(47). Some results, although not significant, indicated a trend toward improved long-term neurodevelopmental status in the vitamin A–treated group (47).

Childhood

The postweaning life stage has long been recognized as a vulnerable period with respect to the development of vitamin A deficiency. Increased risk at this time appears to be the result of a combination of low accrual of vitamin A from breast milk (low concentration, volume, or short duration of breast feeding) and a vitamin A–poor postweaning diet (48). In children in well-nourished populations, plasma retinol levels rise gradually with age (49). Serum retinol was determined in NHANES III (1988–1994) for males and females from 4 to >71 y old. As shown in Figure 2.2, 4- to 8-y-old girls and boys had equivalent distributions of serum retinol, with median (50th percentile) values of 1.19 µmol/L for both sexes. When subdivided by race/ethnicity, non-Hispanic White children had slightly higher median values (1.22 µmol/L) compared with non-Hispanic Black and Mexican-American children (1.12 µmol/L each). The median values were higher for 9- to 13-y-old children, and a difference between girls and boys, with boys having higher levels, began to become evident. Similar to the values for 4- to 8-y-olds, the median retinol concentrations were higher in 9- to 13-y-old non-Hispanic White children than in non-Hispanic Black and Mexican-American children (49).

Evidence that vitamin A supplementation is effective in reducing child mortality in 6-mo- to 6-y-old children in high-risk populations is now available from sev-

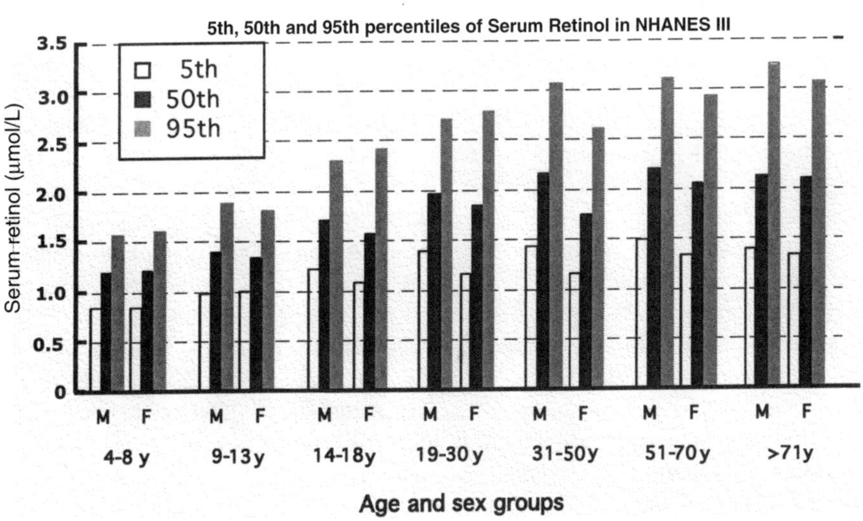

Fig. 2.2. Serum retinol distributions by age and sex in the National Health and Nutrition Examination Survey, NHANES III, 1998–1994. Data shown are the 5th, 50th, and 95th percentiles for all individuals in the age-sex groups shown. [See (49) for additional data.]

eral randomized controlled studies conducted in various regions of the developing world; several meta-analyses of these data were published [see (50,51) for reviews]. In brief, vitamin A provided either in high-dose form (60–120 mg retinol) at 4- to 6-mo intervals, or weekly at a level similar to the RDA proved effective in reducing all-cause mortality by ~23%. Now that a consensus has been reached that vitamin A is an effective means of reducing mortality, research has shifted to studies designed to compare the effectiveness of various modes of delivering vitamin A, and to multimicronutrient supplementation in at-risk populations.

Adolescence, Adulthood and Aging

The period from adolescence through old age appears to be a time of maintenance with respect to vitamin A functions. It is typically a period in which tissue vitamin A accumulates slowly, differing in extent according to dietary intake. A nutritional deficiency of vitamin A in this age range is extremely rare. In experimental animals, a prolonged period of time consuming a vitamin A–free diet is required to induce low plasma retinol and vitamin A deficiency in adult animals whose tissues stores of vitamin A are already substantial. Based on retinol kinetic studies, vitamin A catabolism is limited (52,53). In affluent human populations, adolescent and adult plasma retinol concentrations are higher than those in children. The divergence in serum retinol values between males and females becomes more marked around puberty, and serum retinol levels in men remain slightly higher than those in women through the 5th decade. Serum retinol values for women > 51–70 y old are similar to those for men of the same age, and serum levels remain relatively constant into the 8th decade of life [(49) and Fig. 2.2].

The capacity for vitamin A storage is maintained across the life span, and tissues levels may accumulate substantially when dietary vitamin A exceeds needs. In a study of rats fed diets containing a marginal, adequate, or nontoxic but excessive level of retinol, from weaning until 2–3, 8–10, or 18–20 mo of age, liver vitamin A accumulation depended on both diet and age (54). LRAT and CYP26, enzymes implicated in retinoid homeostasis, were expressed at very low levels in the liver of rats fed a marginal vitamin A diet. In contrast, rats fed the vitamin A–supplemented diet had significantly elevated levels of LRAT and CYP26, and both of these increased gradually with age and vitamin A storage (55,56). Although diet-dependent differences in vitamin A storage and the expression of LRAT and CYP26 were not significant in 2- to 3-mo-old rats, the effects of different levels of vitamin A intake were highly significant in middle-aged rats, and increased further in old rats. Overall, tissue retinoids, mainly as RE, accumulated slowly and significantly with aging as long as dietary vitamin A intake was sufficient.

In elderly humans, liver vitamin A levels also appear to be well correlated with dietary vitamin A intake. Ribaya-Mercado *et al.* (57) assessed the dietary vitamin A intake of Filipino elders 60–88-y old using three nonconsecutive 24-h dietary recalls and determining liver vitamin A concentrations by a deuterated-retinol dilution method. Total-body vitamin A or liver vitamin A, but not serum retinol, correlated with dietary vitamin A (total RAE, retinol and β-carotene intake/d). The relation between intake and storage observed in this community study was consistent with a large body of experimental data in animals indicating that tissue vitamin A stores can be predicted from vitamin A intake, but not from serum retinol concentrations.

Risk and Upper Levels

At the adult stage of life, is the risk no longer one of developing vitamin A deficiency but, conversely, of consuming too much retinol? Epidemiologic studies focusing on bone health suggested that this may be the case. Data from animal, human, and laboratory research support an association between a high intake of vitamin A and a loss of bone mineral density (BMD), a risk factor for osteoporosis. In growing rats fed high levels of vitamin A, the epiphyseal growth plate of the proximal tibia disappeared in a dose-dependent manner (58), and rarification of bone (reduced density and increased porosity) preceded the presence of actual fractures (59). In another study, vitamin A reduced total bone ash and bone resorption, regardless of cotreatment with vitamin D_3 (60).

The results of epidemiologic and clinical studies suggest that a loss of BMD may occur with intakes of retinol that do not greatly exceed the RDA. The risk of hip fracture was elevated twofold in older Swedish men and women whose retinol intake was >1500 µg/d compared with those whose intake was <500 µg/d (61). A longitudinal study of Swedish men showed a significant dose-dependent increase in bone fractures in men with the highest levels of serum retinol (62). In a short-term metabolic study of healthy Swedish men and women (n = 9), a single dose of vitamin A in an amount similar to that in a serving of liver caused a significant blunting of the plasma calcium increase produced by the administration of 1,25-dihydroxy vitamin D (63), consistent with the idea that a high intake of vitamin A may antagonize the effect of vitamin D. A study of relatively affluent and active community-living older Americans in Rancho Bernardo, California showed that the use of retinol-containing supplements was relatively high, i.e., 50% of women and 39% of men reported use of supplemental retinol (64). Female supplement users in this population showed a change in BMD that was inversely associated with retinol intake (64). An analysis of data for >72,000 postmenopausal women in the Nurses' Health Study suggests that the risk of hip fracture is higher when retinol intake exceeds 3000 µg/d (65) (see section on Upper Level, below).

Overall, the emerging results from human studies suggest that a chronically elevated intake of vitamin A, on the order of 3000 µg/d (~4 times the RDA), may increase the risk of osteoporotic bone disease and fracture, at least in older men and women. Additional studies to evaluate the effect of vitamin A and its interactions with vitamin D, calcium, and other nutrients on BMD and fracture risk appear to be warranted. It is yet to be determined whether an elevated intake of vitamin A early in adult life might be a risk factor for bone disease and fractures in old age.

If too much retinol is a risk factor for bone health, yet supplement use is prevalent in the affluent elderly, would it be prudent to recommend that most of vitamin A consumed by the elderly be in the form of provitamin A carotenoids, and not retinol? Before such a recommendation could be considered evidence-based, we must gain a better understanding of whether the capacity to utilize carotenoids is maintained during aging. A recent study (66) addressed the question of the capacity to absorb β-carotene, α-carotene, lutein, and lycopene, measured by the postprandial chylomicron response to meals of vegetables containing these carotenoids, in older (60–75 y) French subjects (n = 8) compared with 8 young (20–35 y) subjects. There was no major effect of age on the postprandial area under the curve (AUC) for these carotenoids, implying that absorption of these carotenoids was not impaired with age, although the AUC for lycopene was 40% lower in the old subjects. Numerous conditions could, however, affect the bioavailability of carotenoids present in fibrous foods, including dentition, the efficiency of mastication, hydrochloric acid production, and digestive enzyme release, all conditions that are more likely to be impaired in elderly than in younger individuals. These physiologic conditions may be underappreciated factors in an individual's ability to effectively utilize the carotenoids that are present in their diet.

The Upper Level (UL)—Recommendations for Limiting the Intake of Preformed Retinol. An excess of vitamin A has serious and potentially irreversible effects, as observed by the teratogenic effects of excess vitamin A in pregnancy and numerous reports of impaired liver function in individuals who have consumed an excess of vitamin A either chronically or acutely at high dosage. The IOM report of 2002 established, for the first time, a UL for preformed vitamin A (1). Upper Level values for vitamin A by life stage group are shown in Table 2.1. The UL is defined as the highest intake of a nutrient that, consumed over time, is likely to pose no risk of adverse health effects in nearly all healthy individuals; it is meant to be a guideline for safe levels of consumption. In establishing the UL for vitamin A, the critical adverse effects used were risk of birth defects (teratogenesis) in women of reproductive age, and liver abnormalities for men and women >50 y old. Because these effects are seen only after ingestion of excessive retinol, not carotenoids, the UL for vitamin A applies only to the intake of preformed vitamin A (retinol from foods, fortified foods, and supplements). The UL is not meant to apply to individuals taking vitamin A under medical supervision. The UL established by the IOM is consistent with the American Teratology Society's recommendation that the daily intake of vitamin A be <3000 μg in pregnant women. It is noteworthy that the UL values for vitamin A for some age-sex groups are no more than fourfold above the RDA.

Summary and Speculation

In this brief tour of the life cycle, we observed that human neonates and their animal counterparts begin life with only a small endowment of vitamin A. Therefore, they depend on receiving an adequate intake of vitamin A from breast milk or a replacement formula to develop a "margin of safety," that is, to increase the vitamin A concentrations of their liver and to supply adequate retinol to other tissues. In humans and animals, plasma retinol levels are lower at birth than in early childhood, even when maternal vitamin A status is adequate. Thus "physiological low vitamin A status" appears to be the norm for neonates. But does the *norm* represent what is *optimal* for the health of the neonate? We do not know at this point. The ability of a perinatal vitamin A supplement to improve the survival of neonates in an at-risk population, as shown recently in Tamil Nadu, India, suggests that the gap between vitamin A deficiency and vitamin A adequacy could be very narrow at this time of life. In this study, vitamin A was most beneficial to infants of low birth weight. Premature very-low-birth weight infants, even in the United States, have lower plasma retinol levels, lower liver vitamin A concentrations, and a higher rate of morbidity and mortality than term infants. Would routine supplementation with vitamin A improve outcomes in this vulnerable group? Extending this question to term infants, "Would there be measurable benefit from routinely supplementing all newborns with vitamin A?" A prophylactic dose of vitamin K soon after birth is considered part of standard care in many affluent countries—might the provision of a "safety dose" of vitamin A be of advantage to high-risk infants, or even to neonates presumed to be healthy? As vitamin A research soon enters its second century, many questions concerning the health implications of dietary vitamin A remain to be addressed.

In the elderly, the risk appears to be that of consuming an excess of vitamin A, specifically in the form of retinol. Additional research is warranted to better understand the ability of the elderly to digest, absorb, metabolize, and maintain retinoid homeostasis from diets rich in fruits and vegetables, or supplements containing β-carotene instead of retinol. With additional data on the relation of dietary vitamin A to health status across the life span, it may be possible to recommend "tailored patterns" of vitamin A intake that more closely suit the shifting need for vitamin A from early life through old age.

References

1. Institute of Medicine (2002) Dietary Reference Intakes for Vitamin A, Vitamin K, Arsenic, Boron, Chromium, Copper, Iodine, Iron, Manganese, Molybdenum, Nickel, Silicon, Vanadium, and Zinc, National Academy Press, Washington, D.C.
2. Tanumihardjo, S.A. (2002) Factors Influencing the Conversion of Carotenoids to Retinol: Bioavailability to Bioconversion to Bioefficacy, *Int. J. Vitam. Nutr. Res. 72:* 40–45.
3. Edwards, A.J., Nguyen, C.H., You, C.S., Swanson, J.E., Emenhiser, C., and Parker, R.S. (2002) α- and β-Carotene from a Commercial Puree Are More Bioavailable to Humans than from Boiled-Mashed Carrots, as Determined Using an Extrinsic Stable Isotope Reference Method, *J. Nutr. 132:* 159–167.
4. Livera, G., Rouiller-Fabre, V., Pairault, C., Levacher, C., and Habert, R. (2002) Regulation and Perturbation of Testicular Functions by Vitamin A, *Reproduction 124:* 173–180.

5. Morales, A., and Cavicchia, J.C. (2002) Spermatogenesis and Blood-Testis Barrier in Rats After Long-Term Vitamin A Deprivation, *Tissue Cell 34:* 349–355.
6. Harney, J.P., Masaratt, A., Vedeckis, W.V., and Bazer, F.W. (1994) Porcine Conceptus and Endometrial Retinoid-Binding Proteins, *Reprod. Fertil. Dev. 6:* 211–219.
7. McCaffery, P.J., Adams, J., Maden, M., and Rosa-Molinar, E. (2003) Too Much of a Good Thing: Retinoic Acid as an Endogenous Regulator of Neural Differentiation and Exogenous Teratogen, *Eur. J. Neurosci. 18:* 457–472.
8. Tickle, C. (2002) The Early History of the Polarizing Region: from Classical Embryology to Molecular Biology, *Int. J. Dev. Biol. 46:* 847–852.
9. Dickman, E.D., Thaller, C., and Smith, S.M. (1997) Temporally-Regulated Retinoic Acid Depletion Produces Specific Neural Crest, Ocular and Nervous System Defects, *Development 124:* 3111–3121.
10. Antipatis, C., Grant, G., and Ashworth, C.J. (2000) Moderate Maternal Vitamin A Deficiency Affects Perinatal Organ Growth and Development in Rats, *Br. J. Nutr. 84:* 125–132.
11. Ulven, S.M., Gundersen, T.E., Weedon, M.S., Landaas, V.O., Sakhi, A.K., Fromm, S.H., Geronimo, B.A., Moskaug, J.O., and Blomhoff, R. (2000) Identification of Endogenous Retinoids, Enzymes, Binding Proteins, and Receptors During Early Postimplantation Development in Mouse: Important Role of Retinal Dehydrogenase Type 2 in Synthesis of all-*trans*-Retinoic Acid, *Dev. Biol. 220:* 379–391.
12. Sakai, Y., Meno, C., Fujii, H., Nishino, J., Shiratori, H., Saijoh, Y., Rossant, J., and Hamada, H. (2001) The Retinoic Acid-Inactivating Enzyme CYP26 Is Essential for Establishing an Uneven Distribution of Retinoic Acid Along the Anterio-Posterior Axis Within the Mouse Embryo, *Genes Dev. 15:* 213–225.
13. Wellik, D.M., Norback, D.H., and DeLuca, H.F. (1997) Retinol Is Specifically Required During Midgestation for Neonatal Survival, *Am. J. Physiol. 272:* E25–E29.
14. Makover, A., Soprano, D.R., Wyatt, M.L., and Goodman, D.S. (1989) An *in Situ*-Hybridization Study of the Localization of Retinol-Binding Protein and Transthyretin Messenger RNAs During Fetal Development in the Rat, *Differentiation 40:* 17–25.
15. Thomas, T., Southwell, B.R., Schreiber, G., and Jaworowski, A. (1990) Plasma Protein Synthesis and Secretion in the Visceral Yolk Sac of the Fetal Rat: Gene Expression, Protein Synthesis and Secretion, *Placenta 11:* 413–430.
16. West, K.P., Jr., Katz, J., Khatry, S.K., LeClerq, S.C., Pradhan, E.K., Shrestha, S.R., Conner, P.B., Dali, S.R., Christian, P., Pokhrel, R.P., Sommer, A., and the N-S Group (1999) Double Blind, Cluster Randomised Trial of Low Dose Supplementation with Vitamin A or β-Carotene on Mortality Related to Pregnancy in Nepal, *Br. Med. J. 318:* 570–575.
17. Christian, P., West, K.P., Jr., Khatry, S.K., Katz, J., LeClerq, S.C., Kimbrough-Pradhan, E., Dali, S.M., and Shrestha, S.R. (2000) Vitamin A or β-Carotene Supplementation Reduces Symptoms of Illness in Pregnant and Lactating Nepali Women, *J. Nutr. 130:* 2675–2682.
18. Christian, P. (2003) Micronutrients and Reproductive Health Issues: An International Perspective, *J. Nutr. 133:* 1969S–1973S.
19. Levin, M.S., Li, E., Ong, D.E., and Gordon, J.I. (1987) Comparison of the Tissue-Specific Expression and Developmental Regulation of Two Closely Linked Rodent Genes Encoding Cytosolic Retinol-Binding Proteins, *J. Biol. Chem. 262:* 7118–7124.
20. Ghyselinck, N.B., Båvik, C., Sapin, V., Mark, M., Bonnier, D., Hindelang, C., Dierich, A., Nilsson, C.B., Håkansson, H., Sauvant, P., Azaïs-Braesco, V., Frasson, M., Picaud, S., and

Chambon, P. (1999) Cellular Retinol-Binding Protein I Is Essential for Vitamin A Homeostasis, *EMBO J. 18:* 4903–4914.

21. Saari, J.C., Nawrot, M., Garwin, G.G., Kennedy, M.J., Hurley, J.B., Ghyselinck, N.B., and Chambon, P. (2002) Analysis of the Visual Cycle in Cellular Retinol-Binding Protein Type I (CRBPI) Knockout Mice, *Investig. Ophthalmol. Vis. Sci. 43:* 1730–1735.
22. E, X.P., Zhang, L., Lu, J.Y., Tso, P., Blaner, W.S., Levin, M.S., and Li, E. (2002) Increased Neonatal Mortality in Mice Lacking Cellular Retinol-Binding Protein II, *J. Biol. Chem. 277:* 36617–36623.
23. Batten, M.L., Imanishi, Y., Maeda, T., Tu, D.C., Moise, A.R., Bronson, D., Possin, D., Van Gelder, R.N., Baehr, W., and Palczewski, K. (2004) Lecithin-Retinol Acyltransferase Is Essential for Accumulation of all-*trans*-Retinyl Esters in the Eye and in the Liver, *J. Biol. Chem. 279:* 10422–10432.
24. Gluckman, P.D., Sizonenko, S.V., and Bassett, N.S. (1999) The Transition from Fetus to Neonate—An Endocrine Perspective, *Acta Paediatr. Suppl. 88:* 7–11.
25. Marshall-Clarke, S., Reen, D., Tasker, L., and Hassan, J. (2000) Neonatal Immunity: How Well Has It Grown Up? *Immunol. Today 21:* 35–41.
26. Dann, W.J. (1932) The Transmission of Vitamin A from Parents to Young in Mammals, *Biochem. J. 26:* 1072–1080.
27. Ismadi, S.D., and Olson, J.A. (1982) Dynamics of the Fetal Distribution and Transfer of Vitamin A Between Rat Fetuses and Their Mother, *Int. J. Vitam. Nutr. Res. 52:* 111–118.
28. Davila, M.E., Norris, L., Cleary, M.P., and Ross, A.C. (1985) Vitamin A During Lactation: Relationship of Maternal Diet to Milk Vitamin A Content and to the Vitamin A Status of Lactating Rats and Their Pups, *J. Nutr. 115:* 1033–1041.
29. Olson, J.A. (1984) Serum Level of Vitamin A and Carotenoids as Reflectors of Nutritional Status, *J. Natl. Cancer Inst. 73:* 1439–1444.
30. Dahro, M., Gunning, D., and Olson, J.A. (1983) Variations in Liver Concentrations of Iron and Vitamin A as a Function of Age in Young American Children Dying of the Sudden Infant Death Syndrome as Well as of Other Causes, *Int. J. Vitam. Nutr. Res. 53:* 13–18.
31. Stoltzfus, R.J., Hakimi, M., Miller, K.W., Rasmussen, K.M., Dawiesah, S., Habicht, J.-P., and Dibley, M.J. (1993) High Dose Vitamin A Supplementation of Breast-Feeding Indonesian Mothers: Effects on the Vitamin A Status of Mother and Infant, *J. Nutr. 123:* 666–675.
32. Humphrey, J.H., and Rice, A.L. (2000) Vitamin A Supplementation of Young Infants, *Lancet 356:* 422–424.
33. Canfield, L.M., Clandinin, M.T., Davies, D.P., Fernandez, M.C., Jackson, J., Hawkes, J., Goldman, W.J., Pramuk, K., Reyes, H., Sablan, B., Sonobe, T., and Bo, X. (2003) Multinational Study of Major Breast Milk Carotenoids of Healthy Mothers, *Eur. J. Nutr. 42:* 133–141.
34. Ross, A.C., Pasatiempo, A.M., and Green, M.H. (2004) Chylomicron Margination, Lipolysis, and Vitamin A Uptake in the Lactating Rat Mammary Gland: Implications for Milk Retinoid Content, *Exp. Biol. Med. 229:* 46–55.
35. Bonet, M.L., Oliver, J., Pico, C., Felipe, F., Ribot, J., Cinti, S., and Palou, A. (2000) Opposite Effects of Feeding a Vitamin A-Deficient Diet and Retinoic Acid Treatment on Brown Adipose Tissue Uncoupling Protein 1 (UCP1), UCP2 and Leptin Expression, *J. Endocrinol. 166:* 511–517.
36. Massaro, D., and Massaro, G.D. (2002) Pre- and Postnatal Lung Development, Maturation, and Plasticity—Invited Review: Pulmonary Alveoli: Formation, the "Call for Oxygen," and Other Regulators, *Am. J. Physiol. 282:* L345–L358.

37. Dirami, G., Massaro, G.D., Clerch, L.B., Ryan, U.S., Reczek, P., and Massaro, D. (2004) Lung Retinol Storing Cells Synthesize and Secrete Retinoic Acid, an Inducer of Alveolus Formation, *Am. J. Physiol. 286:* L249–L256.
38. McGowan, S.E. (2002) Contributions of Retinoids to the Generation and Repair of the Pulmonary Alveolus, *Chest 121:* 206S–208S.
39. Maden, M., and Hind, M. (2003) Retinoic Acid, a Regeneration-Inducing Molecule, *Dev. Dyn. 226:* 237–244.
40. Shenai, J.P., Kennedy, K.A., Chytil, F., and Stahlman, M.T. (1987) Clinical Trial of Vitamin A Supplementation in Infants Susceptible to Bronchopulmonary Dysplasis, *J. Pediatr. 111:* 269–277.
41. Tyson, J.E. for the National Institute of Child Health and Human Development Neonatal Research Network (2000) Vitamin A Supplementation for Extremely-Low-Birth-Weight Infants, *J. Pediatr. 136:* 124–125.
42. Darlow, B.A., and Graham, P.J. (2002) Vitamin A Supplementation for Preventing Morbidity and Mortality in Very Low Birthweight Infants (Cochrane Review), *Cochrane Database Syst. Rev.:* CD000501.
43. Ambalavanan, N., Wu, T.J., Tyson, J.E., Kennedy, K.A., Roane, C., and Carlo, W.A. (2003) Comparison of Three Vitamin A Dosing Regimens in Extremely-Low-Birth-Weight Infants, *J. Pediatr. 142:* 656–661.
44. Rahmathullah, L., Tielsch, J.M., Thulasiraj, R.D., Katz, J., Coles, C., Devi, S., John, R., Prakash, K., Sadanand, A.V., Edwin, N., and Kamaraj, C. (2003) Impact of Supplementing Newborn Infants with Vitamin A on Early Infant Mortality: Community Based Randomised Trial in Southern India, *Br. Med. J. 327:* 254–257.
45. Humphrey, J.H., Agoestina, T., Wu, L., Usman, A., Nurachim, M., Subardja, D., Hidayat, S., Tielsch, J., West, K.P., Jr., and Sommer, A. (1996) Impact of Neonatal Vitamin A Supplementation on Infant Morbidity and Mortality, *J. Pediatr. 128:* 489–496.
46. Humphrey, J.H., Agoestina, T., Juliana, A., Septiana, S., Widjaja, H., Cerreto, M.C., Wu, L.S.F., Ichord, R.N., Katz, J., and West, K.P., Jr. (1998) Neonatal Vitamin A Supplementation: Effect on Development and Growth at 3 y of Age, *Am. J. Clin. Nutr. 68:* 109–117.
47. Ambalavanan, N., Tyson, J.E., Kennedy, K.A., Hansen, N., Vohr, B.R., Wright, L.L., Carlo, W.A. (2004) Does Vitamin A Supplementation Affect Long-Term Neurodevelopment in Extremely Low Birth Weight (ELBW) Infants? *Pediatr. Res.* Abstract.
48. Stoltzfus, R.J., and Humphrey, J.H. (2002) Vitamin A and the Nursing Mother-Infant Dyad: Evidence for Intervention, *Adv. Exp. Med. Biol. 503:* 39–47.
49. Ballew, C., Bowman, B.A., Sowell, A.L., and Gillespie, C. (2001) Serum Retinol Distributions in Residents of the United States: Third National Health and Nutrition Examination Survey, 1988–1994, *Am. J. Clin. Nutr. 73:* 586–593.
50. Beaton, G.H., Martorell, R., Aronson, K.A., Edmonston, B., McCabe, G., Ross, A.C., and Harvey, B. (1994) Vitamin A Supplementation and Child Morbidity and Mortality in Developing Countries, *Food Nutr. Bull. 15:* 282–289.
51. Sommer, A., and West, K.P., Jr. (1996) *Vitamin A Deficiency: Health, Survival, and Vision*, Oxford University Press, New York.
52. Green, M.H., and Green, J.B. (1994) Vitamin A Intake and Status Influence Retinol Balance, Utilization and Dynamics in Rats, *J. Nutr. 124:* 2477–2485.

53. Green, M.H., and Green, J.B. (1996) Quantitative and Conceptual Contributions of Mathematical Modeling to Current Views on Vitamin A Metabolism, Biochemistry, and Nutrition, *Adv. Food Nutr. Res. 40:* 3–23.
54. Dawson, H.D., Yamamoto, J., Zolfaghari, R., Rosales, F., Dietz, J., Shimada, T., Li, N.-Q., and Ross, A.C. (2000) Regulation of Hepatic Vitamin A Storage in a Rat Model of Controlled Vitamin A Status During Aging, *J. Nutr. 130:* 1280–1286.
55. Zolfaghari, R., Wang, Y., Sancher, A., Chen, Q., and Ross, A.C. (2002) Cloning and Molecular Expression Analysis of Large and Small Lecithin:Retinol Acyltransferase mRNAs in the Liver and Other Tissues of Adult Rats, *Biochem. J. 368:* 621–631.
56. Yamamoto, Y., Zolfaghari, R., and Ross, A.C. (2000) Regulation of CYP26 (Cytochrome P_{450}RAI) mRNA Expression and Retinoic Acid Metabolism by Retinoids and Dietary Vitamin A in Liver of Mice and Rats, *FASEB J. 14:* 2119–2127.
57. Ribaya-Mercado, J.D., Solon, F.S., Fermin, L.S., Perfecto, C.S., Solon, J.A.A., Dolnikowski, G.G., and Russell, R.M. (2004) Dietary Vitamin A Intakes of Filipino Elders with Adequate or Low Liver Vitamin A Concentrations as Assessed by the Deuterated-Retinol-Dilution Method: Implications for Dietary Requirements, *Am. J. Clin. Nutr. 79:* 633–641.
58. Kodaka, T., Takaki, H., Soeta, S., Mori, R., and Naito, Y. (1998) Local Disappearance of Epiphyseal Growth Plates in Rats with Hypervitaminosis A, *J. Vet. Med. Sci. 60:* 815–821.
59. Johansson, S., Lind, P.M., Hakansson, H., Oxlund, H., Orberg, J., and Melhus, H. (2002) Subclinical Hypervitaminosis A Causes Fragile Bones in Rats, *Bone 31:* 685–689.
60. Rohde, C.M., Manatt, M., Clagett-Dame, M., and DeLuca, H.F. (1999) Vitamin A Antagonizes the Action of Vitamin D in Rats, *J. Nutr. 129:* 2246–2250.
61. Melhus, H., Michaëlsson, K., Kindmark, A., Bergström, R., Holmberg, L., Mallmin, H., Wolk, A., and Ljunghall, S. (1998) Excessive Dietary Intake of Vitamin A Is Associated with Reduced Bone Mineral Density and Increased Risk for Hip Fracture, *Ann. Intern. Med. 129:* 770–778.
62. Michaëlsson, K., Lithell, H., Vessby, B., and Melhus, H. (2003) Serum Retinol Levels and the Risk of Fracture, *N. Engl. J. Med. 348:* 287–294.
63. Johansson, S., and Melhus, H. (2001) Vitamin A Antagonizes Calcium Response to Vitamin D in Man, *J. Bone Miner. Res. 16:* 1899–1905.
64. Promislow, J.H.E., Goodman-Gruen, D., Slymen, D.J., and Barrett-Connor, E. (2002) Retinol Intake and Bone Mineral Density in the Elderly: The Rancho Bernardo Study, *J. Bone Miner. Res. 17:* 1349–1358.
65. Feskanich, D., Singh, V., Willett, W.C., and Colditz, G.A. (2002) Vitamin A Intake and Hip Fractures Among Postmenopausal Women, *J. Am. Med. Assoc. 287:* 47–57.
66. Cardinault, N., Tyssandier, V., Grolier, P., Winklhofer-Roob, B.M., Ribalta, J., Bouteloup-Demange, C., Rock, E., and Borel, P. (2003) Comparison of the Postprandial Chylomicron Carotenoid Responses in Young and Older Subjects, *Eur. J. Nutr. 42:* 315–323.

Chapter 3

The Essential Role of Vitamin A in Signal Transduction

Beatrice Hoyos and Ulrich Hammerling

Immunology Program, Memorial Sloan-Kettering Cancer Center, New York, NY 10021

Introduction

Most cells in the body are endowed with an elaborate biochemistry to take up retinol from plasma (1), convert it to retinyl esters for storage (2), hydrolyze these esters on demand, and modify the recovered retinol to 14-hydroxy-*retro*-retinol and 13,14-dihydroxyretinol (3,4). Although the capacity to alter retinol biosynthetically to the di- and trihydroxylated forms is nearly universal and in fact evolutionarily goes back to insects, vertebrate cells that produce retinoic acid (RA) are rare, even though the latter is considered to be the dominant bioactive form of vitamin A (Fig. 3.1). Moreover, the nanomolar quantities of RA required to activate the cognate retinoic acid receptor (RAR) or retinoid X receptor (RXR) (5,6) seem out of proportion with the steady retinol supply in the micromolar range that all tissues receive from the bloodstream. What is all this retinol used for? This paradox has stubbornly persisted because no role for vitamin A other than that of prohormone has come to light, despite early indications in the literature that RA does not substitute for retinol in all cases. Recent evidence for a cytoplasmic role for retinoids is mounting (7), but to assign effector function to retinol itself is as new as it is opportune for resolving the paradox.

Serine/threonine kinases have been known for some time to harbor retinol receptors in their activation domains, the zinc-fingers (7,8). Because retinol freely partitions from extracellular abundance to high-affinity binding sites inside cells (9), the protein kinase C (PKC) and cRaf family members exist normally as complexes with retinol. Bound retinol, *per se*, does not appear to affect these kinases, which remain as inactive enzymes in the cytoplasm. Nor do they play any role during triggering by classic second messengers, despite the coincidence that they bind the same activation modules. Lipid second messengers, diacylglycerol (DAG) and phorbol esters, bind the zinc-fingers of most PKC isoforms (10), and GTP/ras docks at the cRaf zinc-finger (11,12). Both use sites distinct from retinol binding (8). Reactive oxygen intermediates (ROI) have emerged as a new class of second messenger (13). PKC and cRaf are among those signaling molecules that respond positively to ROI (14,15). Responses in terms of target destination, substrate choice, and downstream effect appear indistinguishable from those elicited by classic second messengers, with the proviso, however, that retinol must be physically associated with the kinases. In its absence, as would be the case in vitamin A

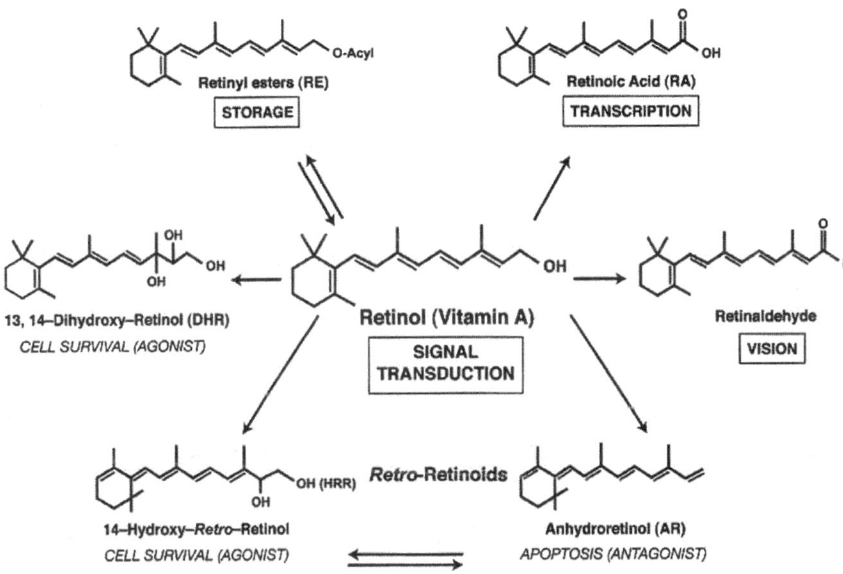

Fig. 3.1. The four different branches of vitamin A physiology.

malnutrition, both PKC and cRaf activations by the redox pathway are strongly impeded (16,17). To what extent this malfunction of serine/threonine kinases contributes to disease is presently unclear. Vitamin A deficiency leads to a complex syndrome affecting growth, development, reproduction, and immunity, among others. The multiplicity of affected primary signaling molecules and downstream paths suggests that vitamin A depletion causes system-wide transmission failure, which is in tune with the complexity of organelle damage observed with cell lines, the severity of infectious disease processes in animals, and ultimately their death.

The underlying mechanisms of vitamin A action are not understood. Although the number and diversity of target molecules interacting directly with retinol is likely to increase in the coming years, a plausible paradigm is in the making on how serine/threonine kinases are regulated. The essential features of this paradigm will be presented, and the potential effect on redox regulation of cells discussed.

Role of Vitamin A in the Immune System

Vitamin A was discovered 100 years ago during the course of then popular nutritional studies in animals. Depending on the severity of deprivation, animals displayed symptoms of multiple and progressive organ failure. In its most severe form, vitamin A deficiency leads to death. In the 1930s, biochemists identified the structure of retinol and several metabolites, notably retinaldehyde (RAL) and retinoic acid (RA). When RA was found to reverse several deficiency manifestations, such as developmental defects, growth retardation, weight loss, and skin abnormalities, the opinion took hold that vitamin A represented solely a precursor that gave rise to the true mediators, *trans* and 19-*cis*-RA, which proved indispensable for transcription control of a variety of genes. However, a number of deficiency symptoms could be reversed only by vitamin A itself; these included immune system dysfunction and male sterility, suggesting a direct role for retinol as a biological mediator. When a specific mechanism of action for retinol failed to materialize, the magnificent progress in understanding vision and the transcriptional role of RA (18) began to dominate vitamin A research. The idea of a direct function of vitamin A faded. In the past few years, reports on cytoplasmic action of a number of synthetic retinoids (19), as well as retinol (3,20,21), have renewed interest in vitamin A as a primary mediator. Because the influence of vitamin A on the immune system is well researched, it is presented as a case study below. It should be stressed, however, that the effect of vitamin A is much broader, due to the basic nature of its role in cell biology, as will become apparent in this review.

According to the WHO, an estimated 1.5 million children are blinded from severe vitamin A deficiency. A second category of 15 million children in ~40 countries with less severe vitamin A deficiency suffer from clinical xerophthalmia (night-blindness) and are at risk of adverse consequences for their healthy development and survival. These severe cases are but the tip of the iceberg because a third category without clinical xerophthalmia, numbering in the hundreds of millions, have suboptimal blood levels of vitamin A and are at heightened risk from morbidity associated with common childhood infections such as measles [reviewed in (22)]. The data suggest that vitamin A deficiency does not affect the frequency of infection, but does adversely affect the severity of clinical symptoms. Hence, immune system suppression is the most likely consequence of vitamin A malnutrition. Although other adverse effects such as stunted growth and development are also discussed, the field trials have not yielded definitive answers, primarily because vitamin A deficiency is always coupled with confounding factors, including general malnutrition, low socioeconomic status, and endemic infection with microorganisms and parasites. When vitamin A deficiency was induced in laboratory animals, the immune deficiency, the developmental abnormalities, and stunted growth were confirmed to be part of the syndrome. Additionally, keratosis of the skin and male infertility occurred with high frequency.

Observations in nutritionally depleted rats suggested as early as 1930 that vitamin A had a role in protection against opportunistic infections. Rearing vitamin A–depleted rats under germfree conditions (23,24), and, later on, under antibiotic protection (25) significantly reduced mortality without reversing the reduced growth characteristic of such animals. Thus, resistance to infection was linked to vitamin A status (26–28). When followed up with clearance experiments, it became evident that the elimination of pathogens, from gram-negative bacteria to helminthes, required normal vitamin A levels. Importantly, RA was insufficient to restore resistance to infection in vitamin A–depleted rodents, indicating a role in host defense for vitamin A itself, or a metabolite other than RA. Classic immunity was suspected as the beneficiary of vitamin A action. This view was supported by evidence, some raised in the mid-1920s by Wolbach and Howe (29) and confirmed later by numerous investigators [reviewed in (30)], that lymphopoiesis was severely compromised in vitamin-deprived rats. These rats had widespread histopathologic changes in multiple organs and tissues, but none as striking as the complete atrophy of the thymus and lymph nodes in some animals (29). We know today, although it was not known then, that the observed lack of cellularity in these glands meant a dysfunctional immune system, offering a ready explanation for opportunistic pathogens in vitamin A–deficient animals.

In contrast to rats, the histopathologic and immunological consequences of low vitamin A status in mice and birds were less dramatic overall; they depended on the regimen and, moreover, were variable from strain to strain (31). Thymus, spleen, and lymph node development were unaffected until late in vitamin A deprivation when the two last-mentioned tended to be enlarged, with signs of accumulation of cell debris and enhanced phagocytic activity. However, the T-helper/cytotoxic and B-cell components in lymphoid organs remained in stable and normal proportions (32,33). Similarly, circulating leukocytes were at normal levels. Inconsistent with the relatively normal distribution of lymphocytes, cell-mediated immune responses were frequently impaired. These included acute as well as memory responses by cytotoxic T cells to viral antigens (34) and by delayed-type hypersensitivity cells (31). *In vitro* parameters of cellular immunity yielded a variable picture, i.e., depending on the lymphoid tissue origin, proliferative responses to mitogens were reported to be reduced, unchanged, or even increased (35,36).

Humoral immune responses were consistently impaired in vitamin A–deficient animals. Despite overt signs of hypergammaglobulinemia, antibody responses of rats to tetanus toxoid as well as experimental protein antigens were greatly reduced, but promptly returned to normal after repletion with vitamin A (37–39). Antibody responses to T-cell independent type II polysaccharide responses were diminished in rats with low vitamin A status (33), whereas type I lipopolysaccharide antigen elicited normal antibody levels (37).

Despite clear end points in the assessment of immunity in vitamin A–deficient animals, the underlying cause for dysfunction has not been identified. Neither a particular cell type nor particular cell product (cytokine, chemokine) has been so impaired that this might be considered the pivotal block in the adaptive immune response. It is not even known whether the adaptive or the innate arms of the immune system, or both, are affected. The consensus in the field is that despite deprivation of a single micronutrient, i.e., vitamin A, the pathology of the immune system that ensues is multifactorial in origin, a view first proposed by Nauss and Newberne (40). At the cellular level, multiple cell types orchestrating the immune response are almost certainly affected. At the control level of each cell type, both signaling processes and differentiation events are presumed to depend on adequate vitamin A. The former were thought to require vitamin A itself as a cofactor, whereas the latter were compromised when RA became limiting due to vitamin A deficiency. A reductionist approach was required to distinguish between these two major branches of vitamin A action.

Receptor Sites for Vitamin A in the Cytoplasm

Although still adhering to the tenet that vitamin A constituted a profactor, Buck *et al.* (41) conducted a broad biochemical search for metabolites that supported biological effects, mainly growth and cell survival, that were unrelated to transcriptional regulation. They identified two new natural retinoids, 14-hydroxy-*retro*-retinol, (HRR) and 13,14-dihydroxyretinol (DHR) that effectively restored normal growth properties to vitamin A–depleted cultures of transformed B lymphocytes (3,4). Both retinoids were produced not only by B lymphocytes but in essence by all cell lines tested, with tissue origins as diverse as liver, skin, colon, lung, brain, kidney, heart (O'Connell and Hammerling, unpublished results). Additionally, a limited survey indicated that they were present in invertebrate cell lines as well (42). Cells were able to maintain production of these hydroxylated retinoids after exterior sources of vitamin A were removed, falling back on stored retinyl esters until the latter were consumed, at which time cells went into growth arrest and suffered apoptosis (43). Because hydroxylated retinoids did not bind or activate any of the RAR or RXR transcription factors (Buck, unpublished results), a search for receptors elsewhere in the cell began.

Instrumental in focusing the search on the cytoplasm were observations on anhydroretinol, another naturally occurring metabolite of retinol (44,45). This retinoid had the property of antagonizing the growth-promoting effects of retinol in B-lymphocyte cultures. The end result, i.e., apoptosis, was the same as that achieved by nutritional vitamin A deprivation; however, the kinetics of its occurrence were greatly accelerated. As mentioned above, apoptosis by nutritional depletion is slow and asynchronous because cells contain intracellular stores of vitamin A (41,43). By contrast, anhydroretinol imposed an instant state of vitamin A deficiency. The pharmacologic analysis of this phenomenon indicated that retinol and anhydroretinol acted as mutually reversible inhibitors. The most direct explanation was that they competed for the same receptor site(s), and that when at high concentration, anhydroretinol had displaced all retinol, vital survival signals were compromised (46). It followed that vitamin A itself acted as an essential survival factor. The conventional wisdom that RA was the requisite mediator was also challenged by the fact that B lymphoblastoid cells neither produced discernible amounts of RA, nor were they rescued by external RA supplementation (41). Furthermore, anhydroretinol-induced apoptosis proceeded whether the cells were transcriptionally or translationally competent because neither actinomycin D nor cycloheximide blocked apoptosis (46).

An important reason to search for retinol targets outside the nucleus and to focus on signal transduction molecules came from results with herbimycin A. This drug, thought to block heat-shock proteins and attendant signaling paths, suppressed the action of anhydroretinol (46). The idea that anhydroretinol in some way disrupted vital signal pathways was subsequently verified. Of biological relevance was not so much the capacity of anhydroretinol to induce apoptosis by interfering with signaling as the converse conclusion that vitamin A was vital for maintenance of normal cell survival signals.

This simple concept was confounded by the observations that anhydroretinol also suppressed the growth-promoting effects of HRR and DHR in a mutually reversible manner. This seemed inconsistent with the existence of a single receptor species and the idea of retinoid binding specificity that was demonstrated to be so exquisite for the RA family. Nevertheless, a large number of cytoplasmic retinol receptors were in fact identified (7,8). What reconciled the simple binding competition model with the existence of multiple receptor molecules was the finding that the latter shared highly conserved retinoid binding domains, whereas the rest of the protein structures were quite diverse.

Molecular Identification of Serine/Threonine Kinases as Vitamin A Receptors

Using a solid-phase agarose matrix with covalently linked *retro*-retinoid, Imam *et al.* (8) succeeded in isolating signal transduction molecules from cytoplasmic extracts; the only disconcerting finding was the unexpectedly large number of proteins binding the matrix. Although some proteins undoubtedly bound nonspecifically *via* hydrophobic interaction, others, such as members of the PKC family, were effectively displaced by free retinol, thus fulfilling the criteria for specific binding. Still the abundance and diversity of proteins was not explained until later when it was learned that entire protein assemblies, and not monomolecular species, were captured by the matrix. Thus, bound 14-3-3 proteins or heat shock proteins were legitimate, copurifying components of signalosomes (47,48), although the retinoid binding sites belonged to members of the serine/threonine kinases with which they are known to associate in the cytoplasm.

Once it became clear that cRaf and multiple isoforms of PKC possessed an affinity for the retinoid matrix, the question arose whether these molecules had their own unique binding sites or shared a common motif. Results of homology comparisons of the genes favored the second possibility because cysteine-rich subdomains of considerable homology were present in all serine/threonine kinases. Most PKC isoforms had two in tandem, except the atypical ones, with one copy. CRaf, Raf-A, and Raf-B possessed one copy. When these cysteine-rich domains spanning 50 amino acids were expressed as fusion proteins in bacteria, the predictions that they harbored the retinol binding site were verified. Specific binding of retinol was demonstrated in qualitative terms *in vitro*, based on the following four independent fluorometric assays: (i) quenching of the intrinsic protein fluorescence by bound retinol; (ii) fluorescence resonance energy transfer from tryptophan donor to retinol acceptor; (iii) enhancement of retinol fluorescence emission; and

(iv) development of a vibronic fine structure coupled with a red-shifted UV absorption spectrum. Titrations of retinol in the first three assays yielded Scatchard plots and binding affinities in the tens of nanomoles. Each of the aforementioned serine/threonine kinases possessed at least one binding site for retinol. However, the PKC θ and μ isoforms harbored two, one associated with each of its tandem cysteine-rich domains. In other PKC isoforms, either the first cysteine-rich domain or the second [called C1A or C1B, respectively (49)] carried the binding site, with no clear systematic pattern emerging.

The cysteine-rich domains are instrumental for regulation of kinase activity. They represent the primary sites of interaction with the second messenger molecules that trigger these kinases to undergo cytoplasmic-to-membrane translocation and to acquire catalytic competence. Most PKC isoforms respond to DAG, or to phorbol esters, its pharmacologic mimetic (10), whereas cRaf interacts with GTP/ras through a critical half-site (11,12). It was of interest to determine whether the retinol binding sites overlapped the phorbol binding sites. Using the PKCα C1A fusion protein as a model, it was determined that they were separate (8). By scanning mutagenesis, Hoyos *et al.* (unpublished data) recently confirmed that the extinction of retinol binding in the cRaf cysteine-rich domain did not affect the ability to interact with GTP/ras, another indication that the retinol binding site was distinct. Finally, introducing the presumed anchor sites gleaned from the mutagenesis study of cRaf into the nonbinding PKCα C1B domain conferred high affinity binding; these contact amino acids were separate in kind and geographic location from those mediating phorbol ester binding (50).

All cells in the body are bathed in plasma containing a level of vitamin A kept constant by the liver (~2 μmol). With its lipid-like properties, retinol can pass through cell membranes and partition to intracellular high-affinity sites (9). At an average of 10 nmol binding affinity (7,8), sites on serine/threonine kinases could be expected to be constitutively complexed with retinol. This prediction was confirmed by radioisotope tracer methodology. Cellular cRaf immunoprecipitated from cytoplasm was saturated with retinol (7). It is noteworthy that this biochemistry differs from that of RA because the latter is supplied in a time- and space-restricted manner from endocrine or paracrine sources. These considerations presaged that retinol would not act as a hormone because constant supply and constitutive occupancy of binding sites were antithetical to hormonal action. In fact, the binding alone of retinol to threonine/kinases has no discernible consequences, whereas its absence causes serious malfunction of signal processes.

As mentioned above, retinol is metabolized ubiquitously to HRR. This retinoid can substitute for retinol as a growth promoter and survival factor (3). Fluorometric assays indicated that HRR bound with similar affinity to all cysteine-rich domains that possessed affinity for retinol, but to none of the retinol non-binders. Anhydroretinol and RA also had the same binding affinity and pattern as HRR and retinol, indicating a lack of binding specificity on the part of natural retinoids. These initially perplexing results were reconciled as follows: First, the assumption that retinoids contacted the protein binding surface by their shared β-ionone ring and/or polyene structure readily explained why they all displayed the same affinities. Second, pharmacologic mutual inhibition, such as that observed between AR and retinol, or AR and HRR (see above), was similarly explained by head-first orientation in the binding groove. Third, the biological agonist activity was dependent on the presence of hydroxyl groups in the tail of the retinoid. The absence of hydroxyls, as in anhydroretinol, or their replacement by carboxyl groups, explained the conversion to antagonist (although it begged the question why the hydroxyl groups are important for agonist function). A limited structure/function analysis confirmed that amino, amide, and ester groups were not compatible with agonist function. Fourth, interference by RA was not an issue (except in artificial situations) because this retinoid seldom reaches sufficiently high concentrations to compete with retinol.

Regulatory Function of Retinol for Serine/Threonine Kinases

To understand the function of the retinol cofactor, it is necessary to recall the basics of serine/threonine kinase regulation. Like src, insulin receptor kinase, and many other enzymes, serine/threonine kinases obey the paradigm of autoinhibition, i.e., when present in the cytoplasm in tightly folded form, they are inactive because their regulatory domains sterically hinder the catalytic domain (51). Activation necessitates retraction of the regulatory domain and unplugging of the substrate binding site to achieve release from autoinhibition. Although secondary protein modifications [phosphorylations at preordained sites (52)], translocation to membranes, as well as protein/protein interactions (e.g., with GTP ras in case of cRaf) are required in preparation for full kinase competence (53), the conserved cysteine-rich domains feature prominently as sites of initiation. These structures harbor the binding sites for DAG second messenger in most PKC isoforms (54,55) and GTPras contact sites in cRaf (12). The retinol binding sites are also embedded in these cysteine-rich domains, but are distinct from the former (7), as discussed above.

In addition to the classic lipid second messenger, several (by extrapolation, perhaps all) members of the conventional, novel, and atypical PKC families, can be fully activated by oxidation. Gopalakrishna and Anderson (14) and later Knapp and Klann (56) demonstrated activation of several isoforms *in vitro* by ROS. The latter authors suggested that attack on sulfhydryl groups of the cysteine-rich domain by superoxide radicals was the likely activating signal. Konishi *et al.* (15) found that the PKCα, δ, and ζ isoforms became catalytically active after treatment of cells with hydrogen peroxide. Irradiation with UV light has long been known to cause activation of cRaf, and what links this finding with PKC is the involvement of ROS that are elicited by UV-B and are presumed to serve as second messengers in this alternative pathway of serine/threonine kinase activation (16). What connects these signal molecules further is the requirement for retinol to promote the redox pathway (17).

Loading cells with retinol seemed to have no consequences *per se* on the functionality of serine/threonine kinases. PKC family members or cRaf showed no signs of becoming activated by retinol alone, neither undergoing translocation nor acquiring catalytic competence. Additionally, the classic activation paths were unaffected, whether cells had adequate or depleted vitamin A stores. Neither the phorbol ester pathway nor the PTK/ras axis leading to cRaf activation was negatively affected by vitamin A deficiency. However, the alternate activation signal *via* reactive oxygen, known to lead to full activation of all serine/threonine kinase members, was found to require vitamin A. The activation of the cRaf kinase by hydrogen peroxide was first noted to be suboptimal if vitamin A was deficient. Subsequently, activation by the more physiologic UV pathway that depended on ROS generation was found inoperative in vitamin A–deficient cells, but retinol readily restored the UV response to normal (16). In keeping with the antagonistic action in cells, anhydroretinol was unable to substitute for retinol. In fact, even when cells had an adequate supply of vitamin A, anhydroretinol was suppressive, provided that the nominal amount added to the culture medium exceeded that of retinol. On the other hand, HRR proved a potent agonist cofactor, promoting cRaf kinase activation by the redox pathway as efficiently as retinol. Identical results were obtained with the PKCα isoforms (17). Activation by hydrogen peroxide proceeded normally as long as the cells had adequate vitamin A or HRR, whereas much less kinase activity was obtained in cells loaded with anhydroretinol (8). In conclusion, the adverse properties of retinol vs. anhydroretinol, or HRR vs. anhydroretinol, first observed in cell biological experiments, were recapitulated at the level of the kinases, reflecting precisely the pharmacological "reversible inhibition" raised in cell survival assays.

It can be argued that the effects of retinol were indirect and depended on the conversion to a metabolite, such as HRR. Judging from experiments in which cells were first made vitamin A deficient and then reconstituted with retinol, this scenario was unlikely because the restoration to full responsiveness was shown to occur within minutes of exposure to retinol. This time span was long enough for uptake and saturation of receptor sites, but too brief for a meaningful biochemical conversion. Although a cell-free system of kinase activation would rigorously exclude the possibility of metabolic conversion to an active retinoid, the available data strongly favor the notion that retinol itself acts as a biological mediator. The physical binding to recombinant cRaf and PKC fragments supports this argument.

The Cysteine-Rich Region as a Redox-Sensitive, Retinol-Regulated Hinge: The Linchpin Hypothesis

When considered in the context of redox regulation, the function of retinol is best described as that of a catalyst. The presence of bound retinol was important for the efficient oxidative activation of the kinases. In its absence, or if replaced by anhydroretinol, the redox mechanism appeared impaired, although the fate of the kinases in that case was unclear, i.e., whether they simply stayed inactive or were targeted for degradation because of inappropriate oxidation. It must be more than coincidental that the very domains binding retinol and supposedly benefiting from controlled oxidation contain six cysteine residues—the most likely primary targets for oxidation (56). A further clue suggesting the utility of the cysteine-rich domains is that they are organized into zinc-finger structures. Although these are generally perceived as rigid structures, reflecting historically the ability by numerous transcription factors endowed with zinc-fingers to intercalate into DNA in cognate fashion, they might in reality constitute flexible hinges. The "galvanization" (57) of a tetrad of cysteine/histidine residues by a central zinc ion lends great strength to this regional module, but its Achilles' heel is the dependence on the thiolate anions for ionic bonding. Oxidation of only a single cysteine would by necessity disrupt this zinc-coordination center. Such a chemical reaction is on the one hand an attractive means for induction of conformation change, i.e., it is energetically cheap; the cost is the transfer of a mere two electrons to oxygen; it is reversible by reduction; it is versatile because different combinations of cysteines and histidines, as well as different spacings of these framework residues, can make for numerous structures. On the downside, it is difficult to see how reactions with ROS may be directed to preordained cysteine residues. This perceived lack of specificity has hindered general acceptance of the redox network as a "public" second messenger system (13), despite much evidence to the contrary that it has an important place in cell biology for the coordinate mobilization of complex resources. To make up for the deficit in selectivity of ROS, it was suggested that retinol binding evolved to channel oxidation reactions to the preferred cysteine groups (17).

Countless zinc-coordinated structures are encountered in nature, and their diversity is reflected in names such as zinc-fingers, ring-fingers, zinc-butterflies. By some estimates, 1% of the human genome is set aside to code for such zinc proteins. Several thousand are known to date in prokaryotes and eukaryotes combined. The reason why zinc ions possess such formidable properties derives from their electron configuration. As a transition metal ion with an 8d electron shell, zinc can form strong ionic bonds with oxygen-, nitrogen-, and sulfur-containing amino acids to produce symmetric tetrahedral structures of great strength. It thus shares features in common with carbon, but is distinct from the latter in that the noncovalent nature of the bonds permits rapid rearrangement, and is favored in situations in which transient bonds are desirable. In addition, unlike iron or copper ions, the zinc ion is redox inactive. For these reasons, zinc ions are deployed in the active sites of many enzymes as a stabilizer of transition states (58,59), as well as in zinc-fingers of signal transduction molecules.

The best-studied example of a zinc-finger hinge concerns the prokaryotic heat shock protein, Hsp 33 (60,61). Like mammalian PKC, this enzyme is inactivated by autoinhibition of its catalytic domain by the regulatory domain. Activation requires the retraction of the regulatory domain to permit dimerization and substrate access. At the fulcrum of this unfolding process lies a zinc-finger module that responds reversibly to changing redox levels of the microenvironment. Under oxidant conditions, thiols are converted to intramolecular disulfide, resulting in the release of zinc ions, rearrangement of the zinc-finger structure, and catalytic competence (62). This process is entirely reversed when reducing conditions prevail.

The zinc-finger domains in PKC and Raf are functional paralogs of their bacterial counterparts, although they are very different in composition and structure. They are organized into two zinc-coordination centers comprising three cysteine and one histidine residue each, which are recruited from noncontiguous portions of a 50-amino acid stretch located at the N-terminus of the regulatory domain. The consensus motif of this dual zinc-finger, $HX_{12}CX_2CX_nCX_2CX_4HX_2CX_7C$, (where C is cysteine, H is histidine, X is any other amino acid) is represented in 11 PKC isoforms, 3 Raf family members, and other signaling molecules including KSR, *n*-chimerin and Unc-13 (63) and, moreover, is conserved from yeast to drosophila, to humans. Because of its importance as the site of interaction with lipid second messengers in PKC or with activated ras in cRaf, the NMR and crystal structures were elucidated (50,64,65). They show two anti-parallel β-sheets formed by three noncontiguous (βAβDβE) and two (βBβC) strands. The last C-terminal strand is connected to an α-helix that folds back toward the N-terminus. The phorbol ester site was mapped to a region between the two loops that connect βA to βB, and βC to βD, respectively (50). CRaf is lacking the second loop (65), accounting for its inability to bind phorbol ester. The retinol binding site has not been defined crystallographically, but according to scanning mutagenesis results of the cRaf zinc-finger, the β-ionone ring was predicted to dock to the rear of the βA/βE strands, whereas the tail likely pointed downward to the second zinc coordination center (Hoyos, unpublished results).

In analogy to the Hsp33 zinc-finger, the PKC as well as the cRaf zinc-fingers give up their zinc ions in response to oxidation, although the last-mentioned two, but not the first, require retinol as a cofactor, as discussed. The release of zinc ions in response to ROS was demonstrated with full-length PKC *in vitro* (56) and was confirmed with recombinant proteins *in vitro* (16) as well as in intact cells (66), using intravital staining with a zinc-sensitive fluorescent probe, TSQ. The kinetics of zinc mobilization and ROS dose-response characteristics closely followed those of PKC kinase activations. For the stated reasons that cystine cannot sustain zinc chelation, the release of zinc after oxidation is not unexpected. Unexpectedly, however, phorbol ester and DAG also caused zinc release *in vitro*, and phorbol did so *in vivo* (66). Phorbol ester is not known to evoke ROS, especially not *in vitro*. Hence, another mechanism related to specific lipid binding (because neither cRaf nor the phorbol nonbinding PKC ζ zinc-fingers responded), but otherwise as yet unknown, must be invoked to explain the disassembly of PKC zinc-fingers by lipid second messenger. It is biologically relevant that two different signals, ROS and lipid second messengers, converge on the same structure, the zinc-finger, and trigger the release of zinc (albeit by different mechanisms), setting up the opportunity for conformational change.

Causality between zinc-release and kinase function was established. The zinc content of PKC α, isolated by immunoprecipitation from phorbol or ROS-activated normal lymphocytes was significantly lower than that of resting cells, as predicted, and was inversely correlated with the phosphotransfer activity measured in the immunoprecipitates (66). The dual responsiveness of serine-threonine kinases is thereby explained at the molecular and biochemical level. This mechanism supports a model whereby the inactive form is kept clamped into a globular cytoplasmic protein by the closed zinc-finger module (51). Activation by either mechanism can be likened to pulling the linchpin (the central zinc ions), which triggers the unfolding of the protein and the exposure of previously hidden protein/protein or protein/lipid interaction surfaces. Recognition of specific anchoring sites can subsequently guide the kinase to the final membrane destination for substrate phosphorylation. Secondary modifications by phosphorylations of regulatory sites and cofactors, such as Ca ions, that might fine-tune the strength and duration of kinase action all form part of the activation cycle.

Model of Vitamin A Action as Cofactor in Redox Regulation

There is no theory at present, or even any biochemical or bioenergetic details, on how retinol might perform its function as a redox catalyst in kinase activation. At best guess, retinol is presumed to facilitate electron transfer from cysteine to an oxygen acceptor, as proposed in the accompanying illustration (Fig. 3.2). It is not known which cysteine residues are targeted by ROS, or even whether one or both zinc-coordination centers are involved. On preliminary evidence, Korichneva *et al.* (66) suggested that ROS and phorbol ester target opposite zinc-coordination centers. This conclusion was based on the observations that phorbol ester caused the release of one zinc-equivalent and that mild oxidation also freed only one zinc-equivalent, whereas the combined treatment resulted in the release of two equivalents. As an extrapolation of these observations, it is proposed in the illustration that redox activation is targeted to zinc-coordination center 2, mainly because the retinol binding site was mapped to a region near that center, whereas the phorbol binding site is proximal to center 1.

How vitamin A might work in signal transduction is part of a broader question: how retinoids and their next of kin, the carotenes, might function. Is there a common denominator? The literature is silent on this subject, except that retinoids and carotenes tend to show up in instances in which electrons are handled as follows: (i) Vision (67): The underlying principle is that upon capturing a photon, the π electrons of retinaldehyde are elevated to an excited state. Their return to baseline triggers the isomerization from *cis* to *trans*. Freed energy is converted to a nerve impulse. (ii) Photosynthesis *via* bacteriorhodopsin (68): A similar excited state drives photoisomerization of retinaldehyde from *trans* to *cis*. Energy stored in *cis*-retinaldehyde drives a proton pump. (iii) Photosynthesis in chloroplasts: β-Carotene, integrated in the light harvesting complexes, absorbs photons, resulting in an excited state.

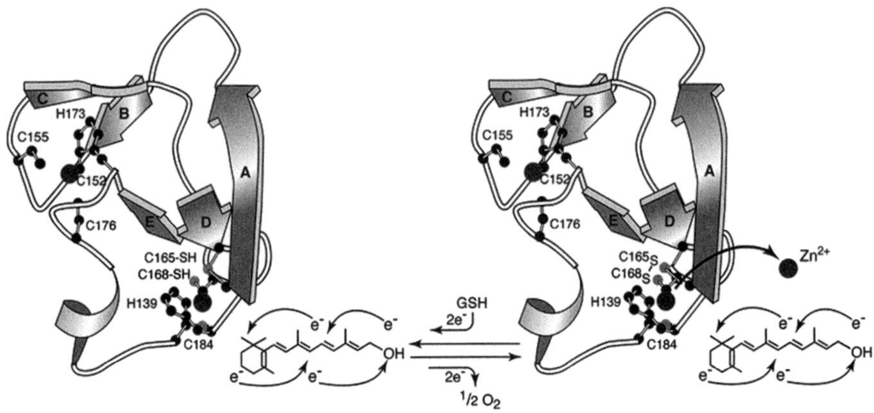

Fig. 3.2. Schematic representation of the catalyst role of retinol. Shown is the nuclear magnetic resonance (NMR) structure of the cRaf zinc-finger domain according to Mott *et al.* (65) with its two zinc-coordination centers, each formed by one histidine and three cysteine residues. H173, C152, C155, C176 form center I, whereas H139, C 165, C168, C184 form center II. The retinol binding site has not been mapped with certainty, but according to scanning mutagenesis (B. Hoyos, unpublished) lies at the rear of the β-sheet A with its ionone ring anchored proximally to F146. A docking study has the hydroxylated tail pointing toward center II (S. Jiang and B. Hoyos, unpublished). The function of retinol as an electron transfer device is depicted schematically, but whether this involves a one- or two-electron mechanism is not known. The oxidation of as yet unidentified cysteine residues is believed to result in disulfide formation, as indicated for the hypothetical C165-C168 cystine. Zn^{2+} ions are released as a result of oxidation, enabling a conformation shift. This process is theoretically reversible by reduction, leading to the restoration of the zinc-finger module.

Stored energy is transmitted to chlorophyll by resonance energy transfer. In no case is there evidence that electrons, although readily elevated to higher orbit, are completely abstracted, as would be required for electron transfer during redox reactions.

α-Tocopherol was proposed to mediate lipid peroxidation by an oxygen radical mechanism (69). The oxidation of unsaturated fatty acid in LDL complexes involves a tocopherol radical intermediate that transfers successively two oxygen radicals creating lipid peroxide, with little consumption of tocopherol. The structural similarities between vitamin E and A suggest that the latter could function in a similar mode. Like tocopherol, vitamin A contains a crucial hydroxyl group. Although it is not known whether it is suitable for electron abstraction, this hydroxyl group is notably important for the function of retinol as a presumptive redox catalyst. The structure-function studies described above indicate that the terminal hydroxyl group is inviolate; substitutions or deletion lead to either an inactive or, as a result of blocking the binding site, to an antagonistic retinoid.

Biological Significance of Vitamin A in Oxygen Radical Biology

In the history of biology, the study of pathologic conditions has in numerous cases led to insight into normal physiology. In this vein, the effects of anhydroretinol on cells, however dramatic, are less important for their (without exception) negative effect as for the flip side that retinol is needed for normal function of these processes. As reported, anhydroretinol caused a long litany of cell injuries: the depolymerization of F-actin, resulting in the destruction of the cytoskeleton (70); the depolarization of mitochondria (71); enhanced production of ROS (72); a general loss of proliferation potential (45,72); apoptosis in several cell types (46). In all cases, displacement of retinol from crucial receptors was the likely reason because the effects were readily reversed by retinol. Vitamin A deprivation produced similar phenomena, and more: Fibroblasts were found to lose their adhesion properties upon retinol depletion, and cell blebbing occurred as an indication of osmotic turmoil. All of these can theoretically be explained by defective signaling by kinases deprived of bound retinol. The resulting complex suite of cell injuries occurring in vitamin A–depleted cells or whole animals is consistent with the existence of multiple retinol-dependent kinases and attendant signal pathways, but whether it is always the same hand that deals the bad card is debatable. More likely, disruptions of multiple signal paths add up to irreversible damage.

This picture of extreme pleomorphism can be augmented by reports that synthetic retinoids produced similar effects. 4-HPR, a drug used successfully in clinical trials, works by a cytoplasmic pathway that converges on mitochondria, causes loss of membrane potential, release of large amounts of ROS together with cytochrome C, activation of caspases, and culminates in apoptosis (19,73). Anhydroretinol causes a strikingly similar sequence of mitochondrial damage (71), but whether 4-HPR and anhydroretinol are targeted to the same or different upstream receptors is not resolved.

It is often said that ROS are elicited in response to, and possibly in defense of, cell stress and injury. New insight gained from the study of retinol as redox cofactor suggests an amended picture in which ROS are indispensable and active at all times. This view emerges from the fact that the function of retinol in signal transduction, at least as it concerns PKC and Raf pathways, was found to be inseparable from redox mechanisms. To reiterate, whether retinol occupied PKC or Raf receptors had no adverse effects *per se* on classic signal processes. On the other hand, the alternate redox-mediated signal paths proved highly sensitive to retinol deprivation. Given this close link, one must wonder: Are ROS not a constant companion? If so, ROS must act also as silent partners in experimental situations in which ostensibly only the vitamin A status, but not the redox status, was manipulated, although a secondary surge of ROS production could follow retinol depletion. The pathologic consequences of disrupting signal processes regulating basic household functions by anhydroretinol could not have been more dramatic. As before, the flip side argument can be made that an intact signal apparatus depending on redox regulation is required for continuous, possibly subtle but nevertheless indispensable adjustments of the PKC and Raf families of kinases. It follows that to drive these kinases, cells must generate a basal level of ROS to ensure the maintenance of a redox signal network that is instrumental for cell survival.

Oxygen joins a group of inorganic gases with surprising messenger properties, comprised of NO (74,75), ROS (13), CO_2 (76), and CO (77). These messengers probably originated in primordial organisms as feedback devices for adaptation of their basic metabolism to the changing microenvironment. This basic strategy of "sniffing at the tailpipe" for monitoring and adjustments of the cell's inner workings seems to have evolved in higher organisms into sophisticated sensor/actuator systems. In the PKC and Raf family of molecules, the zinc-finger domains serve as reactive oxygen sensors, in addition to their role as receptors for classic second messengers. The chemical signal received is converted into conformational change that regulates (in concert with other factors) the covalently attached catalytic domain. The role of retinol can be viewed as a device for the efficient and accurate channeling of ROS to the zinc-finger domains.

References

1. Roberts, A.B., and Sporn, M.B. (1984) Cellular Biology and Biochemistry of the Retinoids, in *The Retinoids* (Sporn, M.B., Roberts, A.B., and Goodman, D.S., eds.), Academic Press, Orlando, Vol. 2, pp. 209–286.
2. Blaner, W.S., and Olson, J.A. (1994) *Retinol and Retinoic Acid Metabolism*, Raven Press, New York.
3. Buck, J., Derguini, F., Levi, E., Nakanishi, K., and Hammerling, U. (1991) Intracellular Signaling by L4-Hydroxy-*retro*-retinol, *Science 254:* 1654–1656.
4. Derguini, F., Nakanishi, K., Hämmerling, U., Chua, R., Eppinger, T., Levy, E., and Buck, J. (1995) 13,14-Dihydroxy-retinol, a New Bioactive Retinol Metabolite, *J. Biol. Chem. 270:* 18875–18880.
5. Evans, R.M., and Hollenberg, S.M. (1988) Zinc Fingers: Guilt by Association, *Cell 52:* 1–3.
6. Green, S., and Chambon, P. (1988) Nuclear Receptors Enhance Our Understanding of Transcription Regulation, *Trends Genet. 4:* 309–314.
7. Hoyos, B., Imam, A., Chua, R., Swenson, C., Tong, G.C., Levi, E., Noy, N., and Hammerling, U. (2000) The Cysteine-Rich Regions of the Regulatory Domains of Raf and PKC as Retinoid Receptors, *J. Exp. Med. 192:* 835–846.
8. Imam, A., Hoyos, B., Swenson, C., Levi, E., Chua, R., Viriya, E., and Hammerling, U. (2000) Retinoids as Ligands and Coactivators of Protein Kinase C Alpha, *FASEB J. 10:* 1096/fj.1000–0329fje. Published online.
9. Noy, N., and Xu, Z.-J. (1990) Interactions of Retinol with Binding Proteins: Implications for the Mechanism of Uptake by Cells, *Biochemistry 29:* 3883–3886.
10. Kazanietz, M.G., Bustelo, X.R., Barbacid, M., Kolch, W., Mischak, H., Wong, G., Pettit, G.R., Bruns, J.D., and Blumberg, P.M. (1994) Zinc Finger Domains and Phorbol Ester Pharmacophore: Analysis of Binding to Mutated Form of Protein Kinase C Zeta and the Vav and c-Raf Proto-Oncogene Products, *J. Biol. Chem. 269:* 11590–11594.
11. Ghosh, S., Xie, W.-Q., Quest, A.F.G., Mabrouk, G.M., Strum, J.C., and Bell, R.M. (1994) The Cysteine-Rich Region of Raf-1 Kinase Contains Zinc, Translocates to Liposomes, and Is Adjacent to a Segment That Binds GTP-Ras, *J. Biol. Chem. 269:* 10000–10007.

12. Luo, Z., Diaz, B., Marshall, M.S., and Avruch, J. (1997) An Intact Raf Zinc Finger Is Required for Optimal Binding to Processed Ras and for Ras-Dependent Raf Activation *in situ*, *Mol. Cell. Biol. 17:* 46–53.
13. Nathan, C. (2003) Specificity of a Third Kind: Reactive Oxygen and Nitrogen Intermediates in Cell Signaling, *J. Clin. Investig. 111:* 769–778.
14. Gopalakrishna, R., and Anderson, W. (1989) Ca^{2+}- and Phospholipid-Independent Activation of Protein Kinase C by Selective Oxidative Modification of the Regulatory Domain, *Proc. Natl. Acad. Sci. USA 86:* 6758–6762.
15. Konishi, H., Tanaka, M., Takemura, Y., Matsuzaki, H., Ono, Y., Kikkawa, U., and Nishizuka, Y. (1997) Activation of Protein Kinase C by Tyrosine Phosphorylation in Response to H_2O_2, *Proc. Natl. Acad. Sci. USA 94:* 11233–11237.
16. Hoyos, B., Imam, A., Swenson, C., Chua, R., Levi, E., Viriya, E., and Hammerling, U. (2002) Activation of cRaf Kinase by UV Light: Regulation by Retinoids, *J. Biol. Chem. 277:* 23949–23957.
17. Imam, A., Hoyos, B., Swenson, C., Levi, E., Chua, R., Viriya, E., and Hammmerling, U. (2001) Retinoids as Ligands and Coactivators of Protein Kinase C Alpha, *FASEB J. 15:* 28–30.
18. Mangelsdorf, D.J., Umesono, K., and Evans, R.M. (1994) *The Retinoid Receptors*, Raven Press, New York.
19. Lotan, R. (2003) Receptor-Independent Induction of Apoptosis by Synthetic Retinoids, *J. Biol. Regul Homeost. Agents 17:* 13–28.
20. Garbe, A., Buck, J., and Hammerling, U. (1992) Retinoids Are Important Cofactors in T Cell Activation, *J. Exp. Med. 176:* 109–117.
21. Chen, Y., Derguini, F., and Buck, J. (1997) Vitamin A in Serum Is a Survival Factor for Fibroblasts, *Proc. Natl. Acad. Sci. USA 94:* 10205–10208.
22. Underwood, B.A. (1994) *Vitamin A in Human Nutrition*, Raven Press, New York.
23. Bieri, J.G., McDaniel, E.G., and Rogers, W.E.J. (1969) Survival of Germfree Rats Without Vitamin A, *Science 163:* 574–575.
24. Raica, N.J., Stedham, M.A., Herman, Y.F., and Sauberlich, H.E. (1970) *Vitamin A Deficiency in Germ-Free Rats*, University of Wisconsin Press, Madison.
25. Anzano, M.A., Lamb, A.J., and Olson, J.A. (1979) Growth, Appetite, Sequence of Pathological Signs and Survival Following the Induction of Rapid, Synchronous Vitamin A Deficiency in the Rat, *J. Nutr. 109:* 1419–1431.
26. Green, H.N., and Mellanby, E. (1928) Vitamin A as Anti-Infective Agent, *Br. Med. J. 2:* 691–696.
27. Green, H.N., and Mellanby, E. (1930) Carotene and Vitamin A: The Anti-Infective Action of Carotene, *Br. J. Exp. Pathol. 11:* 81–89.
28. Lassen, H.C.A. (1930) Vitamin A Deficiency and Resistance Against a Specific Infection. A Preliminary Report, *J. Hygiene 30:* 300–310.
29. Wolbach, S.B., and Howe, P.R. (1925) Tissue Changes Following Deprivation of Fat-Soluble A Vitamin, *J. Exp. Med. 42:* 753–777.
30. Ross, A.C., and Hammerling, U. (1994) Retinoids and the Immune System, in *The Retinoids: Biology, Chemistry and Medicine* (Sporn, M.B., Roberts, A.B., and Goodman, D.S., eds.) pp. 521–543, Academic Press, Orlando.
31. Smith, S.M., Levy, N.L., and Hayes, C.E. (1987) Impaired Immunity in Vitamin A Deficient Mice, *J. Nutr. 117:* 857–865.

32. Nauss, K.M., Phua, C.-C., Ambrogi, L., and Newberne, P.M. (1985) Immunological Changes During Progressive Stages of Vitamin A Deficiency in the Rat, *J. Nutr. 115:* 909–918.
33. Pasatiempo, A.M.G., Taylor, C.E., and Ross, A.C. (1991) Vitamin A Status and the Immune Response to Pneumococcal Polysaccharide: Effects of Age and Early Stages of Retinol Deficiency in the Rat, *J. Nutr. 121:* 556–562.
34. Sijtsma, S.R., Rombout, J.H., West, C.E., and Van Der Zipp, A.J. (1990) Vitamin A Deficiency Impairs Cytotoxic T Lymphocyte Activity in Newcastle Disease Virus-Infected Chickens, *Vet. Immunol. Immunopathol. 26:* 191–201.
35. Nauss, K.M., Mark, D.A., and Suskind, R.M. (1979) The Effect of Vitamin A Deficiency on the *In Vitro* Cellular Immune Response of Rats, *J. Nutr. 109:* 1815–1823.
36. Chandra, R.K., and Au, B. (1981) Single-Nutrient Deficiency and Cell-Mediated Immune Responses. III. Vitamin A, *Nutr. Res. 1:* 181–185.
37. Pasatiempo, A.M.G., Kinoshita, M., Taylor, C.E., and Ross, A.C. (1990) Antibody Production in Vitamin A Depleted Rats Is Impaired After Immunization with Bacterial Polysaccharide or Protein Antigens, *FASEB J. 4:* 2518–2527.
38. Kinoshita, M., Pasatiempo, A.M.G., Taylor, C.E., and Ross, A.C. (1991) Immunological Memory to Tetanus Toxoid Is Established and Maintained in the Vitamin A Depleted Rat, *FASEB J. 5:* 2473–2481.
39. Smith, S.M., and Hayes, C.E. (1987) Contrasting Impairments in IgM and IgG Responses of Vitamin A Deficient Mice, *Proc. Natl. Acad. Sci. USA 84:* 5878–5882.
40. Nauss, K.M., and Newberne, P.M. (1985) Local and Regional Immune Function of Vitamin A Deficient Rats with Ocular Herpes Simplex Virus (HSV) Infections, *J. Nutr. 115:* 1316–1324.
41. Buck, J., Ritter, G., Dannecker, L., Katta, V., Cohen, S., Chait, B., and Hammerling, U. (1990) Retinol is Essential for Growth of Activated Human B Cells, *J. Exp. Med. 171:* 1624–1630.
42. Grun, F., Noy, N., Hammerling, U., and Buck, J. (1996) Purification, Cloning and Bacterial Expression of Retinyl Dehydratase from *Spodoptera frugiperda*, *J. Biol. Chem. 271:* 16135–16138.
43. Eppinger, T.M., Buck, J., and Hammerling, U. (1993) Growth Control or Terminal Differentiation: Endogenous Production and Differential Activities of Vitamin A Metabolites in HL-60 Cells, *J. Exp. Med. 178:* 1995–2005.
44. Shantz, E.M., Cawley, J.D., and Embree, N.D. (1943) Anhydro-("Cyclized") Vitamin A, *J. Am. Chem. Soc. 65:* 901.
45. Buck, J., Grun, F., Kimura, S., Noy, N., Derguini, F., and Hammerling, U. (1993) Anhydroretinol: A Naturally Occurring Inhibitor of Lymphocyte Physiology, *J. Exp. Med. 178:* 675–680.
46. O'Connell, M.J., Chua, R., Hoyos, B., Buck, J., Chen, Y.Q., Derguini, F., and Hammerling, U. (1996) Retro-Retinoids in Regulated Cell Growth and Death, *J. Exp. Med. 184:* 549–555.
47. Stancato, L.F., Chow., Y.-H., Hutchison, K.A., Perdew, G.H., Jove, R., and Pratt, W.B. (1993) Raf Exists in a Native Heterocomplex with hsp90 and p50 That Can Be Reconstituted in a Cell-Free System, *J. Biol. Chem. 268:* 21711–21716.
48. Stancato, L.F., Chow, Y.H., Owens-Grillo, J.K., Yem, A.W., Deibel, M.R., Jove, R., and Pratt, W.B. (1994) The Native vRaf.hsp90.p50 Heterocomplex Contains a Novel Immunophilin of the FK506 Binding Class, *J. Biol. Chem. 269:* 22157–22160.

49. Hurley, J.H., Newton, A.C., Parker, P.J., Blumberg, P.M., and Nishizuka, Y. (1997) Taxonomy and Function of C1 Protein Kinase C Homology Domains, *Protein Sci. 6:* 477–480.
50. Zhang, G., Kazanietz, M.G., Blumberg, P., and Hurley, J.H. (1995) Crystal Structure of the Cys2 Activator-Binding Domain of Protein Kinase C Delta in Complex with Phorbol Ester, *Cell 81:* 917–924.
51. Hurley, J., and Gobler, J. (1997) Protein Kinase C and Phospholipase C: Bilayer Interactions and Regulations, *Curr. Opin. Struct. Biol. 7:* 557–565.
52. Newton, A. (2002) Regulation of the ABC Kinases by Phosphorylation: Protein Kinase C as a Paradigm, *Biochem. J. 370:* 361–371.
53. Marshall, M. (1995) Interactions Between Ras and Raf: Key Regulatory Proteins in Cellular Transformation, *Mol. Reprod. Dev. 42:* 493–499.
54. Burns, D.J., and Bell, R.M. (1991) Protein Kinase C Contains Two Phorbol Ester Binding Domains, *J. Biol. Chem. 266:* 18330–18338.
55. Ono, Y., Fuji, T., Igarashi, K., Takayoshi, K., Tanaka, C., Kikkawa, U., and Nishizuka, Y. (1989) Phorbol Ester Binding to Protein Kinase C Requires a Cysteine-Rich Zinc-Finger-Like Sequence, *Proc. Natl. Acad. Sci. USA 86:* 4868–4871.
56. Knapp, L., and Klann, E. (2000) Superoxide-Induced Stimulation of Protein Kinase C *via* Thiol Modification and Modulation of Zinc Content, *J. Biol. Chem. 275:* 24136–24145.
57. Berg, J.M., and Shi, Y. (1996) The Galvanization of Biology: A Growing Appreciation for the Roles of Zinc, *Science 271:* 1081–1085.
58. Coleman, J.E. (1992) Zinc Proteins: Enzymes, Storage Proteins, Transcription Factors, and Replication Proteins, *Annu. Rev. Biochem. 61:* 897–905.
59. Kiefer, L.L., and Fierke, C.A. (1994) Functional Characterization of Human Carbonic Anhydrase Ii Variants with Altered Zinc Binding Sites, *Biochemistry 33:* 15233–15240.
60. Jakob, U., Muse, W., Eser, M., and Bardwell, J.C. (1999) Chaperone Activity with a Redox Switch, *Cell 96:* 341–352.
61. Hoffmann, J.H., Linke, K., Graf, P.C., Lilie, H., and Jakob, U. (2004) Identification of a Redox-Regulated Chaperone Network, *EMBO J. 23:* 160–168.
62. Graumann, J., Lilie, H., Tan, X., Tucker, C., Hoffmann, J.H., Vijayalakshmi, J., Saper, M., Bardwell, J.C.A., and Jakob, U. (2001) Activation of the Redox-Regulated Molecular Chaperone Hsp33—A Two-Step Mechanism, *Structure 9:* 377–387.
63. Kazanietz, M.G., Barchi, J.J., Omichinski, J.G., and Blumberg, P.M. (1995) Low Affinity Binding of Phorbol Esters to Protein Kinase C and Its Recombinant Cysteine-Rich Region in the Absence of Phospholipids, *J. Biol. Chem. 270:* 14679–14684.
64. Hommel, U., Zurini, M., and Luyten, M. (1994) Solution Structure of a Cysteine Rich Domain of Rat Protein Kinase C, *Nat. Struct. Biol. 1:* 383–388.
65. Mott, H., Carpenter, J., Zhong, S., Ghosh, S., Bell, R., and Campbell, S. (1996) The Solution Structure of the Raf-1 Cysteine-Rich Domain: A Novel Ras and Phospholipid Binding Site, *Proc. Natl. Acad. Sci. USA 93:* 8312–8317.
66. Korichneva, I., Hoyos, B., Chua, R., Levi, E., and Hammerling, U. (2002) Zinc Release from Protein Kinase C as the Common Event During Activation by Lipid Second Messenger of Reactive Oxygen, *J. Biol. Chem. 277:* 44327–44331.
67. Wald, G. (1968) Molecular Basis for Visual Excitation, *Science 162:* 230–232.
68. Haupts, U., Haupts, C., and Oesterhelt, D. (1995) The Photoreceptor Sensory Rhodopsin I as a Two-Photon-Driven Proton Pump, *Proc. Natl. Acad. Sci. USA 92:* 3834–3838.

69. Upton, J.M., Terentis, A.C., and Stocker, R. (1999) Tocopherol-Mediated Peroxidation of Lipoproteins: Implications of Vitamin E as a Potential Antiatherogenic Supplement, *FASEB J. 13:* 977–994.
70. Korichneva, I., and Hammerling, U. (1999) F-Actin as a Functional Target for Retro-Retinoids: A Potential Role in Anhydroretinol-Triggered Cell Death, *J. Cell. Sci. 112:* 2521–2528.
71. Korichneva, I., Waka, J., and Hammerling, U. (2003) Regulation of the Cardiac Mitochondrial Membrane Potential by Retinoids, *J. Pharm. Exp. Ther. 305:* 1–8.
72. Chen, Y., Buck, J., and Derguini, F. (1999) Anhydroretinol Induces Oxidative Stress and Cell Death, *Cancer Res. 59:* 3985–3990.
73. Sabichi, A.L., Xu, H., Fischer, S., Zou, C., Ang, X., Steele, V.E., Kelloff, G.J., Lotan, R., and Clifford, J.L. (2003) Retinoid Receptor-Dependent and Independent Biological Activities of Novel Fenretinide Analogues and Metabolites, *Clin. Cancer Res. 9:* 4606–4613.
74. Palmer, R.M., Ferige, A.G., and Moncada, S. (1987) Nitric Oxide Release Accounts for the Biological Activity of Endothelium-Derived Relaxing Factor, *Nature 327:* 524–526.
75. Ignarro, L.J., Buga, G.M., Wood, K.S., Byrns, R.E., and Chaudhuri, G. (1987) Endothelium-Derived Relaxation Factor Produced and Released from Artery and Vein Is Nitric Oxide, *Proc. Natl. Acad. Sci. USA 84:* 9265–9269.
76. Chen, Y., Cann, M.J., Litvin, T.N., Iourgenko, V., Sinclair, M.L., Levin, L.R., and Buck, J. (2000) Soluble Adenylyl Cyclase as an Evolutionarily Conserved Bicarbonate Sensor, *Science 289:* 559–560.
77. Ryter, S.W., Morse, D., and Choi, A.M. (2004) Carbon Monoxide: To Boldly Go Where NO Has Gone Before [Review], Sci. STKE 230, RE6.

Chapter 4

Chemical and Metabolic Oxidation of Carotenoids

Frederick Khachik

Department of Chemistry and Biochemistry, Joint Institute for Food Safety and Applied Nutrition (JIFSAN), University of Maryland, College Park, MD 20742

Introduction

To date, a wide range of carotenoids have been isolated, identified, and quantified from the extracts of fruits and vegetables commonly consumed in the United States (1–6). These studies revealed that 40–50 carotenoids may be available from the diet and absorbed, metabolized, or utilized by the human body (4). However, among these, only 13 *all-E* (*trans*)- and 12 *Z* (*cis*)-carotenoids are routinely found in human serum and milk (7–11). In addition, 1 *Z* (*cis*)- and 8 *all-E* (*trans*)-carotenoid, which are metabolites resulting from the two major dietary carotenoids, lutein and lycopene, were also characterized (7,8,10,11). This brings the total number of carotenoids and their metabolites detected in humans to 34 (11). These dietary carotenoids include lutein, α- and β-cryptoxanthin, lycopene, ζ-carotene, α- and β-carotene, phytofluene, and phytoene; they have been detected in μg–ng/g quantities in human lung, liver, breast, and cervical tissues (12). The correlation between dietary carotenoids and the carotenoids routinely found in extracts from human serum/plasma revealed that only selected groups of carotenoids make their way into the human bloodstream. For example, carotenoid epoxides such as lutein epoxide (taraxanthin), antheraxanthin, violaxanthin, and neoxanthin, which are major constituents of all green vegetables (1,4,5) have not been detected in human serum (7,11). Among serum carotenoids, lutein, zeaxanthin, and lycopene are not only absorbed intact but appear to undergo a series of metabolic transformations, resulting in the formation of carotenoid by-products (13–16). To explain the evidence in support of the metabolism of carotenoids in humans, the biosynthetic pathways that lead to the formation of dietary carotenoids and their oxidation products in plants will be discussed. Due to susceptibility of carotenoids to oxidative degradation, the observed biological activity of these compounds in certain cell culture studies has been difficult to interpret. Therefore, in an attempt to establish a model for monitoring the stability of carotenoids, a sample of purified lutein was exposed to air at room temperature for 2 wk and the oxidation products of this carotenoid were tentatively identified by HPLC-ultraviolet/visible absorption spectrophotometry (UV/Vis)-mass spectrometry (MS).

Materials and Methods

Technical-grade (3R,3′R,6′R) lutein (80% pure) containing ~5% (3R,3′R)-zeaxanthin was obtained from Kemin Foods LC (Des Moines, IA) and was further purified to 97% by two consecutive crystallizations according to our published procedure (17). A purified sample of (3R,3′R,6′R)-lutein (1 g) was kept in a brown bottle exposed to air at room temperature for 2 wk; at various intervals, the stability of this sample was monitored by HPLC-UV/Vis-MS. The separations were carried out on a silica-based nitrile bonded column (25 cm × 4.6 mm i.d.; 5-μm particle size), employing a mixture of 75% hexane, 25% dichloromethane, 0.25% methanol, and 0.10% *N,N*-diisopropylethylamine as eluent at a flow rate of 0.70 mL/min. The column was protected with a Brownlee nitrile-bonded guard cartridge (3 cm × 4.6 mm i.d.; 5-μm particles). The monitoring wavelength for lutein (λ_{max} = 448 nm) and zeaxanthin (λ_{max} = 454 nm) was selected at 450 nm. However, the chromatograms were also monitored at 440, 420, 380, 340, 290, and 250 nm by the photodiode-array detector to determine the presence of possible impurities and carotenoid oxidation products.

The mass spectra were obtained by interfacing the HPLC system into a Hewlett-Packard Model 5989A particle beam MS. The flow rate with the HPLC/MS system was 0.7 mL/min. Eluate from the HPLC was divided in a ratio of 1:2 with the smaller amount entering the particle beam interface, which was operated at a desolvation temperature of 45°C. Electron capture negative ionization (ECNI) was achieved using methane at a pressure of 160 Pa (1.2 torr) and a source temperature of 250°C. Spectra were collected from *m/z* 100 to *m/z* 700 using a scan cycle time of 1.5 s.

To monitor the formation of lutein/zeaxanthin oxidation products, a solution of these carotenoids at a concentration of 0.023 mg/mL was first analyzed and then concentrated by 10- and 20-fold and reexamined by HPLC-UV/Vis-MS, respectively.

Discussion

Dietary Carotenoids in Fruits, Vegetables, and Human Serum and Tissues

The major dietary carotenoids in fruits and vegetables can be classified into five categories as follows: (i) hydrocarbon carotenoids, (ii) monohydroxycarotenoids, (iii) dihydroxycarotenoids, (iv) carotenoid acyl esters, and (v) carotenoid epoxides (4,6). Certain hydrocarbon carotenoids such as α-carotene, β-carotene, and γ-carotene are converted to vitamin A in humans and are also absorbed intact, whereas others such as lycopene, neurosporene, ζ-carotene, phytofluene, and phytoene are not vitamin A active and are mainly absorbed intact. Among the latter group, only lycopene appears to undergo metabolic transformation (14,18,19). Monohydroxycarotenoids in fruits and vegetables are mainly α-cryptoxanthin and β-cryptoxanthin, and only the latter is among the provitamin A active carotenoids. The dihydroxycarotenoids in fruits and vegetables, (3R,3′R,6′R)-lutein and (3R,3′R)-zeaxanthin, are found in most green and yellow/orange fruits and vegetables (20). Carotenoid acyl esters constitute straight-chain fatty acid esters of mono- and dihydroxycarotenoids such as α-cryptoxanthin, β-cryptoxanthin, (3R,3′R,6′R)-lutein, and (3R,3′R)-zeaxanthin, which are found in certain fruits and vegetables (2–4). Carotenoid acyl esters have not been detected in human serum. However, in the presence of pancreatic secretions, this group of carotenoids undergoes hydrolysis to regenerate the parent hydroxycarotenoids, which are then absorbed. The major carotenoid epoxides in foods are the following: lutein epoxide (taraxanthin), antheraxanthin, violaxanthin, and neoxanthin. With the exception of carotenoid epoxides, which have not been detected in human serum and tissues, all of the other carotenoids described above are present in human serum (7,8), milk (11), organs (12), skin (21), and ocular tissues (22,23).

Carotenoid Metabolites in Humans

In addition to dietary carotenoids, a number of carotenoid metabolites that are absent in foods were also isolated and identified from human serum, organs, and ocular tissues (7,8,11,12,21–23). The chemical structures of these metabolites are shown in Figure 4.1. The first group of these metabolites appears to result from a series of oxidation-reduction and double-bond isomerization reactions of dietary (3R,3′R,6′R)-lutein and (3R,3′R)-zeaxanthin. This is supported by two separate human studies in which supplementation with either (3R,3′R,6′R)-lutein or (3R,3′R)-zeaxanthin increased the plasma concentration of these metabolites (13,14). Two dehydration products of dietary lutein, i.e., (3R,6′R)-3′,4′-didehydro-β,γ-caroten-3-ol and (3R,6′R)-2′,3′-didehydro-β,ε-caroten-3-ol (Fig. 4.1), were also isolated from the extracts of human plasma and fully characterized (7,10,11). These carotenoids are nonenzymatic by-products of dietary lutein and are presumably formed in the presence of acids in the human digestive system. The metabolism of lycopene appears to be rather complex and involves the oxidation of lycopene at the C-5 double bond, resulting in the formation of lycopene 5,6-epoxide. This epoxide is highly unstable and rapidly undergoes rearrangement at ambient temperature to form two diastereomeric 2,6-cyclolycopene-1,5-oxides. Although these bicyclic lycopene oxides have not been detected in human plasma, their ring opening products, which are shown in Figure 4.1, have been isolated and fully characterized (18,19). A supplementation study with lycopene conducted in healthy humans demonstrated that an increase in plasma lycopene concentration is directly associated with an increase in the plasma concentration of the oxidative metabolites of lycopene (16). For a detailed examination of the transformation of dietary carotenoids to their metabolites, see publications by Khachik *et al.* (12–15,19).

The role and function of carotenoids in human ocular tissues was also a focus of intense research in the past decade. In the early 1990s, a number of epidemiologic studies suggested a beneficial role for carotenoids, more specifically, lutein and zeaxanthin, as well as antioxidant vitamins in the prevention of neovascular

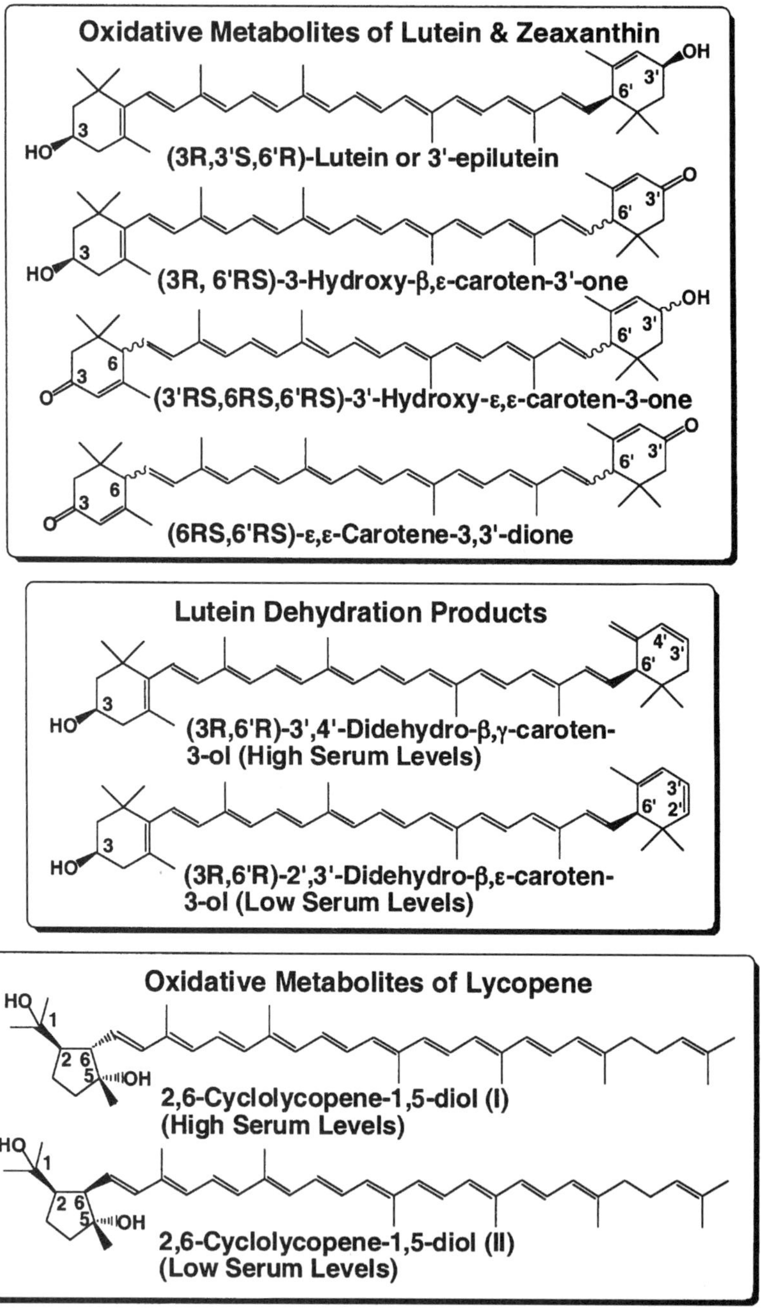

Fig. 4.1. The chemical structures of metabolites of lutein, zeaxanthin, and lycopene.

age-related macular degeneration (AMD) (24–26). One of the underlying hypotheses for the protective role of carotenoids in AMD was based on the ability of these carotenoids to act as antioxidants that can protect the human retina from photooxidation (27,28). The possible role of lutein and zeaxanthin in protection from AMD was studied extensively by a number of investigators (29–33). In 1997, we provided preliminary evidence for the photoprotective role of two dietary carotenoids, lutein and zeaxanthin, in the human retina as antioxidants in the prevention of AMD (22). This was accomplished by isolating, identifying the chemical structure, and measuring the concentrations of lutein, zeaxanthin, and their oxidation products in the retinas of 11 human donor eyes and that of one monkey. Although lutein, zeaxanthin, and a direct oxidation product of lutein were the major carotenoids in the retina, 11 minor carotenoids were also identified. On the basis of these findings, we proposed and postulated a series of metabolic reactions by which (3R,3′R,6′R)-lutein and (3R,3′R)-zeaxanthin in the retina may be transformed into their oxidation products to protect the macula against photooxidation and prevent AMD. More recently, we identified and quantified lutein, zeaxanthin, and their oxidative metabolites in the pooled extracts from various tissues of the human eye [neural retina, retinal pigment epithelium (RPE/choroid), ciliary body, iris, lens] (23). Lycopene and a diverse range of carotenoids were also identified and quantified in the human ciliary body and RPE/choroid. Our recent data also revealed the presence of lycopene in the human iris, but no detectable levels were found in the lens (34). The presence of carotenoids and their oxidation products in the iris and lens is consistent with the hypothesized role of these compounds in the prevention and treatment of cataract. For a review of the biological importance of lycopene in human health see the publication by Stahl and Sies (35).

Chemical Oxidation of Purified Carotenoids

In an attempt to evaluate the susceptibility of certain dietary carotenoids such as (3R,3′R,6′R)-lutein and lycopene, the chemical oxidation of these carotenoids was extensively studied. These studies allowed the identification of the oxidative metabolites of carotenoids that were isolated from human plasma and tissues and provided valuable insight into the metabolic transformations in humans. The metabolic oxidation of carotenoids in humans is dependent on many factors and could potentially take a course that is entirely different from that of chemical oxidation of purified carotenoids, which is normally achieved with an oxidizing reagent in organic solvents. Nonetheless, in several cases, we showed that the chemical oxidation of purified carotenoids under controlled conditions resulted in the formation of the same oxidation products of these compounds observed in humans. For example, allylic oxidation of (3R,3′R,6′R)-lutein with nickel peroxide yielded (3R,6′R)-3-hydroxy-β,ε-caroten-3′-one, which was also isolated from human plasma (7,11) as well as human ocular tissues (22,23,34). The nature of the oxidizing reagent also plays an important role in the outcome of the reaction. When lutein is allowed to react with *m*-chloroperoxybenzoic acid under controlled conditions, lutein 5,6-epoxide is obtained as the main product (36,37). However, as mentioned earlier, carotenoid epoxides such as lutein 5,6-epoxide, which are present in all green fruits and vegetables, are absent in human plasma and tissues. The chemical reaction of carotenoids with oxidizing reagents can also result in the oxidative cleavage of the polyene chain of these compounds and yield a number of apocarotenals and apocarotenones. The degree to which these breakdown products of carotenoids are formed is dependent on the reaction conditions such as temperature as well as the nature and the mole equivalence of the oxidizing reagent. To date, none of the oxidative cleavage products of carotenoids have been detected in human plasma and tissues. Lycopene is another example of a dietary carotenoid that proved to be an effective scavenger for reactive oxygen species protecting against oxygen-mediated cytotoxicity and genotoxicity by scavenging singlet oxygen and other reactive oxygen species (38,39). The chemical oxidation of lycopene by hydrogen peroxide and *m*-chloroperoxybenzoic acid allowed the identification of two metabolites of this carotenoid, i.e., 2,6-cyclolycopene-1,5-diols I and II, which were isolated and identified from the extract of tomato products and human plasma (Fig. 4.1) (18,19,40). More recently, Yokota *et al.* (41) investigated the reactivity of lycopene toward peroxynitrile in an *in vitro* model and proposed a mechanism for the peroxynitrile scavenging action of lycopene.

It is interesting to note that in the chemical oxidation of lycopene, the 1,2-double bond and the 5,6-double bond are most susceptible to oxidation and lead to the formation of lycopene 1,2-epoxide and lycopene 5,6-epoxide, respectively. Although the 1,2-epoxide is quite stable, the 5,6-epoxide undergoes rearrangement and yields two diasteromeric bicyclic oxides identified as 2,6-cyclolycopene-1,5 oxides I and II. The products of the acid-catalyzed ring opening of these lycopene oxides were shown to be identical to the metabolites of lycopene identified in human plasma (18,19).

The chemical and metabolic oxidation of lutein and lycopene described above indicates that under controlled conditions, the functional groups in the end-groups (hydroxyl group and double bond) of these carotenoids are most susceptible to oxidation as opposed to oxidative cleavage of the polyene chain. Although the details and mechanisms of the metabolic oxidation of lutein and lycopene in humans remain unexplored and may take an entirely different course from that of chemical oxidation, the nature of the oxidation products of these carotenoids produced by both processes is remarkably similar.

Biological Activity of Dietary Carotenoids

The evidence for nutritional significance of carotenoids in the prevention of chronic diseases such as cancer, cardiovascular disease, and macular degeneration was obtained from various interdisciplinary studies; these may be classified as follows: (i) epidemiologic studies; (ii) carotenoid distribution in fruits, vegetables, human serum, and tissues; (iii) *in vitro* studies of chemopreventive properties; and (iv) *in vivo* studies with rodents. For more than two decades, it was suggested that one mechanism by which carotenoids exert their biological activity in disease prevention was by functioning as an antioxidant. Preliminary support for this mechanism of action was obtained by isolation and characterization of the oxidation products of carotenoids from human serum and tissues. However, in addition to their antioxidant mechanism of action, carotenoids can exert their biological activity in disease prevention *via* several other mechanisms. These include the following: gap junctional intercellular communications, antiinflammatory and antitumor promoting property, and induction of the detoxication (phase 2) enzymes (15). The observed biological activities of carotenoids in *in vitro* models are often difficult to interpret because of the susceptibility of these compounds to air oxidation, which may result in the formation of carotenoid breakdown products in cell cultures. Therefore, in certain cases in which the cell cultures containing carotenoids are incubated for extended periods of time, it may be argued that the breakdown products of these compounds may be responsible for the observed biological activity or lack thereof. In addition, during extraction, isolation, and manipulation of carotenoids in the laboratory, the integrity of these compounds may be affected, resulting in carotenoid artifacts that may be misidentified as carotenoid metabolites. To address this problem, an attempt should be made to investigate the stability of the carotenoid in question and establish the identity of the possible by-products that might be formed from it as a result of air oxidation. Because (3R,3′R,6′R)-lutein is one of the major dietary carotenoids that appears to undergo extensive metabolic transformation in humans, the stability of this carotenoid as a result of prolonged exposure to air was studied extensively and the results are described below.

Air Oxidation of Lutein as a Model for Monitoring the Stability of Carotenoids

As discussed earlier, in the chemical oxidation of carotenoids, two types of products are normally observed with these compounds. In the first type, the polyene chain of the carotenoid remains intact and the end-group is modified by oxidation. In this case, oxidation of a double bond or a hydroxyl group in the carotenoid end-group results in the formation of a carotenoid epoxide or an α,β-unsaturated cyclic ketone, respectively. For example, one of the most likely oxidation products of (3R,3′R,6′R)-lutein is (3R,6′R)-3-hydroxy-β,ε-caroten-3′-one (3′-oxolutein), which can be formed by allylic oxidation of the hydroxyl group in the ε-ring of this carotenoid. Alternatively, epoxidation of the double bonds in the β- and/or ε-end groups of lutein can also yield lutein 5,6-epoxide or lutein 4′,5′-epoxide, and/or lutein 5,6,4′,5′-diepoxide. Although chemical oxidation of lutein by various reagents has not yet revealed products that are formed by a combination of allylic oxidation and epoxidation, this possibility should also be considered.

The second type of carotenoid oxidation involves the oxidative cleavage of the polyene chain of the carotenoid and can result in the formation of a number of apocarotenals, apocarotenones, and apocarotenoic acids. Under more severe conditions, a third possibility exists in which the oxidative cleavage may also be accompanied by oxidative modification of the carotenoid end-group. Fortunately, regardless of the mode of oxidation, the nature and identity of the oxidation products of a carotenoid such as lutein can be tentatively established by a combination of UV/Vis-MS. Therefore the HPLC-UV/Vis photodiode array detection-MS technique was employed to monitor the stability of a purified sample of (3R,3′R,6′R)-lutein (98%) [containing ~5% (3R,3′R)-zeaxanthin] that was kept in a brown bottle and exposed to air at room temperature for 2 wk. The purity of the lutein that was employed as the starting material was fully established by HPLC-UV/Vis-MS and ^1H NMR spectroscopy. For this purpose, a high concentration of the purified lutein was examined by HPLC-UV/Vis-MS in an attempt to detect the presence of minute amounts of lutein oxidation products and/or impurities. However, the highest concentration of lutein that could be injected into the HPLC system without column overloading and disruption in the separation profile was estimated to be 0.46 mg/mL of HPLC eluent. Although lutein was above the detection limit of the photodiode array detector at this concentration, no impurities and/or oxidation products of this carotenoid could be detected by HPLC before the stability study.

In wk 1 of lutein exposure to air, there was no noticeable change in the HPLC-UV/Vis-MS profile, and no apparent oxidation products of this carotenoid could be detected by monitoring the chromatograms between 250 and 600 nm. However, after 2 wk, HPLC revealed the presence of a number of unknown peaks that eluted before lutein on a silica-based nitrile bonded column. These were identified by HPLC-UV/Vis-MS as a number of apocarotenals and apocarotenones, resulting from oxidative cleavage of the polyene chain of lutein. In addition, (3R,6′R)-3-hydroxy-β,ε-caroten-3′-one (3′-oxolutein), which is among the major oxidation products of lutein in human plasma, was also identified; this carotenoid is formed by allylic oxidation of the hydroxyl group in the ε-ring of lutein. The HPLC profile and the mass ion chromatogram (ECNI) of lutein after 2 wk of exposure to air at room temperature is shown in Figures 4.2A and B, respectively. The chemical structures of the identified apocarotenals and apocarotenones are shown in Figure 4.3, in their order of chromatographic elution.

Several characteristics of the HPLC-UV/Vis-MS profiles of lutein and zeaxanthin were also employed for distinguishing apocarotenoids with a β-group from those with an ε-end group. For example, the mass spectrum of lutein showed the molecular parent ion peak at m/z = 568 Da as well as an ion peak at m/z = 550 Da due to the loss of water from the molecular parent ion. Although the mass spectrum of zeaxanthin also showed the molecular parent ion at 568 Da, the molecular ion due to the loss of water from the parent ion peak in this compound was not observed. This is because the dehydration of the allylic hydroxyl group in lutein is activated by the neighboring double bond, whereas in zeaxanthin, in which the hydroxyl groups are nonallylic, this activation is absent. This pattern of ionization was used consistently to differentiate between 3-hydroxy-ε- and 3-hydroxy-β-end

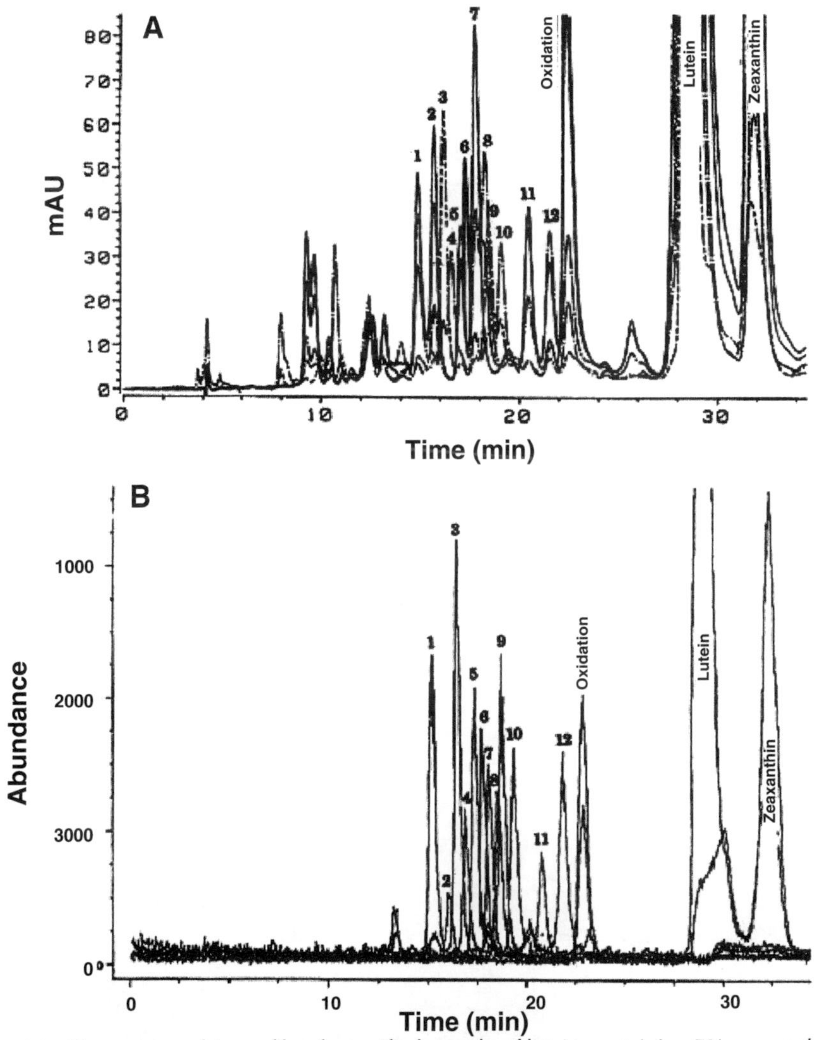

Fig. 4.2. Chromatographic profile of a purified sample of lutein containing 5% zeaxanthin after exposure to air for 2 wk at ambient temperature. Panel A: HPLC profile on a silica-based nitrile bonded column (conditions described in text) simultaneously monitored by a photodiode array detector between 250 and 600 nm; panel B: mass ion chromatogram (electron capture negative ionization, ECNI). For peak identification see Figure 4.3.

groups of apocarotenoids. Similarly, the photodiode array detector also provided the absorption spectra of lutein and zeaxanthin in the HPLC eluent. Lutein exhibited a main absorption maximum at 448 nm, which was 6 nm lower than the main absorption maxima of zeaxanthin at 454 nm. This hypsochromic shift from the absorption maximum of zeaxanthin to lutein is due to the isolated double bond in

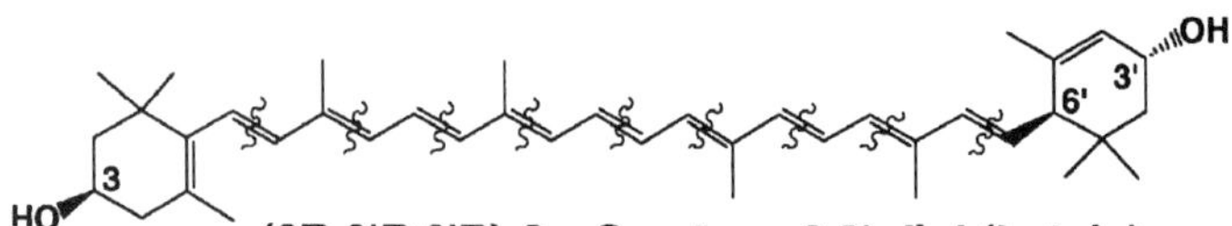

(3R,3'R,6'R)-β,ε-Carotene-3,3'-diol (Lutein)

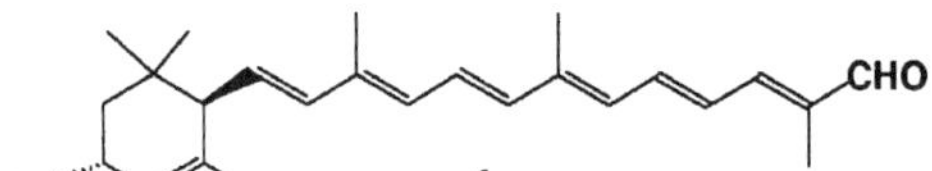

1

$C_{25}H_{34}O_2$
Mass = 366, 348 (M-18)
λ_{max} = 416 nm

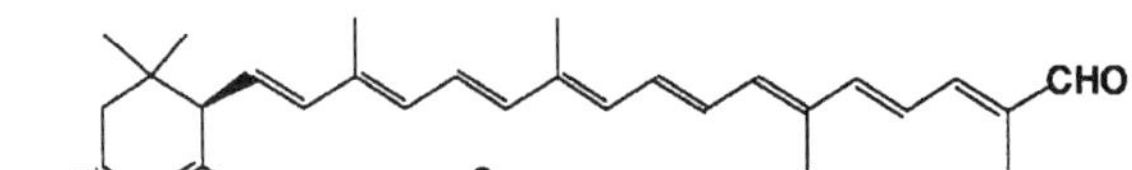

2

$C_{30}H_{40}O_2$
Mass = 432, 414 (M-18)
λ_{max} = 450 nm

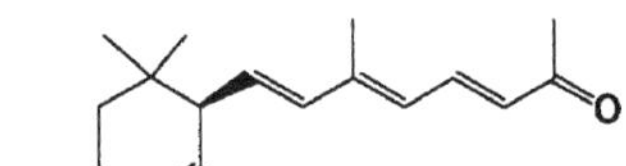

3

$C_{18}H_{26}O_2$
Mass = 274, 256 (M-18)
λ_{max} = 322 nm

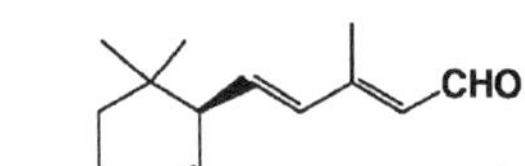

4

$C_{15}H_{22}O_2$
Mass = 234, 216(M-18)
λ_{max} = 280 nm

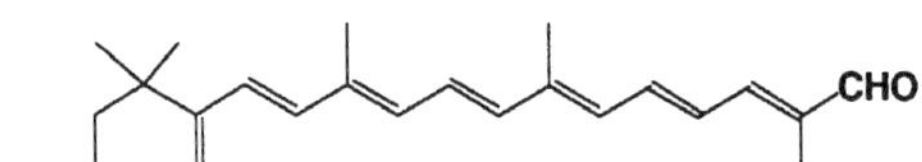

5

$C_{25}H_{34}O_2$ Mass = 366
λ_{max} = 420 nm

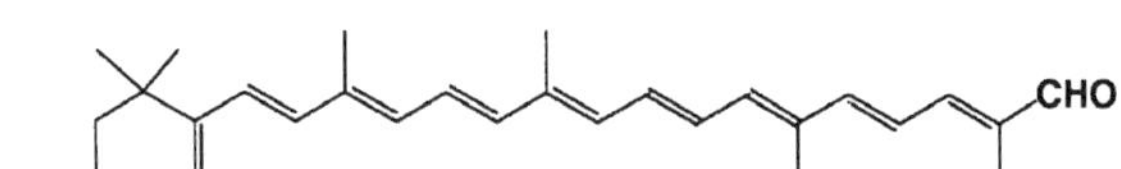

6

$C_{30}H_{40}O_2$ Mass = 432
λ_{max} = 455 nm

(Continued) →

Fig. 4.3. The chemical structure of lutein and its oxidation products, apocarotenals and apocarotenones, formed by 2 wk of exposure of this carotenoid to air at ambient temperature. The mass spectrometric and ultraviolet/visible data obtained by HPLC-UV/Vis-MS are also shown for each compound. The HPLC and mass ion chromatograms are shown in Figure 4.2.

the ε-end group of lutein, which is not in conjugation with the polyene chain system in this compound. This 6-nm difference between the wavelengths of the main absorption maxima of lutein and zeaxanthin was used effectively to distinguish apocarotenoids with an ε-end group from those with a β-end group. As shown in the HPLC (Fig. 4.2A) and the mass ion (Fig. 4.2B) chromatograms, in addition to

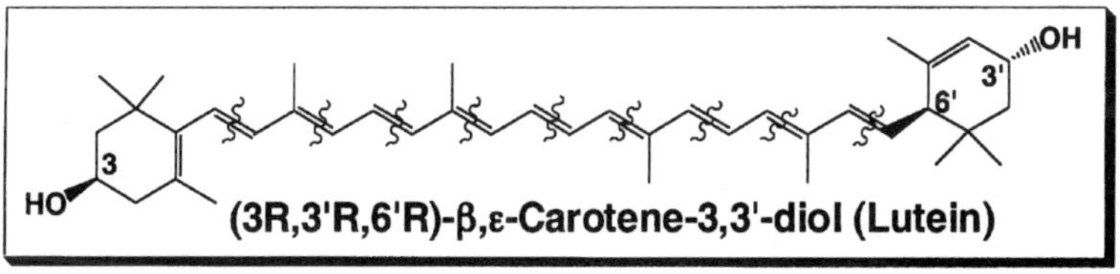

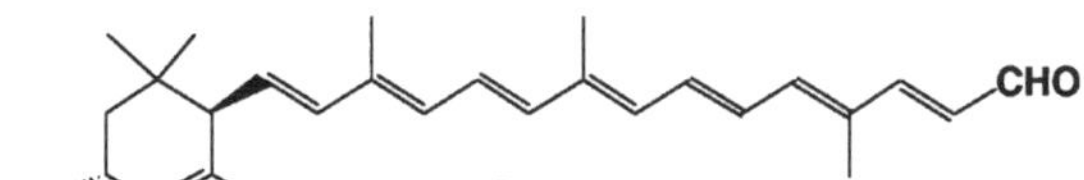

7

$C_{27}H_{36}O_2$
Mass = 392, 374 (M-18)
λ_{max} **= 438 nm**

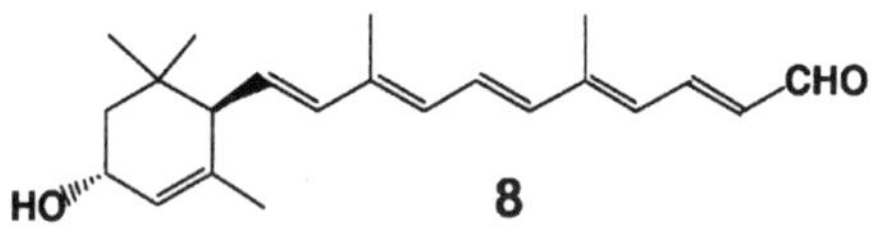

8

$C_{22}H_{30}O_2$
Mass = 326, 308 (M-18)
λ_{max} **= 382 nm**

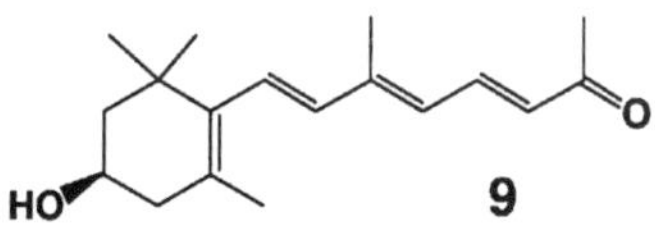

9

$C_{18}H_{26}O_2$ **Mass = 274**
λ_{max} **= 338 nm**

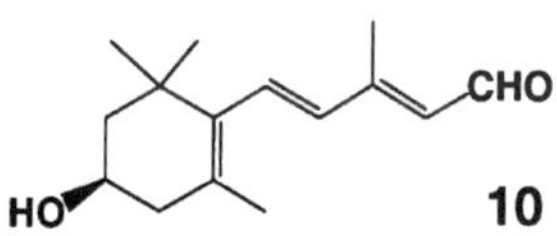

10

$C_{18}H_{26}O_2$ **Mass = 234**
λ_{max} **= 286 nm**

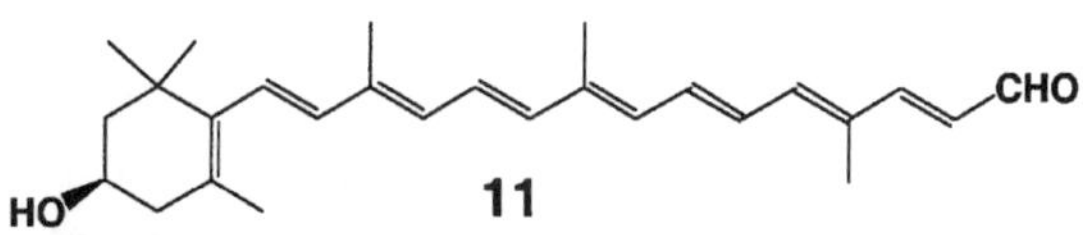

11

$C_{27}H_{36}O_2$ **Mass = 392**
λ_{max} **= 444-446 nm**

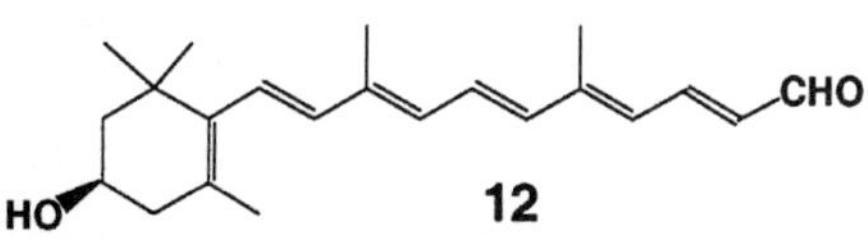

12

$C_{22}H_{30}O_2$ **Mass = 326**
λ_{max} **= 390 nm**

Fig. 4.3. (Continued)

the formation of the oxidative cleavage products, lutein also undergoes allylic oxidation to form 3′-oxolutein as a result of exposure to air at room temperature. It is interesting to note that apocarotenoids that contain an allylic hydroxyl group (compounds 1, 2, 3, 4, 7, and 8) can be readily distinguished from their isomeric counterparts with a nonallylic hydroxyl group (compounds 5, 6, 9, 10, 11, and 12) by evaluation of both the mass and UV/Vis spectra of these compounds. This is because in MS analysis by ECNI, apocarotenoids with an allylic hydroxyl group in an ε-ring exhibit ions due to the loss of water from their molecular parent ions, whereas this is not the case for apocarotenoids with a nonallylic hydroxy group in a β-ring. In addition, apocarotenoids with an allylic hydroxyl group in an ε-ring exhibit absorption maxima at wavelengths 4–8 nm lower than those with a non-allylic hydroxyl group in a β-ring. This is due to the lack of conjugation of the isolated double bond in apocarotenoids with an ε-end group with polyene chain double bonds as opposed to their isomeric counterparts with a β-end group. These findings are also consistent with the UV/Vis and MS data described earlier for lutein and zeaxanthin.

The results from this study clearly show that *in vitro* cell culture studies with lutein, which are normally carried out within 24–48 h of incubation, would not be expected to be accompanied by complications due to the formation of apocarotenoids. However, the matrix to which carotenoids are exposed could also play an important role in either preventing or facilitating the oxidation of these compounds. Therefore, it is advisable to monitor the stability of carotenoids in a given cell culture medium by HPLC-UV/Vis-MS to ensure that the structural integrity of these compounds remains intact. Because the stability of carotenoids is greatly dependent on their chemical structure and their ability to quench singlet oxygen and other oxidizing species, similar studies are warranted for other carotenoids of interest (38).

Summary

Although the number of dietary carotenoids is in excess of 40, only 13 major *all-E* (*trans*)- and 12 *Z* (*cis*)-carotenoids are absorbed, metabolized, or utilized by the human body. There are also 9 carotenoid metabolites that have been isolated and characterized from human plasma and tissues. These metabolites result from modification of the carotenoid end-group by a series of oxidation-reduction, double-bond isomerization, and rearrangement reactions. Under controlled conditions, the chemical oxidation of carotenoids such as lutein and lycopene can yield the same metabolites as those observed in humans. Although metabolic and controlled chemical oxidation of carotenoids involve two completely different processes, the nature of the products is remarkably similar.

Under more severe conditions, chemical oxidation of carotenoids can also result in the formation of apocarotenoids. The outcome of cell culture studies and laboratory operations in which carotenoids are exposed to air for an extended period of time has been difficult to interpret because of the susceptibility of these compounds to air oxidation. Therefore, to gain a better understanding of the metabolic and chemical oxidation of carotenoids in various *in vitro* models, the integrity of these air-sensitive compounds should be established systematically. Because (3R,3′R,6′R)-lutein is one of the major dietary carotenoids that appears to undergo extensive metabolic transformations in humans, the stability of this carotenoid as a result of prolonged exposure to air was examined extensively in this report. This study revealed that lutein exposure to air after 2 wk at ambient temperature is accompanied by oxidative cleavage of the polyene chain of the carotenoid and results in the formation of a number of apocarotenoids that are absent in human plasma and tissues. This air oxidation is found to be a very slow and gradual process and does not appear to interfere with *in vitro* cell culture studies and laboratory operations. However, the results from this study cannot be generalized to include other carotenoids that may exhibit a different degree of susceptibility toward air oxidation.

Acknowledgments

The author thanks the Joint Institute for Food Safety and Applied Nutrition (JIFSAN), University of Maryland (UM), Food and Drug Administration (FDA), and Kemin Foods, LC (Des Moines, IA) for their financial support.

References

1. Khachik, F., Beecher, G.R., and Whittaker, N.F. (1986) Separation, Identification, and Quantification of the Major Carotenoid and Chlorophyll Constituents in the Extracts of Several Green Vegetables by Liquid Chromatography, *J. Agric. Food Chem. 34:* 603–616.
2. Khachik, F., and Beecher, G.R. (1988) Separation and Identification of Carotenoids and Carotenol Fatty Acid Esters in Some Squash Products by Liquid Chromatography. Part I: Quantification of Carotenoids and Related Esters by HPLC, *J. Agric. Food Chem. 36:* 929–937.
3. Khachik, F., Beecher, G.R., and Lusby, W.R. (1989) Separation, Identification, and Quantification of the Major Carotenoids in Extracts of Apricots, Peaches, Cantaloupe, and Pink Grapefruit by Liquid Chromatography, *J. Agric. Food Chem. 37:* 1465–1473.
4. Khachik, F., Beecher, G.R., Goli, M.B., and Lusby, W.R. (1991) Separation, Identification, and Quantification of Carotenoids in Fruits, Vegetables and Human Plasma by High Performance Liquid Chromatography, *Pure Appl. Chem. 63:* 71–80.
5. Khachik, F., Goli, M.B., Beecher, G.R., Holden, J., Lusby, W.R., Tenorio, M.D., and Barrera, M.R. (1992) The Effect of Food Preparation on Qualitative and Quantitative Distribution of Major Carotenoid Constituents of Tomatoes and Several Green Vegetables, *J. Agric. Food Chem. 40:* 390–398.
6. Khachik, F., Beecher, G.R., Goli, M.B., and Lusby, W.R. (1992) Separation and Quantification of Carotenoids in Foods, in *Methods in Enzymology*, Packer, L., ed., Academic Press, New York, vol. 213A, pp. 347–359.
7. Khachik, F., Beecher, G.R., Goli, M.B., Lusby, W.R., and Smith, J.C. (1992) Separation and Identification of Carotenoids and Their Oxidation Products in Extracts of Human Plasma, *Anal. Chem. 64:* 2111–2122.
8. Khachik, F., Beecher, G.R., Goli, M.B., Lusby, W.R., and Daitch, C.E. (1992) Separation and Quantification of Carotenoids in Human Plasma, in *Methods in Enzymology*, Packer, L., ed., Academic Press, New York, vol. 213A, pp. 205–219.
9. Khachik, F., Englert, G., Daitch, C.E., Beecher, G.R., Lusby, W.R., and Tonucci, L.H. (1992) Isolation and Structural Elucidation of the Geometrical Isomers of Lutein and Zeaxanthin in Extracts from Human Plasma, *J. Chromatogr. Biomed. Appl. 582:* 153–166.
10. Khachik, F., Englert, G., Beecher, G.R., and Smith, J.C., Jr. (1995) Isolation, Structural Elucidation, and Partial Synthesis of Lutein Dehydration Products in Extracts from Human Plasma, *J. Chromatogr. Biomed. Appl. 670:* 219–233.

11. Khachik, F., Spangler, C.J., Smith, J.C., Jr., Canfield, L.M., Pfander, H., and Steck, A. (1997) Identification, Quantification, and Relative Concentrations of Carotenoids, and Their Metabolites in Human Milk and Serum, *Anal. Chem. 69:* 1873–1881.
12. Khachik, F., Askin, F.B., and Lai, K. (1998) Distribution, Bioavailability, and Metabolism of Carotenoids in Humans, in *Phytochemicals: Today's Knowledge for Tomorrow's Products*, Omaye, S., and Bidlack, W., eds., Technomic Publishing, Lancaster, PA, pp. 77–96.
13. Khachik, F., Beecher, G.R., and Smith, J.C., Jr. (1995) Lutein, Lycopene, and Their Oxidative Metabolites in Chemoprevention of Cancer, *J. Cell. Biochem. 22:* 236–246.
14. Khachik, F., Steck, A., and Pfander, H. (1997) Bioavailability, Metabolism, and Possible Mechanism of Chemoprevention by Lutein and Lycopene in Humans, in *Food Factors for Cancer Prevention*, Ohigashi, H., Osawa, T., Terao, J., Watanabe, S., and Yoshikawa, T., eds., Springer-Verlag, Tokyo, pp. 542–547.
15. Khachik, F., Bertram, J.S., Huang, M.T., Fahey, J.W., and Talalay, P. (1999) Dietary Carotenoids and Their Metabolites as Potentially Useful Chemopreventive Agents Against Cancer, in *Antioxidant Food Supplements in Human Health*, Packer, L., Hiramatsu, M., and Yoshikawa, T., eds., Academic Press, Tokyo, Chapter 14, pp. 203–229.
16. Paetau, I., Khachik, F., Brown, E.D., Beecher, G.R., Kramer, T.R., Chittams, J., and Clevidence, B.A. (1998) Chronic Ingestion of Lycopene-Rich Tomato Juice or Lycopene Supplements Significantly Increases Plasma Concentrations of Lycopene and Related Tomato Carotenoids in Humans, *Am. J. Clin. Nutr. 68:* 1187–1195.
17. Khachik, F., Steck, A., and Fander, H. (1999) Isolation and Structural Elucidation of (13Z,13′Z,3R,3′R,6′R)-Lutein from Marigold Flowers, Kale, and Human Plasma, *J. Agric. Food Chem. 47:* 455–461.
18. Khachik, F., Steck, A., Niggli, U.A., and Pfander, H. (1998) Partial Synthesis and Structural Elucidation of the Oxidative Metabolites of Lycopene Identified in Tomato Paste, Tomato Juice and Human Serum, *J. Agric. Food Chem. 46:* 4874–4884.
19. Khachik, F., Pfander, H., and Traber, B. (1998) Proposed Mechanisms for the Formation of the Synthetic and Naturally Occurring Metabolites of Lycopene in Tomato Products and Human Serum, *J. Agric. Food Chem. 46:* 4885–4890.
20. Humphries, J.M., and Khachik, F. (2003) Distribution of Lutein, Zeaxanthin and Related Geometrical Isomers in Fruit, Vegetables, Wheat and Pasta Products, *J. Agric. Food Chem. 51:* 1322–1327.
21. Scholz, T.A., Hata, T.R., Pershing, L.K., Gellermann, W., McClane, R., Alexeeva, M., Irmakov, I., and Khachik, F. (2000) Non-Invasive Raman Spectroscopic Detection of Carotenoids in Human Skin, *J. Investig. Dermatol. 115:* 441–448.
22. Khachik, F., Bernstein, P., and Garland, D.L. (1997) Identification of Lutein and Zeaxanthin Oxidation Products in Human and Monkey Retinas, *Investig. Ophthalmol. Vis. Sci. 38:* 1802–1811.
23. Bernstein, P.S., Khachik, F., Carvalho, L.S., Muir, G.J., Zhao, D.Y., and Katz, N.B. (2001) Identification and Quantitation of Carotenoids and Their Metabolites in the Tissues of the Human Eye, *Exp. Eye Res. 72:* 215–223.
24. The Eye-Disease Case Control Study Group (1992) Risk Factors for Neovascular Age-Related Macular Degeneration, *Arch. Ophthalmol. 110:* 1701–1708.
25. The Eye-Disease Case Control Study Group (1993) Antioxidant Status and Neovascular Age-Related Macular Degeneration, *Arch. Ophthalmol. 111:* 104–109.

26. Seddon, J.M., Ajani, U.A., Sperduto, R.D., Hiller, R., Blair, N., Burton, T.C., Farber, M.D., Gragoudas, E.S., Haller, J., Miller, D.T., Yannuzzi, L.A., and Willett, W. (1994) Dietary Carotenoids, Vitamin A, C, and E, and Advanced Age-Related Macular Degeneration, *J. Am. Med. Assoc. 272:* 1413–1420.
27. Snodderly, D.M. (1995) Evidence for Protection Against Age-Related Macular Degeneration by Carotenoids and Antioxidant Vitamins, *Am. J. Clin. Nutr. 162:* 1448S–1461S.
28. Schalch, W., Dayhaw-Barker, P., and Barker, F.M. (1999) The Carotenoids of Human Macula, in *Nutritional and Environmental Influences on the Eye*, Taylor, A., ed., CRC Press, Boca Raton, FL, pp. 215–250.
29. Landrum, J.T., Bone, R.A., and Kilburn, M.D. (1997) The Macular Pigment: A Possible Role in Protection from Age-Related Macular Degeneration, *Adv. Pharmacol. 38:* 537–556.
30. Snodderly, D.M., Auran, J.D., and Delori, F.C. (1984) The Macular Pigment, II: Spatial Distribution in Primate Retinas. *Investig. Ophthalmol. Vis. Sci. 25:* 674–685.
31. Bone, R.A., Landrum, J.T., Friedes, L.M., Gomez, C.M., Kilburn, M.D., Menendez, E., Vidal, I., and Wang, W. (1997) Distribution of Lutein and Zeaxanthin Stereoisomers in the Human Retina, *Exp. Eye Res. 64:* 211–218.
32. Snodderly, D.M., Handelman, G.J., and Adler, A.J. (1991) Distribution of Individual Macular Pigment Carotenoids in Central Retina of Macaque and Squirrel Monkeys, *Investig. Ophthalmol. Vis. Sci. 32:* 268–279.
33. Beatty, S., Boulton, M., Henson, D., Koh, H.H., and Murray, I.J. (1999) Macular Pigment and Age-Related Macular Degeneration, *Br. J. Ophthalmol. 83:* 867–877.
34. Khachik, F., Moura, F.F., Zhao, D.Y., Aebischer, C.P., and Bernstein, P.S. (2002) Transformations of Selected Carotenoids in Plasma, Liver, and Ocular Tissues of Humans and in Nonprimate Animal Models, *J. Investig. Ophthalmol. Vis. Sci. 43:* 3383–3392.
35. Stahl, W., and Sies, H. (1996) Lycopene: A Biologically Important Carotenoid for Humans? *Arch. Biochem. Biophys. 336:* 1–9.
36. Karrer, P., and Jucker, E. (1945) Partial Synthesis of Flavoxanthin, Chrysanthemaxanthin, Antheraxanthin, Violaxanthin, Mutatoxanthin, and Auroxanthin, *Helv. Chim. Acta 28:* 300–315.
37. Subbarayan, C., Jungalwala, F.B., and Cama, H.R. (1965) Applicability of Partition Ratios, Mercuric Chloride Complexes, and Chromatographic Behavior for the Identification of Carotenoids, *Anal. Biochem. 12:* 275–281.
38. DiMacio, P., Kaiser, S., and Sies, H. (1989) Lycopene as the Most Efficient Biological Carotenoid Singlet Oxygen Quencher, *Arch. Biochem. Biophys. 274:* 532–538.
39. Lu, Y., Etoh, H., Watanabe, N., Ina, K., Ukai, N., Oshima, S., Ojima, F., Sakamoto, H., and Ishiguro, Y. (1995) A New Carotenoid: Hydrogen Peroxide Oxidation Product from Lycopene, *Biosci. Biotechnol. Biochem. 59:* 2153–2155.
40. Yokota, T., Etoh, H., Ukai, N., Oshima, S., Sakamoto, H., and Ishiguro, Y. (1997) 1,5-Dihydroxyiridanyl Lycopene in Tomato Puree, *Biosci. Biotechnol. Biochem. 61:* 549–550.
41. Yokota, T., Ohtake, T., Ishikawa, H., Inakuma, T., Ishiguro, Y., Terao, J., Nagao, A., and Etoh, H. (2003) Quenching of Peroxynitrile by Lycopene *In Vitro*, *Chem. Lett. 33:* 80–81.

Chapter 5

Gap Junctional Intercellular Communication: Carotenoids and Retinoids

Wilhelm Stahl and Helmut Sies

Institut für Biochemie und Molekularbiologie I, Heinrich-Heine-Universität Düsseldorf, D-40001 Düsseldorf, Germany

Introduction

Epidemiologic studies established a positive correlation between eating a balanced diet with ample amounts of fruits and vegetables and a lowered risk of cancer (1,2). The cancer-preventing properties of such a diet were associated with an increased intake of micronutrients including antioxidants such as polyphenols, vitamins E and C, and carotenoids (3). Although recent intervention studies with single antioxidants, especially β-carotene (4), yielded disappointing results, their possible role in cancer prevention remains to be evaluated. Cooperative interactions within an optimized mixture leading to synergistic antioxidant efficacy may explain why supplementation with single compounds failed to achieve protection. Revealing the biological properties of micronutrients is essential for understanding their contribution to human health as bioactive components of a healthy diet. Antioxidant activity of micronutrients represents only one aspect of their spectrum of bioactivities. It was shown that vitamin E, polyphenols, and carotenoids affect cell growth and differentiation, interfere with the progression of the cell cycle, and play a role in the regulation of gene expression and post-transcriptional modification of gene products (5). Carotenoids and their cleavage products, the apo-carotenoids and retinoids, influence cellular signaling pathways. They also modulate the direct exchange of signals between neighboring cells, triggering gap junctional intercellular communication (GJIC).

Gap Junctional Intercellular Communication

Microdomains of the plasma membrane containing arrays of channels that directly link the cytosol of adjacent cells are denoted as gap junctions. The channels provide a pathway that permits small molecules to shuttle from one cell to another (6). The upper limit in molecular weight for passage through gap junctions is ~1000 Da, and compounds such as cyclic AMP, inositol triphosphate, glucose-6-phosphate, or nucleotides may pass through.

Each channel is composed of two connexons (half-channels), which are connected in the intermembrane gap. Connexons are hexagonal tubes formed from six connexin subunits. Several subtypes of connexins have been identified, and the pattern differs among various organisms and organs (7). Gap junctions are seen in cells in contact with other cells in a tissue. Most cell types express multiple connexin isoforms; therefore, a spectrum of heteromeric hemichannels and heterotypic gap junctions may be formed, providing the structural basis for the selectivity of signaling *via* GJIC. Connexins share a common structure of four membrane-spanning domains (8). The channels are dynamic structures within the membrane and present in an open or closed configuration (9). High levels of intracellular calcium or low intracellular pH are stimuli for rapid closure. Multiple other pathways for the regulation of GJIC are known, including effects on the rate of transcription of connexin genes as well as stabilization of connexin mRNA. Connexins may also be modified post-translationally. Phosphorylation is a common modification of these proteins.

Gap junction–dependent signaling involves electric and metabolic coupling important for coordinating the activity of groups of cells . Electric coupling is of particular relevance in cardiac and smooth muscle. Cardiomyocytes and smooth muscle cells react to changes in intracellular ion levels, especially Ca^{2+}, that are spread *via* gap junctions. Thus, depolarization of one group of muscle cells is rapidly mediated to adjacent cells, leading to well-coordinated contractions of muscles.

In heart, gap junctions are assembled from different connexin combinations providing cell-to-cell pathways for the precise patterns of current that govern heart rhythm. Changes in the organization of gap junctional communication have been associated with heart diseases in which arrhythmia plays a role (11). This remodeling of gap junctional organization may be due to changes in channel permeabilities or expression of connexin patterns. A decrease in the expression of ventricular connexin43 levels was found in congestive heart failure .

Metabolic coupling *via* gap junctions is involved in the maintenance of cell homeostasis, control of proliferation, transmission of developmental signals, and regulation of apoptosis (13). There is substantial evidence that downregulation of GJIC provides a tumor-promoting stimulus (14). Normal contact-inhibited cells are connected *via* functional gap junctions, whereas many tumors or transformed cells exhibit diminished communication. GJIC is inhibited after the activation of some oncogenes, and a number of nongenotoxic carcinogens and tumor-promoting agents inhibit GJIC (15). The latter include phorbol ester (12-*O*-tetradecanoylphorbol-13-acetate),2,3,7,8-tetrachlorodibenzo-*p*-dioxin (TCDD), and polychlorinated biphenyls. Disruption of GJIC between initiated and normal cells interrupts the transfer of growth control signals imposed by normal neighboring cells, finally leading to clonal expansion of the transformed cell and tumor formation (16). Although dysregulation of GJIC or connexin expression was correlated with decreased growth control, restoration of GJIC is related to increased growth control and decreased tumorigenicity (17). It was also speculated that apoptotic signals are mediated through gap junctions (9). Thus, interruption of GJIC may release tumor cells from undergoing apoptosis. In this context, it is interesting to note that several compounds that exhibit antitumor properties stimulate this pathway of signal exchange.

It was also postulated that GJIC may play a role in cancer treatment in both radiotherapy and chemotherapy (18,19). Toxic metabolites generated for cancer treatment may be transferred *via* gap junctions into neighboring cells less affected by direct exposure. However, downregulation of GJIC is an adaptive, protective response after exposure to toxic agents (14). At the cellular level, the transport of toxic compounds such as reactive oxygen species or signals mediating toxicity can be interrupted.

Carotenoids: Structure-Activity

Carotenoids are pigments that play a major role in the protection of plants against photooxidative processes (20); they are efficient antioxidants, scavenging singlet molecular oxygen and peroxyl radicals. In humans, carotenoids are part of the antioxidant defense system and are thought to protect against degenerative diseases. Apart from their antioxidant activity, carotenoids modulate GJIC, which was discussed as a possible mechanism underlying the protective effects of these compounds toward the development of cancer (see above). However, no correlation was found between the antioxidant effects of selected carotenoids and their capability of stimulating GJIC (21,22), indicating that the two properties are not directly related to each other regarding mechanisms mediating the bioactivity of carotenoids.

β-Carotene, α-carotene, lutein, and β-cryptoxanthin are major carotenoids in human blood and tissues. Experiments in cell culture systems showed that all of these carotenoids stimulate GJIC (16,22). Provitamin A carotenoids are metabolized *in vitro* and *in vivo* to retinal, which is finally oxidized to yield retinoic acid, a potent stimulator of GJIC. However, stimulation of intercellular communication *via* gap junctions is not restricted to compounds with provitamin A activity.

A series of structurally different carotenoids was investigated to evaluate the structure-activity relation with respect to their effects on GJIC (22). Natural and synthetic carotenoids carrying different 5- and 6-membered ring substituents at the end of the conjugated double bond system were included in this study. The location and nature of the substituent at the 6-membered ring had little influence on the activity of different carotenoids. Echinenone, canthaxanthin, β-cryptoxanthin, and 4-hydroxy-β-carotene induce GJIC as does retro-dehydro-β-carotene, a structural analog of β-carotene. Carotenoids carrying a ring system composed of 5 carbon atoms, like the dinorcanthaxanthin are less active. The 6-carbon ring carotenoid canthaxanthin is about twice as active as its 5-membered analog. No stimulatory effects on GJIC were found for capsorubin and violerythrin (22).

Structure-activity relations regarding the effects of linear (acyclic) carotenoids and related compounds on GJIC are difficult to interpret (see Fig. 5.1). Lycopene, another important carotenoid in humans, moderately induces gap junctional com-

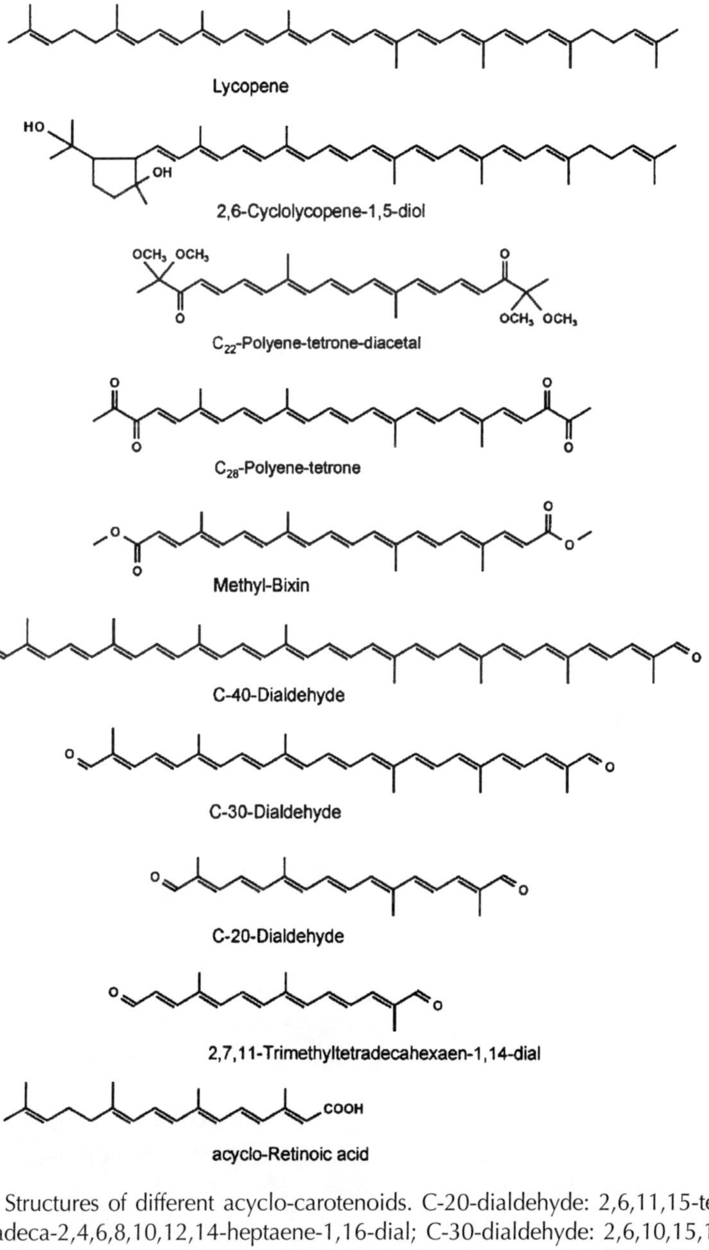

Fig. 5.1. Structures of different acyclo-carotenoids. C-20-dialdehyde: 2,6,11,15-tetramethylhexadeca-2,4,6,8,10,12,14-heptaene-1,16-dial; C-30-dialdehyde: 2,6,10,15,19,23-hexamethyl-tetracosane-2,4,6,8,10,12,14,16,18,20,22-undecaene-1,24-dial; C-40-dialdehyde: 2,6,10,14,19,23,27,31-octamethyl-dotriacontane-2,4,6,8,10,12,14, 16,18,20, 22,24,26,28,30-pentadecaene-1,31-dial. C-22 polyene-tetrone-diacetal: 8,13-dimethyl-2,2,19,19-tetramethoxy-eicosa-4,6,8,10,12,14,16-heptaene-3,18-dione.

munication (23,24), whereas the open-ring compound methyl-bixin was inactive. GJIC was also not stimulated by the open-ring polyenic C-20, C-30, and C-40-dialdehydes (see Fig. 5.1) with 7, 11, or 15 carbon-carbon double bonds in conjugation, respectively. However, stimulation was reported for other long-chain polyenes, C-22 polyene-tetrone-diacetal and C-28 polyenetetrone, which also efficiently inhibited carcinogen-induced neoplastic transformation *in vitro* (25). After complete *in vitro* oxidation of lycopene with hydrogen peroxide/osmium tetroxide, a product that efficiently stimulated GJIC was isolated (26). The active compound was identified by gas chromatography coupled with mass spectrometry, ultraviolet/visible-, and infrared spectrophotometry as a C-17 dialdehyde 2,7,11-trimethyltetradecahexaene-1,14-dial. At present, there is only limited information on the metabolism of lycopene in humans, and it is not known whether the compounds mentioned above are formed *in vivo*. However, it was shown that epimeric 2,6-cyclolycopene-1,5-diol appears in human blood, probably derived from lycopene-5,6-epoxide, which is present in tomatoes (27) but may also be formed *in vivo* upon metabolic oxidation of lycopene (28). After treatment of human keratinocytes with this metabolite, an increased expression of connexin43 was observed (29). The lycopene derivative was more active than the parent compound, and it is speculated that 2,6-cyclolycopene-1,5-diol is partially responsible for lycopene activity on GJIC.

Lycopene may be cleaved at the center of the molecule by the enzyme 15,15′-β-carotene-oxygenase, yielding acyclo-retinal and further, acyclo-retinoic acid, a retinoic acid analog. Lycopene, acyclo-retinoic acid and retinoic acid were tested for their effects on GJIC *in vitro* (24). GJIC increased significantly when the cells were exposed to retinoic acid or lycopene. Acyclo-retinoic acid affected GJIC only at very high levels, indicating that it is not a major mediator of lycopene effects on intercellular communication. It should be noted, however, that the studies described here were performed under various conditions in different cell models. Thus, they are difficult to compare and should be interpreted with caution.

Little is known about the effects of carotenoids on GJIC *in vivo*. β-Carotene, α-carotene, and lycopene were investigated at different dose levels in rats (30). After application of the compounds for 5 d at dose levels of 0.5, 5, or 50 mg/kg body weight, GJIC was measured in the liver. Compared with controls, no effect was found for any of the three compounds at the level of 0.5 mg carotenoid/kg body weight. Induction of GJIC was determined after treatment with 5 mg carotenoid/kg body weight. However, GJC was inhibited when the rats were treated with 50 mg carotenoid/kg body weight. Inhibition was most pronounced when lycopene was administered.

Canthaxanthin Cleavage Products

The effect of cleavage products on GJIC was investigated with canthaxanthin-derived compounds. After decomposition of canthaxanthin, fractions were isolated using HPLC and tested in a cell communication assay. Two of the active components were identified as *all-trans* and 13-*cis* 4-oxo-retinoic acid (31), apparently derived from central cleavage of the parent carotenoid. Both isomers stimulated intercellular communication *via* gap junctions. Furthermore, a simultaneous increase in the expression of connexin43 mRNA was observed.

Apo-canthaxanthinoic acids that differed in chain length, which might be formed after excentric cleavage of canthaxanthin and subsequent oxidation, were synthesized and investigated for their effects on GJIC. 11-Apo-canthaxanthin-11-oic acid and 13-apo-canthaxanthin-13-oic acid were not active, whereas the activity of 14′-apo-canthaxanthin-14′-oic acid was below that of 4-oxo-retinoic and the positive control retinoic acid (32). The effects of 4-oxo-retinoic acid and retinoic acid were comparable. The data indicate that the presence of four conjugated double bonds in the side chain provides optimal activity. Decreasing the number of double bonds leads to inactive compounds. Similarly, addition of a further double bond decreases activity.

Retinoic acid and its analogs exert their biological effects acting as ligands of nuclear receptors of the retinoic acid receptor (RAR) and retinoid X receptor (RXR) family (16,33). The ligand-activated receptor forms dimers, which bind to retinoic acid responsive elements (RARE) at the DNA, acting as transcription factors involved in the regulation of gene expression. There is increasing evidence that unliganded retinoid receptors are involved in gene repression (34).

A construct of a RARβ-promoter connected with a lacZ reporter gene was used to test the activity of carotenoids and their oxidation products (32,35). After activation by its ligand, the receptor activates the transcription of β-galactosidase mRNA, which finally leads to an increase in β-galactosidase enzyme activity as a measure for activation by a suitable ligand. 4-Oxo-retinoic acid is accepted as a ligand of the receptor, and increased β-galactosidase activity was found after treatment of the cells. However, the short-chain analogs of 4-oxo-retinoic acid, 11-apo-canthaxanthin-11-oic acid and 13-apo-canthaxanthin-13-oic acid, had no effect. A slight, but significant induction was observed with 14′-apo-canthaxanthin-14′-oic acid. Stimulatory effects were more pronounced at higher dose levels.

Possible Interactions with Other Micronutrients

In addition to retinoic acid and other retinoids, a number of biologically occurring low-molecular-weight molecules serve as ligands for nuclear receptors. These include 1α,25-dihydroxy-vitamin D_3 or the thyroid hormones 3,3′,5-triiodo-L-thyronine and L-thyroxine (36,37). The active metabolite of vitamin D, 1α,25-dihydroxy-vitamin D_3, is known for its role in the regulation of cell growth and differentiation and is an important factor in calcium homeostasis. It is a ligand for the vitamin D receptor, a nuclear acting transcription factor. The activated receptor binds as a heterodimer with the RXR receptor to the vitamin D response element (VDRE), regulating transcription.

Interestingly, vitamin D_3 and vitamin D_2 induce gap junctional intercellular communication at nanomole levels as shown in mouse fibroblasts (38). The effects on GJIC are similar to those observed after incubation with carotenoids or retinoic acid. Applying vitamin D_3 at high (nonphysiologic) concentrations (5 μM) suppresses GJIC, which is reversible upon exposure to *all-trans* retinoic acid after vitamin D_3 is removed from the medium. When the intercellular communication between cells was prestimulated with *all-trans* retinoic acid, replacement of the retinoid by vitamin D_3 was followed by a rapid decrease in GJIC. The inductory effects of vitamin D_3 at low concentrations are rather slow compared with the rapid inhibition of GJIC at high vitamin D levels. This implies two different vitamin D–dependent mechanisms. One tends to activate, possibly by regulation of connexin gene expression; the other mechanism is inhibitory and might be operative through changes in cellular calcium concentration.

In human skin fibroblasts, 1α,25-dihydroxyvitamin D_3 also induces GJIC (39). However, in cells devoid of a functional nuclear vitamin D receptor, the compound had no effect. Parallel to the increase in cell-cell communication, increases in connexin43 protein and connexin43 mRNA levels were observed that were dependent on the presence of a functional vitamin D receptor.

In addition to retinoic acid and vitamin D, the ligands of the thyroid hormone receptor, 3,3′,5-triiodo-L-thyronine and L-thyroxine, stimulated GJIC (40). The increase in GJC is preceded by an increase in connexin mRNA levels and is accompanied by elevated connexin43 protein levels. It is interesting to note that several ligands of nuclear receptors influence GJIC. Interactions among the signaling pathways of retinoic acid, vitamin D, and thyroid hormones may play a role in the regulation of this system of intercellular communication.

Another group of micronutrients, which has also been suggested to be a part of the dietary prevention system for degenerative diseases, is composed of the polyphenols, e.g., flavonoids and isoflavonoids. Although their antioxidant activities are thought to play an important role in protection, other biological activities have been attributed to these compounds. Epicatechin and genistein at levels of 4–40 μM induce GJIC in rat liver epithelial cells (41). Treatment with these compounds prevents inhibition of GJIC mediated by the tumor promoter 12-*O*-tetradecanoylphorbol-13-acetate (TPA) (41,42). In communicating cells, most of the gap junction protein connexin43 was located in the plasma membrane, whereas in cells exposed to the tumor promoter, considerably less protein was found in the membrane. Such a delocalization of connexin43 proteins was not observed when TPA was coincubated with epicatechin or genistein.

Conclusions

Carotenoids were shown *in vitro* and in animal models to influence molecular and cellular processes *via* mechanisms not related to their antioxidant activity. There is substantial evidence that the effects of carotenoids on regulatory and signaling pathways are important with respect to their cancer-preventing properties and that effect on GJIC is of interest. Although the functional effects of carotenoids were demonstrated in various cell culture systems, there are limited and somewhat controversial data on their mechanism of action. Further research is required regarding the role of metabolites and cleavage products in this pathway. Because a number of other vitamins, micronutrients, and signaling molecules also affect GJIC, interactions among signaling pathways are likely to be operative.

Acknowledgments

H. Sies is a Fellow of the National Foundation for Cancer Research (NFCR), Bethesda, MD. Supported by Deutsche Forschungsgemeinschaft, SFB 575/B4.

References

1. Riboli, E., and Norat, T. (2003) Epidemiologic Evidence of the Protective Effect of Fruit and Vegetables on Cancer Risk, *Am. J. Clin. Nutr. 78:* 559S–569S.
2. Willett, W.C. (2001) Diet and Cancer: One View at the Start of the Millennium, *Cancer Epidemiol. Biomark. Prev. 10:* 3–8.
3. Sies, H. (1997) *Antioxidants in Disease Mechanisms and Therapy*, Academic Press, London.
4. Albanes, D. (1999) β-Carotene and Lung Cancer: A Case Study, *Am. J. Clin. Nutr. 69 (Suppl.):* 1345S–1350S.
5. Stahl, W., Ale-Agha, N., and Polidori, M.C. (2002) Non-Antioxidant Properties of Carotenoids, *Biol. Chem. 383:* 553–558.
6. Goodenough, D.A., Goliger, J.A., and Paul, D.L. (1996) Connexins, Connexons, and Intercellular Communication, *Annu. Rev. Biochem. 65:* 475–502.
7. Willecke, K., Eiberger, J., Degen, J., Eckardt, D., Romualdi, A., Guldenagel, M., Deutsch, U., and Sohl, G. (2002) Structural and Functional Diversity of Connexin Genes in the Mouse and Human Genome, *Biol. Chem. 383:* 725–737.
8. Evans, W.H., and Martin, P.E. (2002) Gap Junctions: Structure and Function (Review), *Mol. Membr. Biol. 19:* 121–136.
9. De Maio, A., Vega, V.L., and Contreras, J.E. (2002) Gap Junctions, Homeostasis, and Injury, *J. Cell Physiol. 191:* 269–282.
10. Jongsma, H.J., and Wilders, R. (2000) Gap Junctions in Cardiovascular Disease, *Circ. Res. 86:* 1193–1197.
11. Severs, N.J. (2001) Gap Junction Remodeling and Cardiac Arrhythmogenesis: Cause or Coincidence? *J. Cell Mol. Med. 5:* 355–366.
12. Severs, N.J., Dupont, E., Coppen, S.R., Halliday, D., Inett, E., Baylis, D., and Rothery, S. (2004) Remodelling of Gap Junctions and Connexin Expression in Heart Disease, *Biochim. Biophys. Acta 1662:* 138–148.
13. Trosko, J.E., and Chang, C.C. (2000) Modulation of Cell-Cell Communication in the Cause and Chemoprevention/Chemotherapy of Cancer, *Biofactors 12:* 259–263.
14. Chipman, J.K., Mally, A., and Edwards, G.O. (2003) Disruption of Gap Junctions in Toxicity and Carcinogenicity, *Toxicol. Sci. 71:* 146–153.
15. Trosko, J.E., Yotti, L.P., Warren, S.T., Tsushimoto, G., and Chang, C.-C. (1982) Inhibition of Cell-Cell Communication by Tumor Promoters, in *Cocarcinogenesis and Biological Effects of Tumor Promoters*, Hecker, E., Fusenig, N.E., Kunz, W., Marks, F., and Thielmann, H.W., eds., Raven Press, New York, vol. 7, pp. 564–585.

16. Bertram, J.S. (1999) Carotenoids and Gene Regulation, *Nutr. Rev. 57:* 182–191.
17. King, T.J., Fukushima, L.H., Yasui, Y., Lampe, P.D., and Bertram, J.S. (2002) Inducible Expression of the Gap Junction Protein Connexin43 Decreases the Neoplastic Potential of HT-1080 Human Fibrosarcoma Cells *In Vitro* and *In Vivo*, *Mol. Carcinog. 35:* 29–41.
18. Little, J.B. (2003) Genomic Instability and Bystander Effects: A Historical Perspective, *Oncogene 22:* 6978–6987.
19. Edwards, G.O., Botchway, S.W., Hirst, G., Wharton, C.W., Chipman, J.K., and Meldrum, R.A. (2004) Gap Junction Communication Dynamics and Bystander Effects from Ultrasoft X-Rays, *Br. J. Cancer 90:* 1450–1456.
20. Demmig-Adams, B., and Adams, W.W., III (2002) Antioxidants in Photosynthesis and Human Nutrition, *Science 298:* 2149–2153.
21. Zhang, L.X., Cooney, R.V., and Bertram, J.S. (1992) Carotenoids Up-Regulate Connexin43 Gene Expression Independent of Pro-Vitamin A or Antioxidant Properties, *Cancer Res. 52:* 5707–5712.
22. Stahl, W., Nicolai, S., Briviba, K., Hanusch, M., Broszeit, G., Peters, M., Martin, H.D., and Sies, H. (1997) Biological Activities of Natural and Synthetic Carotenoids: Induction of Gap Junctional Communication and Singlet Oxygen Quenching, *Carcinogenesis 18:* 89–92.
23. Zhang, L.-X., Cooney, R.V., and Bertram, J.S. (1991) Carotenoids Enhance Gap Junctional Communication and Inhibit Lipid Peroxidation in C3H/10T1/2 Cells: Relationship to Their Cancer Chemopreventive Action, *Carcinogenesis 12:* 2109–2114.
24. Stahl, W., von Laar, J., Martin, H.D., Emmerich, T., and Sies, H. (2000) Stimulation of Gap Junctional Communication: Comparison of Acyclo-Retinoic Acid and Lycopene, *Arch. Biochem. Biophys. 373:* 271–274.
25. Pung, A., Franke, A., Zhang, L.-X., Ippendorf, M., Martin, H.D., Sies, H., and Bertram, J.S. (1993) A Synthetic C22 Carotenoid Inhibits Carcinogen-Induced Neoplastic Transformation and Enhances Gap Junctional Communication, *Carcinogenesis 14:* 1001–1005.
26. Aust, O., Ale-Agha, N., Zhang, L., Wollersen, H., Sies, H., and Stahl, W. (2003) Lycopene Oxidation Product Enhances Gap Junctional Communication, *Food Chem. Toxicol. 41:* 1399–1407.
27. Khachik, F., Beecher, G.R., and Smith, J.C. (1995) Lutein, Lycopene, and Their Oxidative Metabolites in Chemoprevention of Cancer, *J. Cell. Biochem. S 22:* 236–246.
28. Khachik, F., Spangler, C.J., Smith, J.C., Canfield, L.M., Steck, A., and Pfander, H. (1997) Identification, Quantification, and Relative Concentrations of Carotenoids and Their Metabolites in Human Milk and Serum, *Anal. Chem. 69:* 1873–1881.
29. Khachik, F., Bertram, J.S., Huang, M.-T., Fahey, J.W., and Talalay, P. (1999) Dietary Carotenoids and Their Metabolites as Potentially Useful Chemoprotective Agents Against Cancer, in *Antioxidant Food Supplements in Human Health*, Packer, L., Hiramatsu, M., and Yoshikawa,T., eds., pp. 203–229, Academic Press, San Diego.
30. Krutovskikh, V.A., Asamoto, M., Takasuka, N., Murakoshi, M., Nishino, H., and Tsuda, H. (1997) Differential Dose-Dependent Effects of α-, β-Carotenes and Lycopene on Gap-Junctional Intercellular Communication in Rat Liver *In Vivo*, *Jpn. J. Cancer Res. 88:* 1121–1124.
31. Hanusch, M., Stahl, W., Schulz, W.A., and Sies, H. (1995) Induction of Gap Junctional Communication by 4-Oxoretinoic Acid Generated from Its Precursor Canthaxanthin, *Arch. Biochem. Biophys. 317:* 423–428.

32. Teicher, V.B., Kucharski, N., Martin, H.D., van der Saag, P., Sies, H., and Stahl, W. (1999) Biological Activities of Apo-Canthaxanthinoic Acids Related to Gap Junctional Communication, *Arch. Biochem. Biophys. 365:* 150–155.
33. Lippman, S.M., and Lotan, R. (2000) Advances in the Development of Retinoids as Chemopreventive Agents, *J. Nutr. 130:* 479S–482S.
34. Weston, A.D., Blumberg, B., and Underhill, T.M. (2003) Active Repression by Unliganded Retinoid Receptors in Development: Less Is Sometimes More, *J. Cell Biol. 161:* 223–228.
35. Nikawa, T., Schulz, W.A., van den Brink, C.E., Hanusch, M., van der Saag, P., Stahl, W., and Sies, H. (1995) Efficacy of all-*trans*-β-Carotene, Canthaxanthin, and all-*trans*, 9-*cis*, and 4-Oxoretinoic Acids in Inducing Differentiation of an F9 Embryonal Carcinoma RARβ-lacZ Reporter Cell Line, *Arch. Biochem. Biophys. 316:* 665–672.
36. Lin, R., and White, J.H. (2004) The Pleiotropic Actions of Vitamin D, *BioEssays 26:* 21–28.
37. Yen, P.M. (2001) Physiological and Molecular Basis of Thyroid Hormone Action, *Physiol Rev. 81:* 1097–1142.
38. Stahl, W., Nicolai, S., Hanusch, M., and Sies, H. (1994) Vitamin D Influences Gap Junctional Communication in C3H/10T1/2 Murine Fibroblast Cells, *FEBS Lett. 352:* 1–3.
39. Clairmont, A., Tessmann, D., Stock, A., Nicolai, S., Stahl, W., and Sies, H. (1996) Induction of Gap Junctional Intercellular Communication by Vitamin D in Human Skin Fibroblasts Is Dependent on the Nuclear Vitamin D Receptor, *Carcinogenesis 17:* 1389–1391.
40. Stock, A., Sies, H., and Stahl, W. (1998) Enhancement of Gap Junctional Communication and Connexin43 Expression by Thyroid Hormones, *Biochem. Pharmacol. 55:* 475–479.
41. Ale-Agha, N., Stahl, W., and Sies, H. (2002) (-)-Epicatechin Effects in Rat Liver Epithelial Cells: Stimulation of Gap Junctional Communication and Counteraction of Its Loss Due to the Tumor Promoter 12-*O*-Tetradecanoylphorbol-13-acetate, *Biochem. Pharmacol. 63:* 2145–2149.
42. Kang, K.S., Kang, B.C., Lee, B.J., Che, J.H., Li, G.X., Trosko, J.E., and Lee, Y.S. (2000) Preventive Effect of Epicatechin and Ginsenoside Rb_2 on the Inhibition of Gap Junctional Intercellular Communication by TPA and H_2O_2, *Cancer Lett. 152:* 97–106.

Chapter 6

Raman Detection of Carotenoids in Human Tissue

Werner Gellermann[a], Jeff A. Zidichouski[b], Carsten R. Smidt[b], and Paul S. Bernstein[c]

[a]Department of Physics, University of Utah, Salt Lake City, UT 84112; [b]Pharmanex Research Laboratories, Provo, UT 84601; [c]Moran Eye Center, University of Utah, Salt Lake City, UT 84132

Introduction

Raman spectroscopy is a highly specific form of vibrational spectroscopy that can be used to identify and quantify chemical compounds. Carotenoid molecules are especially suitable for Raman measurements because they can be excited with light overlapping their visible absorption bands; under these conditions, they exhibit a very strong resonance Raman scattering (RRS) response, with an enhancement factor of ~5 orders of magnitude relative to nonresonant Raman spectroscopy (1). This allows one to detect in a noninvasive manner the characteristic vibrational energy levels of the carotenoids through their corresponding spectral fingerprint signature, even in complex biological systems.

Motivated to find a noninvasive, objective, *in vivo* method for the detection of carotenoid antioxidants in human tissue, we started developing RRS for this purpose ~6 years ago. Using excised eyecups as test samples, we applied the technology initially to the detection of macular pigment (MP). Comprised of the carotenoids, lutein and zeaxanthin, and bound to the macular tissue in very high concentrations, the species are thought to play a major role in the prevention of age-related macular degeneration (2). Since that time, we extended the technology to *in vivo* MP measurements in laboratory and clinical settings (3–6). Independent trials using the ocular Raman technology are now in progress at several sites worldwide. Recently, we began to develop RRS spectroscopy for the detection of carotenoid antioxidants in human skin and mucosal tissue (7–10), tissues in which other carotenoid species such as lycopene and β-carotene are thought to play an important protective role, such as in the protection of skin from UV and short-wavelength visible radiation. The carotenoids, lutein and lycopene, may also have protective functions for cardiovascular health, and lycopene may play a role in the prevention of prostate cancer. It is conceivable that skin levels of these species are correlated with corresponding levels in the internal tissues.

Optical Properties and Resonance Raman Scattering of Carotenoids

Carotenoids are π-electron congugated carbon-chain molecules and are similar to polyenes with regard to their structure and optical properties. Distinguishing features are the number of conjugated carbon double bonds (C=C bonds), the number of attached methyl side groups, and the presence and structure of attached end groups. The molecular structures of some of the most important carotenoid species found in human tissue, along with their absorption spectra, are shown in Figure 6.1. They include β-carotene, zeaxanthin, lycopene, lutein, and phytofluene. The electronic absorptions are strong in each case, occur in broad bands (~100 nm width), and shift to a longer wavelength with an increasing number of carbon double bonds of the corresponding molecule. The absorption of phytofluene (5 conjugated C=C bonds) is positioned in the near UV; β-carotene, lutein and zeaxanthin (11,10, and 11 C=C bonds, respectively) are centered at ~440 nm, and lycopene (11 bonds) peaks at ~450 nm. All show a clearly resolved vibronic substructure due to weak electron-phonon coupling, with a spacing of ~1400 cm^{-1}. A strong electric-dipole allowed absorption transitions to occur between the molecules' delocalized π-orbitals from the 1 ^1A_g singlet ground state to the 1 ^1B_u singlet excited state (see inset of Fig. 6.1).

In all cases, optical excitation within the absorption band leads to only very weak luminescence bands (not shown). The extremely low quantum efficiency of the luminescence is caused by the existence of a second excited singlet state, a 2 ^1A_g state, which lies below the 1 ^1B_u state (see Fig. 6.1A, inset). After excitation of the 1 ^1B_u state, the carotenoid molecule relaxes very rapidly, within ~200–250 fs (11), *via* nonradiative transitions, to this lower 2 ^1A_g state from which electronic emission to the ground state is parity-forbidden (dashed, downward pointing arrows in inset of Fig. 6.1). The low 1 $^1B_u \rightarrow 1\ ^1A_g$ luminescence efficiency (10^{-5} to 10^{-4}) and the absence of 2 $^1A_g \rightarrow 1\ ^1A_g$ fluorescence of the molecules allows one to detect the RRS response of the molecular vibrations (shown as a solid, downward pointing arrow in the inset of Fig. 6.1) without potentially masking fluorescence signals. Specifically, RRS detects the stretching vibrations of the polyene backbone as well as the methyl side groups (1).

For tetrahydrofuran solutions of the carotenoids of Figure 6.1, we obtained the RRS spectra shown in Figure 6.2. β-Carotene, zeaxanthin, lycopene, and lutein all have strong and clearly resolved Raman signals superimposed on a weak fluorescence background, with three prominent Raman Stokes lines appearing at ~1525 cm^{-1} (C=C stretch), 1159 cm^{-1} (C-C stretch), and 1008 cm^{-1} (C-CH_3 rocking motions) (1). In the shorter-chain phytofluene molecule, only the C=C stretch appears, and it is shifted significantly to higher frequencies (by ~40 cm^{-1}).

Raman scattering does not require resonant excitation, in principle, and is therefore useful to simultaneously detect the vibrational transitions of all Raman active compounds in a given sample. However, off-resonant Raman scattering is a very weak optical effect, requiring intense laser excitation, long signal acquisition times, and highly sensitive, cryogenically cooled detectors. Also, in biological systems, the spectra tend to be very complex due to the diversity of compounds present. This scenario changes drastically if the compounds exhibit absorption bands due to electronic dipole transitions of the molecules, particularly if these are located in the visible wavelength range. When illuminated with monochromatic light

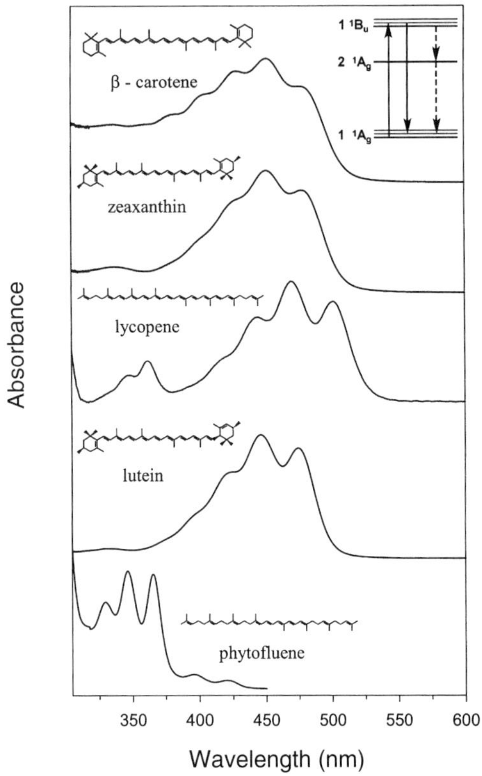

Fig. 6.1. Absorption spectra, molecular structure, and energy level scheme of major carotenoid species found in human tissue, including β-carotene, zeaxanthin, lycopene, lutein, and phytofluene. Carotenoid molecules feature an unusual even parity excited state (see inset). As a consequence, electric-dipole absorption transitions are allowed in these molecules, but spontaneous emission is forbidden. The resulting absence of any strong fluorescence in carotenoids is the main reason why Resonance Raman scattering, shown as a solid, downward pointing arrow (optical transition) in the inset, can be used as a noninvasive way of carotenoid detection in human tissue. All carotenoid molecules feature a linear, chain-like conjugated carbon backbone consisting of alternating carbon single (C–C) and double bonds (C=C) with varying numbers of conjugated C=C double bonds, and a varying number of attached methyl side groups. β-Carotene, lutein, and zeaxanthin feature additional ionone rings as end groups. In β-carotene and zeaxanthin, the double bonds of these ionone rings add to the effective C=C double bond length of the molecules. Lutein and zeaxanthin have an OH group attached to the ring. Lycopene has 11 conjugated C=C bonds, β-carotene 11, zeaxanthin 11, lutein 10, and phytofluene 5. The absorptions of all species occur in broad bands in the blue/green spectral range, with the exception of phytofluene, which, as a consequence of the shorter C=C conjugation length absorbs in the near UV. Note also, that a small (~10 nm) spectral shift exists between lycopene and lutein absorptions. The spectral shifts can be explored with RRS to selectively detect some of the carotenoids existing as a mixture in human tissues.

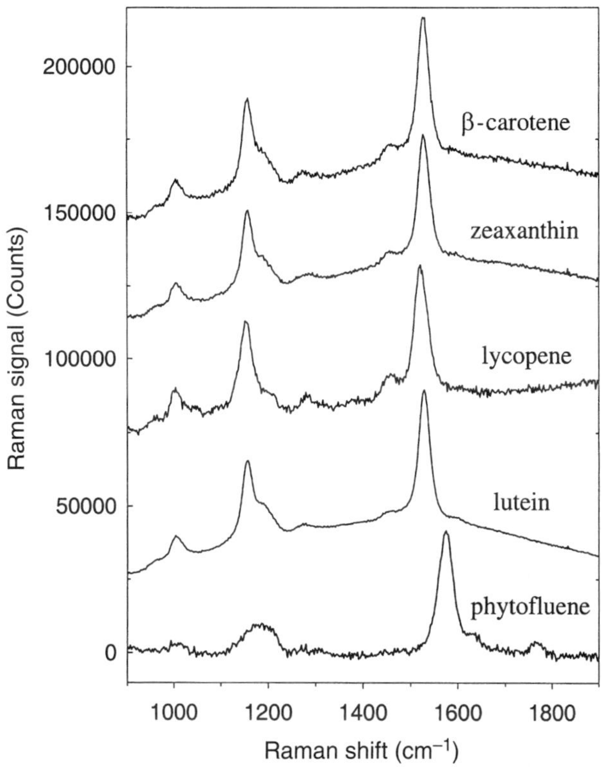

Fig. 6.2. Resonance Raman spectra of β-carotene, zeaxanthin, lycopene, lutein, and phytofluene solutions, showing the three major "spectral fingerprint" Raman peaks of carotenoids originating from rocking motions of the methyl components ($C–CH_3$) and stretch vibrations of the carbon–carbon single bonds (C–C) and double bonds (C=C). In all carotenoids except phytofluene, these peaks appear at 1008, 1159, and 1525 cm^{-1}, respectively. In phytofluene, the C=C stretch frequency is shifted by ~40 cm^{-1} to higher frequencies. Note the large contrast between Raman response and broad background signal, which is due to the inherently weak fluorescence of carotenoids.

overlapping one of these absorption bands, the Raman scattered light will exhibit a substantial resonance enhancement. In the case of carotenoids, 488 nm argon laser light provides an extraordinarily high resonant enhancement of the Raman signals on the order of 10^5 (1). No other biological molecules found in significant concentrations in human tissues exhibit similar resonant enhancement at this excitation wavelength; thus, *in vivo* carotenoid RRS spectra are remarkably free of confounding Raman responses.

Raman scattering is a linear spectroscopy, meaning that the Raman scattering intensity (I_S) scales linearly with the intensity of the incident light (I_L), as long as the scattering compound can be considered to be optically thin. Furthermore, at fixed incident light intensity, the Raman response scales with the population density of the scatterers $N(E_i)$ in a linear fashion with the Raman scattering cross section $\sigma_R(i \rightarrow f)$ (a fixed constant whose magnitude depends on the excitation and collection geometries) as long as the scatterers can be considered to be optically thin (3).

$$I_S = N(E_i) \times \sigma_R \times I_L$$

In optically thick media, as in geometrically thin but optically dense tissue, a deviation from the linear Raman response of the I_S vs. concentration N can occur, of course, e.g., due to self-absorption of the Stokes Raman line by the strong electronic absorption. In general, this can be taken into account, albeit over a limited concentration range, by calibrating the Raman response with suitable tissue phantoms.

RRS spectroscopy has an additional advantage over ordinary Raman spectroscopy because it influences the Raman response through a judicious choice of the excitation wavelength. This allows one to selectively enhance the Raman response of one carotenoid species over another in a mixture of compounds. For example, exciting a mixture of phytofluene and lutein at 450 nm would result in a RRS response only from lutein, thus allowing the selective quantification of lutein in this mixture.

In complex biological tissues, several carotenoid species are usually present. For quantification of the composite RRS response, it is therefore important to account for individual RRS responses of the excited species. Because the RRS response generally follows the absorption line shape, the individual RRS depends on the extent of the overlap of the excitation laser with the absorption. In the case of equal Raman scattering cross sections, realized when exciting all carotenoids at their respective absorption maxima, the RRS responses should be additive. To verify this assumption, we measured RRS spectra for solutions of β-carotene, lycopene, and a mixture of both. The results are shown in Figure 6.3 for solutions with carotenoid concentrations that are higher than typical physiologic concentrations encountered in human tissue; the RRS response for the carotenoid is roughly equal to the sum of the responses for the individual concentrations.

Raman Detection of Macular Pigments

Macular Pigments and Existing Measuring Methods. It has been hypothesized that the macular carotenoid pigments, lutein and zeaxanthin (12,13), might play a role in the treatment and prevention of age-related macular degeneration (AMD) (14,15). In the United States, this leading cause of blindness affects ~30% of people >70 y old. Supportive epidemiologic studies showed that there is an inverse correlation between high dietary intakes and blood levels of lutein and zeaxanthin and risk of advanced AMD (16,17). It was also demonstrated that MP levels can be altered through dietary manipulation (18,19) and that carotenoid pigment levels are lower in eyes from patients with AMD at autopsy (20,21). Randomized, placebo-controlled, prospective clinical study data to support this hypothesis are not yet available, however.

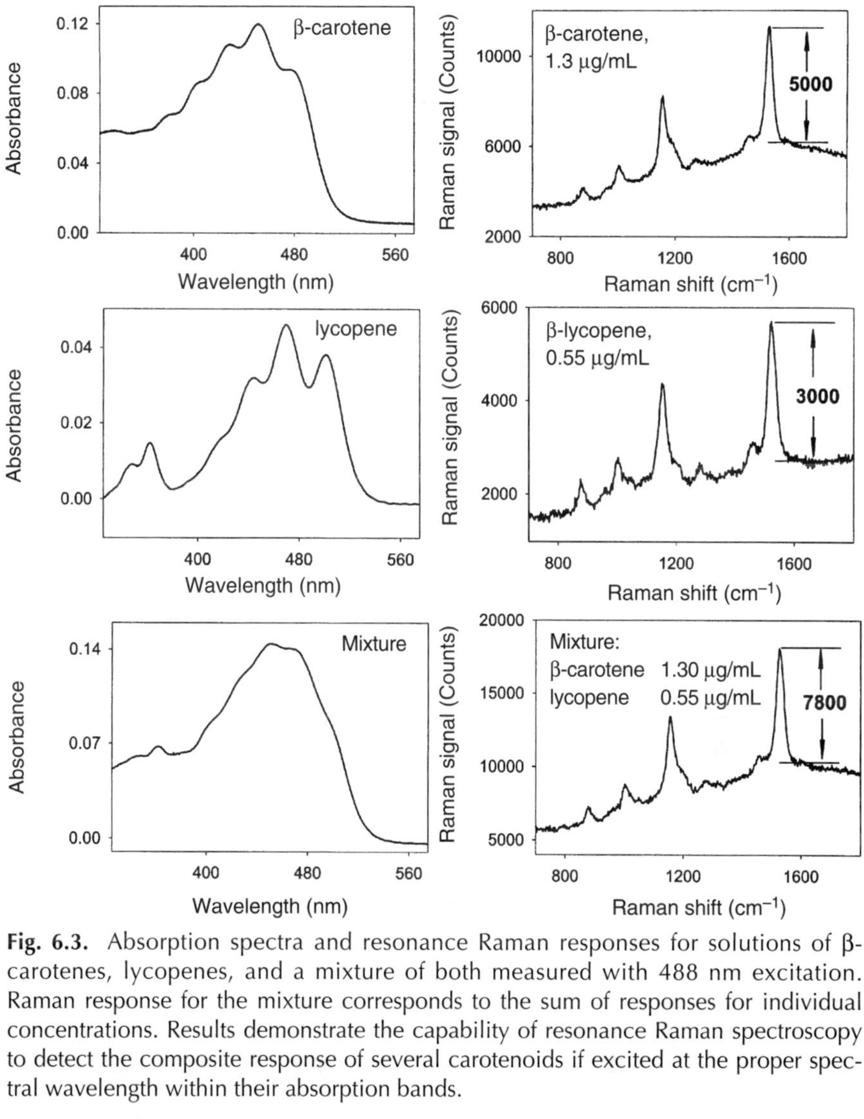

Fig. 6.3. Absorption spectra and resonance Raman responses for solutions of β-carotenes, lycopenes, and a mixture of both measured with 488 nm excitation. Raman response for the mixture corresponds to the sum of responses for individual concentrations. Results demonstrate the capability of resonance Raman spectroscopy to detect the composite response of several carotenoids if excited at the proper spectral wavelength within their absorption bands.

Spectroscopic studies of tissue sections of primate maculae (the central 5–6 mm of the retina, see Fig. 6.4A) indicate that there are very high concentrations of carotenoid pigments, shown as a shaded area in Figure 6.4B, in the Henle fiber layer of the fovea and smaller amounts in the inner plexiform layer (22) (Fig. 6.4B). The mechanisms by which these two macular pigments, derived exclusively from dietary sources such as green leafy vegetables and orange and yellow fruits and vegetables, might protect against AMD remain unclear. They are known to be

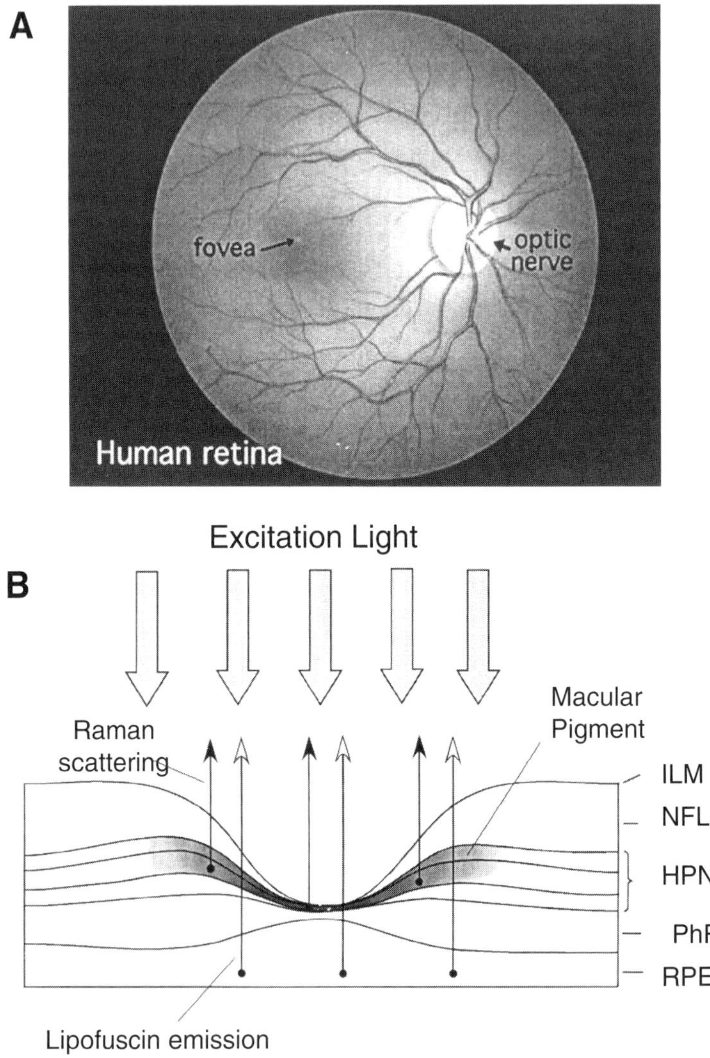

Fig. 6.4. (A) Fundus photograph of healthy human retina, showing optic nerve head (bright spot at left) and macula [dark shaded area outlining macular pigment (MP) distributions]. (B) Schematic representation of retinal layers participating in light absorption, transmission, and scattering of excitation and emission light. ILM: inner limiting membrane, NFL: nerve fiber layer; HPN: Henle fiber, plexiform and nuclear layers; PhR: photoreceptor layer; RPE: retinal pigment epithelium. In Raman scattering, the scattering response originates from the MP, which is located anteriorly to the photoreceptor layer. The influence of deeper fundus layers such as the RPE is avoided. In autofluorescence spectroscopy, light emission of deeper fundus layers such as lipofuscin emission from the RPE, can be stimulated on purpose to generate an intrinsic "light source" for single-path absorption measurements of anteriorly located MP layers (see text).

excellent free radical scavenging antioxidants in a tissue at high risk of oxidative damage due to the high levels of light exposure, and abundant highly unsaturated lipids (14,15,23,24). In addition, because these molecules absorb in the blue-green spectral range, they act as filters that may attenuate photochemical damage and/or image degradation caused by short-wavelength visible light reaching the retina (25).

There is considerable interest in the noninvasive measurement of macular carotenoid levels in the elderly to determine whether low levels of MP are associated with an increased risk of AMD (26). Currently, the most commonly used noninvasive method for measuring human MP levels is a subjective psychophysical heterochromatic flicker photometry test involving color intensity matching of a light beam aimed at the fovea and another aimed at the perifoveal area (27). However, this method is rather time consuming and requires an alert, cooperative subject with good visual acuity; it may also exhibit a high intrasubject variability when macular pigment densities are low or if significant macular pathology is present (28). Thus, the usefulness of this method for assessing macular pigment levels in the elderly population most at risk for AMD is severely limited. Nevertheless, researchers have used flicker photometry to investigate important questions such as variation of macular pigment density with age and diet. In a recent flicker photometry study, for example, the pigment density was found to increase slightly with age (29), whereas two other studies found the opposite trend (26,30).

A number of objective techniques for the measurement of MP in the human retina were explored recently as alternatives to the subjective psychophysical tests. The underlying optics principles of these techniques are based on either fundus reflection or fundus fluorescence (autofluorescence) spectroscopy. In traditional fundus reflectometry, which uses a light source with a broad spectral continuum, reflectance spectra of the bleached retina are separately measured for fovea and perifovea. The double-path absorption of MP is extracted from the ratio of the two spectra by reproducing its spectral shape in a multiparameter fitting procedure using appropriate absorption and scattering profiles of the various fundus tissue layers traversed by the source light (31–34). One of the imaging variants of fundus reflectometry uses a TV-based imaging fundus reflectometer with sequential, narrow bandwidth light excitation over the visible wavelength range and digitized fundus images (35). Another powerful variant uses a scanning laser ophthalmoscope (36,37), employing raster-scanning of the fundus with discrete laser excitation wavelengths to produce highly detailed information about the spatial distribution of MP (and photopigments) (38–42). In autofluorescence spectroscopy, lipofuscin in the retinal pigment epithelium is excited with light within and outside the wavelength range of macular pigment absorption. Measuring the lipofuscin fluorescence levels for fovea and perifovea, an estimate of the single-pass absorption of MP can be obtained (42,44). An imaging variant of this method is described below.

Autofluorescence measurements of MP levels were recently carried out in a group of 159 subjects (ages 15–80 y; normal retinal status) and compared with reflectometry. In a small subgroup, they were also compared with heterochromatic flicker photometry measurements (44). The mean MP optical densities for the whole age group were 0.48 ± 0.16 for autofluorescence (using a 2° test field), 0.23 ± 0.07 for reflectometry (2° test field), and 0.37 ± 0.26 for the psychophysical measurement (0.8° test field). Furthermore, autofluorescence measured densities reproducibly to within 9% of the mean density, superior to the reproducibility obtained by reflectometry (22%) and flicker photometry (15–35% depending on age, training, and experience of subjects). In another recent study, comparing reflectometry, scanning laser ophthalmoscopy (SLO), and flicker photometry in the measurement of macular lutein uptake, SLO was superior to spectral fundus reflectance, whereas psychophysical measurements yielded widely varying results (34,45).

Ocular Raman Measurements. We investigated Raman scattering as a new approach (2) for the measurement of MP levels in living eyes (46,47). The technique is objective as well as noninvasive, appears to be fast and quantitative, and its specificity for carotenoids means that it could be used for patients with a variety of ocular pathologies.

In vivo, RRS spectroscopy in the eye takes advantage of several favorable anatomical properties of the tissue structures encountered in the light scattering pathways. First, the major site of macular carotenoid deposition in the Henle fiber layer is on the order of only 100 μm in thickness (22). This provides a chromophore distribution very closely resembling an optically thin film having no significant self-absorption of the illuminated or scattered light. Second, the ocular media (cornea, lens, vitreous) are generally of sufficient clarity not to attenuate the signal, and they should require appropriate correction factors only in cases of substantial pathology such as visually significant cataracts. Third, because the macular carotenoids are situated anteriorly in the optical pathway through the retina (see Fig. 6.4B), the illuminating light and the back-scattered light never encounter any highly absorptive pigments such as photoreceptor rhodopsin and RPE melanin, whereas the light unabsorbed by the macular carotenoids and the forward- and side-scattered light will be efficiently absorbed by these pigments (3). In contrast, emission of lipofuscin used in autofluorescence (AF)-based measurements of MP has to traverse the photoreceptor layer (see Fig. 6.4B).

Our initial "proof of principle" studies of ocular carotenoid RRS employed a laboratory-grade high resolution Raman spectrometer, flat-mounted human cadaver retinas, human eyecups, and a few whole frog eyes. We were able to record characteristic carotenoid RRS spectra from these tissues with a spatial resolution of ~100 μm, and we were able to confirm linearity of response by extracting and analyzing tissue carotenoids by HPLC after completion of the Raman measurements (2). For *in vivo* experiments and clinical use, we developed a new instrument with lower spectral resolution but highly improved light throughput. The latest instrument version, shown schematically in Figure 6.5A, consists of a low-power air-cooled argon laser, which projects a 1-mm diameter, 1.0-mW 488-nm spot onto the

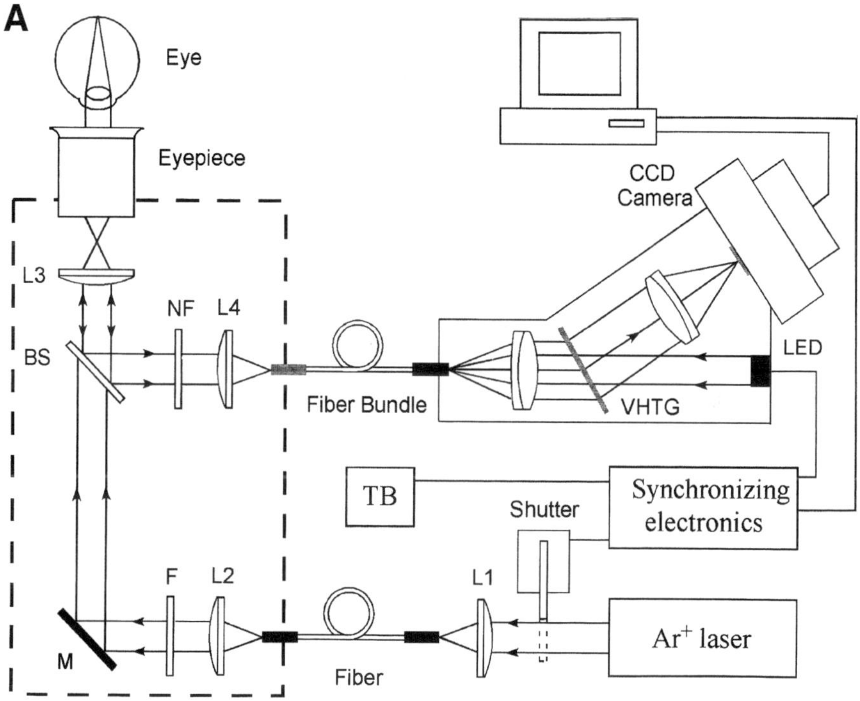

Fig. 6.5. (A) Schematic diagram of a macular pigment resonance Raman detector designed for human clinical studies. Light from an argon laser is routed *via* optical fiber into an optical probe head (outlined by dashed line) and from there into the eye of a subject where it is projected as a ~1 mm diameter spot onto the retina. L1–L4: lenses; F: laser line filter; BS: dichroic beam splitter; NF: notch filter; VHF: volume holographic grating; LED: red light-emitting diode for visualization of fiber bundle. Raman scattered light is collected in back-scattering geometry with lens L3, split off by dichroic beam splitter BS, and sent into a spectrograph *via* fiber bundle for light dispersion. A CCD array is used to detect the spectrally dispersed light. The instrument is interfaced to a personal computer for light exposure control, data acquisition, and processing. Typical settings are 1 mW of 488 nm laser light for 0.2 s with a 1-mm spot size on the macula. (B) Subject looking into the optical probe head of the instrument. (C) Typical Raman spectra from the retina of a healthy volunteer, measured with dilated pupil (8 mm), and displayed on the computer monitor of the instrument. Left panel: Spectrum obtained after a single measurement, clearly showing the carotenoid Raman signals superimposed upon a broad fluorescent background. Right panel: Same enlarged spectrum obtained after fitting background with a 4th-order polynomial and subtracting it from the original spectrum. For quantitation of MP concentration, the software displays the Raman response of the strongest peak, corresponding to the C=C stretch, as an intensity score.

(Continued)

B

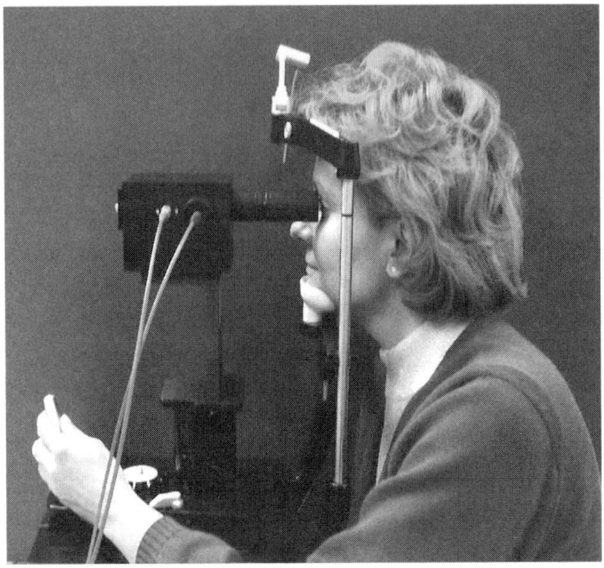

C

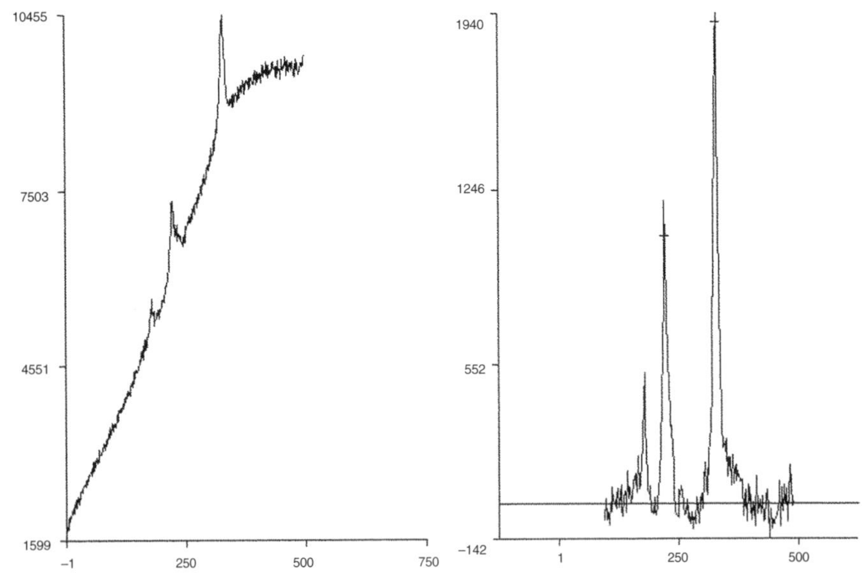

Fig. 6.5. (A) (B) and (C) (Continued).

foveal region for 0.2 s through a pharmacologically dilated pupil (4). Light backscattered at 180° is collimated with a lens, the light scattered at the laser excitation wavelength is rejected by a high efficiency band-pass filter, and the remainder is routed *via* fiber optics to a Raman spectrograph/CCD camera combination. The instrument is interfaced to a personal computer equipped with custom-designed software that can subtract background fluorescence and quantify the intensity of the Raman peaks. Living human subjects were asked to fixate on a suitable target to ensure self-alignment (see Fig. 6.5B), whereas in the monkey experiments, we used an additional video monitoring system and red laser aiming beam to confirm foveal targeting. The instrument could be calibrated against solutions of lutein and zeaxanthin in optically thin 1-mm quartz cuvettes placed at the focal point of a lens whose power duplicated the refractive optics of the human eye. Detector response was linear up to optical densities of nearly 0.8, well past carotenoid levels normally encountered in the human macula. Using living monkey eyes, a linearity of response at "eye safe" laser illumination levels could again be established relative to HPLC analysis (3).

Typical RRS spectra, measured from the macula of a healthy human volunteer with a dilated pupil, are shown in Figure 6.5C. The left panel of this figure shows a typical spectrum obtained from a single measurement and clearly reveals carotenoid Raman signals superimposed on a weak and spectrally broad fluorescence background. The background is caused in part by weak intrinsic fluorescence of carotenoids, and in part by the short-wavelength emission tail of lipofuscin fluorophors deposited in the retinal pigment epithelial layer. The ratio between the intensities of the carotenoid C=C Raman peak and the fluorescence background is high enough (~0.25) that it is easily possible to quantify the amplitude of the C=C peak after digital background subtraction (right panel of Fig. 6.5C). Carotenoid RRS spectra collected from the living human macula were indistinguishable from the spectra originating from RRS spectra solutions of pure lutein or zeaxanthin solutions (6).

Clinical Results. We measured the macular carotenoid pigment levels of hundreds of human subjects at the Moran Eye Center of the University of Utah using RRS spectroscopy. Most patients can readily perform the test with acceptable intrasession and intersession repeatability of ±10% as long as they have preserved central fixation (typically a visual acuity ≥20/80). The subjects noted a central afterimage from the laser illumination after each measurement that they described as similar to that of a camera flash. This afterimage usually faded within 1–2 min. Subjects with dense media opacities such as visually significant cataracts or with poor pupillary dilation of <6 mm were usually excluded from measurements in our clinical studies because we found their measurements to be artifactually low.

When we measured a large population of normal subjects, none of whom were consuming supplements containing substantial amounts of lutein or zeaxanthin, we found a striking decline of average macular carotenoid levels with age (3), a phenomenon sometimes observed with other MP measurement techniques (26). Part of this decline can be explained by "yellowing" of the crystalline lens with age, which would attenuate some of the illuminating and back-scattered light, but we also found consistently low MP levels even in patients who had previously had cataract surgery with implantation of optically clear prosthetic intraocular lenses (pseudophakia) (3). In addition, we noted that patients with unilateral cataracts after trauma or retinal detachment repair typically had very similar RRS carotenoid levels in the normal and in the pseudophakic eye. Thus, we concluded that there is a decline in macular carotenoids that reaches a low steady state just at the time when the incidence and prevalence of AMD begins to rise dramatically. The conclusion is further confirmed by our recent HPLC analysis of MP in 49 excised donor eyes in which we found a decline of MP levels with age (unpublished data). These results also emphasize the importance of ensuring that populations are properly age-matched when using RRS spectroscopy in case-control studies.

We compared several different populations with significant macular pathology vs. age-matched controls. AMD patients who did not regularly consume lutein or zeaxanthin supplements had 32% lower macular carotenoid Raman measurements than age-matched controls (6). Interestingly, AMD patients who had been consuming high-dose lutein supplements (≥4 mg/d) for at least 3 mo after their diagnosis of AMD had macular carotenoid levels that on average would be considered normal for their age. These results are very supportive of the hypothesis that low macular carotenoid levels may be a risk factor for AMD and that macular pigment levels can be modified through supplements even in an elderly population with significant macular pathology. We also observed similar reductions in macular carotenoid levels relative to age-matched controls in smaller populations of subjects with macular dystrophies such as Stargardt disease (5) and early-onset macular drusen. On the other hand, mean levels of patients with retinitis pigmentosa or choroideremia, inherited dystrophies that affect the peripheral retina while sparing the fovea until very late in the disorder, were in the same range as those of age-matched controls (5).

Spatial Imaging of Macular Pigments. The current version of the clinical RRS instrument measures total carotenoids in the 1-mm illuminated area centered on the fovea, but it would be very useful to also have information on the spatial distributions of the macular carotenoids. To achieve this *in vivo* spatial mapping, we developed a resonance Raman imaging system, which is shown schematically in Figure 6.6A. It uses parallel CCD camera arrays in conjunction with a narrow band-pass tunable filter. A topographic pseudocolor map of MP distribution at <50 μm resolution is obtained by subtracting two images: one taken with the filter rotated to transmit the 1525 cm^{-1} C=C carotenoid Raman peak, and the other with the filter slightly rotated to transmit light just a few wave numbers away. A fraction of the excitation light is split off by a beam splitter to a second CCD camera to record an image of the excitation laser intensity distribution. Currently, the procedure takes 2 s. Steady fixation of the measured eye is ensured by fixation of the fellow eye onto

A

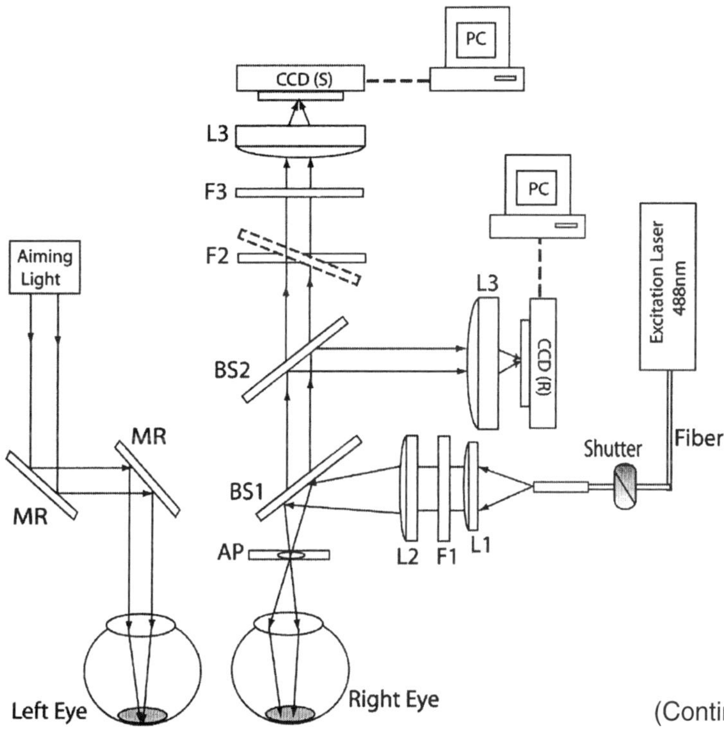

(Continued)

Fig. 6.6. (A) Schematic diagram used for resonance Raman imaging of the macular pigment (MP) distribution of volunteer subjects. Light from blue and green excitation laser wavelengths is projected sequentially onto the retina of a subject as a ~5 mm diameter spot while the fellow eye is looking at a fixation target. The Raman scattered light is collimated by the eye lens and imaged simultaneously onto the 2-D arrays of two separate digital imaging CCD cameras, using a beam splitter, QBS, for the generation of two separate light paths. One of the cameras images the spatial distribution of the excitation light (reference), and the other images light levels in the spectral range of the C=C stretch Raman peak. A rotatable, narrow-band filter (F2) is used to selectively realize C=C on-peak and off-peak transmission for the scattered light. After the registration of an on-peak and an off-peak image, the two images are digitally subtracted (after alignment using vessel landmarks) and displayed as topographic pseudocolor images or surface plots showing the spatial distribution of MP concentrations. L1–L3: lenses; F1: laser line filter; BS: dichroic beam splitters; F2: tunable filter; F3: band-pass filter. (B) Pseudocolor surface plot of MP distribution in a living human eye measured without a Raman imaging instrument. Raman intensities are coded according to the intensity scale shown on the upper right-hand side of the image. Note the roughly circularly symmetric MP distribution in the central area of the fovea but the appearance of side bumps in peripheral areas. (C) A line scan of the above image, obtained by plotting intensities along a horizontal meridional line running through the center of the macular pigment distribution. The width at half maximum is ~150 μm in this individual.

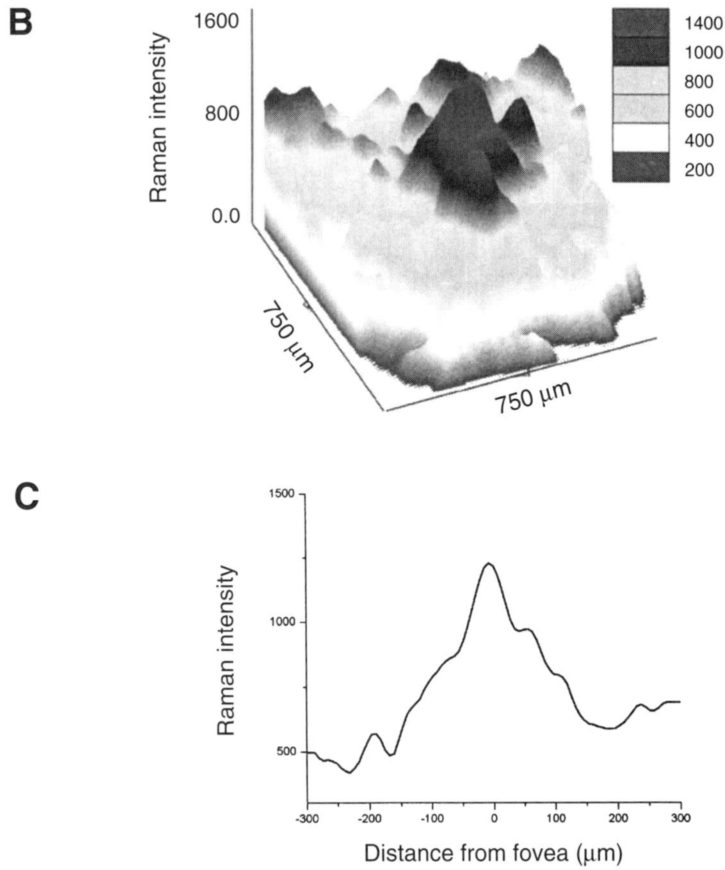

Fig. 6.6. (Continued).

an alignment target provided by a red laser beam that is colinear with the Raman excitation beam. The RRS image of the macula of a volunteer is shown in Figure 6.6B, and a line plot registering the pixel intensities along a horizontal meridional is displayed in Figure 6.6C. Note the roughly circularly symmetric MP distribution in the central area of the fovea but the appearance of side bumps in peripheral areas. The integral of the Raman response over the area illuminated in this imaging technique can be correlated with the previous single-spot resonance Raman method or with extraction and HPLC analysis in the case of measurements on excised eye-cups. For *in vivo* imaging, however, the required light intensities approach the limits of safe ocular exposure at the present stage of development. We are hopeful that further improvements will allow us to reduce exposure levels and thus to include Raman imaging in future clinical studies of the possible role of lutein and zeaxanthin in the maintenance of macular health.

As a promising objective alternative to MP Raman imaging, we investigated the potential merits and drawbacks of lipofuscin fluorescence excitation spectroscopy (AF spectroscopy). AF-based MP measurements were proposed by Delori some time ago as an indirect way to measure the integrated single-path absorption of MP (48); recently, AF was expanded to an imaging configuration using a customized scanning laser ophthalmoscope (49). In AF, the emission of lipofuscin, located in the retinal pigment epithelial layer, is excited at two wavelengths, i.e., one wavelength that overlaps both the MP and lipofuscin absorption and another, longer wavelength, that lies outside the MP absorption range, but that still excites the lipofuscin emission. The MP absorption is then derived from the difference in lipofuscin emission intensities. In our own realization of AF-based MP measurements, we used a novel, very simple imaging approach based on an imaging CCD camera, two laser light sources, and a light delivery and collection module (see schematics in Fig. 6.7A). Digital MP images of a subject are recorded indirectly by detecting the lipofuscin fluorescence of the retinal pigment epithelium in its long-wavelength emission range (λ >730 nm) upon sequential excitation with 488 and 532 nm light; the spatial extent of MP and its topographic concentration distribution are obtained by digital image subtraction. In Figure 6.7B, we show the digital difference images obtained for two volunteer subjects, displaying gray-scale coded spatial MP distributions and line plots of transmission and absorptions derived

A

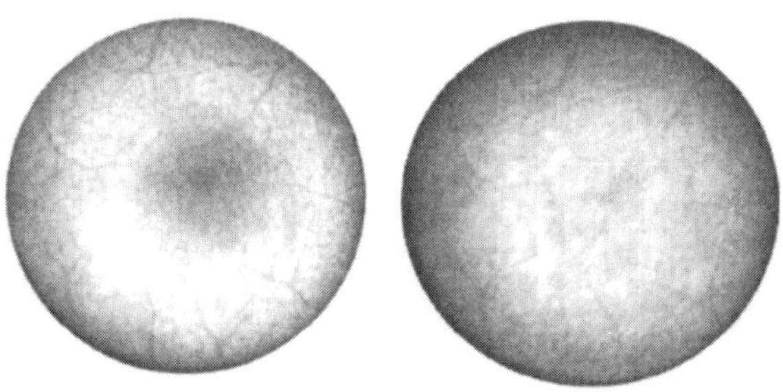

(Continued)

Fig. 6.7. (A) Digital fundus images of a human subject obtained by detecting lipofuscin fluorescence of the retinal pigment epithelium in its long-wavelength emission range (λ >730 nm) upon laser excitation with 488 nm (image A) and 532 nm (image B). The spatial extent of macular pigment (MP) and its topographic concentration distribution can be obtained by digitally subtracting image B, serving as a reference pixel intensity map, from image A, which has pixel areas with reduced intensities due to absorption of the lipofuscin emission by MP (central shaded area). (B) Digital difference images obtained for two subjects, showing spatial MP distribution, and line plots of transmission and absorptions derived from difference images by evaluating pixel intensities along horizontal meridional horizontal lines. (C) Comparison of MP intensities obtained for both subjects with autofluorescence and resonance Raman detection techniques, shown as bar graphs for the Raman and autofluorescence responses integrated over the macular region.

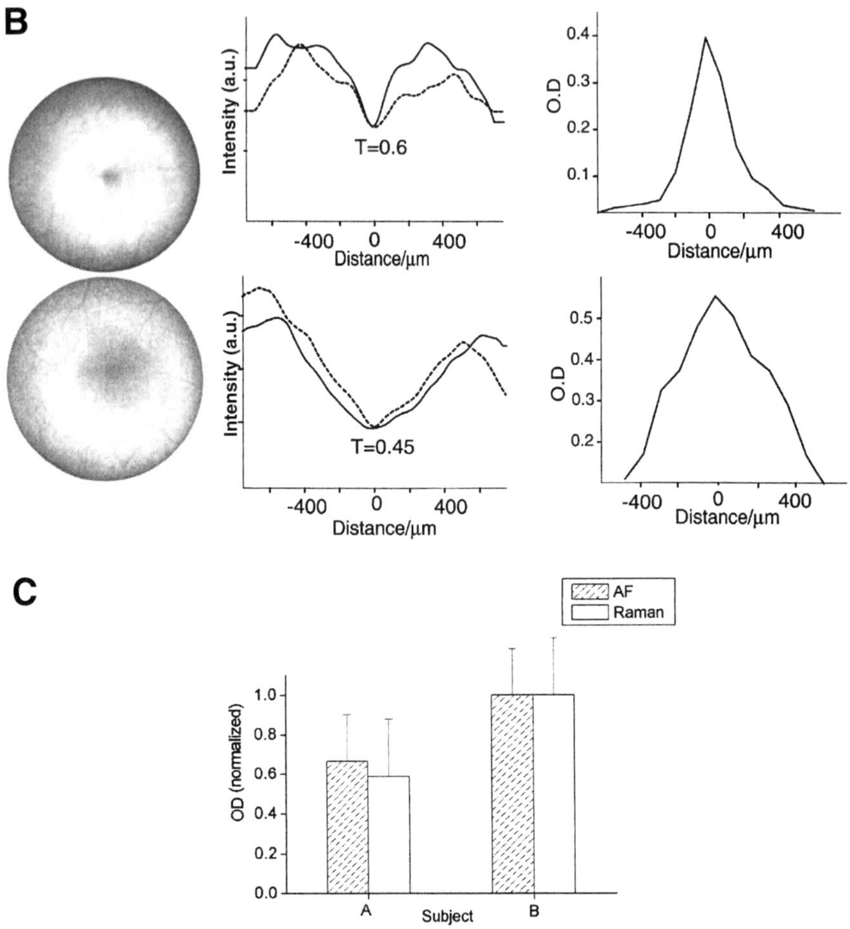

Fig. 6.7. (Continued).

from the difference images by evaluating the corresponding pixel intensities along horizontal meridional horizontal lines. As can be clearly seen from these data, the spatial width, symmetries, and concentrations of MP vary significantly in these subjects. Similar effects were seen in ~6 other subjects measured to date in this way.

To compare AF-derived MP imaging results with Raman measurements, we integrated the AF image pixel intensities over the macular area and compared these average MP results with those obtained using the integral Raman measurement method described above. The results are plotted as bar graphs in Figure 6.7C and show that the MP levels are roughly in agreement in these subjects. Experiments are now in progress to investigate the differences between the two techniques in a larger subject base. If there are no differences, the AF-based MP method might well be a viable highly sensitive imaging methodology, at least in subjects having no macular pathology.

Raman Detection of Carotenoids in Skin

Properties and Function of Skin Carotenoids. Carotenoid molecules play an important role in the skin's antioxidant defense system (10). The six most concentrated carotenoid antioxidants in human skin are lycopene, α-carotene, β-carotene, β-cryptoxanthin, phytoene, and phytofluene, with lycopene and the carotenes accounting for ~60–70% of total carotenoid content (7,50). They are thought to act as scavengers for free radicals (51), singlet oxygen (52,53), and other harmful reactive oxygen species (54), which are all formed, e.g., by metabolic processes or by excessive exposure of skin to the UV components of sunlight. If unbalanced by a lack of antioxidants, the destructive effects of reactive oxygen species and free radicals can lead to skin malignancies and disease. In animal models, carotenoids were shown to inhibit carcinoma formation in the skin (55). It was shown that skin carotenoid levels are strongly and significantly correlated with carotenoid levels in plasma (50). As is found in plasma, skin carotenoid levels are lower in smokers than in nonsmokers. Carotene levels in skin increase with supplementation (56), and supplemental β-carotene is used to treat patients with erythropoietic protoporphyria, a photosensitive disorder (57). Supplemental carotenoids were also shown to delay erythema in normal healthy subjects exposed to UV light (58–60). There is limited evidence that they may be protective against skin malignancies (7), but more research is required to confirm these findings. Because carotenoids are lipophilic molecules, they are well placed in the skin to act as chain-breaking antioxidants, protecting epidermal polyunsaturated fatty acids from oxygen peroxidation (61).

Other dermal antioxidants such as superoxide dismutase, glutathione peroxidase, α-tocopherol, ascorbic acid, and melanins work in collaboration with carotenoids to provide skin with a defensive mechanism against free radical attack and oxidative stress (62). Because these molecules work as a network, definitive measurement of a subset of these antioxidants provides an indication of the relative strength of the whole system. The effectiveness of this protective network can be diminished either by excessive generation of free radicals or by insufficient antioxidant molecules being supplied to the skin. The result is a state of oxidative stress in which important skin constituents are exposed to free radical damage and the associated deleterious structural and chemical changes. If an individual is measured and found to have a lower than normal level of carotenoids in the skin, that person's antioxidant defense system would likely be relatively ill equipped to balance oxidative processes compared with an individual having higher levels of antioxidants. Skin antioxidant measurements provide an opportunity for intervention strategies such as increasing the dietary intake of fruits and vegetables, smoking cessation, and/or prescribing dietary antioxidant supplements.

For many decades, the standard technique for measuring carotenoids has been HPLC. This time-consuming and expensive chemical method works well for the measurement of carotenoids in serum, but it is difficult to perform in skin tissue because it requires biopsies of relatively large tissue volumes. Additionally, serum antioxidant measurements are more indicative of short-term dietary intakes of antioxidants rather than steady-state accumulations in body tissues exposed to external oxidative stress factors such as smoking and UV-light exposure. Nevertheless, the scientific basis of carotenoid function in the human body has been extensively studied for >30 years using the HPLC methodology.

Skin Raman Measurements. Recently, we extended RRS to carotenoid measurements in skin and oral mucosal tissue (8,63). This method is an appealing alternative to reflectance due to its high sensitivity and specificity, which obviate the need for complex correction models. In addition, this method allows one to measure absolute carotenoid levels in these tissues; thus, the method does not have to rely on induced concentration changes. Although absolute levels of carotenoids are much lower in the skin relative to the macula of the human eye, the laser power can be much higher and acquisition times much longer to compensate. Because the bulk of the skin carotenoids are in the superficial layers of the dermis (7), the thin-film Raman equation given above is still valid. Background fluorescence of the tissue can be quite high, but baseline correction algorithms are still adequate to yield carotenoid resonance Raman spectra with excellent signal-to-noise ratios. The Raman method exhibits excellent precision and reproducibility (7,8). Deep melanin pigmentation likely interferes with penetration of the laser beam; thus, measurements are best performed on the palm of the hand where pigmentation is usually quite light even in darkly pigmented individuals. Also, the stratum corneum layer in the palm of the hand is thicker (~1–2 mm) than the penetration depth of blue light (~several hundred μm), and it is bloodless. Therefore, using this tissue site, one realizes measuring conditions of a fairly homogeneous uniform layer with well-defined absorption and scattering conditions. In fact, simple reflection spectroscopy, which is less specific and accordingly not as precise, supports these assumptions because it was already demonstrated that the stratum corneum can be used to monitor changes in carotenoid levels of the human body upon supplementation (56), and it was also shown that dermal carotenoid levels measured at various tissue sites are highly correlated with serum carotenoid levels (58).

An image of the skin carotenoid Raman detection set-up is shown in Figure 6.8A. The main parts of the instrument are an argon ion laser, spectrograph, and the light delivery and collection module. A typical measurement involves the placement of the palm of the hand against the window of the module and exposing the palm for ~1 min at laser intensities of ~10 mW in a 2-mm diameter spot. Carotenoid Raman signals are detected with a 2-D CCD camera integrated with the spectrograph (left side of image). Typical skin carotenoid Raman spectra measured *in vivo* are shown in Figure 6.8B. The spectrum shown at left was obtained directly after laser exposure, and reveals a broad, featureless, and strong fluorescence background of skin with three superimposed sharp Raman peaks characteristic of the

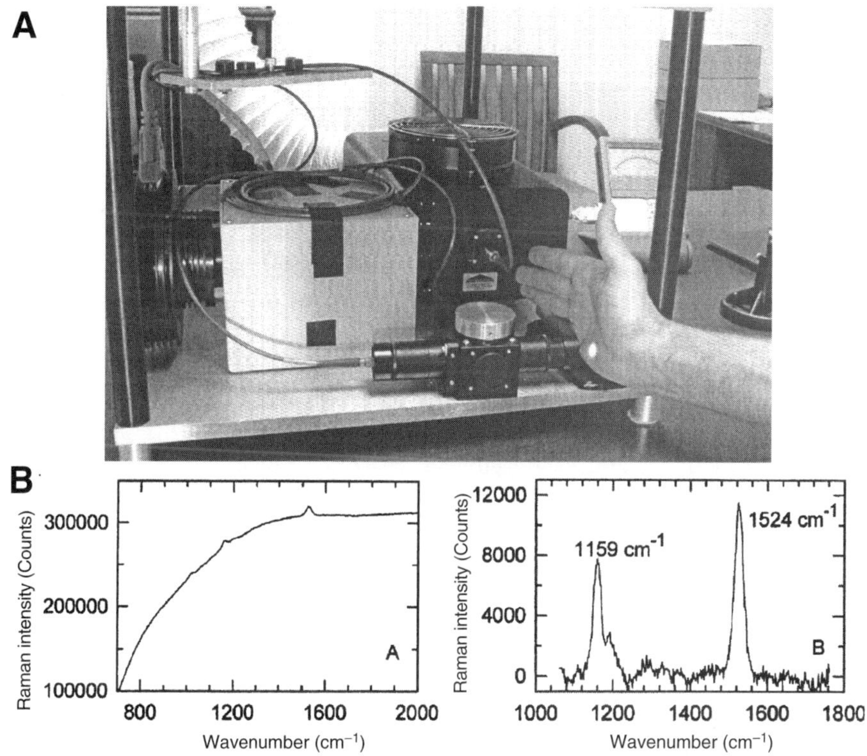

Fig. 6.8. (A) Image of skin carotenoid resonance Raman detector, showing argon laser, spectrograph, light delivery/collection module, and excitation laser spot on the palm of the hand of a subject. A typical measurement involves the placement of the palm of the hand against the window of the module and exposing the palm for about 1 min at laser intensities of ~10 mW in a 2-mm diameter spot. Carotenoid Raman signals are detected with a 2-D CCD camera integrated with the spectrograph (left side of image), and processed similarly to the ocular Raman instrument. (B) Typical skin carotenoid Raman spectra measured *in vivo*. The spectrum shown on top is the spectrum obtained directly after exposure, and reveals broad, featureless, and strong autofluorescence background of skin with superimposed sharp Raman peaks characteristic of carotenoid molecules. The spectrum at the bottom is a difference spectrum obtained after fitting the fluorescence background with a 4th-order polynomial and subtracting it from the top spectrum. The main characteristic carotenoid peaks are clearly resolved with a good signal-to-noise ratio at 1159 and 1524 cm^{-1}.

carotenoid molecules at 1015, 1159, and 1524 cm^{-1}. The spectrum shown at the right represents the Raman spectrum obtained after fitting the fluorescence background with a 5th-order polynomial and subtracting it from the unprocessed spectrum, thus revealing the frequency region of the two strongest carotenoid peaks, at 1159 and 1524 cm^{-1} at an expanded scale.

We found that as with reflectometry results reported in the literature (58), relatively high levels of skin carotenoids are measured by the Raman method on the forehead and on the palm of the hand, whereas other body areas are significantly lower (7,64). To verify the validity of the skin Raman measurements, we conducted a study involving a group of 104 healthy men and women (volunteers) in which we compared HPLC-derived carotenoid levels of serum from fasting subjects with Raman skin levels measured in the inner palm. They were highly correlated ($P < 0.001$) with a correlation coefficient of 0.78 (65). The main result of that study is shown in Figure 6.9A. Measurements of large populations with the Raman device demonstrated widely varying concentrations of carotenoid levels in the palm of the hand (10). Field studies were recently conducted in which a population of 1375 healthy subjects could be screened within a period of several weeks (65). The results, shown in Figure 6.9B, demonstrate a bell-shaped distribution of skin carotenoid concentrations with a large variation throughout the population. Analysis of the data confirmed a pronounced positive relationship between self-reported fruit and vegetable intake (a source of carotenoids) and skin Raman response (Fig. 6.10A). Furthermore, the study showed that people with habitual high sunlight exposure have significantly lower skin carotenoid levels than people with little sunlight exposure, independent of their carotenoid intake or dietary habits (Fig. 6.10B), and that smokers had dramatically lower levels of skin carotenoids than nonsmokers (Fig. 6.10C). When analyzed by a chemical assay based on urinary malondialdehyde excretion, an indicator of oxidative lipid damage, people with high oxidative stress had significantly lower skin carotenoid levels than people with low oxidative stress. Again, this relation was not confounded by dietary carotenoid intakes, which were similar in the two groups. These observations provide evidence that skin carotenoid resonance Raman readings might be useful as a surrogate marker for general antioxidant status (10), which was suggested recently for plasma carotenoids as well (66). The recent availability of a commercial, portable resonance Raman instrument for skin carotenoids (BioPhotonic Scanner™, Pharmanex LLC, Provo, UT) facilitates further epidemiologic and nutrition studies.

Studies are also underway to determine whether low skin Raman measurements may be associated with an increased risk of various skin cancers. Initial studies demonstrated that lesional and perilesional Raman carotenoid intensities of cancerous and precancerous skin lesions are significantly lower than in the region-matched skin of healthy subjects (7).

Selective Measurement of Lycopene. In all previous Raman measurements of dermal carotenoids, we measured the composite level of the long-chain carotenoid species because they are all excited simultaneously under the conditions used, and all contribute to the overall Raman response. The increased conjugation length in lycopene compared with the other carotenoids in skin produces a small (~10 nm) but distinguishable red shift of the absorption band that can be used to measure lycopene independently of the other carotenoids (9). There is considerable interest

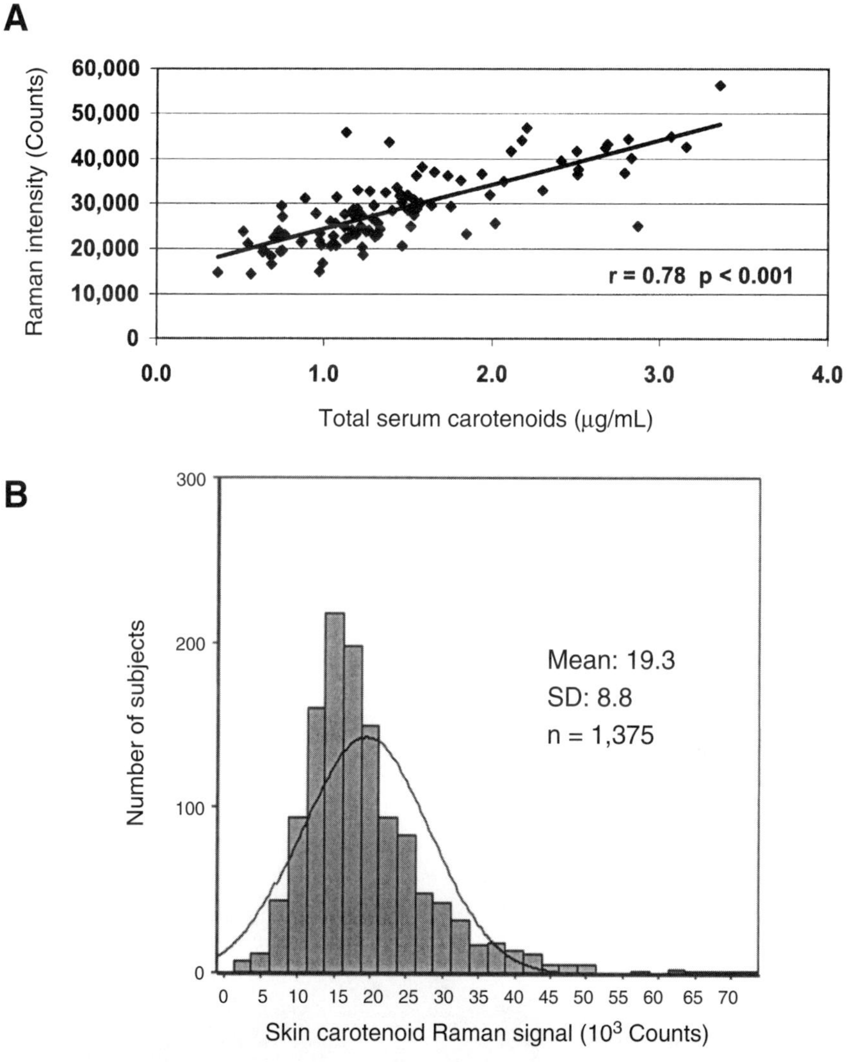

Fig. 6.9. (A) Correlation of skin resonance Raman intensity measured in the inner palm of the hand with serum carotenoids determined by HPLC, obtained for a group of 104 healthy men and women. Note the high correlation coefficient of $r = 0.78$ ($P < 0.001$). (B) Histogram of skin carotenoid resonance Raman response measured in the palm of the hand for 1375 subjects, showing the wide distribution of skin carotenoid levels in a large population.

in a specific role for lycopene in the prevention of prostate cancer and other diseases (60,67), and a noninvasive biomarker for lycopene consumption would be of tremendous utility. As seen in Figure 6.11A for solutions of lycopene and β-carotene, the resonance Raman response has approximately the same strengths

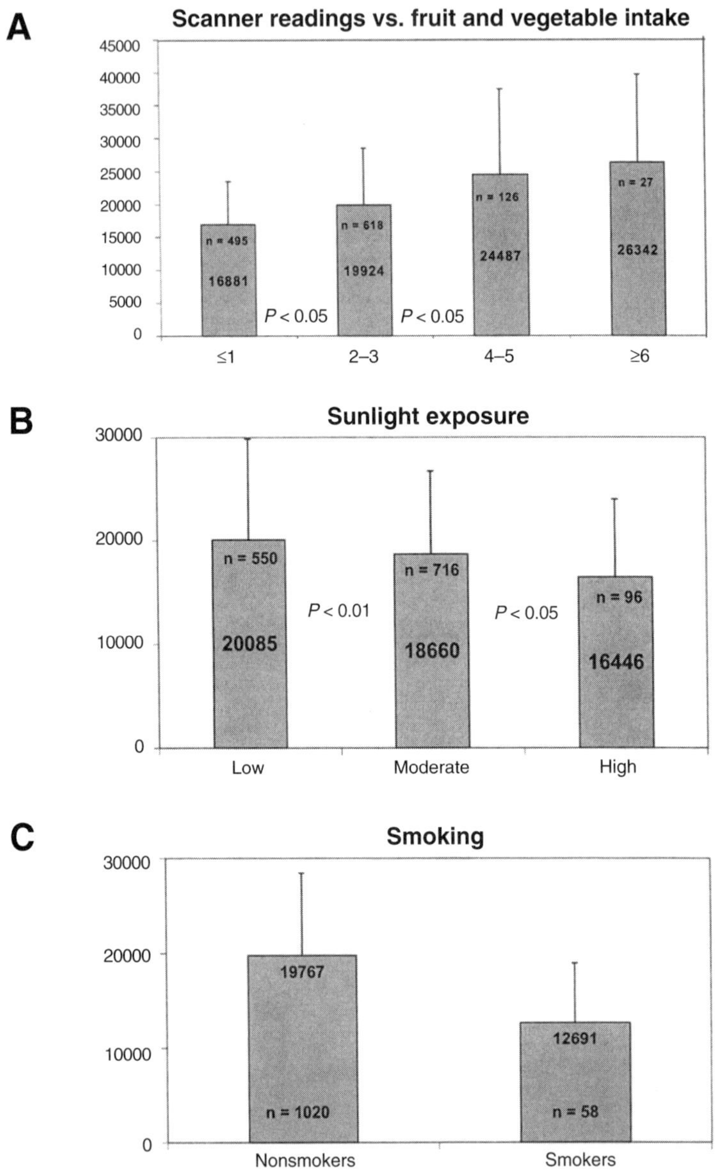

Fig. 6.10. (A) Resonance Raman intensity (counts) vs. reported number of daily servings of fruits and vegetables consumed, demonstrating the increase in skin carotenoid concentrations with increased fruit and vegetable uptake. Values are means ± SD. (B) Resonance Raman intensity (counts) vs. self-reported exposure of skin to sunlight, showing the decrease in skin carotenoid levels with increased sun exposure. (C) Resonance Raman intensity (counts) in nonsmokers and cigarette smokers, showing the ~30% decrease of skin carotenoid levels in smokers.

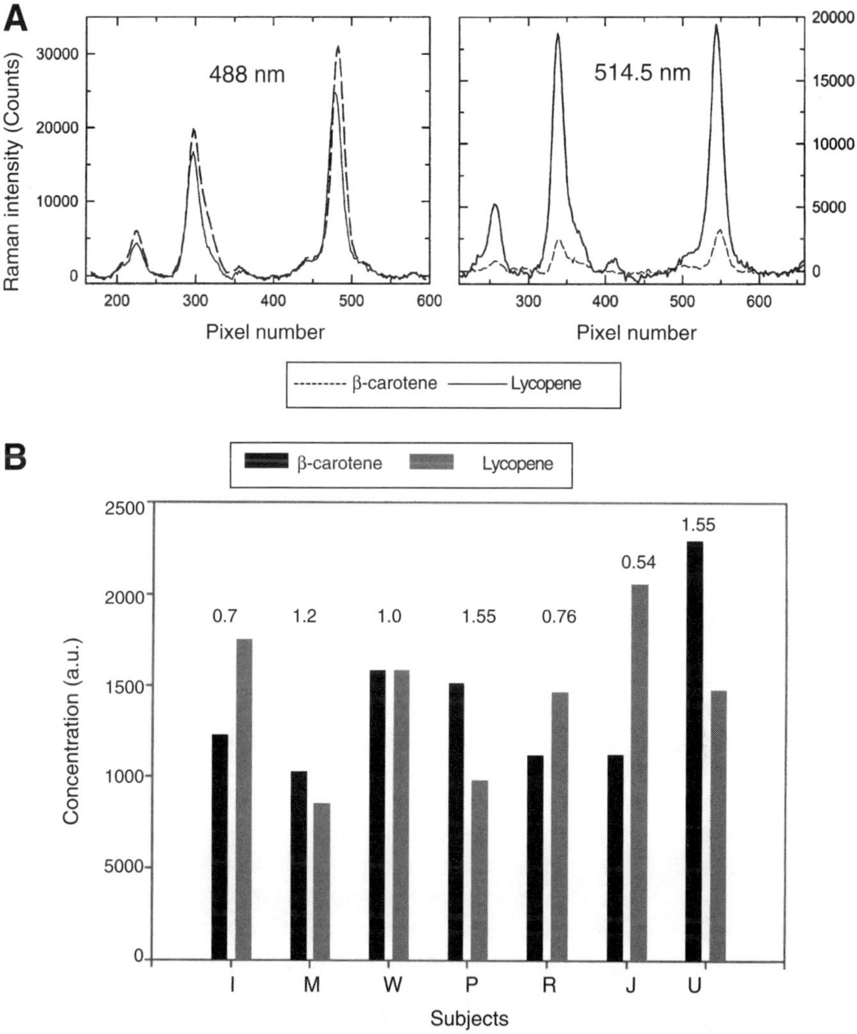

Fig. 6.11. (A) Resonance Raman spectrum of an acetone solution of lycopene (solid lines) and β-carotene (dotted lines), measured under argon laser excitation at 488 nm (upper panel) and 514.5 nm (bottom panel). Both solutions had identical carotenoid concentrations. Raman spectra were recorded using identical excitation power and sensitivity-matched instruments. The strongest Raman peaks correspond to the stretch vibrations of the carbon single and double bonds of the molecule (at ~1159 and ~1525 cm^{-1}, respectively). Note the strongly reduced Raman response of C=C stretch for lycopene compared with β-carotene under 514.5 nm excitation. (B) β-Carotene and lycopene skin Raman levels measured with selective resonant Raman spectroscopy for 7 subjects. Note the strong intersubject variability of the β-carotene to lycopene ratios (indicated above bar graphs).

under 488 nm excitation. However, under excitation at 514.5 nm, the response is ~6 times higher for lycopene. Therefore, it is possible to measure the individual responses in a mixture of both by measuring in addition to the 488-nm response the 514.5-nm response, and taking into account the relative carotenoid contributions in the resulting spectra (9). In initial experiments with 7 volunteers, we measured the skin Raman response from the stratum corneum of the palm of the hand using such a dual-wavelength carotenoid Raman detection instrument and demonstrated large variations in individual skin lycopene levels. The most important result of this experiment is summarized in Figure 6.11B, which indicates the individual lycopene and carotene levels together with the lycopene/carotene ratio for each subject. Obviously, there is a strong, almost threefold variation in the carotene to lycopene ratio in the subjects measured, ranging from 0.54 to 1.55. This means that substantially different carotenoid compositions exist in human skin, with some subjects exhibiting almost twice the concentration of lycopene compared with carotene, and other subjects showing the opposite effect. This behavior could reflect different dietary patterns in terms of the intake of lycopene or lycopene-containing vegetables, or it could point toward differing abilities among subjects to accumulate these carotenoids in the skin.

Conclusions

Resonance Raman spectroscopy is a highly specific, sensitive, and precise noninvasive optical method allowing one to rapidly assess macular and dermal carotenoid content in large populations with excellent correlation of dermal to serum levels. To our knowledge, there are no serious confounding factors for the technology, and it has exciting application potential. As the technology is further validated and enters common clinical usage, it is likely to play an important role in the early diagnosis of individuals at risk for many debilitating disorders such as age-related macular degeneration and other degenerative diseases. In the nutritional supplement industry, it is already being used as an objective, portable device for monitoring the effect of carotenoid-containing supplements on skin tissue carotenoid levels. In cancer epidemiology, it may serve as a noninvasive novel biomarker for fruit and vegetable intake, replacing costly plasma carotenoid measurements with inexpensive and rapid skin Raman measurements. Last, due to its capability to selectively detect lycopene, the technology may be useful in investigating a specific role for lycopene in the prevention of prostate cancer and other diseases.

Acknowledgments

The authors acknowledge significant contributions to this work by I.V. Ermakov, M.R. Ermakova, M. Sharifzadeh, and D.-Y. Zhao. This research was supported by grants from Spectrotek, L.C., the National Eye Institute (grants R29-EY 11600, STTR 1 R41 EY 1234–01, and STTR 2 R42 EY 1234–02), Research to Prevent Blindness, and by the State of Utah (Center of Excellence for Biomedical Optics grant).

References

1. Koyama, Y. (1995) Resonance Raman Spectroscopy, in *Carotenoids, Vol. 1B, Spectroscopy*, Britton, G., Liaaen-Jensen, S., and Pfander, H., eds., Birkhäuser, Basel, pp. 135–146.
2. Bernstein, P.S., Yoshida, M.D., Katz, N.B., McClane, R.W., and Gellermann, W. (1998) Raman Detection of Macular Carotenoid Pigments in Intact Human Retina, *Investig. Ophthalmol. Vis. Sci. 39:* 1003–2011.
3. Gellermann, W., Ermakov, I.V., Ermakova, M.R. McClane, R.W., Zhao, D.-Y., and Bernstein, P.S. (2002) *In Vivo* Resonant Raman Measurement of Macular Carotenoid Pigments in the Young and the Aging Human Retina, *J. Opt. Soc. Am. A 19:* 1172–1186.
4. Ermakov, I.V., Ermakova, M.R., Gellermann, W., and Bernstein, P.S. (2004) Macular Pigment Raman Detector for Clinical Applications, *J. Biomed. Opt. 9:* 139–148.
5. Zhao, D.-Y., Wintch, S.W., Ermakov, I.V., Gellermann, W., and Bernstein, P.S. (2003) Resonance Raman Measurement of Macular Carotenoids in Retinal, Choroidal, and Macular Dystrophies, *Arch. Ophthalmol. 121:* 967–972.
6. Bernstein, P.S., Zhao, D.-Y., Wintch, S.W., Ermakov, I.V., McClane, R.W., and Gellermann, W. (2002) Resonance Raman Measurement of Macular Carotenoids in Normal Subjects and in Age-Related Macular Degeneration Patients, *Ophthalmology 109:* 1780–1787.
7. Hata, T.R., Scholz, T.A., Ermakov, I.V., McClane, R.W., Khachik, F., Gellermann, W., and Pershing, L.K. (2000) Non-Invasive Raman Spectroscopic Detection of Carotenoids in Human Skin, *J. Investig. Dermatol. 115:* 441–448.
8. Ermakov, I.V., Ermakova, M.R., McClane, R.W., and Gellermann, W. (2001) Resonance Raman Detection of Carotenoid Antioxidants in Living Human Tissues, *Opt. Lett. 26:* 1179–1181.
9. Ermakov, I.V., Ermakova, M.R., Gellermann, W., and Lademann, J. (2004) Non-Invasive Selective Detection of Lycopene and Beta-Carotene in Human Skin Using Raman Spectroscopy, *J. Biomed. Opt. 9:* 332–338.
10. Gellermann, W., Ermakov, I.V., Scholz, T.A., and Bernstein, P.S. (2002) Noninvasive Laser Raman Detection of Carotenoid Antioxidants in Skin, *Cosmet. Dermatol. 15:* 65–68.
11. Shreve, A.P., Trautman, J.K., Owens, T.G., and Albrecht, A.C. (1991) Determination of the S2 Lifetime of β-Carotene, *Chem. Phys. Lett. 178:* 89.
12. Bone, R.A., Landrum, J.T., and Tarsis, S.L. (1985) Preliminary Identification of the Human Macular Pigment, *Vision Res. 25:* 1531–1535.
13. Handelman, G.J., Snodderly, D.M., Adler, A.J., Russett, M.D., and Dratz, E.A. (1992) Measurement of Carotenoids in Human and Monkey Retinas, *Methods Enzymol. 213:* 220–230.
14. Schalch, W., Dayhaw-Barker, P., and Barker, F.M. (1999) The Carotenoids of the Human Retina, in *Nutritional and Environmental Influences on the Eye*, Taylor A., ed., CRC, Boca Raton, FL, pp. 215–250.
15. Snodderly, D.M. (1995) Evidence for Protection Against Age-Related Macular Degeneration by Carotenoids and Antioxidant Vitamins, *Am. J. Clin. Nutr. Suppl. 62:* 1448S–1461S.
16. Eye Disease Case Control Study Group (1993) Antioxidant Status and Neovascular Age-Related Macular Degeneration, *Arch. Ophthalmol. 111:* 104–109.

17. Seddon, J.M., Ajani, U.A., Sperduto, R.D., Hiller, R., Blair, N., Burton, T.C., Farber, M.D., Gragoudas, E.S., Haller, J, Miller, D.T., Yannuzzi, L.A., and Willet, W. (1994) Dietary Carotenoids, Vitamins A, C, and E, and Advanced Age-Related Macular Degeneration, *J. Am. Med. Assoc. 272:* 1413–1420.
18. Hammond, B.R., Johnson, E.J., Russell, R.M., Krinsky, N.I., Yeum, K.-J., Edwards, R.B., and Snodderly, D.M. (1997) Dietary Modification of Human Macular Pigment Density, *Investig. Ophthalmol. Visual Sci. 38:* 1795–1801.
19. Landrum, J.T., Bone, R.A., Joa, H., Kilburn, M.D., Moore, L.L., and Sprague, K.E. (1997) A One Year Study of the Macular Pigment: The Effect of 140 Days of a Lutein Supplement, *Exp. Eye Res. 65:* 57–62.
20. Landrum, J.T., Bone, R.A., and Kilburn, M.D. (1997) The Macular Pigment: A Possible Role in Protection from Age-Related Macular Degeneration, *Adv. Pharmacol. 38:* 537–556.
21. Bone, R.A., Landrum, J.T., Mayne, S.T., Gomez, C.M., Tibor, S.E., and Twaroska, E.E. (2000) Macular Pigment in Donor Eyes with and Without AMD: A Case-Control Study, *Investig. Opthalmol. Visual Sci. 42:* 235–240.
22. Snodderly, D.M., Auran, J.D., and Delori, F.C. (1984) The Macular Pigment, I: Absorbance Spectra, Localization, and Discrimination from Other Yellow Pigments in Primate Retinas, *Investig. Ophthalmol. Visual Sci. 25:* 660–673.
23. Bernstein, P.S., and Katz, N.B. (2000) The Role of Ocular Free Radicals in Age-Related Macular Degeneration, in *Environmental Stressors: Effects on Lung, Skin, Eye and Immune System Function,* Fuchs, J., and Packer, L., eds., Marcel Dekker, New York, pp. 423–456.
24. Beatty, S., Koh, H.H., Henson, D., and Boulton, M. (2000) The Role of Oxidative Stress in the Pathogenesis of Age-Related Macular Degeneration, *Surv. Ophthalmol. 45:* 115–134.
25. Reading, V.M., and Weale, R.A. (1974) Macular Pigment and Chromatic Aberration, *J. Am. Optom. Assoc. 64:* 231–234.
26. Beatty, S., Murray, I.J., Henson, D.B., Carden, D., Koh, H.-H., and Boulton, M.E. (2001) Macular Pigment and Risk for Age-Related Macular Degeneration in Subjects from a Northern European Population, *Investig. Ophthalmol. Visual Sci. 42:* 439–446.
27. Hammond, B.R., Wooten, B.R., and Snodderly, M.R. (1997) Individual Variations in the Spatial Profile of Human Macular Pigment, *J. Opt. Soc. Am. A 14:* 1187–1196.
28. Snodderly, D.M., and Hammond, B.R. (1999) *In Vivo* Psychophysical Assessment of Nutritional and Environmental Influences on Human Ocular Tissues: Lens and Macular Pigment, in *Nutritional and Environmental Influences on the Eye*, Taylor A., ed., CRC, Boca Raton, FL, pp. 251–273.
29. Werner, J.S., Bieber, M.L., and Schefrin, B.E. (2000) Senescence of Foveal and Parafoveal Cone Sensitivities and Their Relations to Macular Pigment Density, *J. Opt. Soc. Am. A 17:* 1918–1932.
30. Hammond, B.R., and Caruso-Avery, M. (2000) Macular Pigment Optical Density in a Southwestern Sample, *Investig. Ophthalmol. Visual Sci. 41:* 1492–1497.
31. van Norren, D., and Tiemeijer, L.F. (1986) Spectral Reflectance of the Human Eye, *Vision Res. 26:* 313–320.
32. Delori, F.C., and Pflibsen, K.P (1989) Spectral Reflectance of the Human Ocular Fundus, *Appl. Opt. 28:* 1061–1077.
33. van de Kraats, J., Berendshot, T.T.J.M., and van Norren, D. (1996) The Pathways of Light Measured in Fundus Reflectometry, *Vision Res. 36:* 2229–2247.

34. Berendshot, T.T.J.M., van de Kraats, J., and van Norren, D. (1999) Three Methods to Measure Macular Pigment Compared in a Lutein Intake Study, *Investig. Ophthalmol. Visual Sci. 39:* S314 (1999).
35. Kilbride, P.E., Alexander, K.R., Fishman, M., and Fishman, G.A. (1989) Human Macular Pigment Assessed by Imaging Fundus Reflectometry, *Vision Res. 29:* 663–674.
36. Webb, R.H., Hughes, G.W., and Pomerantzeff, O. (1980) Flying Spot TV Ophthalmoscope, *Appl. Opt. 19:* 2991–2997.
37. Webb, R.H., Hughes, G.W., and Delori, F.C. (1987) Confocal Scanning Laser Ophthalmoscope, *Appl. Opt. 26:* 1492–1499.
38. van Norren, D., and van de Kraats, J. (1989) Imaging Retinal Densitometry with a Confocal Scanning Ophthalmoscope, *Vision Res. 29:* 1825–1830.
39. Elsner, A.E., Burns, S.A., Delori, F.C., and Webb, R.H. (1990) Quantitative Reflectometry with the SLO, in *Laser Scanning Ophthalmoscopy and Tomography*, Nasemann, J.E., and Burk, R.O., eds., Quintessenz-Verlag, Berlin, pp. 109–121.
40. Elsner, A.E., Burns, S.A., Hughes, G.W., and Webb, R.H. (1992) Reflectometry with a Scanning Laser Ophthalmoscope, *Appl. Opt. 31:* 3697–3710.
41. Elsner, A.E., Burns, S.A., Beausencourt, E., and Weiter, J.J. (1998) Foveal Cone Photopigment Distribution: Small Alterations Associated with Macular Pigment Distribution, *Investig. Ophthalmol. Visual Sci. 39:* 2394–2404.
42. Schweitzer, D., Hammer, M., and Scibor, M. (1998) Imaging Spectrometry in Ophthalmology-Principle and Applications in Microcirculation and in Investigation of Pigments, *Investig. Ophthalmol. Visual Sci. 39:* 2001–2011.
43. Delori, F.C. (1993) Macular Pigment Density Measured by Reflectometry and Fluorophotometry, in *Ophthalmic and Visual Optics and Noninvasive Assessment of the Visual System*, vol. 3 of 1993 OSA Technical Digest Series, Optical Society of America, Washington, DC, pp. 240–243.
44. Delori, F.C., Goger, D.G., Hammond, B.R., Snodderly, D.M, and Burns, S.A. (2001) Macular Pigment Density Measured by Autofluorescence Spectrometry: Comparison with Reflectometry and Heterochromatic Flicker Photometry, *J. Opt. Soc. Am. A 18:* 1212–1230.
45. Berendshot, T.T.J.M., Goldbohm, R.A., Kloepping, W.A.A., van de Kraats, J., van Norel, J., and van Norren, D. (2000) Influence of Lutein Supplementation on Macular Pigment, Assessed with Two Objective Techniques, *Investig. Ophthalmol. Visual Sci. 41:* 3322–3326.
46. Ermakov, I.V., McClane, R.W., Gellermann, W., and Bernstein, P.S. (2001) Resonant Raman Detection of Macular Pigment Levels in the Living Human Retina, *Opt. Lett. 26:* 202–204.
47. Bernstein, P.S., Gellermann, W., and McClane, R.W., U.S. Patent 5,873,831 (1999).
48. Delori, F.C. (1994) Spectrophotometer for Noninvasive Measurement of Intrinsic Fluorescence and Reflectance of the Ocular Fundus, *Appl. Opt. 33:* 7439–7452.
49. Robson, A.G., Moreland, J.D., Pauleikoff, D., Morrissey, T., Holder, G.E., Fitzke, F.W., Bird, A.D., and van Keijk, F.J.G.M.D. (2003) Macular Pigment Density and Distribution: Comparison of Fundus Autofluorescence with Minimum Motion Photometry, *Vision Res. 43:* 1765–1775.
50. Peng, Y.M., Peng, Y.S., Lin, Y., Moon, T., Roe, D.J., and Ritenbaugh, C. (1995) Concentrations and Plasma-Tissue-Diet Relationships of Carotenoids, Retinoids, and Tocopherols in Humans, *Nutr. Cancer 23:* 233–246.

51. Böhm, F., Tinkler, J.H., and Truscott, T.J. (1995) Carotenoids Protect Against Cell Membrane Damage by the Nitrogen Dioxide Radical, *Nat. Med. 1:* 98–99.
52. Foote, C.S., and Denny, R.W. (1968) Chemistry of Singlet Oxygen. VIII. Quenching by β-Carotene, *J. Am. Chem. Soc. 90:* 6233–6235.
53. Farmillo, A., and Wilkinson, F. (1973) On the Mechanism of Quenching of Singlet Oxygen in Solution, *Photochem. Photobiol. 18:* 447– 450.
54. Conn, P.F., Schalch, W., and Truscott, T.G. (1991) The Singlet Oxygen and Carotenoid Interaction, *J. Photochem. Photobiol. B: Biol. 11:* 41–47.
55. Chen, L.C., Sly, L., Jones, C.S., De Tarone, R., and Luca, L.M. (1993) Differential Effects of Dietary β-Carotene on Papilloma and Carcinoma Formation Induced by an Initiation-Promotion Protocol in SENCAR Mouse Skin, *Carcinogenesis 14:* 713–717.
56. Prince, M.R., and Frisoli, J.K. (1993) Beta-Carotene Accumulation in Serum and Skin, *Am. J. Nutr. 175:* 175–181.
57. Mathews-Roth, M.M. (1993) Carotenoids in Erythropoietic Protoporphyria and Other Photosensitivity Diseases, *Ann. N.Y. Acad. Sci. 691:* 127–138.
58. Stahl, W., Heinrich, U., Jungmann, H., Sies, H., and Tronnier, H. (2000) Carotenoids and Carotenoids Plus Vitamin E Protect Against Ultraviolet Light-Induced Erythema in Humans, *Am. J. Clin. Nutr. 71:* 795–798.
59. Lee, J., Jiang, S., Levine, N., and Watson, R. (2000) Carotenoid Supplementation Reduces Erythema in Human Skin After Simulated Solar Radiation Exposure, *Proc. Soc. Exp. Biol. Med. 223:* 170–174.
60. Giovannucci, E. (1999) Tomatoes, Tomato-Based Products, Lycopene, and Cancer: Review of the Epidemiological Literature, *J. Natl. Cancer Inst. 91:* 317–331.
61. Krinsky, H.I., and Deneke, S.M. (1982) The Interaction of Oxygen and Oxyradicals with Carotenoids, *J. Natl. Cancer Inst. 69:* 205–210.
62. Steenvorden, D.P.T., and Beijerbergen van Henegouwen, G.M.J. (1997) The Use of Endogenous Antioxidants to Improve Photoprotection, *J. Photochem. Photobiol. B: Biology 41:* 1–10.
63. Gellermann, W., Bernstein, P.S., and McClane, R.W., U.S. Patent 6,205,354 B1 (2001).
64. Alaluf, S., Heinrich, U., Stahl, W., Tronnier, H., and Wiseman, S. (2002) Dietary Carotenoids Contribute to Normal Human Skin Color and UV Photosensitivity, *J. Nutr. 132:* 399–403.
65. Smidt, C.R., Gellermann, W., and Zidichouski, J.A. (2004) Non-Invasive Raman Spectroscopy Measurement of Human Carotenoid Status, *FASEB J. 18:* A 480 (Abstr.).
66. Svilaas, A., Sakhi, A.K., Andersen, L.F., Svilaas, T., Ström, E.C., Jacobs, D.R., Jr., Ose, L., and Blomhoff, R. (2004) Intakes of Antioxidants in Coffee, Wine, and Vegetables Are Correlated with Plasma Carotenoids in Humans, *J. Nutr. 134:* 562–567.
67. Rao, V., and Heber, D. (2002) Lycopene and the *Prevention of Chronic Diseases, Nutrition and Health Conference Report, Vol. 1, No. 1*, Caledonian Science Press, Barcelona.

Chapter 7

Macular Carotenoids in Eye Health

Richard A. Bone[a] and John T. Landrum[b]

[a]Department of Physics and [b]Department of Chemistry and Biochemistry, Florida International University, Miami, FL 33199

Introduction

The central few millimeters of human and other primate retinas is called the macula, named for the concentration of carotenoids that results in a distinct yellow spot, or "*macula lutea*." The existence of this macular pigment in the living eye was in dispute until as recently as ~50 years ago. Its presence in autopsy eyes was considered by some to be a *post-mortem* artifact (1,2). Now the question is not whether it exists in living eyes, but what functions it serves. Based upon epidemiologic findings, there is a growing consensus that it has a significant influence on the etiology and risk of at least one, and possibly two, diseases of the eye (3).

The macular carotenoids were characterized by HPLC of the compounds themselves and their derivatives on several different stationary phases (4). They were also identified by MS and UV-visible spectroscopy (4,5). They consist largely of lutein (L) and stereoisomers of zeaxanthin. These are oxygenated carotenoids that are known collectively as "xanthophylls." The stereoisomers of zeaxanthin in the retina are 3*R*,3′*R*-zeaxanthin [hereafter "zeaxanthin" (Z)], 3*R*,3′*S*-zeaxanthin ["meso-zeaxanthin" (MZ)], together with much smaller amounts of 3*S*,3′*S*-zeaxanthin ["SS-zeaxanthin" (SZ)] (5). The structures of the major retinal carotenoids, L, Z, and MZ, are shown in Figure 7.1. HPLC chromatograms of macular pigment also reveal the presence of small amounts of lutein and zeaxanthin oxidation products as well as geometric isomers of these carotenoids (6). Although visually discernible only in the center of the macula, lutein and zeaxanthin are present throughout the retina, although at much lower concentrations (7,8) (see Fig. 7.2.) Lutein and zeaxanthin are also found in other ocular tissues including the lens, ciliary body, iris, and retinal pigment epithelium (RPE)/choroid (9). In the retina, the relative amounts of the major carotenoids change with eccentricity from the fovea (a small depression in the center of the retina) (10). In the fovea, the ratio of L:Z:MZ is ~1.3:1:0.8, whereas in the peripheral retina it is ~2.4:1:0.2. Thus zeaxanthin and meso-zeaxanthin are dominant in the fovea, and lutein is dominant in the peripheral retina. This is the basis of the hypothesis that lutein is partially converted to meso-zeaxanthin in the retina by a process that becomes more effective toward the foveal center (10). In this context, it should be noted that lutein and zeaxanthin are found in blood serum, having been derived directly from the diet. Meso-zeaxanthin is not common in human food sources (11) and consequently is not found in the serum.

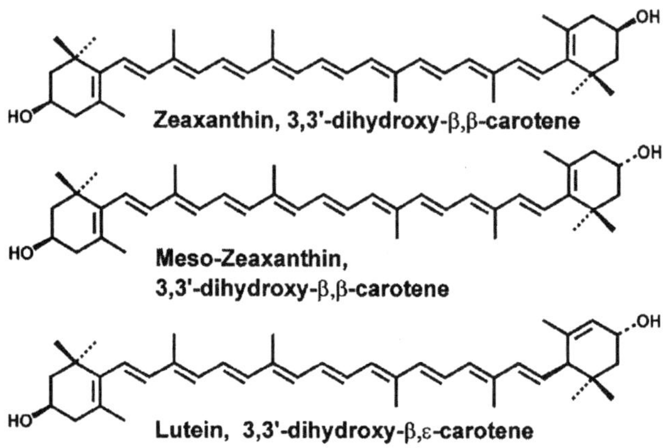

Fig. 7.1. The structures of the major carotenoids found in the human retina.

Microspectrophotometry of transverse retinal sections shows high concentrations of macular pigment in the Henle fibers, the photoreceptor axons of the foveal region (12). The preferred orientations of the macular pigment molecules perpendicular to these radially oriented fibers is the basis for an entoptical phenomenon called Haidinger's brushes (13) and supports the suggestion that the carotenoids are incorporated within the cell membranes where their preferred packing arrangement is one in which they span the membrane. Haidinger's brushes refers to a shadowy hourglass-shaped figure seen at the fixation point while viewing a blue surface through a polarizing filter. The explanation of the brushes lies in the fact that lutein and zeaxanthin possess an elongated structure (see Fig. 7.1) and preferentially absorb light that is polarized parallel to their long axes (14). Separated rod outer segments were also shown to contain lutein and zeaxanthin (15,16), and although the same may be true of cones, their outer segments are not readily separable for carotenoid analysis. As will be discussed below, carotenoids in Henle fibers may protect posterior tissues (photoreceptor outer segments, Bruch's membrane and the RPE) passively by screening out blue light. In addition, they may provide active protection for the outer segments through their known antioxidant activity (17).

Lutein and zeaxanthin are detectable in small amounts in the fetal and neonatal retina, and reach concentrations comparable to those of adults by the age of ~2 y (7). Averaged results from most studies show little tendency for macular pigment to increase or decrease throughout the remaining life span (7,18–22). The exception is when macular pigment is detected by resonance Raman spectroscopy because the Raman signal declines markedly with age (23).

Age-Related Macular Degeneration (AMD)

AMD is the leading cause of vision loss in the United States (24). It is characterized, ultimately, by photoreceptor loss, or dysfunction, in the macula and a corre-

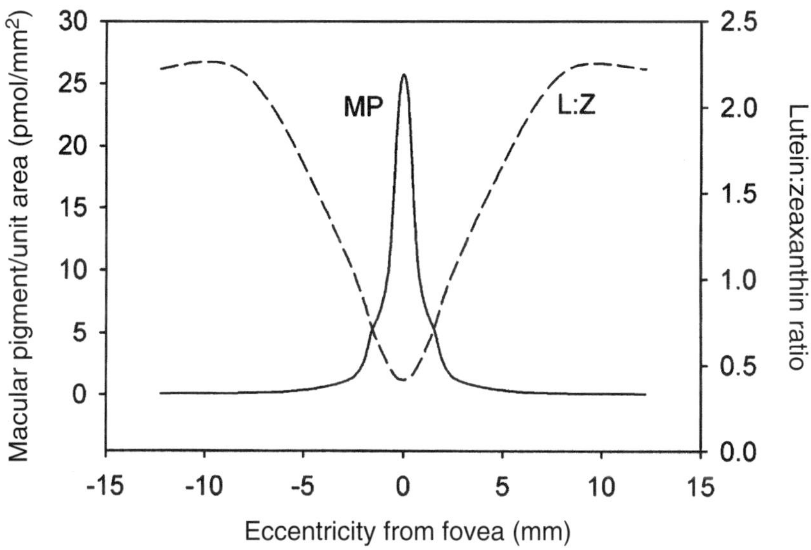

Fig. 7.2. Based on HPLC analysis of human retinas, the total amount of lutein and zeaxanthin declines rapidly with eccentricity from the center of the fovea. In addition, the ratio of lutein to zeaxanthin increases from ~1:2 in the fovea to 2:1 in the periphery.

sponding blind spot in the visual field. Approximately 20% of the population >70 y old is affected. Risk factors for this disease, other than age, are both genetic and environmental (25). The former includes a family history of the disease and race, with Whites appearing to be at greater risk than Blacks. Environmental factors include smoking, cardiovascular disease, poor antioxidant status, and, possibly, light exposure. Epidemiologic evidence indicates that low levels of lutein and zeaxanthin in the serum and diet are associated with an increased risk of the advanced form of the disease, neovascular AMD (26,27). A study of autopsy eyes reported that, on average, there were lower levels of carotenoids throughout the retinas of subjects with AMD compared with controls who did not have the disease (28). Although AMD itself could be responsible for the decreased amounts of carotenoids observed in the macula, it would not be expected to have a significant effect on the peripheral retina. Similarly, Beatty *et al.* (29) compared the macular pigment optical density in subjects with healthy maculae, and in the healthy eye of subjects with advanced AMD in the fellow eye. Those in the latter group had significantly less macular pigment than those with healthy maculae.

The processes involved in the development of AMD are not fully understood. A somewhat simplistic model involving oxidative stress was proposed (30). Photoreceptor outer segments are continuously being renewed, at a rate that may depend on light exposure (31). The photoreceptors grow outward toward the back of the retina, and the older disks at the outer ends are phagocytosed by RPE cells and exocytosed into the choroidal circulation. However, the elimination process is imperfect, and insoluble lipophilic material accumulates as whitish clumps called "drusen," an early indicator of susceptibility to AMD (32). The drusen are found between the RPE cell layer and Bruch's membrane. As drusen grow, or coalesce, normal processes that maintain the photoreceptors are interrupted, eventually resulting in cell death. These processes include the transfer of nutrients between the choroidal capillaries and photoreceptors through the intervening RPE and Bruch's membrane, and the transfer of vitamin A back and forth between the photoreceptors and RPE as part of the visual cycle. In addition, RPE cells accumulate lipofuscin, a fluorescent material consisting partly of the photosensitizer, A2E (33). Photooxidation of polyunsaturated fatty acids in outer segments may accelerate the outer segment renewal process and hence the build-up of drusen.

The macular carotenoids may interfere with these processes in two ways. By reducing the incident blue light flux on the outer segments and RPE, the macular pigment contained in the Henle fibers may reduce photooxidative stress, thereby slowing the turnover of outer segments and the concomitant accumulation of drusen. The carotenoids in the outer segments themselves may be protective as antioxidants by quenching singlet oxygen and other reactive species. It is noteworthy that zeaxanthin, at least under some conditions, is a more efficient quencher of singlet oxygen *in vitro* than lutein (34), and that lutein appears to be converted to a stereoisomer of zeaxanthin, MZ, in the eye (10). As a result, antioxidant defenses may be improved. The other environmental risk factors for AMD referred to above are also probably related to oxidative stress. Smoking raises the level of prooxidants in the blood (35) and lowers the level of carotenoids (36), and A2E, exposed to blue light, initiates the formation of singlet oxygen and other reactive oxygen species. The phototoxicity of A2E is significantly reduced in the presence of lutein (37).

Thus, plausible mechanisms exist that could account for the reduced risk of advanced AMD observed in epidemiologic studies involving lutein and zeaxanthin. Further evidence for a protective role of macular pigment is provided by a number of studies. In one study, monkeys were fed a carotenoid-free diet (38). The monkeys, not surprisingly, lacked macular pigment. They also developed drusen at an earlier age than those fed a normal, carotenoid-containing diet, and "window defects," which reflect a loss of melanin in the RPE, a characteristic of early maculopathy. One of the advanced forms of AMD, geographic atrophy, is characterized by a sparing of the central foveal region, where macular pigment density peaks, until the disease is well advanced (39,40). The same central sparing is apparent in annular macular degeneration ("bull's eye maculopathy") (41).

Cataract

The human lens, like the retina, accumulates the carotenoids lutein and zeaxanthin exclusively (42). The total amount is only ~4 ng/lens, and the two carotenoids are present in roughly equal amounts. The carotenoids do not appear to be uniformly distributed throughout the lens but are more concentrated in the cortical and epithelial layers (43). It is possible that the concentration of lutein and zeaxanthin in these more metabolically active layers is sufficiently high for physiologic antioxidant activity to take place. Thus, the carotenoids may participate in the deactivation of reactive oxygen species, which might otherwise promote damage to lens proteins leading to cataract. The observation that cataracts occur most frequently in the nucleus of the lens is possibly related to the lower levels of antioxidants in this region compared with the cortex and epithelium. An association has emerged from epidemiologic studies between dietary intake of lutein and zeaxanthin and a reduced risk of cataract. Jacques *et al.* (44) found a strong, significant, inverse association between carotenoid serum levels and cataract, whereas dietary intake did not reach a level of significance. In a more recent study, intake was found to be significant in relation to nuclear cataract (45). In another study, the risk of cataract extraction was reduced by ~20% for men and women in the highest quintile of lutein and zeaxanthin intake compared with those in the lowest quintile (46,47). Importantly, careful analysis of the carotenoid intake shows that only lutein and zeaxanthin had a strong significant association, and other carotenoids did not.

In Vivo *Measurement of Macular Pigment*

The evidence that has accumulated for the protective function of lutein and zeaxanthin in the eye emphasizes the need for noninvasive methods of determining the concentration and distribution of these carotenoids in the relevant tissues. Although no such methods are currently available for the lens, several methods, both subjective and objective, were developed for the macula. One of these, based on reflectometry, is under development in our laboratory, and will be described in detail after a review of the existing methodology.

Heterochromatic Flicker Photometry (HFP). HFP is currently the most popular method of measuring macular pigment optical density (MPOD). It is a subjective method that is based on the reduced sensitivity of the central region of the retina to blue light resulting from prereceptor filtering by the macular pigment. With the instrument used in our studies, the subject views a 1.5° visual stimulus that alternates between a wavelength of 460 nm (blue, corresponding to the peak absorbance by the macular pigment) and 540 nm (green, beyond the absorption envelope of the macular pigment) (13). Subjects adjust the intensity of the blue component of the stimulus until the apparent luminosity matches that of the green component as judged by an absence, or minimization, of perceived flicker. The match is made while viewing the stimulus centrally, and again at 8° eccentricity from the center of the fovea where the MPOD is negligible. The increase of blue light intensity needed for the central match compared with the peripheral match indicates the amount of light absorbed by the macular pigment. The MPOD is calculated as the log ratio of the intensity settings for the central and peripheral match.

It is also possible to use HFP to obtain the distribution of MPOD in the retina. This is accomplished by the use of a centrally viewed stimulus consisting of a thin annulus. Thus, instead of providing an averaged MPOD over the area of the retina corresponding to the circular stimulus, the annular stimulus provides the MPOD at an eccentricity corresponding to the radius of the annulus. By employing annuli of different radii, the MPOD profile can be generated. Alternatively, a very small circular stimulus can be viewed at different eccentricities.

Motion Photometry. This method is similar to HFP. The stimulus consists of alternating blue and green bars that move across the visual field (48). The subject adjusts the relative intensities of the two colors until the perception of motion is minimized. This occurs when the luminances are matched. As with HFP, a reference measurement is made with the stimulus imaged at an extrafoveal location. Again, it is possible to determine the MPOD distribution by viewing the stimulus at different eccentricities.

Fluorescence of Lipofuscin. Autofluorescence of chromophores that are present beneath the retina can be used to determine macular pigment absorbance. Lipofuscin is a fluorescent material found in the RPE, and therefore posterior to the macular pigment. It can be excited both by a wavelength that is absorbed by the macular pigment, e.g., 470 nm, or by a wavelength that is transmitted, e.g., 550 nm (49). The resulting fluorescence is detected at a wavelength (710 nm was found to be useful) outside the absorbance range of the macular pigment. If the intensity of the emitted fluorescent light is measured for each exciting wavelength, the differential absorbance of the macular pigment at the two exciting wavelengths can be determined after correcting for the fluorescence efficiency of the lipofuscin chromophores at the two wavelengths. In this technique, the probe wavelength (470 or 550 nm) makes only a single pass through the retina in contrast to the reflectometric methods which will be described below.

Resonance Raman Spectroscopy. When lutein and zeaxanthin are excited by light in the wavelength range 450–550 nm, relatively strong resonance Raman signals are emitted that are dependent on the amount of carotenoid present. Originally, an argon laser (488 nm) was employed as the excitation source (50), but the more recent instrumentation makes use of a filtered mercury arc lamp. Similarly, the Raman spectrometer that was used as the detector on early instruments has been superseded by a CCD camera, thereby permitting the capture of density maps of the macular pigment in the retina (51).

Reflectometry. Like the two previous methods, reflectometry provides an objective means of determining MPOD (49), and has the distinction of being the earliest such method. Brindley and Willmer (52) first observed that the reflectance spectrum of the bleached, central retina differed from that of the peripheral retina.

Bleaching minimizes the absorption of light by photopigments in the rods and cones; thus, the differences in the spectra were ascribed to the macular pigment. Later investigators developed models of the retina that, in addition to macular pigment, included the absorbing properties of the lens, blood and melanin, and scattering in the ocular media (53,54). Imaging reflectometry was introduced by Kilbride *et al.* (55). With a modified retinal camera, they captured two images of the bleached retina, one at 462 nm and the other at 559 nm. The two images were aligned and digitally subtracted after first logarithmically transforming the pixel values. The resulting image represented an MPOD map. Using a similar technique, Chen *et al.* (56) found that the macular pigment distribution broadened with age. Bour *et al.* (57) used reflectometry to examine macular pigment in children, but with a film camera replacing the video camera. Wüstmeyer *et al.* (58) obtained images at 488 and 514 nm using a scanning laser ophthalmoscope, rather than a retinal camera, and determined that macular pigment levels were lower in subjects with AMD than in controls without the disease.

We further refined this imaging technique using a digital fundus camera that we modified for macular pigment measurements. The camera has the advantage of being "nonmydriatic," i.e., pupil dilation using topical drugs is not required. Fundus cameras typically provide either color images of the retina, or fluorescein angiograms. For the latter, the camera is provided with excitation and blocking filters so that retinal blood vessels stand out clearly in the image as a result of a fluorescein injection. In our modification, the excitation filter, which is between the camera's flash-lamp and the subject's eye, was replaced with a triple bandpass interference filter. We also removed the blocking filter, which is between the subject's eye and the CCD elements of the camera. The new interference filter restricts the spectral responses of the camera's three charge-coupled devices to narrow wavelength bands centered at 460, 530, and 610 nm, respectively. Image analysis software (ImagePro Plus) is used to extract the three corresponding grayscale bitmap images from a captured color image of a subject's retina. The three bitmaps quantify the distribution of light reflected from the retina at each of the three wavelengths, respectively. The light that forms the images is attenuated by absorbing pigments in the light path, i.e., macular pigment, rod and cone photopigments, melanin, hemoglobin, and lens pigments. By assuming that the last-mentioned three provide uniform attenuation of light across the central area of the retina, a distribution map of the macular pigment optical density, D_{MP}, at 460 nm is obtained as follows from a linear combination of log-transformed bitmaps, L_B, L_G, and L_R:

$$D_{MP} = -0.525L_B + 0.355L_G - 0.882L_R \qquad [1]$$

The numerical coefficients in Equation 1 represent different combinations of the known extinction coefficients of macular pigment and rod and cone photopigments at the three wavelengths. (Other linear combinations provide the optical density distributions of rod and cone photopigments.) Line-scans through the foveal region of the D_{MP} map reveal the optical density distribution of macular pigment along the line. An example is shown in Figure 7.3. The peak optical density in this case, relative to baseline, is ~0.3 absorbance units (AU).

Although further validation experiments are in progress, the method shows promise for the speed and simplicity with which a subject's MPOD can be determined. Using a macro allows the required mathematical operations to be performed with a single computer keystroke, and the MPOD distribution image is produced instantaneously. In previous reflectometry methods, separate images were obtained at two different wavelengths, and these images had to be carefully aligned before digital subtraction of one from the other. Alignment is unnecessary with the current method because the three monochromatic images are derived from a single color image and are therefore perfectly registered with each other. Furthermore, the other methods required pupil dilation and photopigment bleaching to eliminate photopigment contributions to the MPOD map, steps that are unnecessary with the current method. For clinical applications in which time is an important consideration, this technique is a dramatic improvement over other methods. It appears likely that it will also be competitive on the basis of instrumental cost.

Supplementation

The evidence for a protective role of macular carotenoids has prompted the commercial development of lutein supplements. Lutein has been added to many brand-name multivitamins and may soon emerge as a food supplement. This has taken place with little knowledge of the influence of dose, which generally ranges from

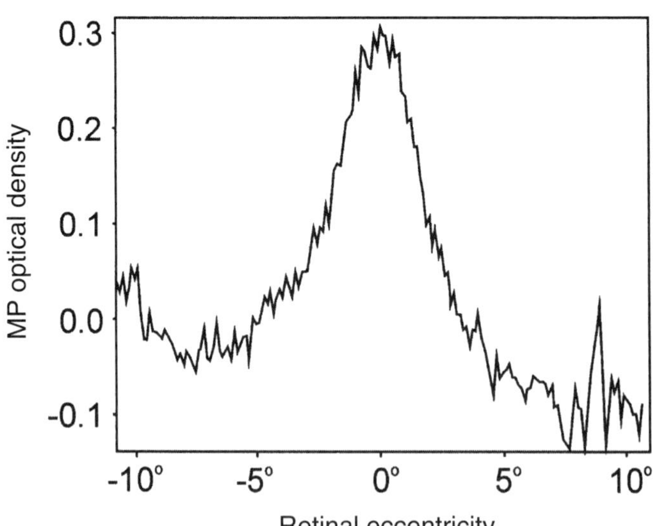

Fig. 7.3. Distribution of macular pigment optical density (relative) revealed by reflectometry. The distribution is determined along a line passing through the center of the fovea (0°).

~0.25 to ≥20 mg. To evaluate the effects of different doses, we conducted several supplementation studies (59). Most recently, we compared lutein doses of 5, 10, and 20 mg/d with a placebo. Healthy, nonsmoking subjects were assigned to one of these dosage groups on a double-blind basis, and took the supplement or placebo for 120 d. Blood samples were drawn before supplementation and at 2-wk intervals throughout the supplementation period. Lutein extracted from serum was quantified by HPLC using standard methods (59).

MPOD in the central 1.5° of the retina was measured before supplementation and 2 times/wk throughout the supplementation period using the HFP technique described earlier. An important admission criterion was the subject's ability to generate consistent and precise measurements. In a single session, the subject made 5 matches for each viewing condition (central and peripheral) from which the average MPOD was calculated. The subject was required to achieve an accompanying SEM not exceeding 0.020 AU. Measurements were repeated for both eyes.

Figure 7.4 shows a typical result for lutein supplementation on the serum levels of this carotenoid. Such plots are characterized by a rise in lutein concentration to a plateau and an exponential decline after the supplement is discontinued. Figure 7.5 shows an example of the approximately linear increase in MPOD in a subject's eye resulting from a 20 mg/d dose. However, not all subjects responded so robustly. Figure 7.6 is an example of a nonresponsive subject's macular pigment during a

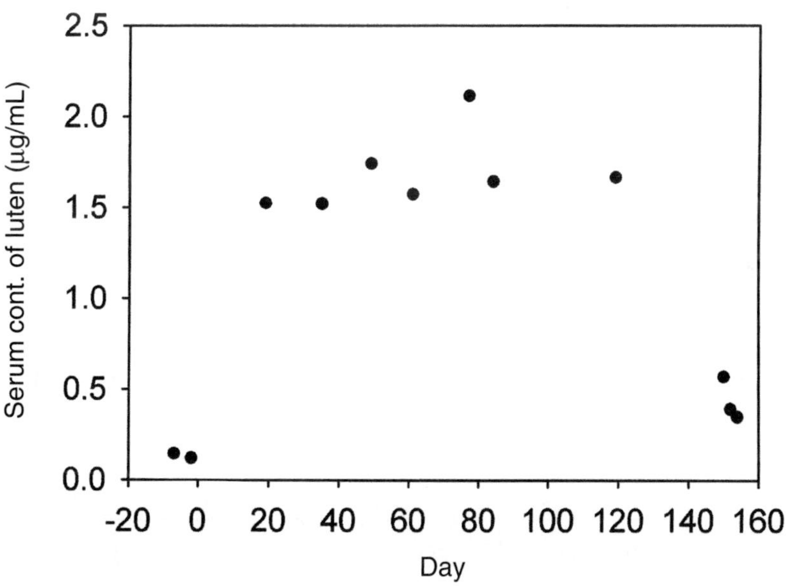

Fig. 7.4. The serum response to lutein supplementation at 20 mg/d for 120 d (d 1 to d 120) in a typical subject. The lutein concentration increased from ~0.1 µg/mL before supplementation to ~1.7 µg/mL during supplementation.

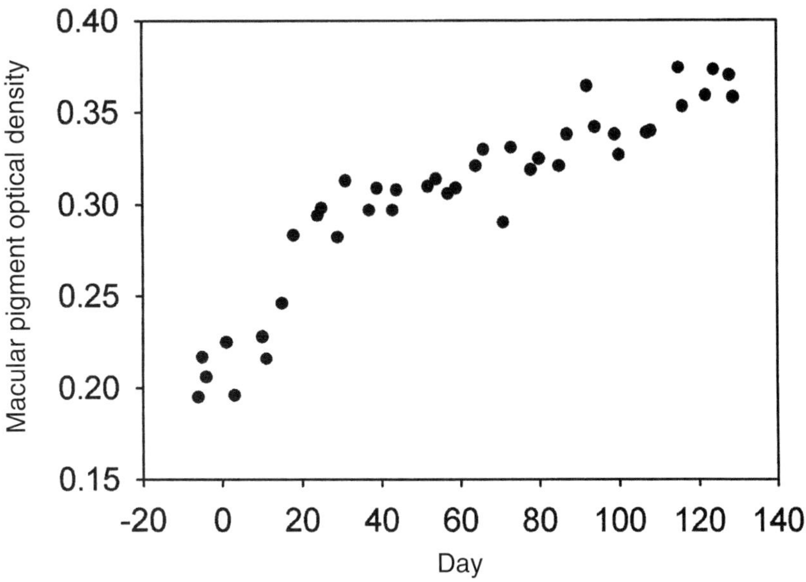

Fig. 7.5. The macular pigment optical density in a responsive subject shows a substantial increase as a result of supplementation with lutein at 20 mg/d for 120 d.

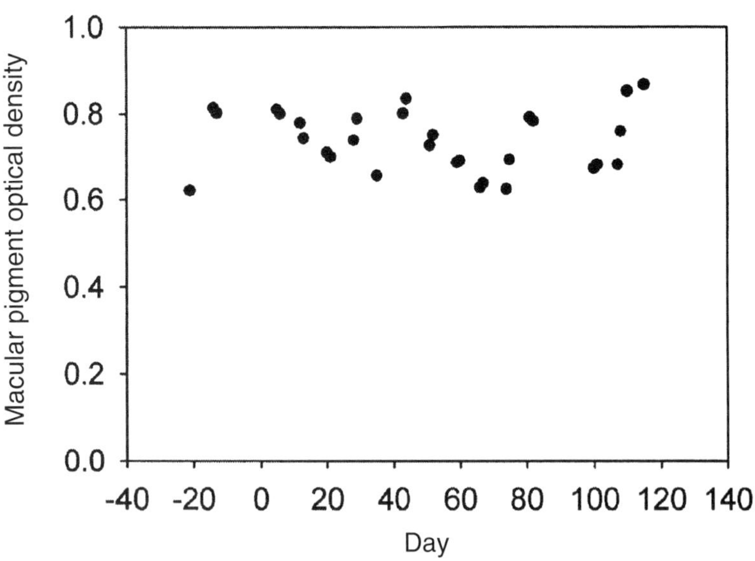

Fig. 7.6. The macular pigment optical density in an unresponsive subject shows little change as a result of supplementation with lutein at 10 mg/d for 120 d.

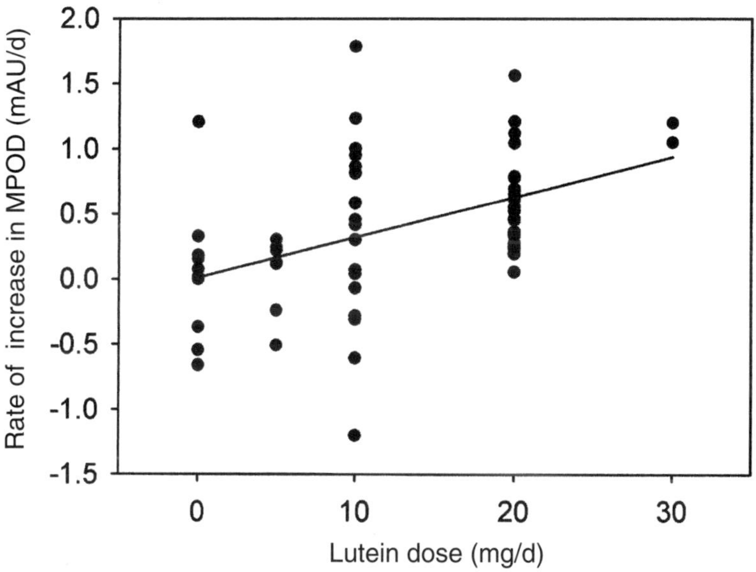

Fig. 7.7. The rate of increase in macular pigment optical density resulting from lutein supplementation varies substantially among subjects, but there is trend toward higher rates for those taking larger doses (20 and 30 mg/d).

supplementation trial. Despite such cases, the average increase in MPOD in subjects generally increases with dose. This trend is shown in Figure 7.7, which includes the results of 10, 7, 18, and 24 subjects who were taking the placebo, 5, 10, and 20 mg/d of lutein, respectively. Each subject's left and right eye data were first averaged, and these averages were used to calculate the mean for each dosage group. The average rates of increase in MPOD that we observed were –0.090 ± 0.432, –0.078 ± 0.596, 0.383 ± 0.720 and 0.490 ± 0.630 mAU/d for the 0, 5, 10, and 20 mg groups, respectively. Also included in Figure 7.7 are data from an earlier study in which 2 subjects were administered 30 mg lutein/d for 180 d. (Hence the choice of rate of increase in MPOD as the dependent variable rather than the absolute change.) The combined data indicate that the rate of increase in MPOD is roughly proportional to the dosage, albeit with a wide variability in individual responses. Doses ≥20 mg were necessary to produce positive rates of increase in MPOD in all subjects. Several subjects in the lower dosage groups had negative rates, and there was little to distinguish their average response from that of the placebo group.

Summary

Evidence from a variety of sources supports the hypothesis that macular pigment protects the central retina from AMD. In addition, the same carotenoids are associated with a reduced risk of cataract. In general, macular pigment density may be increased by dietary modification that increases the intake of lutein or, as described here, by consuming lutein supplements. Subjective methods, such as HFP, for assessing a person's macular pigment density are sometimes beyond the person's capabilities. For such individuals, an objective method is desirable, and our new reflectometry method shows great promise.

Acknowledgments

Supported by National Institutes of Health grants GM08205, GM 61347. The authors thank Cognis-US Corporation and Howard Foundation for lutein supplements.

References

1. Gullstrand, A. (1905) Die Farbe der Macula centralis retinae, *Graefe's Arch. Ophthalmol. 62:* 1–72.
2. Hartridge, H. (1951) Macular Pigment, *Nature 167:* 76–77.
3. Krinsky, N.I., Landrum, J.T., and Bone, R.A. (2003) Biologic Mechanisms of the Protective Role of Lutein and Zeaxanthin in the Eye, *Annu. Rev. Nutr. 23:* 171–201.
4. Bone, R.A., Landrum, J.T., and Tarsis, S.L. (1985) Preliminary Identification of the Human Macular Pigment, *Vision Res. 25:* 1531–1535.
5. Bone, R.A., Landrum, J.T., Hime, G.W., Cains, A., and Zamor, J. (1993) Stereochemistry of the Human Macular Carotenoids, *Investig. Ophthalmol. Vis. Sci. 34:* 2033–2040.
6. Khachik, F., Bernstein, P.S., and Garland, D.L. (1997) Identification of Lutein and Zeaxanthin Oxidation Products in Human and Monkey Retinas, *Investig. Ophthalmol. Vis. Sci. 38:* 1802–1811.
7. Bone, R.A., Landrum, J.T., Fernandez, L., and Tarsis, S.L (1988) Analysis of the Macular Pigment by HPLC: Retinal Distribution and Age Study, *Investig. Ophthalmol. Vis. Sci. 29:* 843–849.
8. Handelman, G.J., Dratz, E.A., Reay, C.C., and van Kuijk, F.J.G.M. (1988) Carotenoids in the Human Macula and Whole Retina, *Investig. Ophthalmol. Vis. Sci. 29:* 850–855.
9. Bernstein, P.S., Khachik, F., Carvalho, L., Muir, G., Zhao, D.-Y., and Katz, N. (2001) Identification and Quantitation of Carotenoids and Their Metabolites in the Tissues of the Human Eye, *Exp. Eye Res. 72:* 215–223.
10. Bone, R.A., Landrum, J.T., Friedes, L.M., Gomez, C.M., Kilburn, M.D., Menendez, E., Vidal, I., and Wang, W. (1997) Distribution of Lutein and Zeaxanthin Stereoisomers in the Human Retina, *Exp. Eye Res. 64:* 211–218.
11. Maoka, T., Arai, A., Shimizu, M., and Matsuno, T. (1984) The First Isolation of Enantiomeric and Meso-Zeaxanthin in Nature, *Comp. Biochem. Physiol. 83B:* 121–124.
12. Snodderly, D.M., Auron, J.D., Delori, F.C. (1984) The Macular Pigment. II. Spatial Distribution in Primate Retinas, *Investig. Ophthalmol. Vis. Sci. 25:* 674–685.
13. Bone, R.A., Landrum, J.T., and Cains, A. (1992) Optical Density Spectra of the Macular Pigment *In Vivo* and *In Vitro*, *Vision Res. 32:* 105–110.
14. Bone, R.A., and Landrum, J.T. (1983) Dichroism of Lutein: A Possible Basis for Haidinger's Brushes, *Appl. Opt. 22:* 775–776.

15. Rapp, L.M., Seema, S.S., and Choi, J.H. (2000) Lutein and Zeaxanthin Concentrations in Rod Outer Segment Membranes from Perifoveal and Peripheral Human Retina, *Investig. Ophthalmol. Vis. Sci. 41:* 1200–1209.
16. Sommerburg, O.G., Siems, W.G., Hurst, J.S., Lewis, J.W., Kliger, D.S., and van Kuijk, F.J.G.M. (1999) Lutein and Zeaxanthin Are Associated with Photoreceptors in the Human Retina, *Curr. Eye Res. 19:* 491–495.
17. Krinsky, N.I. (1989) Antioxidant Function of Carotenoids, *Free Radic. Biol. Med. 7:* 617–635.
18. Delori, F.C., Goger, D.G., Hammond, B.R., Snodderly, D.M., Burns, S.A. (2001) Macular Pigment Density Measured by Autofluorescence Spectrometry; Comparison with Reflectometry and Heterochromatic Flicker Photometry, *J. Opt. Soc. Am. A 18:* 1212–1230.
19. Hammond, B.R., and Caruso-Avery, M. (2000) Macular Pigment Optical Density in a Southwestern Sample, *Investig. Ophthalmol. Vis. Sci. 41:* 1492–1497.
20. Werner, J.S., Bieber, M.L., and Schefrin B.E. (2000) Senescence of Foveal and Parafoveal Cone Sensitivities and Their Relation to Macular Pigment Density, *J. Opt. Soc. Am. A 17:* 1918–1932.
21. Werner, J.S., Donnelly, S.K., and Kliegl, R. (1987) Aging and Human Macular Pigment Density, *Vision Res. 27:* 257–268.
22. Mallerio, J., Ahmadi-Iari, S., van Kuijk, F.J.G.M., Pauleikhoff, D., Bird, A.C., and Marshall, J. (2002) A Portable Instrument for Measuring Macular Pigment with Central Fixation, *Curr. Eye Res. 25:* 37–47.
23. Gellerman, W., Ermakov, I.V., Ermakov, M.R., McClane, R.W., Zhao, D.-Y., and Bernstein, P. (2001) *In Vivo* Resonant Raman Measurement of Macular Carotenoid Pigments in the Young and Aging Human Retina, *J. Opt. Soc. Am. A 19:* 1172–1186.
24. Vingerling, J.R., Klaver, C.C.W., Hofman, A., de Jong, P.T.V.M. (1995) Epidemiology of Age-Related Maculopathy, *Epidemiol. Rev. 17:* 347–360.
25. Age-Related Eye Disease Study Research Group (2000) Risk Factors Associated with Age-Related Macular Degeneration, *Ophthalmology 107:* 2224–2232.
26. Seddon, J.M., Ajani, U.A., Sperduto, R.D., Hiller, R., Blair, N., Burton, T.C., Farber, M.D., Gragoudas, E.S., Haller, J., Miller, D.T., Yannuzzi, L.A., and Willett, W.C. (1994) Dietary Carotenoids, Vitamins A, C, and E, and Advanced Age-Related Macular Degeneration, J. *Am. Med. Assoc. 272:* 1413–1420.
27. Eye Disease Case-Control Study Group (1993) Antioxidant Status and Neovascular Age-Related Macular Degeneration, *Arch Ophthalmol. 111:* 104–109.
28. Bone, R.A., Landrum, J.T., Mayne, S.T., Gomez, C.M., Tibor, S.E., and Twaroska, E.E. (2001) Macular Pigment in Donor Eyes with and Without AMD: A Case-Control Study, *Investig. Ophthalmol. Vis. Sci. 42:* 235–240.
29. Beatty, S., Murray, I.J., Henson, D.B., Carden, D., Koh, H.-H., and Boulton, M.E. (2001) Macular Pigment and Risk for Age-Related Macular Degeneration in Subjects from a Northern European Population, *Investig. Ophthalmol. Vis. Sci. 42:* 439–446.
30. Sun, H., and Nathans, J. (2001) The Challenge of Macular Degeneration, *Sci. Am. 285:* 68–75.
31. Goldman, A.I., Teirstein, P.S., and O'Brien, P.J. (1980) The Role of Ambient Lighting in Circadian Disc Shedding in the Rod Outer Segment of the Rat Retina, *Investig. Ophthalmol. Vis. Sci. 19:* 1257–1267.
32. Klein, R., Davis, M.D., Magli, Y.L., Segal, P., Klein, B.E.K, and Hubbard, L.D. (1991) The Wisconsin Age-Related Maculopathy Grading System, *Ophthalmology 98:* 1128–1134.

33. Suter, M., Reme, C.E., Grimm, C., Wenzel, A., Jaattela, M., Esser, P., Kociok, N., Leist, M., and Richter, C. (2000) Age-Related Macular Degeneration: The Lipofuscin Component A2E Detaches Pro-Apoptotic Proteins from Mitochondria and Induces Apoptosis in Mammalian Retinal Epithelial Cells, *J. Biol. Chem. 275:* 39625–39630.
34. Cantrell, A., McGarvey, D.J., Truscott, T.G., Rancan, F., and Böhm, F. (2003) Singlet Oxygen Quenching by Dietary Carotenoids in a Model Membrane Environment, *Arch. Biochem. Biophys. 412:* 47–54.
35. Church, D.F., and Pryor, W.A. (1985) Free-Radical Chemistry of Cigarette Smoke and Its Toxicological Implications, *Environ. Health Perspect. 64:* 111–126.
36. Ascherio, A., Stampfer, M.J., Colditz, G.A., Rimm, E.B., Litin, L., and Willett, W.C. (1992) Correlations of Vitamin A and E Intakes with the Plasma Concentrations of Carotenoids and Tocopherols Among American Men and Women, *J. Nutr. 122:* 1792–1801.
37. Shaban, H., and Richter, C. (2002) A2E and Blue Light in the Retina: The Paradigm of Age-Related Macular Degeneration, *Biol. Chem. 383:* 537–545.
38. Malinow, M.R., Feeney-Burns. L., Peterson, L.H., Klein, M., and Neuringer, M. (1980) Diet-Related Macular Anomalies in Monkeys, *Investig. Ophthalmol. Vis. Sci. 19:* 857–863.
39. Schatz, H., and McDonald, H.R. (1989) Atrophic Macular Degeneration. Rate of Spread of Geographic Atrophy and Visual Loss, *Ophthalmology 96:* 1541–1551.
40. Sunness, J.S., Bressler, N.M., Tian, Y., Alexander, J., and Applegate, C.A. (1999) Measuring Geographic Atrophy in Advanced Age-Related Macular Degeneration, *Investig. Ophthalmol. Vis. Sci. 40:* 1761–1769.
41. Weiter, J.J., Delori, F., and Dorey, C.K. (1988) Central Sparing in Annular Macular Degeneration, *Am. J. Ophthalmol. 106:* 286–292.
42. Yeum, K.-J., Taylor, A., Tang, G., and Russell, R.M. (1995) Measurement of Carotenoids, Retinoids, and Tocopherols in Human Lenses, *Investig. Ophthalmol. Vis. Sci. 36:* 2756–2761.
43. Jacques, P.F., and Chylack, L.T., Jr. (1991) Epidemiologic Evidence of a Role for the Antioxidant Vitamins and Carotenoids in Cataract Prevention, *Am. J. Clin. Nutr. 53:* 352S–355S.
44. Jacques, P.F., Chylack, L.T., Jr., Hankinson, S.E., Khu, P.M., Rogers, G., Friend, J., Tung, W., Wolfe, J.K., Padhye, N., Willett, W.C., and Taylor, A. (2001) Long-Term Nutrient Intake and Early Age-Related Nuclear Lens Opacities, *Arch. Ophthalmol. 119:* 1009–1019.
45. Yeum, K.-J., Shang, F., Schalch, W., Russell, R.M., and Taylor, A. (1999) Fat Soluble Nutrient Concentrations in Different Layers of Human Cataractous Lens, *Curr. Eye Res. 19:* 502–505.
46. Chasan-Taber, L., Willett, W.C., Seddon, J.M., Stampfer, M.J., Rosner, B., Colditz, G.A., Speizer, F.E., and Hankinson, S.E. (1999) A Prospective Study of Carotenoid and Vitamin A Intakes and Risk of Cataract Extraction in US Women, *Am. J. Clin. Nutr. 70:* 509–516.
47. Brown, L., Rimm, E.B., Seddon, J.M., Giovannucci, E.L., Chasan-Taber, L., Spiegelman, D., Willett, W.C., and Hankinson, S.E. (1999) A Prospective Study of Carotenoids Intake and Risk of Cataract Extraction in US Men, *Am. J. Clin. Nutr. 70:* 517–524.
48. Moreland, J.D., Robson, A.G., Soto-Leon, N., and Kulikowski, J.J. (1998) Macular Pigment and the Colour-Specificity of Visual Evoked Potentials, *Vision Res. 38:* 3241–3245.

49. Delori, F.C., Goger, D.G., Hammond, B.R., Snodderly, D.M., and Burns, S.A. (2001) Macular Pigment Density Measured by Autofluorescence Spectrometry: Comparison with Reflectometry and Heterochromatic Flicker Photometry, *J. Opt. Soc. Am. 18:* 1212–1230.
50. Bernstein, P.S., Yoshida, M.D., Katz, N.B., McClane, R.W., and Gellerman, W. (1998) Raman Detection of Macular Carotenoid Pigments in the Intact Human Retina, *Investig. Ophthalmol. Vis. Sci. 39:* 2003–2011.
51. Gellerman, W., Ermakov, I.V., McClane, R.W., and Bernstein, P.S. (2002) Raman Imaging of Human Macular Pigments, *Opt. Lett. 27:* 833–835.
52. Brindley, G.S., and Willmer, E.N. (1952) The Reflexion of Light from the Macular and Peripheral Fundus Oculi in Man, *J. Physiol. 116:* 350–356.
53. Delori, F.C., and Pflibsen, K.P. (1989) Spectral Reflectance of the Human Ocular Fundus, *Appl. Opt. 28:* 1061–1077.
54. van de Kraats, J., Berendschot, T.T., and van Norren, D. (1996) The Pathways of Light Measured in Fundus Reflectometry, *Vision Res. 36:* 2229–2247.
55. Kilbride, P.E., Alexander, K.R., Fishman, M., and Fishman, G.A. (1989) Human Macular Pigment Assessed by Imaging Fundus Reflectometry, *Vision Res. 29:* 663–674.
56. Chen, S.-F., Chang, Y., and Wu, J.-C. (2001) The Spatial Distribution of Macular Pigment in Humans, *Curr. Eye Res. 23:* 422–434.
57. Bour, L.J., Koo, L., Delori, F.C., Apkarian, P., and Fulton, A.B. (2002) Fundus Photography for Measurement of Macular Pigment Density Distribution in Children, *Investig. Ophthalmol. Vis. Sci. 43:* 1450–1455.
58. Wustemeyer, H., Jahn, C., Nester, A., Barth, T., and Wolf, S. (2002) A New Instrument for the Quantification of Macular Pigment Density: First Results in Patients with AMD and Healthy Subjects, *Graefe's Arch. Clin. Exp. Ophthalmol. 240:* 666–671.
59. Bone, R.A., Landrum, J.T., Guerra, L.H., and Ruiz, C.A. (2003) Lutein and Zeaxanthin Dietary Supplements Raise Macular Pigment Density and Serum Concentrations of These Carotenoids in Humans, *J. Nutr. 133:* 992–998.

Chapter 8

β-Carotene Cleavage Products Impair Cellular and Mitochondrial Functions and May Lead to Genotoxicity

Werner Siems[a], Ingrid Wiswedel[b], Avdulla Alija[c], Nikolaus Bresgen[c], Peter Eckl[c], Claus-Dieter Langhans[d], and Olaf Sommerburg[e]

[a]Loges School of Physical Medicine and Rehabilitation, Bad Harzburg, Germany; [b]Department of Pathological Biochemistry, Institute of Clinical Chemistry and Pathological Biochemistry, Otto-von-Guericke University, Magdeburg, Germany; [c]Institute of Genetics and General Biology, University of Salzburg, Salzburg, Austria; [d]Department of Pediatrics, University of Heidelberg, Heidelberg, Germany; and [e]Department of Pediatrics, University of Ulm, Ulm, Germany

Introduction

Carotenoids are important micronutrients. Of all the known carotenoids, ~50 display provitamin A activity (1,2). Carotenoids are also precursors of retinoids. It was suggested that the antioxidant potency of β-carotene (BC) is transformed by scavenging oxygen radicals, thus protecting against cancer (3,4), cardiovascular, and other diseases (5–10). A number of publications also demonstrated antigenotoxic effects of β-carotene and other carotenoids (11–16). Therefore, intake of BC is recommended, especially in the form of supplements, to prevent and to treat diseases associated with oxidative stress (6,17), such as cancer, UV-mediated skin diseases, neurodegenerative diseases, and cystic fibrosis. The majority of epidemiologic studies consistently showed that increased consumption of foods rich in BC is associated with a reduced risk of lung and some other types of cancer (18). A similar relation was found between levels of BC in plasma and the risk of cancer (18,19).

In contrast, the Alpha-Tocopherol, Beta-Carotene Cancer Prevention (ATBC) study and the Beta-Carotene and Retinol Efficacy Trial (CARET) indicated that supplementation with BC and/or vitamin A in individuals having a high risk of lung cancer increased incidences of lung cancer (20,21). The prooxidant activity of BC and procarcinogenic action in the case of preexisting premalignant lesions are possible reasons for these unexpected effects (20,22–30). It was also reported that low concentrations of retinal (vitamin A aldehyde) and of retinol (vitamin A) cause cellular DNA cleavage and the induction of 8-oxo-7,8-dihydro-2-deoxyguanosine formation in HL-60 and HP-100 cells (31).

Furthermore, it was demonstrated that BC decreases the level of retinoic acid in the lungs and this reduces the inhibitory effect of retinoids on activator protein-1

(AP-1), giving rise to enhanced lung cell proliferation and potentially tumor formation (32). The same authors suggested that BC metabolites are responsible for the carcinogenic response in the lungs of cigarette smokers. These data are supported by the finding that BC supplementation in smokers who also consume alcohol, promotes pulmonary cancer and possibly also cardiovascular diseases (33). According to these investigations Van Popel *et al.* (34) proved the lack of a protective effect of BC on smoking-induced DNA damage in lymphocytes of heavy smokers.

The effects of BC were reported to be modified under certain conditions and concentrations. It was demonstrated that the antioxidant and prooxidant effects of BC are dependent on oxygen tension (35). It was also demonstrated that apo-8′-β-carotenal, a metabolite of BC, acts as a strong inducer of liver cytochromes P_{450} 1A1 and 1A2 (36). In earlier work, we provided evidence that carotenoid cleavage products (CP) inhibit Na^+-K^+-ATPase activity (37). Interestingly, the cleavage products were stronger inhibitors of Na^+-K^+-ATPase activity than the endogenous major lipid peroxidation product 4-hydroxynonenal (HNE) (37). Attempts to use carotenoids for cancer chemoprevention and treatment continue (38,39). However, supplementation with BC even at a high dosage seems to be essential in the treatment of several diseases such as cystic fibrosis (40). BC supplementation is also beneficial for infants fed formula preparations (41). Therefore, the causal mechanism of increased risk of cancer mediated by BC intake has to be elucidated to establish safe conditions for carotenoid supplementation.

Here we address the question whether BCCP are able to attack two important subcellular structures, the mitochondria and the nucleus. The aim of the studies was to investigate whether BCCP may impair mitochondrial function and lead to genotoxicity. In pathophysiologic situations, mitochondria are the main producers of superoxide radical anions and H_2O_2 within the cell (42). Impairment of mitochondrial function, including changes in calcium homeostasis (43) can cause an increase in the formation of "superoxide" (44); this promotes oxidative stress and results in the oxidation of lipids, proteins, and DNA-molecules. Oxidative DNA damage is a hallmark of carcinogenesis. Important parameters for the evaluation of mitochondrial functional status are the state 3 minus state 4 respiration (ADP-stimulated respiration), the mitochondrial glutathione and protein-sulfhydryl (SH) status, and the formation of aldehydes. These experiments were conducted with isolated rat liver mitochondria.

Because the metabolism of xenobiotic substances takes place mainly in the liver and is catalyzed by the enzymes of the phase I (oxidation reactions *via* the cytochrome P_{450} enzyme system) and phase II reactions (conjugation reactions with glutathione, glucuronides, and other water-soluble groups), primary hepatocytes can be considered to be an ideal and highly sensitive test system for the evaluation of the genotoxic potential of mutagens/promutagens. Cultures were exposed to different concentrations of CP and the following endpoints of cyto- and genotoxicity were determined: the mitotic index, the percentage of necrotic and apoptotic cells, micronucleated cells, chromosomal aberrations, and sister chromatid exchange (SCE).

Materials and Methods

Preparation of β-Carotene Cleavage Products (BCCP). The formation and analysis of the BCCP was performed as described by Handelman *et al.* (45). Mixtures of CP from BC, retinal, and β-ionone were produced by mixing samples of a methanolic stock solution of these compounds at a concentration of 0.04 mM, each with 1 mM hypochlorous acid at room temperature (37,46). CP collection was carried out after the "bleaching" reaction was finished, indicated by a stable color of the stock solution. The extraction was carried out twice with hexane. The hexane phases were combined and evaporated completely with nitrogen. The residue was redissolved in aliquots of hexane, adjusted to 1 mM or 0.5 mM stock solutions of CP, which were stored at –80°C. In addition, a blank solution was prepared under exactly the same conditions.

Preparation of Mitochondria (Liver, Brain, Lung). For the experiments freshly isolated liver mitochondria from 180- to 220-g male Wistar rats were used. The mitochondria were prepared in ice-cold medium containing 250 mM sucrose, 1 mM EGTA, and 1% (wt/vol) bovine serum albumin (BSA) at pH 7.4 (isolation medium) using a standard procedure (47). After the initial isolation, Percoll was used for purifying mitochondria from a fraction containing some endoplasmic reticulum, Golgi apparatus, and plasma membranes. The mitochondria were well coupled, as indicated by a respiratory control index that was >5 with glutamate + malate as substrates. Mitochondrial protein concentration was measured by the method of Lowry *et al.* (48) using BSA as a standard and adjusted to 1 mg/mL during incubation experiments. For comparing experiments, mitochondria of brain and lung of rats were also prepared, but the detailed procedures of their preparation will not be described here. Nevertheless, it should be mentioned that the brain and the lung mitochondria that were used for the experiments were also well coupled according to the respiratory control index. In addition, in the experiments with brain and lung mitochondria, the mitochondrial protein concentration was adjusted to 1 mg/mL.

Incubation of Mitochondria and Measurement of Mitochondrial Respiration. For measurements of mitochondrial respiration, aliquots of 5 BCCP solutions (retinal, β-ionone, retinalCP, β-iononeCP, and β-caroteneCP) were transferred into reaction vials and evaporated completely with argon. Then, 2 mL of the incubation medium (containing 10 mM sucrose, 120 mM KCl, 15 mM NaCl, 20 mM TRIS, 2 mM $MgCl_2$, 5 mM NaH_2PO_4, pH 7.4) was added to dissolve carotenoid cleavage products (CCP). The solution was transferred into a thermostat-controlled chamber equipped with a Clark-type electrode. Then the mitochondrial suspension (final concentration in the chamber was adjusted to 1 mg protein/mL) was added. After 3 min of preincubation, 5 mM glutamate and 5 mM malate were added and the state 4 respiration was measured. State 3 respiration was adjusted by adding 200 μM ADP. Uncoupled respiration (state 4u) was accomplished by the addition of 0.1 μM carbonyl cyanide *p*-(tri-fluoromethoxy)phenylhydrazone (FCCP) to the mitochondrial suspension in the presence of hydrogen-supplying substrates. For control experiments, the incubation medium without any carotenoid cleavage products was transferred into the incubation chamber followed by preincubation of mitochondria (1 mg/mL) for 3 min. The addition of substrates and the time schedule were identical to incubations in the presence of BCCP. The reaction temperature was 30°C. Furthermore, incubations of up to 20-min duration in the presence and absence of CP were carried out, and samples were withdrawn for measurements of mitochondrial reduced glutathione (GSH), oxidized glutathione (GSSG), protein-SH, and malondialdehyde (MDA). The incubation and measurement of the respiration of brain and lung mitochondrial suspensions were carried out in an analogous manner at pH 7.4 and 30°C.

Measurement of GSH, GSSG, of Protein-SH Content, MDA, and Mitochondrial Membrane Potential. GSH and GSSG were analyzed by a microtiter plate assay according to Baker *et al.* (49). The ratio of oxidized to total glutathione was calculated as 2 GSSG/(GSH + 2 GSSG). The ratio was calculated on the basis of SH, i.e., the nanomoles of GSSG/mg protein were doubled for calculating the total glutathione and then the ratio was calculated. The content of protein-SH was determined according to Ellman (50). Thiobarbituric acid-MDA conjugates were measured using an HPLC-based method (51,52). The dissipation of the mitochondrial membrane potential was followed at 30°C in a thermostat-controlled chamber equipped with a tetra phenyl phosphonium cation (TPP^+)-sensitive electrode (53).

Hepatocyte Isolation and Culture for Evaluation of Genotoxicity of BCCP. Rat hepatocytes were isolated by the two-step *in situ* collagenase perfusion technique as described by Michalopoulos *et al.* (54). For the preparation, female Fischer 344 rats weighing ~100 g were used. The rats were obtained from Harlan, Winkelman (Germany). They were kept in a temperature- and humidity-controlled room with a 12-h light:dark cycle. Water was freely available. The rats were allowed to acclimate for at least 2 wk before hepatocyte isolation. After the isolation procedure, the hepatocytes were plated at a density of 20,000 viable cells/cm^2 on collagen-coated 60-mm diameter plastic culture dishes. According to Eckl *et al.* (55), the hepatocytes were plated in 5 mL of serum-free minimum essential medium (MEM) containing 1.8 mM calcium, supplemented with pyruvate (1 mM), aspartate (0.2 mM), serine (0.2 mM), and penicillin (100 U)/streptomycin (100 μg/mL). The dishes were incubated at 37°C, 5% CO_2, and 95% relative humidity. After an incubation period of 3 h, the medium was exchanged for fresh MEM and the cultures were returned to the incubator.

Treatment of the Cultures for Genotoxicity Measurements. Approximately 20 h after the first exchange of the medium, the test substances were added to the cultures at concentrations of 0.1, 1, and 5 μM of CP, or 0.1, 1, 5, and 10 μM of apo-8′-β-carotenal and were incubated for 3 h. Then, the medium was aspirated and the plates were washed twice with fresh medium to completely remove the applied substances. Finally, fresh MEM containing 0.4 mM Ca^{2+}, supplemented as described above with the further addition of insulin (0.1 μM) and epidermal growth factor (40 ng/mL) was added. To determine SCE induction, 5-bromo-2′-deoxyuridine (BrdU; 10 μM), was added to three dishes of each concentration. Thereafter, cells were incubated for an additional 48 h. Apo-8′-β-carotenal was a gift from BASF AG, Ludwigshafen, Germany.

Fixation, Staining and Cytogenetic Analysis. Cytogenetic studies were performed as described by Eckl *et al.* (55). After 48 h, colcemide (0.4 μg/mL) was added to three dishes (to which BrdU was added) per concentration, and the cultures were incubated for a further 3 h. No colcemide was added to the cultures for the micronucleus assay. More detailed descriptions of methodological procedures can be found in Alija *et al.* (56).

For the micronucleus assay, cells were fixed in the dishes with methanol:glacial acetic acid (3:1, vol/vol) for 15 min, briefly rinsed with distilled water, and air dried. For the chromosome preparations, cells were harvested by replacing the medium with 2 mL of collagenase solution (0.5 mg collagenase/mL) and incubation for 10 min. The detached cells were collected by centrifugation (44 × *g*), treated with hypotonic KCl solution (0.02 M) for 10 minutes at 37°C, and fixed in freshly prepared methanol: glacial acetic acid (3:1, vol/vol) overnight. Preparations were made by dropping the cell suspension on precleaned frosted slides.

For micronucleus determination, the fixed cells were stained with the fluorescent dye DAPI (4,6′,6-diamidino-2-phenylindol) in Mc Ilvaine buffer (0.2 M Na_2HPO_4 buffer adjusted with 0.1 M citric acid to pH 7) for 30 min in the dark at room temperature. After being washed with Mc Ilvaine buffer, the dishes were rinsed with distilled water followed by air drying. For microscopic observation, fixed and stained cells were mounted in glycerol. To determine the mitotic index, rates of apoptotic and necrotic cells and the number of cells with micronuclei, 1000 cells/dish were analyzed under the fluorescence microscope (Leitz Aristoplan). Nuclear morphology allowed discrimination between apoptotic and necrotic cells, as described by Bresgen *et al.* (57).

The slides for studying chromosomal aberrations and SCE induction were stained with Hoechst 33258 (4.5 μg/mL) in Sörensen phosphate buffer, pH 6.8, for 20 min, rinsed with Sörensen phosphate buffer and exposed to black light (General Electric, F 40 BLB Blacklight) for 15 min on a warming plate kept at 50°C. After removal of the coverslips, the slides were washed briefly with distilled water and stained in 5% Giemsa solution. Well-spread first division metaphases (n = 20) were analyzed for chromosomal aberrations under a phase contrast microscope (Leitz Laborlux 11), and 20 well-spread second division metaphases were analyzed for SCE. The number of aberrations is given per diploid cell, i.e., 42 chromosomes; SCE are reported per chromosome.

Statistical Analysis. Data for the experiments with both mitochondria and primary hepatocyte cultures are presented as means ± standard error (SE). Significant differences were determined using Student's *t* test. A probability of $P < 0.05$ was accepted as significant.

Results

Impairment of Respiration and SH Status of Rat Liver Mitochondria by BCCP. All of the carotenoid CP that were investigated (retinal, β-ionone, retinalCP, β-iononeCP, and β-caroteneCP) strongly inhibited the ADP-stimulated respiration in a dose-dependent manner (Fig. 8.1) (58). Because state 4 respiration was scarcely

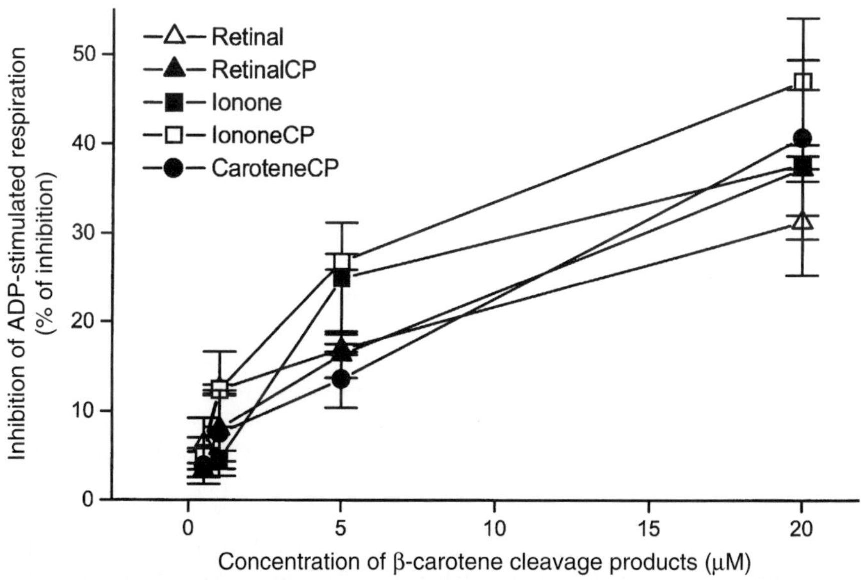

Fig. 8.1. Effects of β-carotene cleavage products (BCCP) toward ADP-stimulated mitochondrial respiration. Rat liver mitochondria were incubated at 30°C in a medium containing 10 mM sucrose, 120 mM KCl, 15 mM NaCl, 20 mM TRIS, 2 mM $MgCl_2$, 5 mM NaH_2PO_4, pH 7.4 (incubation medium). Five types of CP were used: retinal, β-ionone, mixtures of retinal (retinalCP) or β-ionone (iononeCP), or β-carotene cleavage products (caroteneCP). The concentrations of CP were 0.5, 1, 5, or 20 μM, respectively. Inhibition of respiration is presented as the decrease in the difference of respiration after and before ADP-addition (state 3 – state 4 respiration) in the percentage of complete inhibition where 100% inhibition corresponds to a decrease in respiration of 53.4 ± 3.5 nmol O_2/(mg·min). Values are means ± SE, n = 3 independent experiments.

affected by BCCP (data not shown), the data were presented as differences between state 3 and state 4 respirations. The presence of 20 μM CP led to a 30–50% decrease. Low concentrations of CP, such as 1 μM and even 0.5 μM, exerted clear inhibitions. The inhibition by retinal was 12.4 ± 0.5% at 1 μM and 6.3 ± 2.9% at a concentration of 0.5 μM. The ranges of inhibition for the different CP were between 5 and 12% at 1 μM and between 3 and 6% at 0.5 μM. Because blank solutions obtained after extraction of hypochlorous acid did not significantly inhibit respiration, we ruled out the possibility that the inhibition of state 3 respiration by CCP was due to contaminating impurities of hypochlorite.

In addition, dissipation of the mitochondrial membrane potential was measured to distinguish between inhibition of the adenine nucleotide translocator and inhibition of the F_0F_1-ATPase by the decomposition products of carotenoids. After blocking the electron transport within the respiratory chain by cyanide, the F_0F_1-ATPase splits ATP to ADP plus inorganic phosphate, paralleled by pumping protons into the extramitochondrial space. Therefore, inhibition of the F_0F_1-ATPase accelerates the depolarization [detailed data were published in (58)]. It was shown for different carotenoid CP, which inhibited the ADP-induced increase in respiration by ~40% at a concentration of 20 μM, that the CP had no effect on the kinetics of membrane depolarization. This observation supports the suggestion that the inhibition of respiration by decomposition products of carotenoids is caused mainly by impaired adenine nucleotide translocation.

The mitochondrial GSH content decreased rapidly in the presence of retinal and other β-carotene derivatives. Figure 8.2 demonstrates the changes in mitochondrial GSH during incubations over a 20-min period (58). The strongest decreases were observed in the presence of retinal and retinalCP. Both CP at a concentration of 20 μM decreased GSH from 6.21 ± 0.54 nmol/mg protein (control) to 1.71 ± 0.32 (retinal) and 2.51 ± 0.44 (retinalCP). Parallel to this, GSSG content increased. The GSSG increase in combination with a GSH decrease increased the ratio of GSSG/total glutathione. A 20-min incubation of mitochondria in the presence of 20 μM retinal led to a threefold increased ratio compared with controls. Moreover, most of the CP caused a decrease in the total glutathione pool (GSH + GSSG) (data not shown).

Loss of protein-SH occurred during the 20-min incubation with either of the CP mixtures at a concentration of 20 μM (Fig. 8.3). The mitochondrial protein-SH content decreased from 85.9 ± 4.6 nmol SH/mg protein (control) to 67.7 ± 1.9, and 78.2 ± 7.7 nmol SH/mg protein in the presence of 20 μM retinalCP and retinal, respectively. In comparing the loss of sulfhydryl groups from GSH and from proteins, <20% of the share of protein-SH of mitochondria was lost in total. Nevertheless, the absolute protein-SH loss was markedly higher than the absolute loss of GSH. In the presence of 20 μM retinalCP, the total SH loss was 21.9 ± 2.8 nmol/mg protein. Taking into account the final protein concentration of ~1 mg mitochondrial protein/mL of suspension suggests that the bulk of retinalCP may be bound to SH groups.

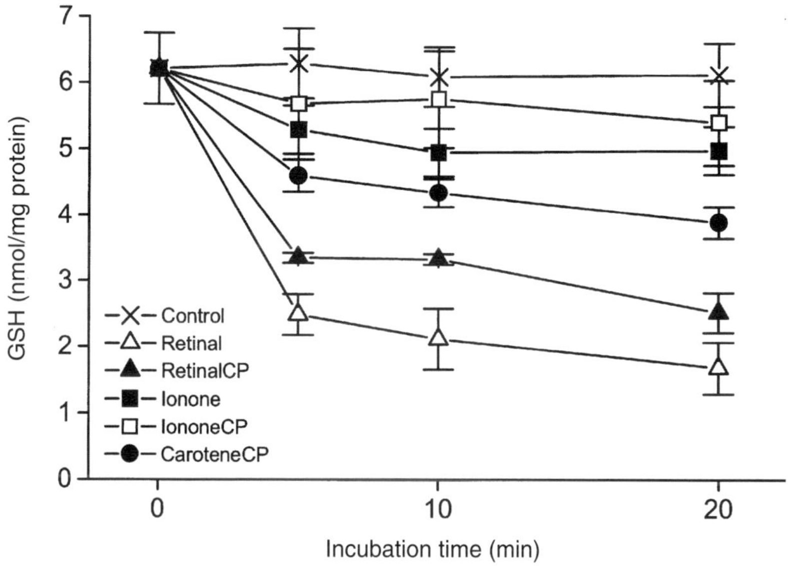

Fig. 8.2. Decrease in mitochondrial glutathione (GSH) by β-carotene cleavage products (BCCP). Rat liver mitochondria were incubated at 30°C for 20 min in the incubation medium in the presence of 20 μM BCCP. Values are means ± SE, n = 8 measurements (control and β-caroteneCP) or n = 4 measurements (retinal, retinalCP, β-ionone, and β-iononeCP).

The formation of MDA was used as a marker of lipid peroxidation after the exposure of mitochondria to CP. The concentrations of MDA were enhanced >10-fold after a 20-min incubation in the presence of 20 μM CP compared with control incubations. However, the amounts of MDA that were formed (up to 120 pmol/mg protein) were relatively small compared with MDA formation rates during iron/ascorbate-induced lipid peroxidation in isolated mitochondria (59).

Brain and Lung Mitochondria are More Sensitive Toward BCCP Than Liver Mitochondria. In additional experiments with rat brain mitochondria and rat lung mitochondria, the inhibition of ADP-stimulated (state 3 – state 4) respiration was stronger than in rat liver mitochondria. The presence of 20 μM CP led to a ~40–60% decrease in brain mitochondria, and in lung mitochondria, to as much as a 40–85% decrease in ADP-stimulated oxygen consumption. Low concentrations of CP, such as 1 μM and even 0.5 μM, clearly inhibited brain and lung mitochondria. The inhibition by retinal was ~10% in brain mitochondria and 15% in lung mitochondria at 1 μM and 6% in brain mitochondria and 10% in lung mitochondria at a concentration of 0.5 μM. At the highest concentration of retinal used (20 μM), the inhibition was 60% in brain mitochondria and 85% in lung mitochondria compared with only 33% in liver mitochondria. Blank solutions obtained after extrac-

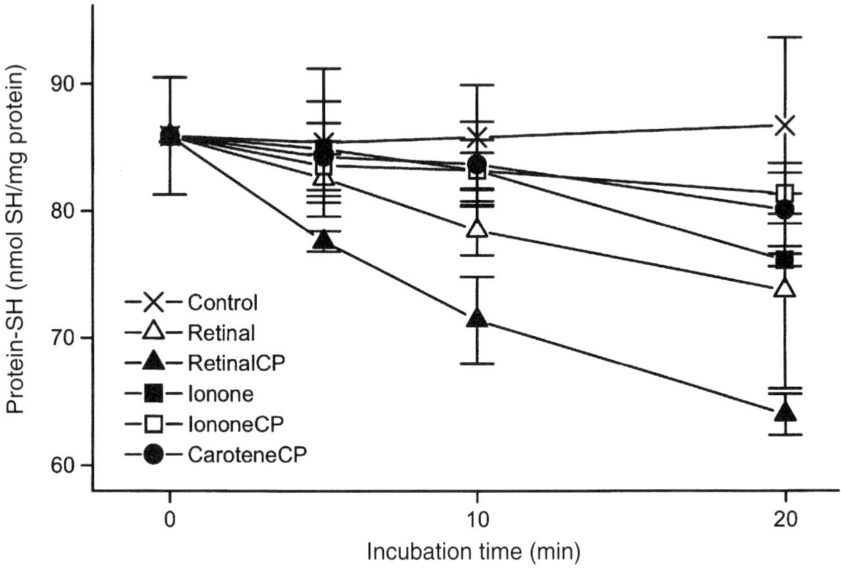

Fig. 8.3. Influence of β-carotene cleavage products (BCCP) on protein-sulfhydryl (SH) content. Rat liver mitochondria were incubated in presence of 20 μM CCP at 30°C in the incubation medium for 20 min. Values are given as nmol SH/mg protein. The initial value was 85.9 ± 4.6 nmol/mg protein (control incubation, n = 8). Values from 8 measurements (control and caroteneCP) or 4 measurements (retinal, retinalCP, β-ionone, and β-iononeCP) are presented.

tion of hypochlorous acid did not inhibit respiration in brain and lung mitochondria. In addition, in examining intramitochondrial GSH and the protein-SH, the brain mitochondria and the lung mitochondria were more sensitive toward BCCP than the liver mitochondria. The sequence of sensitivity toward BCCP from highest to lowest was lung mitochondria > brain mitochondria > liver mitochondria.

Evaluation of Genotoxicity Induction by BCCP in Primary Hepatocyte Cultures. CP and apo-8′-β carotenal had a prominent effect on the rates of micronucelated cells (Fig. 8.4A) (56). Micronucleus induction by BCCP (the mixture of BCCP formed by treatment of BC with hypochlorous acid leading mainly to short-chain CP was used at the one side; apo-8′-β-carotenal as long-chain cleavage product was used at the other side) was significant at concentrations of 100 nM ($P < 0.005$) and 1 μM ($P < 0.05$). At higher concentrations, the efficiency of CP to induce micronucleated cells declined. Similar data were obtained for the long-chain cleavage product apo-8′-β-carotenal, which significantly increased levels of micronucleated cells at concentrations of 100 nM, 1, and 5 μM ($P < 0.005$). In contrast to the mixture of CP, apo-8′-β-carotenal also gave rise to a significant ($P < 0.05$) formation of micronucleated cells at a concentration of 5 μM.

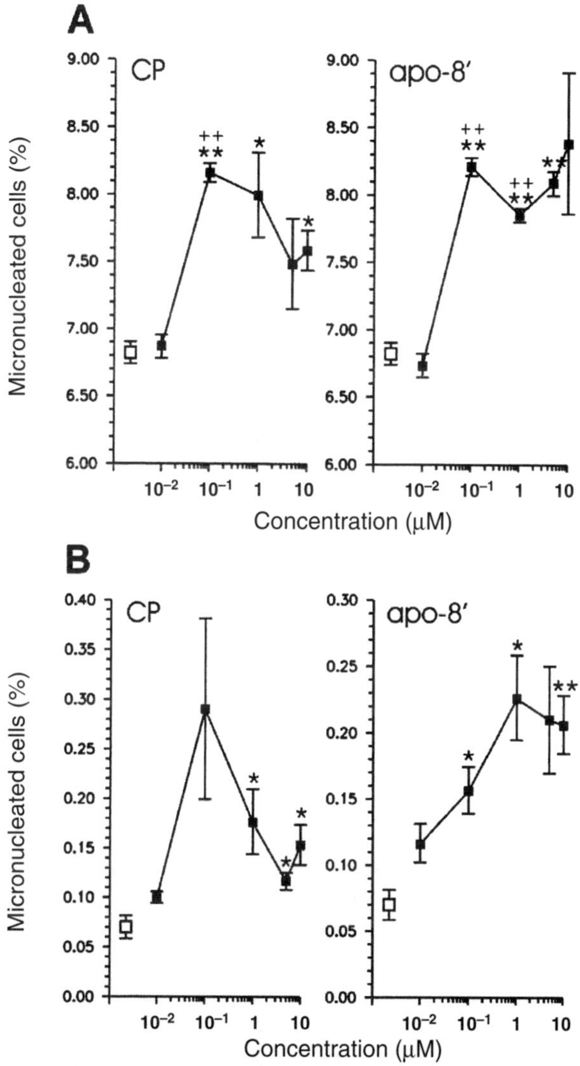

Fig. 8.4. Genotoxic effects β-carotene cleavage products in primary rat hepatocytes. (A) Frequencies of micronucleated cells in control cultures (□) and cultures treated with the mixture of mainly short-chain β-carotene cleavage products (CP) produced by means of hypochlorous acid treatment of β-carotene and with the long-chain product apo-8′-carotenal (■). Data are means ± SE, n = 3 independent experiments. $^*P < 0.05$; $^{**}P < 0.005$ compared with the control. $^{++}P < 0.005$ compared with the preceding concentration. (B) Frequencies of chromosomal aberrations in control cultures (□) and cultures treated with the mixture of mainly short-chain β-carotene cleavage products (CP) and with the long-chain product apo-8′-carotenal (■). Data are means ± SE, n = 3 independent experiments. $^*P < 0.05$; $^{**}P < 0.005$ compared with the control.

Figure 8.4B shows the results on the induction of chromosomal aberrations (56). Although both the mixture of mainly short-chain CP and apo-8′-β-carotenal proved to be potent inducers of chromosomal aberrations, the dose response was different as demonstrated in the figure. Chromosomal aberrations induced by the mixture of mainly short-chain CP had a maximum mean value at a concentration of 100 nM; however, due to large variations among the independent experiments, this was not significant. Significant levels of chromosomal aberrations by the mixture of CP were observed at concentrations of 1 and 5 μM ($P < 0.05$). Apo-8′-β-carotenal significantly increased the levels of chromosomal aberrations at 100 nM, 1 μM ($P < 0.05$), and 10 μM ($P < 0.005$).Chromosomal aberrations increased with apo-8′-β-carotenal concentration. At concentrations ≥1 μM, the level of aberrations remained more or less constant. Both the mixture of CP and apo-8′-β-carotenal gave rise to a concentration-dependent increase in the rate of SCE, which became significant ($P < 0.05$) at a concentration of 10 μM (data not shown), although the mean values of the levels of SCE were higher by 25% at 1 μM mixture of CP and by 40% at 1 μM apo-8′-β-carotenal compared with the control.

Discussion

Oxidative Degradation of β-carotene by Hypochlorous Acid and by Stimulated Neutrophils. Under physiologic conditions, carotenoids are attacked predominantly by oxidative enzymes, such as dioxygenases, epoxidases, hydroxylases, dehydrogenases, and aldehyde oxidases (60,61). Retinal is the primary cleavage product of β-carotene. Two molecules of retinal are formed by 15,15′-dioxygenase, which is the key enzyme in the metabolism of carotene to vitamin A (60–64). The successful cloning and sequencing of cDNA-encoding enzymes with β-carotene 15,15′-dioxygenase activity were reported recently (64,65).

In addition to enzymatic formation, BCCP are formed nonenzymatically. The attack on carotenoids by different free radical species essentially results in the formation of numerous breakdown products. This occurs with pathophysiologic consequences under conditions of oxidative stress, in which increased amounts of cleavage and oxidation products of carotenoids are accumulated. Liebler and his group identified specific products of carotenoid antioxidant reactions, using them as markers for antioxidant function of carotenoids (66–69). Other authors also reported carotenoid enzymatic and nonenzymatic cleavage products (70–72). From reports coming from different laboratories, it is evident that many of these products are carbonyls and epoxides (37,66,68,73). Handelman *et al.* (45) described the cleavage product formation during oxidation of β-carotene in the presence of hypochlorite. This condition mimics the *in vivo* formation of CP in inflammatory regions after activation of polymorphonuclear leucocytes. We adapted this method for the production of BCCP and extended it to the generation of short-chain cleavage products of retinal (retinalCP) and β-ionone (iononeCP). In our further experiments, hypochlorous acid was used as model for the oxidative degradation of BC and for the identification of new CP. Those results were published by Sommerburg *et al.* (2003) (46). In *in vitro* studies, primary cultured human neutrophils were also able to cleave BC after the cells were stimulated with phorbol myristate acid (46). Previously, it was shown that the rate of oxidative BC cleavage by stimulated human neutrophils can be mitigated by various components of the human antioxidative network (Siems, Crifo, Salerno *et al.*, unpublished results). That, of course, would lead to the hope that the potentially toxic effects of the cleavage products of carotenoids could be mitigated or almost completely prevented.

β-Carotene Cleavage Products Exert Prooxidant Effects Toward Enzymes and Mitochondria. A great number of the carotenoid breakdown products were identified as aldehydes. Retinal, the different apo-carotenals, and also a certain number of newly identified short-chain derivatives are of aldehydic nature. Aldehydes react rapidly with sulfhydryl groups and lysyl and histidine residues even at low cellular levels. Recently, while working with the group of van Kuijk, UTMB, Galveston, Texas, we demonstrated the inhibition of the Na^+-K^+-ATPase by a mixture of β-carotene cleavage products derived from hypochlorite treatment (37). We found that BCCP exerted a much higher *in vitro* toxicity than HNE, nonenal, and nonanal, which are also aldehydes and react with nucleophilic groups (73,74). This means that BCCP are very reactive and that they may be of particular high relevance under pathophysiologic conditions. Thus, the depletion of mitochondrial protein-SH and GSH after exposure of isolated rat liver mitochondria to BCCP may be caused by direct reactions of aldehydes with mitochondrial SH-groups. The data presented in this chapter further demonstrate that the decrease in SH-groups under the influence of CCP was paralleled by the inhibition of state 3 respiration (predominantly the ADP-induced increase in respiration) due to impairment of adenine nucleotide translocation. This could be expected because the adenine nucleotide translocator was shown to be sensitive to fatty acid CoA-derivatives and to lipid peroxidation products such as HNE (75,76). Inhibition of electron transfer by the respiratory chain due to the inhibition of adenine nucleotide transport leads to a rise in superoxide radical anion production by the respiratory chain and subsequently to the formation of H_2O_2 and hydroxyl radicals (77). Accordingly, an increase in oxidative stress is induced in mitochondria. This suggestion is in line with our observation that BCCP caused additional MDA formation.

β-Carotene Cleavage Products Also Exert Genotoxic Effects. Many toxicological investigations on β-carotene and carotenoids focused on antimutagenic properties. Antimutagenic effects of β-carotene and carotenoids were described mainly for bone marrow cells of mice (11,12,14,15) or were demonstrated in the *Salmonella typhimurium* test (15,16). The same accounts for the breakdown product apo-8′-β-carotenal, i.e., antimutagenic properties of apo-8′-β-carotenal with respect to BaP and AFB1 were found in the Ames test and by investigating micronucleated polychromatic erythrocytes in bone marrow cells of mice (15).

Additionally, Durnev *et al.* (14) reported antimutagenic properties of apo-8′-β-carotenal when applied at a dose of 50 mg/kg. Lower concentrations did not induce a significant antimutagenic effect against cyclophosphamide- and dioxidine-induced mutagenicity in bone marrow cells taken from mice.

In the studies that were carried out with primary rat hepatocyte cultures in Eckl's group at the University of Salzburg, prominent genotoxic effects of BCCP were observed. The genotoxicity test system of primary rat hepatocytes with its different parameters is highly sophisticated and well accepted (78–82). By using this system, it was shown for the first time that β-carotene breakdown products already possess a genotoxic potential at a concentration of 100 nM, which is in the physiologic range of BCCP concentrations. Apo-8′-β-carotenal was reported to be a strong inducer of liver cytochromes P_{450} 1A1 and 1A2 (36). The data obtained with apo-8′-β-carotenal in primary rat hepatocytes revealed a significant mutagenic potential with respect to micronucleus induction, induction of chromosomal aberrations, and SCE.

Notably, the induction of micronuclei and chromosomal aberrations by CP follows a bell-shaped dose-response curve, with the efficiency decreasing at concentrations > 1 μM (see mean value trend), which was most prominent in the case of the induction of chromosomal aberrations. This also holds true for apo-8′-β-carotenal–induced chromosomal aberrations. Treatment with a mixture of mainly short-chain CP or with the long-chain CP apo-8′-β-carotenal did not influence the mitotic index, and the rates of necrotic and apoptotic cells remained unchanged at any concentration tested; thus, the two substances did not have cytotoxic effects. Therefore, the bell-shaped dose-response curves could be explained by the onset of adaptive mechanisms, i.e., the induction of certain isoforms of cytochrome P_{450} induced at higher concentrations of BCCP.

Because micronuclei are the result of either chromosome breaks or disturbances of the mitotic spindle (83), and chromosomal aberrations result from clastogenic events with and without chromosomal rearrangements (84), these parameters are usually considered to be clear evidence for mutagenicity. On the other hand, SCE may not represent actual damage to chromosomes, but could be considered to be the result of damage repair. This could explain the different shapes of the dose-response curves obtained. As a consequence, it appears to be necessary to determine more than one endpoint in parallel (82).

The observations clearly indicate that BCCP are able to affect the biological material of cell nuclei in cell experiments, i.e., to induce genotoxic effects. This finding could be helpful in explaining adverse side effects such as carcinogenic effects reported in the Beta-Carotene and Retinol Efficacy Trial and the Alpha-Tocopherol, Beta-Carotene-Cancer-Prevention study (20–22). This finding also underlines the potential toxicity of BCCP toward subcellular organelles such as mitochondria and cell nucleus.

Pathophysiologic Effect of BCCP. Where in the human organism can a quantitative significant formation and accumulation of BCCP be expected? Of course, in tissues with a high activity of β-carotene dioxygenase and high levels of the substrate, an elevated formation of retinal can be expected. Furthermore, several pathophysiologically relevant conditions result in the rapid nonenzymatic oxidative cleavage of β-carotene, such as heavy oxidative stress under conditions of cigarette smoking, working with asbestos, severe inflammatory processes, and/or photoirradiation in the skin and eyes (85). The same should be true for hypochlorite-mediated carotenoid cleavage in the neighborhood of activated neutrophils. Hypochlorite released by phagocytic cells is present, at least temporarily, at high concentrations, ranging from 5 to 50 μM in the tissue (86). These levels are high enough to initiate the nonenzymatic cleavage of β-carotene. The total carotenoid concentration in adult subjects is 2.5–77.1 nmol/g liver (mean 21.0 nmol/g tissue), 0.2–12.7 nmol/g kidney (mean 3.1 nmol/g tissue), and 0.1–8.4 nmol/g lung tissue (mean 1.9 nmol/g tissue), with lower values for the tissues of children (87). Thus, carotenoid levels are markedly higher than 1 nmol/g in human tissues. Under the above-mentioned conditions, these levels are high enough to form levels of BCCP in the low micromolar range, and therefore in the range of concentrations that led (*in vitro*) to mitochondriotoxic effects and much higher than the concentrations that led to genotoxic effects.

Our data provide evidence that carotenoid cleavage products deplete mitochondrial sulfhydryl groups and impair oxidative phosphorylation in rat liver mitochondria at the level of the adenine nucleotide translocation. Oxidative stress resulting from the impairment of the mitochondrial energy metabolism in the presence of CCP and indicated by enhanced MDA formation may induce oxidative damage in DNA molecules in the mitochondria and nucleus. Mitochondrial DNA has a pronounced susceptibility to oxidative stress because of the absence of histones and low capacity for DNA repair. Our data also provide clear evidence that carotenoid cleavage products are able to exert genotoxic effects. Oxidative DNA damage, in general, increases the risk of cancer development. Thus, our data on mitochondrio- and genotoxic potential of BCCP may indicate a basic mechanism of the harmful effects of carotenoids in situations of increased oxidative stress.

Acknowledgments

The authors thank the BASF AG for providing apo-8′-β-carotenal for the genotoxicity experiments. Furthermore, the authors thank Prof. F.J.G.M. van Kuijk, Department of Ophthalmology and Visual Sciences at the University of Texas Medical Branch at Galveston, Texas for initiating, starting, and promoting our research on carotenoid degradation products. About 10 years ago, he suggested the toxic potential of carotenoid cleavage products toward enzymes, subcellular organs, and cells. At that moment, we started to collaborate on the influence of BCCP that were produced by bleaching on Na-K-ATPase and compared these effects with the effects of HNE on the same enzyme.

References

1. Olson, J.A., and Krinsky, N.I. (1995) The Colorful, Fascinating World of the Carotenoids: Important Physiologic Modulators, *FASEB J. 9:* 1547–1550.

2. Parker, R.S. (1996) Absorption, Metabolism, and Transport of Carotenoids, *FASEB J. 10:* 542–551.
3. Slaga, T.J. (1995) Inhibition of the Induction of Cancer by Antioxidants, *Adv. Exp. Med. Biol. 369:* 167–174.
4. Burton, G.W., and Ingold, K.U. (1984) β-Carotene: An Unusual Type of Lipid Antioxidant, *Science 224:* 569–573.
5. Stahl, W., and Sies, H. (2003) Antioxidant Activity of Carotenoids, *Mol. Aspects Med. 24:* 345–351.
6. Sies, H., Stahl, W., and Sundquist, A.R. (1992) Antioxidant Functions of Vitamins. Vitamins E and C, β- and Other Carotenoids, *Ann. N.Y. Acad. Sci. 669:* 7–20.
7. Olson, J.A., and Krinsky, N.I. (1995) The Colorful, Fascinating World of the Carotenoids: Important Physiologic Modulators, *FASEB J. 9:* 1547–1550.
8. Halliwell, B. (2000) The Antioxidant Paradox, *Lancet 355:* 1179–1180.
9. Agarwal, S., and Rao, A.V. (2000) Carotenoids and Chronic Diseases, *Drug Metab. Drug Interact. 17:* 189–210.
10. Lu, Q.Y., Hung, J.C., Heber, D., Go, V.L, Reuter, V.E., Cordon-Cardo, C., Scher, H.I., Marshall, J.R., and Zhang, Z.F. (2001) Inverse Associations Between Plasma Lycopene and Other Carotenoids and Prostate Cancer, *Cancer Epidemiol. Biomark. Prev. 10:* 749–756.
11. Raj, A.S., and Katz, M. (1985) Beta-Carotene as an Inhibitor of Benzo(a)pyrene and Mitomycin C Induced Chromosomal Breaks in the Bone Marrow of Mice, *Can. J. Genet. Cytol. 27:* 598–602.
12. Salvadori, D.M., Ribeiro, L.R., Oliveira, M.D., Pereira, C.A., and Becak, W. (1992) Beta-Carotene as a Modulator of Chromosomal Aberrations Induced in Mouse Bone Marrow Cells, *Environ. Mol. Mutagen. 20:* 206–210.
13. Aidoo, A., Lyn-Cook, L.E., Lensing, S., Bishop, M.E., and Wamer, W. (1995) *In Vivo* Antimutagenic Activity of Beta-Carotene in Rat Spleen Lymphocytes, *Carcinogenesis 16:* 2237–2241.
14. Durnev, A.D., Tjurina, L.S., Guseva, N.V., Oreshchenko, A.V., Volgareva, G.M., and Seredenin, S.B. (1998) The Influence of Two Carotenoid Food Dyes on Clastogenic Activities of Cyclophosphamide and Dioxidine in Mice, *Food Chem. Toxicol. 36:* 1–5.
15. Rauscher, R., Edenharder, R., and Platt, K.L. (1998) *In Vitro* Antimutagenic and *In Vivo* Anticlastogenic Effects of Carotenoids and Solvent Extracts from Fruits and Vegetables Rich in Carotenoids, *Mutat. Res. 413:* 129–142.
16. Arriaga-Alba, M., Rivera-Sachez, R., Parra-Cervantes, G., Barro-Moreno, F., Flores-Paz, R., and Garcia-Jimenez, E. (2000) Antimutagenesis of Beta-Carotene to Mutations Induced by Quinolone on Salmonella typhimurium, *Arch. Med. Res. 31:* 156–161.
17. Packer, L. (1993) Antioxidant Action of Carotenoids *In Vitro* and *In Vivo* and Protection Against Oxidation of Human Low-Density Lipoproteins, *Ann. N.Y. Acad. Sci. 691:* 48–60.
18. Ziegler, R.G., Mayne, S.T., and Swanson, C.A. (1996) Nutrition and Lung Cancer, *Cancer Causes Control 7:* 157–177.
19. Peto, R., Doll, R., Buckley, J.D., and Sporn, M.B. (1981) Can Dietary Beta-Carotene Materially Reduce Human Cancer Rates? *Nature 290:* 201–208.
20. Omenn, G.S., Goodman, G.E., Thornquist, M.D., Balmes, J., Cullen, M.R., Glass, A., Keogh, J.P., Meyskens, F.L., Valanis, B., Williams, J.H., Barnhart, S., and Hammar, S. (1996) Effects of a Combination of β-Carotene and Vitamin A on Lung Cancer and Cardiovascular Disease, *N. Engl. J. Med. 334:* 1150–1155.

21. The Alpha-Tocopherol, Beta Carotene Cancer Prevention Study Group (1994) The Effect of Vitamin E and Beta Carotene on the Incidence of Lung Cancer and Other Cancers in Male Smokers, *N. Engl. J. Med. 330:* 1029–1035.
22. Albanes, D., Heinonen, O.P., Taylor, P.R., Virtamo, J., Edwards, B.K., Rautalahti, M., Hartman, A.M., Palmgren, J., Freedman, L.S., Haapakoski, J., Barrett, M.J., Pietinen, P., Malila, N., Tala, E., Liippo, K., Salomaa, E.R., Tangrea, J.A., Teppo, L., Askin, F.B., Takinen, E., Erozan, Y., Greenwald, P., and Huttunen, J.K. (1996) Alpha-Tocopherol and Beta-Carotene Supplements and Lung Cancer Incidence in the Alpha-Tocopherol, Beta-Carotene Cancer Prevention Study: Effects of Base-Line Characteristics and Study Compliance, *J. Natl. Cancer Inst. 88:* 1560–1570.
23. Mayne, S.T., Handelman, G.J., and Beecher, G. (1996) Beta-Carotene and Lung Cancer Promotion in Heavy Smokers—A Plausible Relationship? *J. Natl. Cancer Inst. 88:* 1513–1515.
24. Mayne, S.T. (1996) Beta-Carotene, Carotenoids, and Disease Prevention in Humans, *FASEB J. 10:* 690–701.
25. Wang, X-D., and Russell, R.M. (1999) Procarcinogenic and Anticarcinogenic Effects of Beta-Carotene, *Nutr. Rev. 57:* 263–272.
26. Britton, G. (1995) Structure and Properties of Carotenoids in Relation to Function, *FASEB J. 9:* 1551–1558.
27. Young, A.J., and Lowe, G.M. (2001) Antioxidant and Prooxidant Properties of Carotenoids, *Arch. Biochem. Biophys. 385:* 20–27.
28. Palozza, P., Calviello, G., and Bartoli, G.M. (1995) Prooxidant Activity of β-Carotene Under 100% Oxygen Pressure in Rat Liver Microsomes, *Free Radic. Biol. Med. 19:* 887.
29. Paolini, M., Cantelli-Forti, G., Perocco, P., Peduli, G.F., Abdel-Rahman, S.Z., and Legator, M.S. (1999) Co-Carcinogenic Effect of Beta-Carotene, *Nature 398:* 760–761.
30. Paoloni, M., Antelli, A., Pozzetti, L., Spetlova, D., Perocco, P., Valgimigli, L., Peduli, G.F., and Cantelli-Forti, G. (2001) Induction of Cytochrome P_{450} Enzymes and Over-Generation of Oxygen Radicals in Beta-Carotene Supplemented Rats, *Carcinogenesis 22:* 1483–1495.
31. Murata, M., and Kawanishi, S. (2000) Oxidative DNA Damage by Vitamin A and Its Derivative Via Superoxide Generation, *J. Biol. Chem. 275:* 2003–2008.
32. Wang, X.D., and Russell, R.M. (1999) Procarcinogenic and Anticarcinogenic Effects of β-Carotene, *Nutr. Rev. 57:* 263.
33. Leo, M.A., and Lieber, C.S. (1999) Alcohol, Vitamin A, and Beta-Carotene: Adverse Interactions, Including Hepatotoxicity and Carcinogenicity, *Am. J. Clin. Nutr. 69:* 1071–1085.
34. Van Poppel, G., Kok, F.J., Duijzings, P., and de Vogel, N. (1992) No Influence of Beta-Carotene on Smoking-Induced DNA Damage as Reflected by Sister Chromatid Exchanges, *Int. J. Cancer 28:* 355–358.
35. Zhang, P., and Omaye, S.T. (2001) Antioxidant and Prooxidant Roles for Beta-Carotene, Alpha-Tocopherol and Ascorbic Acid in Human Lung Cells, *Toxicol. In Vitro 15:* 13–24.
36. Gradelet, S., Leclerc, J., Siess, M.H., and Astorg, P.O. (1996) β-Apo-8′-Carotenal, but Not Beta-Carotene, Is a Strong Inducer of Liver Cytochromes P_{450} 1A1 and 1A2 in Rat, *Xenobiotica 26:* 909–919.
37. Siems, W.G., Sommerburg, O., and van Kuijk, F.J.G.M. (2000) Carotenoid Oxidative Degradation Products Inhibit Na^+-K^+-ATPase, *Free Radic. Res. 33:* 427–435.

38. Chen, Y., Buck, J., and Derguini, F. (1999) Anhydroretinol Induces Oxidative Stress and Cell Death, *Cancer Res. 59:* 3985–3990.
39. Pappalardo, G., Maiani, G., Mobarhan, S., Guadalaxara, A., Azzini, E., Raguzzini, A., Salucci, M., Serafini, M., Trifero, M., Illomei, G., and Ferro-Luzzi, A. (1997) Plasma (Carotenoids, Retinol, Alpha-Tocopherol) and Tissue (Carotenoids) Levels After Supplementation with Beta-Carotene in Subjects with Precancerous and Cancerous Lesions of Sigmoid Colon, *Eur. J. Clin. Nutr. 51:* 661–666.
40. Homnick, D.N., Cox, J.H., DeLoof, M.J., and Ringer, T.V. (1993) Carotenoid Levels in Normal Children and in Children with Cystic Fibrosis, *J. Pediatr. 122:* 703–707.
41. Sommerburg, O., Meissner, K., Nelle, M., Lenhartz, H., and Leichsenring, M. (2000) Carotenoid Supply in Breast-Fed and Formula-Fed Neonates, *Eur. J. Pediatr. 159:* 86–90.
42. Kowaltowski, A.J., and Vercesi, A.E. (1999) Mitochondrial Damage Induced by Conditions of Oxidative Stress, *Free Radic. Biol. Med. 26:* 463–471.
43. Schild, L., Keilhoff, G., Augustin, W., Reiser, G., and Striggow, F. (2001) Distinct Ca^{2+} Thresholds Determine Cytochrome C Release or Permeability Transition Pore Opening in Brain Mitochondria, *FASEB J. 15:* 565–567.
44. Turrens, J.F. (1997) Superoxide Production by the Mitochondrial Respiratory Chain, *Biosci. Rep. 17:* 3–8.
45. Handelman, G.J., van Kuijk, F.J.G.M., Chatterjee, A., and Krinsky, N.I. (1991) Characterization of Products Formed During the Autoxidation of Beta-Carotene, *Free Radic. Biol. Med. 10:* 427–437.
46. Sommerburg, O., Langhans, D.C., Arnhold, J., Leichsenring, M., Salerno, C., Crifo, C., Hoffmann, G.F., Dabatin, K.M., and Siems, W.G. (2003) Beta-Carotene Cleavage Products After Oxidation Mediated by Hypochlorous Acid—A Model for Neutrophil-Derived Degradation, *Free Radic. Biol. Med. 35:* 1480–1490.
47. Steinbrecht, I., and Kunz, W. (1970) Anwendung der Cycling-Technik zur stichprobenartigen quantitativen Bestimmung des Reduktionsgrades des NAD- und NADP-Systems von Rattenlebermitochondrien bei kontinuierlich registrierenden Messungen, *Acta Biol. Med. Germ. 25:* 731–747.
48. Lowry, O.H., Rosebrough, N.J., Farr, A.L., and Randal, R.J. (1951) Protein Measurement with the Folin Phenol Reagent, *J. Biol. Chem. 193:* 265–275.
49. Baker, M.A., Cerniglia, G.J., and Zaman, A. (1990) Microtiter Plate Assay for the Measurement of Glutathione and Glutathione Disulfide in Large Numbers of Biological Samples, *Anal. Biochem. 190:* 360–365.
50. Ellman, G.L. (1959) Tissue Sulfhydryl Groups, *Arch. Biochem. Biophys. 82:* 70–77.
51. Buege, J.A., and Aust, S.D. (1978) Microsomal Lipid Peroxidation, *Methods Enzymol. 52:* 302–310.
52. Chirico, S. (1994) High-Performance Liquid Chromatography-Based Thiobarbituric Acid Test, *Methods Enzymol. 233:* 314–318.
53. Kamo, N., Muratsugu, M., Hongoh, R., and Kobatake, Y. (1979) Membrane Potential of Mitochondria Measured with an Electrode Sensitive to Tetra Phenyl Phosphonium and Relationship Between Proton Electrochemical Potential and Phosphorylation Potential in Steady State, *J. Membr. Biol. 49:* 105–121.
54. Michalopoulos, G., Cianciulli, H.D., Novotny, A.R., Kligerman, A.D., Strom, S.C., and Jirtle, R.L. (1982) Liver Regeneration Studies with Rat Hepatocytes in Primary Culture, *Cancer Res. 42:* 4673–4682.

55. Eckl, P.M., Whitcomb, W.R., Michalopoulos, G, and Jirtle, R.L. (1987) Effects of EGF and Calcium on Adult Parenchymal Hepatocyte Proliferation, *J. Cell. Physiol. 132:* 363–366.
56. Alija, A.J., Bresgen, N., Sommerburg, O., Siems, W.G., and Eckl, P.M. (2004) Cytotoxic and Genotoxic Effects of β-Carotene Breakdown Products on Primary Rat Hepatocytes, *Carcinogenesis 25:* 827–831.
57. Bresgen, N., Karlhuber, G., Krizbai, I., Bauer, H., Bauer, H.C., and Eckl, P.M. (2003) Oxidative Stress in Cultured Cerebral Endothelial Cells Induces Chromosomal Aberrations, Micronuclei and Apoptosis, *J. Neurosci. Res. 72:* 327–333.
58. Siems, W.G., Sommerburg, O., Schild, L., Augustin, W., Langhans, C.D., and Wiswedel, I. (2002) Beta-Carotene Cleavage Products Induce Oxidative Stress *In Vitro* by Impairing Mitochondrial Respiration, *FASEB J. 16:* 1289–1291.
59. Wiswedel, I., Hirsch, D., Nourooz-Zadeh, J., Flechsig, A., Lück-Lambrecht, A., and Augustin, W. (2002) Analysis of Monohydroxyeicosatetraenoic Acids and F_2-Isoprostanes as Markers of Lipid Peroxidation in Rat Brain Mitochondria, *Free Radic. Res. 36:* 1–11.
60. Olson, J.A. (1993) Molecular Actions of Carotenoids, *Ann. NY Acad. Sc. 691:* 156–166.
61. Olson, J.A., and Hayaishi, O. (1965) The Enzymatic Cleavage of Beta-Carotene into Vitamin A by Soluble Enzymes of Rat Liver and Intestine, *Proc. Natl. Acad. Sci. USA 54:* 1364–1370.
62. Goodman, D.S., Huang, H.S., and Shiratori, T. (1965) Mechanism of the Biosynthesis of Vitamin A from Beta-Carotene, *J. Biol. Chem. 241:* 1929–1932.
63. Von Lintig, J., and Wyss, A. (2001) Molecular Analysis of Vitamin A Formation: Cloning and Characterization of β-Carotene 15,15′-Dioxygenases, *Arch. Biochem. Biophys. 385:* 47–52.
64. Von Lintig, J., and Vogt, K. (2000) Filling the Gap in Vitamin A Research. Molecular Identification of an Enzyme Cleaving Beta-Carotene to Retinal, *J. Biol. Chem. 275:* 11915–11920.
65. Wyss, A., Wirtz, G., Woggon, W., Brugger, R., Wyss, M., Friedlein, A., Bachmann, H., and Hunziker, W. (2000) Cloning and Expression of Beta,Beta-Carotene 15,15′-Dioxygenase, *Biochem. Biophys. Res. Commun. 271:* 334–336.
66. Liebler, D.C. (1993) Antioxidant Reactions of Carotenoids, *Ann. NY Acad. Sci. 691:* 20–31.
67. Kennedy, T.A., and Liebler, D.C. (1992) Peroxyl Radical Scavenging by β-Carotene in Lipid Bilayers, *J. Biol. Chem. 267:* 4658–4663.
68. Stratton, S.P., Schaefer, W.H., and Liebler, D.C. (1993) Isolation and Identification of Singlet Oxygen Oxidation Products of β-Carotene, *Chem. Res. Toxicol. 6:* 542–547.
69. Liebler, D.C., and McClure, T.D. (1996) Antioxidant Reactions of β-Carotene: Identification of Carotenoid-Radical Adducts, *Chem. Res. Toxicol. 9:* 8–11.
70. Mader, I. (1964) Beta-Carotene: Thermal Degradation, *Science 144:* 533–534.
71. Mordi, R.C., Walton, J.C., Burton, G.W., Hughes, L., Ingold, K.U., Linday, D.A., and Moffatt, D.J. (1993) Oxidative Degradation of β-Carotene and β-Apo-8′-Carotenal, *Tetrahedron 49:* 911–928.
72. Krinsky, N.I. (1998) The Antioxidant and Biological Properties of the Carotenoids, *Ann. NY Acad. Sci. 854:* 443–447.
73. Siems, W.G., Hapner, S.J., and van Kuijk, F.J.G.M. (1996) 4-Hydroxynonenal Inhibits Na^+-K^+-ATPase, *Free Radic. Biol. Med. 20:* 215–223.

74. Uchida, K., Shiraishi, M., Naito, Y., Torii, Y., Nakamura, Y., and Osawa, T. (1999) Activation of Stress Signaling Pathways by the End Product of Lipid Peroxidation. 4-Hydroxy-2-nonenal Is a Potential Inducer of Intracellular Peroxide Production, *J. Biol. Chem. 274:* 2234–2242.
75. Chen, J.J., Bertrand, H., and Yu, B.P. (1995) Inhibition of Adenine Nucleotide Translocator by Lipid Peroxidation Products, *Free Radic. Biol. Med. 19:* 583–590.
76. Shrago, E., Woldegiogis, G., Ruoho, A.E., and DiRusso, C.C. (1995) Fatty Acyl CoA Esters as Regulators of Cell Metabolism, *Prostaglandins Leukot. Essent. Fatty Acids 52:* 163–166.
77. Cadenas, E., and Davies, K.J. (2000) Mitochondrial Free Radical Generation, Oxidative Stress, and Aging, *Free Radic. Biol. Med. 29:* 222–230.
78. Eckl, P.M., Strom, S.C., Michalopoulos, G., and Jirtle, R.L. (1987) Induction of Sister Chromatid Exchanges in Cultured Adult Rat Hepatocytes by Directly and Indirectly Acting Mutagens/Carcinogens, *Carcinogenesis 8:* 1077–1083.
79. Eckl, P.M., and Esterbauer, H. (1989) Genotoxic Effects of 4-Hydroxyalkenals, *Adv. Biosci. 76:* 141–157.
80. Reisenbichler, H., and Eckl, P.M. (1993) Genotoxic Effects of Selected Peroxisome Proliferators, *Mutat. Res. 286:* 135–144.
81. Eckl, P.M., Ortner, A., and Esterbauer, H (1993) Genotoxic Properties of 4-Hydroxyalkenals and Analogous Aldehydes, *Mutation Res. 290:* 183–192.
82. Eckl, P.M. (1995) Aquatic Genotoxicity Testing with Rat Hepatocytes in Primary Culture. II. Induction of Micronuclei and Chromosomal Aberrations, *Sci. Total Environ. 159:* 81–89.
83. Heddle, J.A., (1977) A Rapid *In Vivo* Test for Chromosomal Damage, *Mutat. Res. 18:* 63–69.
84. Brusick, D. (1987) *Principles of Genetics Toxicology*, p. 36, Plenum, New York.
85. Ribaya-Mercado, J.D., Garmyn, M., Gilchrest, B.A., and Russell, R.M. (1995) Skin Lycopene Is Destroyed Preferentially over β-Carotene During Ultraviolet Irradiation in Humans, *J. Nutr. 125:* 1854–1859.
86. Davies, J.M., Horwitz, D.A., and Davies, K.J. (1993) Potential Roles of Hypochlorous Acid and *N*-Chloroamines in Collagen Breakdown by Phagocytic Cells in Synovitis, *Free Radic. Biol. Med. 15:* 637–643.
87. Schmitz, H.H., Poor, C.L., Wellman, R.B., and Erdman, J.W., Jr. (1991) Concentrations of Selected Carotenoids and Vitamin A in Human Liver, Kidney and Lung Tissue, *J. Nutr. 121:* 1613–1621.

Chapter 9

Formation of β-Carotene Cleavage Products in View of the Particular Conditions in Inflamed Lung Tissue

Olaf Sommerburg[a], Claus-Dieter Langhans[b], Costantino Salerno[c], Carlo Crifo[d], and Werner Siems[e]

[a]Department of Pediatrics, University of Ulm, Ulm, Germany; [b]Department of Pediatrics, University of Heidelberg, Heidelberg, Germany; Departments of [c]Clinical Chemistry and [d]Biochemical Sciences, University of Rome, La Sapienza, Italy; and [e]Loges School for Physiotherapy, Bad Harzburg, Germany

Introduction

β-Carotene (BC) has been used as a supplement in food and medicine for many years. The substance is known as a powerful antioxidant and as provitamin A (1). In epidemiologic studies and cell culture experiments, BC had anticarcinogenic effects; consequently, a number of clinical trials were conducted to verify the positive results for certain types of cancer in patients. However, intake of BC failed in the clinical efficacy trials, ATBC and CARET, and even harmful effects occurred when BC and vitamin A were given in high dosages (2,3).

In recent years, it was shown that the negative effects seen in smokers might be related to the oxidative degradation of carotenoids. Rapid BC oxidation leads to the nonenzymatic formation of cleavage products (CP). The major part of these products is comprised of aldehydes or epoxides, which have a high reactivity toward biomolecules (4). Furthermore, some of these CP may demonstrate cell signal activity likely because of their structural similarity to retinoic acid and its precursors. Recent work showed that β-carotene cleavage products (BCCP) might exert carcinogenic action in cells (5,6). Alija *et al.* (5) showed that incubation of rat hepatocytes with BCCP in concentrations as low as 10^{-2} to 1 μM increased micronucleus induction, chromosomal aberrations, and sister chromatid exchanges. From animal studies, it was suggested that tobacco smoke and BCCP might have an effect on cytochrome P_{450}, resulting in retinoic acid catabolism and leading to interference with retinoid signaling (6,7) and alteration of the activity of the transcription factor activator protein-1 (AP-1). This might increase cell proliferation and possibly lead to the development of cancer.

However, it is true that a great number of subjects taking part in the numerous studies testing the efficacy of BC supplementation did not develop cancer. Taking all trials together, the number of subjects who had a negative outcome from BC supplementation was actually small and limited to the patient group of smokers. It seems to be very likely, then, that the special conditions in these patients might have caused the problems.

The relatively high dose of supplemented BC was considered to be one risk factor for the development of cancer. BC is distributed *via* lipoproteins throughout the human body into the different organs and tissues. As shown after the trials, the amounts of BC given in the ATBC and CARET interventions resulted in carotenoid blood levels of 3.0 and 2.1 μg/mL, respectively, compared with the mean blood levels for the U.S. population of 0.05–0.5 μg/mL (8). As known from previous studies, β-carotene accumulates in lung tissue in amounts comparable to that in other organs (9). This leads to the assumption that the BC concentration was extremely high in the lung tissue of the supplemented patients. However, we contend that there are further risk factors leading to an increased rate of BC degradation in lung tissue followed by the formation of increased amounts of BCCP.

Special Conditions in Smokers and Patients with Obstructive Lung Diseases

In the human respiratory system, oxygen enters the lung with a pressure of ~150 mmHg as present in normal air. In the alveoli, an extremely thin barrier between air and capillaries allows oxygen to move from the alveoli into the blood and allows carbon dioxide to move from the blood in the capillaries into the air in the alveoli. Under those conditions, BC present in lung tissue is exposed to a relatively high partial pressure of oxygen. Burton and Ingold (10) reported in 1984 that antioxidant behavior of BC is dependent in part on the partial pressure of oxygen. They showed that BC exhibits good radical-trapping antioxidant activity only at partial pressures of oxygen significantly <150 mmHg. At higher oxygen pressures, the carotenoid loses its antioxidant activity and has an autocatalytic, prooxidant effect, particularly at relatively high concentrations. Although these data were obtained only *in vitro*, the consequences of this effect for *in vivo* situations are still under discussion (11). The lung is the inner organ with the highest partial oxygen pressure. In the alveoli, an average partial oxygen pressure of ~100 mmHg is generally found. In the case of high-dose BC supplementation, this may lead to conditions close to the *in vitro* conditions in which the prooxidative behavior of BC was observed.

Another specialty of the lung is its extraordinary defense system. To understand the necessity of such a powerful defense system, one must consider that a moderately active person breathes ~20,000 L of air every 24 h through the lungs. The immune system of the lungs has to protect the tissue from foreign microorganisms (bacteria, viruses, fungi, yeasts) that infiltrate the lung with every breath. Before coming into contact with the alveolar lining layer, most invading microorganisms are removed by mucociliary clearance. However, the organisms that reach the alveolar compartment *via* the airways first contact the pulmonary epithelium and resident phagocytes (most notably alveolar macrophages). Phagocytes as well as lung epithelial cells are able to release cytokines and chemokines to recruit further defense cells, e.g., neutrophils, from the circulation.

Under normal conditions, those effects might not be relevant. However, several factors may alter the physiologic balance with respect to recruitment and activity of inflammatory cells in lung tissue. One of these factors is smoking behavior. In a recent study, Amin *et al.* (12) investigated bronchial biopsies of asymptomatic smokers (smokers without respiratory problems) and never-smokers. In their study, smokers had an increased thickness of the laminin and tenascin layers in the epithelial layer as a sign of structural changes and disturbance of the epithelial integrity. Furthermore, the number of inflammatory cells was significantly increased in smokers compared with never-smokers. Neutrophils, detected by immunostaining with monoclonal antibodies to interleukin-8, were found in the biopsy specimen of smokers (~12 cells/mm^2; median of 29). In contrast, never-smokers had only ~2 cells/mm^2 (median of 16) (12). These results are in agreement with the work of Eidelman *et al.* (13) who described a fivefold increase in the number of neutrophils in the lung tissue of smokers compared with nonsmokers and others reporting significantly increased numbers of inflammatory cells in general in the lung tissue of smokers (14,15). However, for neutrophils, Seatta *et al.* (16) described a different pattern for heavy smokers and smokers with moderate cigarette consumption. Although the airway wall of heavy smokers was infiltrated with neutrophils, this was not the case in smokers with a moderate consumption of cigarettes. They showed an accumulation of neutrophils only in the lumen of the airways. This disagrees with the findings of Amin *et al.* (12). Nevertheless, for other immunocompetent cells infiltrating the airway wall, the results were similar.

However, smokers are not the only group that has increased numbers of inflammatory cells in the lung. In atopic and nonatopic asthmatics, eosinophils and mast cells, which are able to liberate oxidants as well, were also increased compared with controls (17). However, neutrophils were increased only in nonatopic asthmatics (17). Detailed overviews of the role of inflammatory cells in the pathogenesis of asthma were given in reviews recently published by Adreadis *et al.* (18) and Maddox and Schwartz (19). Both groups of authors reported that increased amounts of eosinophils and neutrophils have the potential to harm host tissue and contribute to inflammatory injury. Primary Sjögren's syndrome, which is connected to rheumatic disorders, was also reported as a disease with increased numbers of neutrophils in lung tissue (2).

Patients with the disorders mentioned above do not necessarily need BC supplementation. However, patients with cystic fibrosis (CF), an inherited multisystem disorder, have very low levels of lipophilic micronutrients in general and might benefit from BC supplementation in high doses (21,22). Most of these patients develop mucous, which clogs the lungs and leads to chronic infections and chronic inflammation of pulmonary tissue. Furthermore, 90% of CF patients have pancreatic exocrine insufficiency. As a result, these patients suffer a malabsorption of lipids and lipid-soluble micronutrients, such as carotenoids. In several studies, it was shown that BC supplementation significantly decreased parameters of oxidative stress and even improved lung function (23–25). Because these patients may benefit from additional BC intake, it is interesting to examine the conditions in a CF lung. In 2001, Hubeau *et al.* (26) published a comprehensive quantitative analysis of inflammatory cells of the airway mucosa of CF patients. Using immunohistochemistry, they found, for example, that elastase-positive neutrophils at the level of the segmental bronchi in CF patients were 7 times higher compared with a specimen from non-CF subjects (CF: 199 ± 54 cells/mm^2 vs. non-CF: 27 ± 7 cells/mm^2) (26). This shows that the risk of oxidative stress due to neutrophils and other phagocytosing cells in CF patients is at a level similar to or even higher than that of smokers.

Neutrophils as a Source of Oxidative Stress in Tissue

Phagocytes, specialized cells that ingest bacteria, are an important part of the immunological defense system in multicellular organisms. In a number of tissues, neutrophils are the most important phagocytes. They are able to release mediators into the extracellular surrounding when they respond to soluble stimuli. However, the enzymatic and chemical reactions involved are dependent on environmental conditions. When they are activated, neutrophils generate a number of prooxidative agents, such as superoxide radicals, H_2O_2, or hypochlorous acid ($HOCl/^{-}OCl$) (27). Extracellular superoxide production might be stimulated by a variety of soluble and particulate stimuli (28–30). As a consequence, hydrogen peroxide is formed by dismutation of the superoxide radical under oxygen consumption (31,32). However, hydrogen peroxide is bactericidal only at high concentrations (31,32), and exogenously generated superoxide does not kill bacteria directly. Therefore, secondary oxidants are formed that amplify the destructive capacity of neutrophils. The hydroxyl radical formed from hydrogen peroxide by the Fenton reaction is one candidate for a secondary oxidant. However, the relevance of this reaction for the action of neutrophils is controversial (35–37). Most of the hydrogen peroxide is consumed by myeloperoxidase (MPO). MPO is a classical heme peroxidase that uses hydrogen peroxidase to oxidize a variety of compounds to yield substrate radicals (38,39). In this way, hypochlorous acid ($HOCl/^{-}OCl$) is formed, representing a very strong nonradical oxidant. $HOCl/^{-}OCl$ is able to damage a high number of different biologically important macromolecules (40–42). Many species of bacteria are killed readily by a myeloperoxidase/hydrogen peroxide/chloride system (43). On the other hand, MPO might also have alternative substrates. Therefore, the products formed in pathophysiologic situations *in vivo* may vary greatly.

As revealed *in vitro*, the reaction rate of $HOCl/^{-}OCl$ with BC is slow compared with other reducing agents, e.g., reduced glutathione (44). However, MPO is a highly cationic protein that quickly becomes attached to membranes after its release from activated neutrophils (45). Thus, an interaction of $HOCl/^{-}OCl$ with more lipophilic constituents such as carotenoids in tissue seems to be realistic under conditions in which neutrophils accumulate extensively. For example, HOCl/⁻OCl is able to destroy carotenoids in LDL (46). Calculations revealed that 2–4 × 10^6 leukocytes/mL produce 100–140 μM HOCl/⁻OCl over a 1-h period (47), leading to realistic BC:HOCl/⁻OCl ratios between 1:100 and 1:10 (48).

β–Carotene Degradation and Cleavage Product Formation by Hypochlorous Acid *In Vitro*

Different experiments were first conducted in methanolic solution to discern which degradation products were formed by an attack of HOCl/⁻OCl on BC. As in earlier studies (4,49,50), these experiments used highly concentrated BC solutions and a high molar ratio NaOCl:BC (100:1) to facilitate product formation for identification. Oxidation of BC was started by adding the entire amount of NaOCl to the methanolic solution while mixing on a vortex. The reaction time was 5 min and was stopped by immediate extraction of the lipophilic compounds. In our results, we could see that degradation of BC occurred very rapidly in the methanolic solution. Therefore, we found mostly short-chain CP in our first experiments. Longer-chain CP, such as apo-carotenals as described earlier (49), were found when the BC concentration in the reaction mixture was increased or when the reaction time was shortened before extracting the lipophilic compounds from the reaction mixture. After these experiments, comprehensive analysis work was done using GC-MS. Peak identification was done by comparison with standard substances. If a potential metabolite was not available as a standard, the substances were identified by comparison with the NIST/EPA/NIH mass spectral library. For identification of a substance, a fit of the MS spectra of the sample and the library data of more than 85%, and a purity of the corresponding total ion chromatogram peak of the sample higher than 70% were required. In this way, we were able to find β-cyclocitral, β-ionone, ionene, 5,6-epoxi-β-ionone, dihydroactinidiolide, and 4-oxo-ionone (Table 9.1) as short-chain CP (51). However, most of them had been found before in BC degradation experiments with different radical-generating systems (49,52–62). Nevertheless, it should be noted that our investigation described most of these products for the first time for BC oxidation by HOCl/⁻OCl. In addition to these CP, other substances that had at least a 75% fit with a suggested metabolite in the MS spectral library and a purity of the total ion chromatogram peak of at least 40% were considered as candidates of BCCP after HOCl/⁻OCl oxidation (51) if it was chemically plausible that they were oxidative metabolites of BC (Table 9.1) .

The short-chain BCCP formed in these reaction mixtures were highly reactive agents. For instance, the products were able to react easily with nucleophilic sites, e.g., with sulfhydryl groups. In previous studies, we demonstrated that these products are able to block enzymes such as Na^+-K^+-ATPase (4) and that these reactions take place when liver mitochondria are incubated with reaction mixtures containing adequate amounts of short-chain BCCP (63).

TABLE 9.1
Identification of Volatile Short-Chain Cleavage Products (CP) Formed After HOCl/⁻OCl–Mediated Degradation of β-Carotene (BC) in Molar Ratios of 1:100 and 1:10 (Carotenoid:NaOCl)[a]

Cleavage products	Chemical structure	Mass spectral data *m/z*
β-Cyclocitral	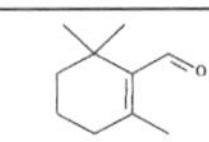	152 (M⁺), 137, 123, 109 (100%), 95, 81, 67
Ionene[b]	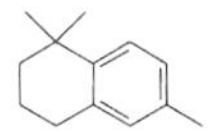	174 (M⁺), 159 (100%), 144, 131, 115, 105, 91, 77
5,6-Epoxi-β-ionone	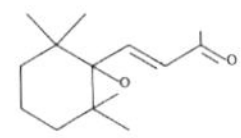	135, 123 (100%), 107, 95
β-Ionone	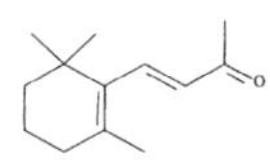	192 (M⁺), 177 (100%) 162, 149, 135, 121, 105, 91
Dihydroactinidiolide	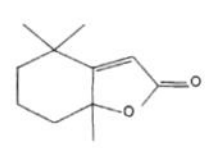	180 (M⁺), 165, 152, 137, 123, 111 (100%), 95, 81, 67
4-Oxo-β-ionone	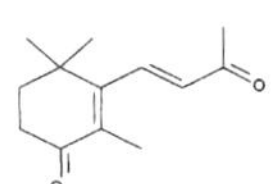	206 (M⁺), 191, 163 (100%), 150, 135, 121, 105, 93, 91
Further candidates		
Hexanedioic acid, mono (2-ethylhexyl)ester	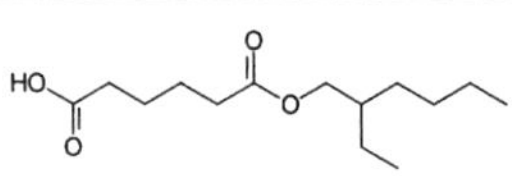	258 , 147, 129 (100%), 111, 83, 70
1,5,5-Trimethyl-6-aceto-methyl-cyclohexene	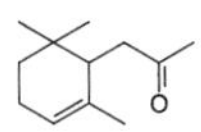	165, 123 (100%), 109, 95, 91, 81, 79, 67
3,7,7-Trimethyl-1-penta-1,3-dienyl-2-oxabicyclo [3.2.0]hept-3-ene	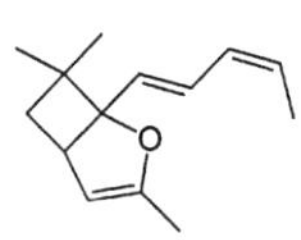	148 (100%), 133, 119, 105, 91, 77
2,6,6-trimethyl-1-Cyclo-hexene-1-acetaldehyde	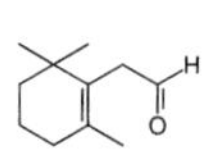	151, 133, 123, 107 (100%), 95, 91, 81, 79, 67
4,6,6-Trimethyl-2(3-methyl-buta-1,3-dienyl)-3-oxa-tricyclo[5.1.0.0(2,4)] octane	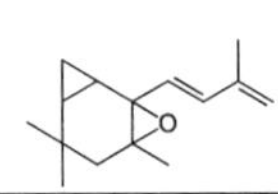	218, 175, 161, 147, 133, 119 (100%), 107, 105, 91, 79

[a]The given substances were found after oxidation of BC in methanol but also after oxidation of BC dispersed in a soybean matrix in aqueous solution. Further candidates for volatile short-chain CP that might be formed after HOCl/⁻OCl–mediated degradation of BC are given in the second section of the table. Conditions of identification analysis are given in the text.

[b]Detected as trimethyltetrahydronaphthalene because the position of substituents is unknown.

However, we had to recognize that the conditions of BC oxidation in methanol were not very physiologic. Therefore, we repeated these experiments in different buffer solutions, which also gave us an opportunity to investigate the oxidation of BC under different pH conditions. However, because BC has essentially no water solubility, for these experiments, BC dispersed in a defined soybean carrier oil (provided by Cognis Australia Pty Limited), termed the matrix, was used. The emulsion was made from a starting material of 30% BC (derived from the algal extract) in soybean oil. This was emulsified into a 30% water/70% glycerol aqueous phase using a glyceryl monooleate emulsifier. The fine emulsion provides the carotenoid (2% wt/vol) in a lipid globule size of ~1 μm or less so that interaction can occur at the cellular level. To act as a control or for dilution of the concentrated BC emulsion, a blank consisting of only the emulsified soybean oil was used. To investigate the influence of the lipid-containing matrix on the degradation rate of BC, two different sample types were provided with 1% of the lipid matrix (vol/vol) and 20% of the lipid matrix (vol/vol). BC concentration in all samples was set at 100 μM; the pH value was adjusted to 5.0 using acetate buffer, and NaOCl was added in a ratio of 100:1 (NaOCl:BC). To establish a possible time dependence, samples were taken after 10 and 30 min and extracted with hexane. As shown in Figure 9.1, BC degradation was slower in general in the samples with 20% of the

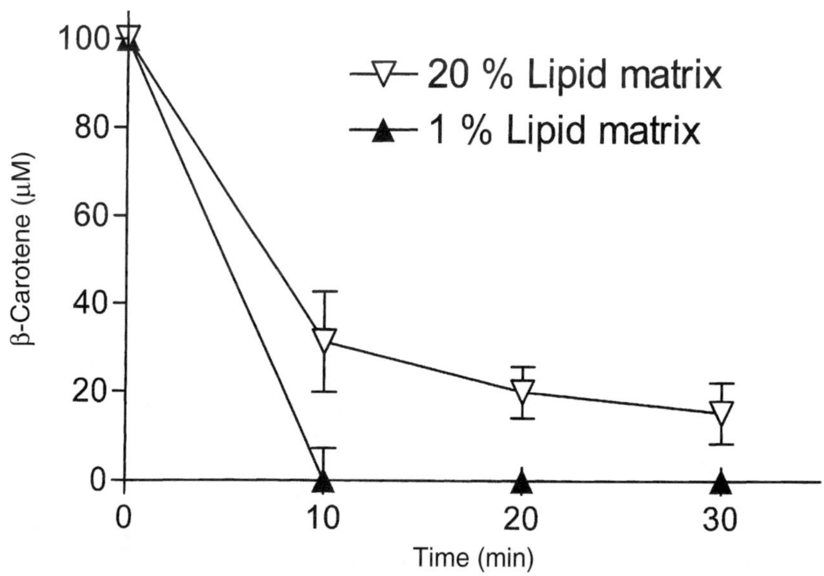

Fig. 9.1. Different pattern of oxidation of 100 μM β-carotene (BC) mediated by HOCl/⁻OCl in a ratio of 100:1 (NaOCl:BC) according to the amount of soybean matrix in the reaction mixture. The amount of the soybean matrix in buffer was 20% (vol/vol) or 1% (vol/vol), respectively. The pH value of the reaction mixture was set to 5.0. Results are given as means ± SEM, n = 5 independent measurements.

soybean matrix than in the samples with 1% of the soybean matrix. Although only 70% of the initial BC was degraded after 10 min in samples with 20% of the soybean matrix, the destruction was already complete in samples with 1% of the soybean matrix. After 30 min of reaction, more than 15% of the initial BC could still be found in samples with 20% of the soybean matrix (51).

In a further experimental approach, we examined the influence of different pH values and the influence of different NaOCl:BC ratios on the formation of long-chain BCCP. For these experiments, only samples with an amount of 20% (vol/vol) of the soybean carrier were used. To prepare these samples, 1 μmol of commercially available crystallized BC was dissolved in dichloromethane and step by step loaded into 2 mL of the blank soybean matrix while the sample was rotated and the organic solvent evaporated. BC concentration was set again to 100 μM in the sample. The pH values were varied to mimic physiologic conditions (pH 7.4) using phosphate buffer, and conditions found in phagosomes (pH 5.0) using acetate buffer. NaOCl was added 10:1 (NaOCl:BC) in the low-ratio samples and 100:1 (NaOCl:BC) in the high-ratio samples. After 10 min of reaction, BC was not degraded completely in any of the samples. In samples with a NaOCl:BC ratio of 10:1, only ~35–45% of the 100 μM BC was oxidized. In contrast, in the samples with a NaOCl:BC ratio of 100:1, ~75% of the BC was degraded (51). The HPLC chromatograms showed a number of long-chain CP formed; however, only a few β-apo-carotenals could be clearly identified. Figure 9.2 shows the accumulation of β-apo8′-carotenal, β-apo12′-carotenal, and β-apo15′-carotenal (retinal) after 10 min of HOCl/$^-$OCl–mediated oxidation. β-Apo4′-carotenal was not formed in any of the samples. For the other β-apo-carotenals, we found different formation rates according to the different NaOCl:BC ratios. Interestingly, no apo8′-carotenal could be detected when NaOCl was added at a ratio of 100:1, obviously due to the rapid degradation of that compound at the high HOCl/$^-$OCl concentration. For apo12′-carotenal and apo15′-carotenal (retinal), a higher HOCl/$^-$OCl concentration increased the formation of these compounds but not their rapid degradation. It was interesting that the pH values of the reaction mixture did not affect the accumulation of long-chain BCCP (51).

After examining all of the results of the different experimental settings (BC oxidation in methanolic and aqueous solution), it became obvious that the broad variety of long-chain products was found only in those samples in which a soybean matrix was used as the lipid matrix. The matrix seemed to slow down the reaction between HOCl/$^-$OCl and BC, probably due to a reduction of the number of HOCl/$^-$OCl molecules attacking BC *via* a parallel reaction of HOCl/$^-$OCl with lipid molecules of the matrix. This was also likely the reason for the incomplete degradation of BC and the lower product yield seen in these samples. However, because these settings meet much more physiologic conditions than the degradation experiments done in organic solution, long-chain CP, such as β-apo-carotenals, must be considered as BCCP that are likely to be formed *in vivo*. Nevertheless, short-chain BCCP were also found in considerable amounts in the reaction mixtures obtained

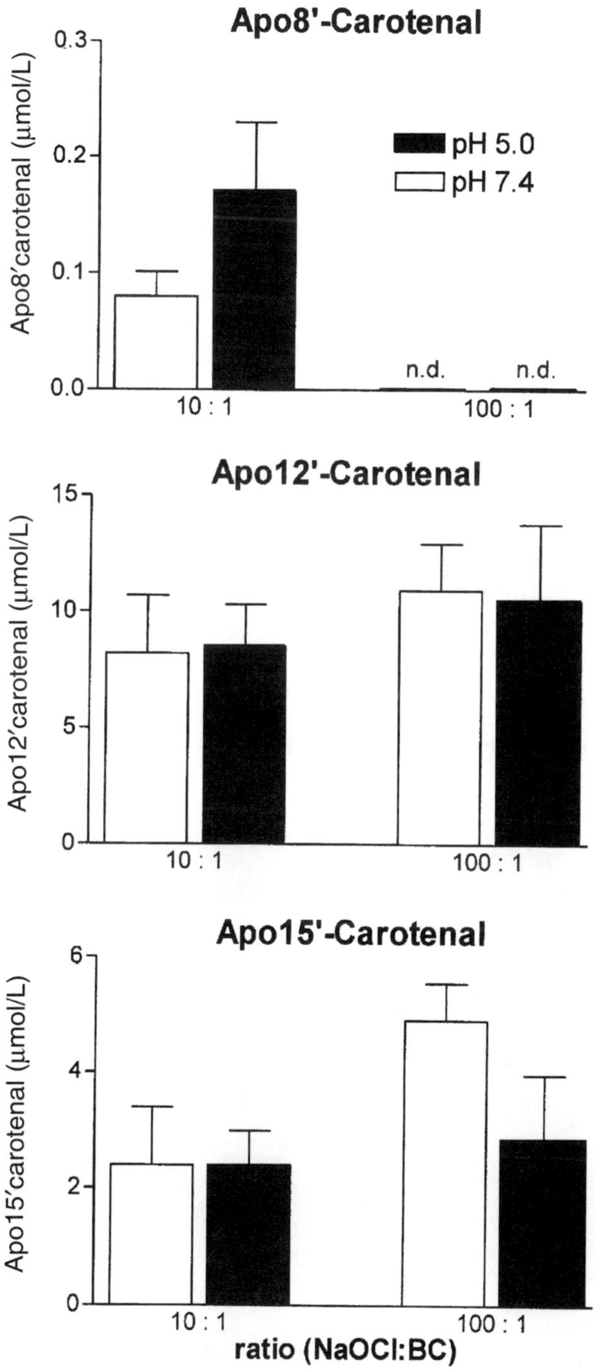

Fig. 9.2. Concentration of apo8′-carotenal, apo12′-carotenal, and apo15′-carotenal (retinal) after 10 min of HOCl/⁻OCl–mediated oxidation of β-carotene (BC, initial concentration 100 µM) dispersed in a lipid matrix. Total amount of lipid matrix consisting of a soybean carrier in the reaction mixture was 20% (vol/vol). NaOCl was added to start the reaction in a ratio of 10:1 or 100:1 (NaOCl:BC). The pH value in the different reaction mixtures was set to 7.4 or to 5.0, respectively. Values are means ± SD, n = 3 independent measurements.

from samples with 1% (vol/vol) and 20% (vol/vol) of the soybean matrix in aqueous solution. However, in these samples, a number of by-products resulting from oxidized lipids were also seen. An overview of the BCCP found under different experimental conditions is given in Table 9.2.

β-*Carotene Degradation and Cleavage Product Formation by Polymorphonuclear Leukocytes (PML)* In Vitro

The next step was to test whether PML are able to oxidize BC. For these experiments leukocytes were purified from heparinized human blood, freshly drawn from healthy donors according to the method of Ferrante and Thong (64). The leukocyte preparations contained 90–98% PML and were apparently free of contaminating erythrocytes. The cells were suspended in 0.5 mM Ca^{2+}-containing PBS at 37°C, pH 7.4, and stimulated in the presence of 1 μg/mL PMA; 5 min after the preincubation of the cells, BC dispersed in a soybean carrier was added to the medium, leading to a final concentration of 1 μM. The cell amount was 10×10^6 cells/mL. Fractions of the culture medium were extracted for measurements of BC absorption at the 0- and 30-min time points. We were able to demonstrate that BC was degraded in a culture medium of activated PML, but not in a medium of nonactivated PML (51). Stimulation of BC significantly decreased BC absorbance after 30

TABLE 9.2
Products Identified After Oxidation of β-Carotene (BC) by HOCl/⁻OCl (in Methanol and Aqueous Solution) and by Polymorphonuclear Leukocytes (PML)[a]

Reaction conditions	Methanol		Aqueous solution				PML	
Amount of lipids (vol/vol)[b]			1%	1%	20%	20%		
Initial β-carotene (μM)	100	100	100	100	100	100	100	10000
NaOCl: β-carotene ratio	100:1	10:1	100:1	10:1	100:1	10:1		
Oxidized β-carotene (%)	100	100	100	100	68	45	82	27
Products identified[c]								
Apo8′-carotenal	–	–	–	–	–	(+)	–	+
Apo12′-carotenal	–	(+)	–	(+)	+	+	–	+
Apo15′-carotenal (retinal)	–	(+)	–	–	+	+	–	+
β-Ionone	+	+	+	+	+	+	+	+
Ionene	(+)	(+)	(+)	(+)	(+)	(+)	(+)	(+)
β-Cyclocitral	+	+	+	+	+	+	+	+
β-Ionone-5,6-epoxide	+	+	+	+	+	+	+	+
Dihydroactinidiolide	+	+	+	+	+	+	+	+
4-Oxo-β-ionone	+	+	+	+	+	–	+	+

[a]BC oxidation by PML was carried out with different initial concentrations of the carotenoid.
[b]The amount of lipids (vol/vol) indicates the amount of soybean carrier in which BC was dispersed to make it soluble in aqueous solution. Symbols: +, found; (+), found in low amounts; –, not found.
[c]The products given for the experiments done in aqueous solution were identified in the reaction mixtures of samples reacted at a pH value of 5.0 AND in the samples reacted at a pH value of 7.4.

min of incubation. To show a time dependence of this reaction and to calculate the degradation rate, further experiments were conducted in which 1 μM of BC was incubated with PML (10×10^6 cells/mL) and the BC remaining in the culture medium was measured after 10, 30, and 60 min. The time dependence is shown in Figure 9.3. The degradation rate of the reaction was 732 pmol/(mL·h) which was equal to 732 pmol/(10×10^6 cells·h).

However, PML-derived degradation of BC is likely mediated by an orchestra of oxidants liberated by these cells. To show the influence of the antioxidant network on the oxidation process, additional inhibition experiments with certain antioxidants were conducted. It was noteworthy that all antioxidants used had a certain inhibitory effect on the degradation process (Siems, W.G., unpublished data). Vitamin E at a concentration of 100 μM was the most effective because it was able to completely block oxidation of BC (Fig. 9.4). Given these results, BC oxidation by PML and consecutive formation of BCCP may be relevant only in cases in which the antioxidant network is depleted for some reason.

After the demonstration of BC degradation by PML, we wanted to determine what kind of CP were formed under these conditions. To perform identification analysis after PML-mediated BC oxidation, additional experiments required that the BC concentration be increased to 100 μM to have sufficient material for HPLC and GCMS analysis. The reaction conditions were kept as before. Fractions of the culture media were taken 0 and 30 min after the addition of BC for CP extraction.

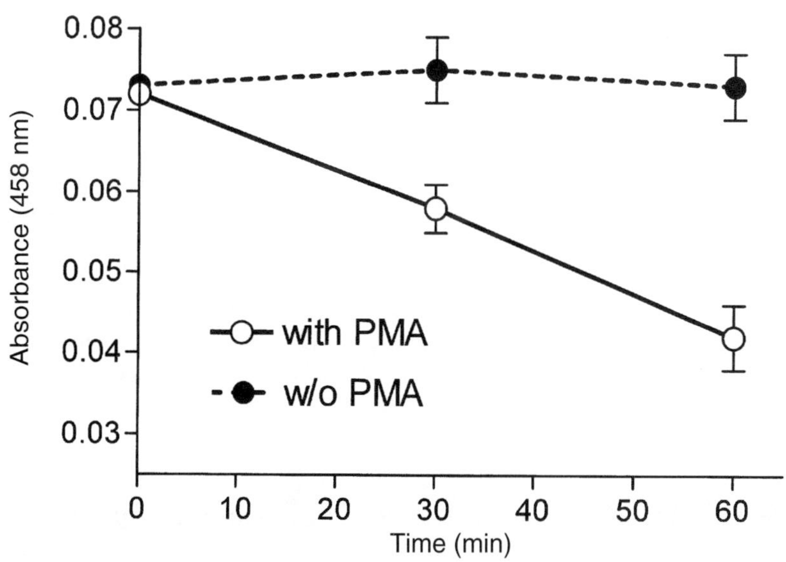

Fig. 9.3. Time-dependent degradation of β-carotene mediated by polymorphonuclear leukocytes (PML). Incubation of 1 μM carotenoid with PML (10×10^6 cells) at 37°C. Values are means ± SEM, n = 6.

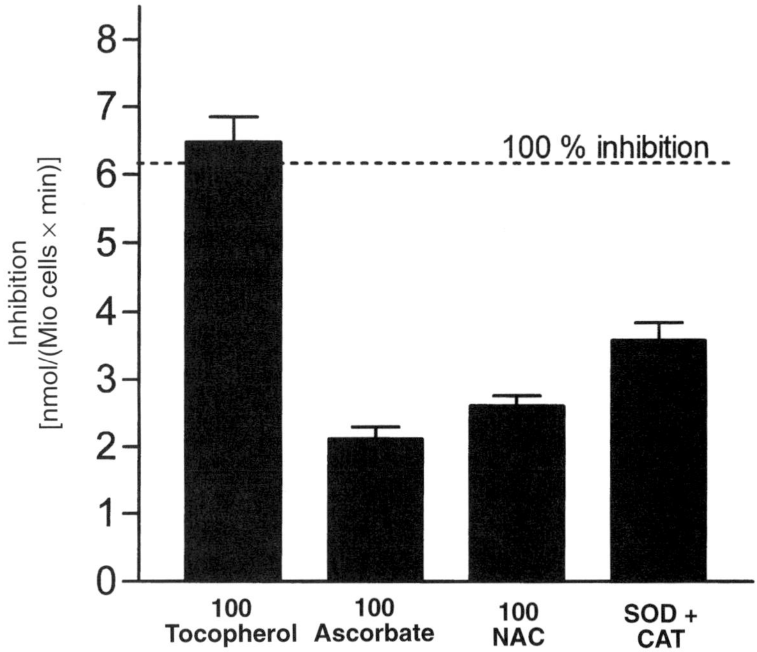

Fig. 9.4. Incubation of activated polymorphonuclear leukocytes (PML) in culture medium containing 2 μM of β-carotene (BC) at 37°C for 1 h. Tocopherol, ascorbate, and *N*-acetylcysteine (NAC) were added, resulting in a final concentration of 100 μM superoxide dismutase (SOD) + catalase (CAT) mean incubation with 50 μg/mL SOD in combination with 20 μg/mL catalase. Data are given as mean ± SEM, n = 6. Abbreviation: Mio, million → 10^6 cells.

Interestingly, after 30 min, BC was degraded completely and only short-chain CP were found. However, to determine whether apo-carotenals are generally formed during PML-mediated BC oxidation, we increased the initial BC concentration to 10 mM in a new experimental approach. Under these reaction conditions, BC degradation was not complete and a number of long-chain CP were detected in the HPLC chromatogram. Among these apo8′-, apo12′-, and apo15′-carotenal (retinal) could be identified (51). The complete pattern of the BCCP formed during oxidation mediated by PML is given in Table 9.2.

Effects of β-Carotene Cleavage Products on PML

To determine the influence of BCCP on PML, we performed experiments in which primary cultured human PML were incubated with different concentrations of BCCP. For these experiments, the above-described BCCP mixture and also several CP available as single compounds were used. In these experiments, stimulation of superoxide production was clearly induced by carotenoid CP with aliphatic chains of different length (retinal, β-ionone, carotenoid CP), but not by the precursor carotenoids lacking the carbonyl moiety (BC or lycopene) (65). However, the stimulatory effect of CP was observed only in cells activated by PMA. Interestingly, there was a significant decrease in lag phase, which was observed by studying the onset time of the respiratory burst. The effect was seen for retinal in a concentration range from 0.04 to 2 μM and for the mixture of BCCP from 0.04 to 40 μM (65). The effect could be attributed to metabolic changes that are triggered by carotenoid CP in PML. The results are consistent with previous findings showing that suboptimal doses of retinal induced minimal O_2^- release but primed guinea pig macrophages to release enhanced amounts of superoxide in response to arachidonate stimulation (66). If this were also true *in vivo*, the presence of BCCP in the given ranges might even amplify oxidative stress mediated by neutrophils.

However, in contrast to previous reports (67,68), we did not observe any activation of NADPH-oxidase activity by the addition of retinal or BCCP alone. The discrepancy can be explained, at least in part, by the fact that nonstimulated PML from peripheral blood were used in our experiments, whereas the previous work (66–68) was performed with phagocytes recruited in guinea pig peritoneal exudate by the injection of colloidal caseinate suspensions, which might have modified the resting state of the cells. At BCCP concentrations slightly higher than those required to stimulate superoxide production, superoxide production by human neutrophils is inhibited. If the chemotactic tripeptide f-MLP is used instead of PMA to trigger neutrophil response, a small inhibition of superoxide production is observed even in the presence of the micromolar concentrations of BCCP used to enhance superoxide production in the presence of PMA.

Inhibition of neutrophil respiratory burst by carotenoids themselves has long been recognized. In fact, it was reported that a high amount of retinal blocks both superoxide release (69–71) and phosphorylation of the 47-kDa component of NADPH-oxidase (72) in intact neutrophils stimulated with PMA and in various fractions derived from neutrophils (72,73). Carotenoids inhibited protein kinase C from various sources (74,75) and were used extensively as an inhibitor of this enzyme in neutrophils (69–71,73). Moreover, carotenoid inhibition of superoxide release from human neutrophils might be due, at least in part, to its ability to scavenge the superoxide anion. Neutrophil apoptosis may be another reason for diminished superoxide production in the presence of carotenoids (65). It was found that chromatin fragmentation in neutrophils can be induced by 20 μM carotenoid breakdown products as well as by cycloheximide, a drug that induces cell apoptosis (76).

Carotenoid concentrations employed in our work are comparable to those present in human tissues (9). However, our model is quite simplistic compared with the *in vivo* situation in which neutrophils are responding to an environment containing cytokines, lipid mediators, bacterial products, and other stimuli. Moreover, there may be important differences in the regulation of cell response and apoptosis between peripheral blood neutrophils and extravasated neutrophils because neutrophil adhesion and the action of local inflammatory mediators are generally believed to modulate neutrophil reactivity and to increase neutrophil lifespan by delaying constitutive apoptosis (77). Nevertheless, it is conceivable that modulation of neutrophil function by carotenoids and especially carotenoid CP may play a critical role because of the opposite effects that we observed by varying carotenoid concentrations in a restricted range. Regulation of superoxide production by relatively low concentrations of carotenoids may contribute to optimize neutrophil response in inflammatory processes but can also lead to increased oxidative stress. On the other hand, because neutrophil apoptosis is a critical step in the resolution of inflammation, premature activation of neutrophil apoptosis mediated by highly concentrated carotenoid CP may greatly increase the potential risk for host tissue injury by toxic agents.

Summary

Neutrophils play an important role in the defense against bacteria and accumulate in the lung tissue of patients with chronic obstructive lung diseases and also in smokers. Oxidants liberated by stimulated neutrophils contribute to a large extent to oxidative stress in the surrounding tissue. Hypochlorous acid oxidized BC, leading to the formation of BCCP. The same was seen when primary cultured human PML were stimulated in BC-containing culture media. However, antioxidants in physiologic concentrations inhibited BC degradation by PML.

Interestingly, once BCCP are formed, they are able to enhance oxidative burst of PML stimulated by PMA in a concentration range of 10^{-2} to 10^{1} μM. If this is true also *in vivo*, an amplification of the prooxidative conditions will occur. It might be possible that the prooxidative effects seen in heavy and/or long-term smokers are the result of the high-dose BC supplementation as seen in the intervention trials, a depleted antioxidant capacity, and increased oxidative stress due to the large pool of stimulated neutrophils in the lung tissue of smokers or patients with an inflammatory lung disease.

However, because carotenoids were shown to be protective in a number of epidemiologic studies, the knowledge of the conditions under which the unexpected results of the intervention trials occurred is essential for the establishment of safe conditions for carotenoid supplementation in disease prevention and clinical therapy.

Acknowledgments

The authors thank BASF AG for providing standard substances for the apo-carotenals used in the studies. Furthermore, the authors gratefully acknowledge Lance Schlipalius and Michael Strahan (Cognis Australia Pty Limited) for providing the soybean oil–based water-dispersible carotenoids.

References

1. Krinsky, N.I. (1998) The Antioxidant and Biological Properties of the Carotenoids, *Ann. N.Y. Acad. Sci. 854:* 443–447.

2. The Alpha-Tocopherol, Beta Carotene Cancer Prevention Study Group (1994) The Effect of Vitamin E and Beta Carotene on the Incidence of Lung Cancer and Other Cancers in Male Smokers, *N. Engl. J. Med. 330:* 1029–1035.
3. Omenn, G.S., Goodman, G.E., Thornquist, M.D., Balmes, J., Cullen, M.R., Glass, A., Keogh, J.P., Meyskens, F.L., Jr., Valanis, B., Williams, J.H., Jr., Barnhart, S., Cherniack, M.G., Brodkin, C.A., and Hammar, S. (1996) Risk Factors for Lung Cancer and for Intervention Effects in CARET, the Beta-Carotene and Retinol Efficacy Trial, *J. Natl. Cancer. Inst. 88:* 1550–1559.
4. Siems, W.G., Sommerburg, O., Hurst, J.S., and van Kuijk, F.J. (2000) Carotenoid Oxidative Degradation Products Inhibit Na^+-K^+-ATPase, *Free Radic. Res. 33:* 427–435.
5. Alija, A.J., Bresgen, N., Sommerburg, O., Siems, W.G., and Eckl, P.M. (2004) Cytotoxic and Genotoxic Effects of Beta Carotene Breakdown Products on Primary Rat Hepatocytes, *Carcinogenesis 25:* 827–831.
6. Wang, X.D., Liu, C., Bronson, R.T., Smith, D.E., Krinsky, N.I., and Russell, M. (1999) Retinoid Signaling and Activator Protein-1 Expression in Ferrets Given Beta-Carotene Supplements and Exposed to Tobacco Smoke, *J. Natl. Cancer Inst. 91:* 60–66.
7. Liu, C., Russell, R.M., and Wang, X.D. (2003) Exposing Ferrets to Cigarette Smoke and a Pharmacological Dose of β-Carotene Supplementation Enhance *In Vitro* Retinoic Acid Catabolism in Lungs *Via* Induction of Cytochrome P_{450} Enzymes, *J. Nutr. 133:* 173–179.
8. Mayne, S.T. (1996) Beta-Carotene, Carotenoids, and Disease Prevention in Humans, *FASEB J. 10:* 690–701.
9. Schmitz, H.H., Poor, C.L., Wellman, R.B., and Erdman, J.J. (1991) Concentrations of Selected Carotenoids and Vitamin A in Human Liver, Kidney and Lung Tissue, *J. Nutr. 121:* 1613–1621.
10. Burton, G.W., and Ingold, K.U. (1984) β-Carotene: An Unusual Type of Lipid Antioxidant, *Science 224:* 569–573.
11. Young, A.J., and Lowe, G.M. (2001) Antioxidant and Prooxidant Properties of Carotenoids, *Arch. Biochem. Biophys. 385:* 20–27.
12. Amin, K., Ekberg-Jansson, A., Lofdahl, C.G., and Venge, P. (2003) Relationship Between Inflammatory Cells and Structural Changes in the Lungs of Asymptomatic and Never Smokers: A Biopsy Study, *Thorax 58:* 135–142.
13. Eidelman, D., Saetta, M.P., Ghezzo, H., Wang, N.S., Hoidal, J.R., King, M., and Cosio, M.G. (1990) Cellularity of the Alveolar Walls in Smokers and Its Relation to Alveolar Destruction. Functional Implications, *Am. Rev. Respir. Dis. 141:* 1547–1552.
14. Saetta, M., Turato, G., Baraldo, S., Zanin, A., Braccioni, F., Mapp, C.E., Maestrelli, P., Cavallesco, G., Papi, A., and Fabbri, L.M. (2000) Goblet Cell Hyperplasia and Epithelial Inflammation in Peripheral Airways of Smokers with Both Symptoms of Chronic Bronchitis and Chronic Airflow Limitation, *Am. J. Respir. Crit. Care Med. 161:* 1016–1021.
15. Di Stefano, A., Capelli, A., Lusuardi, M., Balbo, P., Vecchio, C., Maestrelli, P., Mapp, C.E., Fabbri, L.M., Donner, C.F., and Saetta, M. (1998) Severity of Airflow Limitation Is Associated with Severity of Airway Inflammation in Smokers, *Am. J. Respir. Crit. Care Med. 158:* 1277–1285.
16. Saetta, M. (1999) Airway Inflammation in Chronic Obstructive Pulmonary Disease, *Am. J. Respir. Crit. Care Med. 160:* S17–S20.
17. Amin, K., Ludviksdottir, D., Janson, C., Nettelbladt, O., Bjornsson, E., Roomans, G.M., Boman, G., Seveus, L., and Venge, P. (2000) Inflammation and Structural Changes in

the Airways of Patients with Atopic and Nonatopic Asthma. BHR Group, *Am. J. Respir. Crit. Care Med. 162:* 2295–2301.

18. Andreadis, A.A., Hazen, S.L., Comhair, S.A., and Erzurum, S.C. (2003) Oxidative and Nitrosative Events in Asthma, *Free Radic. Biol. Med. 35:* 213–225.
19. Maddox, L., and Schwartz, D.A. (2002) The Pathophysiology of Asthma, *Annu. Rev. Med. 53:* 477–498.
20. Amin, K., Ludviksdottir, D., Janson, C., Nettelbladt, O., Gudbjornsson, B., Valtysdottir, S., Bjornsson, E., Roomans, G.M., Boman, G., Seveus, L., and Venge, P. (2001) Inflammation and Structural Changes in the Airways of Patients with Primary Sjögren's Syndrome, *Respir. Med. 95:* 904–910.
21. Homnick, D.N., Cox, J.H., Deloof, M.J., and Ringer, T.V. (1993) Carotenoid Levels in Normal Children and in Children with Cystic Fibrosis, *J. Pediatr. 122:* 703–707.
22. Benabdeslam, H., Abidi, H., Garcia, I., Bellon, G., Gilly, R., and Revol, A. (1999) Lipid Peroxidation and Antioxidant Defenses in Cystic Fibrosis Patients, *Clin. Chem. Lab. Med. 37:* 511–516.
23. Winklhofer-Roob, B.M., van't Hof, M.A., and Shmerling, D.H. (1995) Response to Oral Beta-Carotene Supplementation in Patients with Cystic Fibrosis: A 16-Month Follow-Up Study, *Acta Paediatr. 84:* 1132–1136.
24. Rust, P., Eichler, I., Renner, S., and Elmadfa, I. (1998) Effects of Long-Term Oral Beta-Carotene Supplementation on Lipid Peroxidation in Patients with Cystic Fibrosis, *Int. J. Vitam. Nutr. Res. 68:* 83–87.
25. Renner, S., Rath, R., Rust, P., Lehr, S., Frischer, T., Elmadfa, I., and Eichler, I. (2001) Effects of Beta-Carotene Supplementation for Six Months on Clinical and Laboratory Parameters in Patients with Cystic Fibrosis, *Thorax 56:* 48–52.
26. Hubeau, C., Lorenzato, M., Couetil, J.P., Hubert, D., Dusser, D., Puchelle, E., and Gaillard, D. (2001) Quantitative Analysis of Inflammatory Cells Infiltrating the Cystic Fibrosis Airway Mucosa, *Clin. Exp. Immunol. 124:* 69–76.
27. Klebanoff, S.J. (1992) Oxygen Metabolites from Phagocytes, in *Inflammation: Basic Principles and Clinical Correlates*, Gallin, J.L., Goldstein, I.M., and Snyderman, R., eds., Raven Press, New York, pp. 541–588.
28. Klebanoff, S.J. (1992) Phagocytic Cells: Products of Oxygen Metabolism, in *Inflammation: Basic Principles and Clinical Correlates*, Gallin, J.L., Goldstein, I.M., and Snyderman, R., eds., Raven Press, New York, pp. 391–399.
29. Robinson, J.M., and Badwey, J.A. (1995) The NADPH Oxidase Complex of Phagocytic Leukocytes: A Biochemical and Cytochemical View, *Histochem. Cell Biol. 103:* 163– 180.
30. Thomas, M.J., Hedrick, C.C., Smith, S., Pang, J., Jerome, W.G., Willard, A.S., and Shirley, P.S. (1992) Superoxide Generation by the Human Polymorphonuclear Leukocyte in Response to Latex Beads, *J. Leukoc. Biol. 51:* 591–596.
31. Roos, D. (1991) The Involvement of Oxygen Radicals in Microbicidal Mechanisms of Leukocytes and Macrophages, *Klin. Wochenschr. 69:* 975–980.
32. Makino, R., Tanaka, T., Iizuka, T., Ishimura, Y., and Kanegasaki, S. (1986) Stoichiometric Conversion of Oxygen to Superoxide Anion During the Respiratory Burst in Neutrophils. Direct Evidence by a New Method for Measurement of Superoxide Anion with Diacetyldeuteroheme-Substituted Horseradish Peroxidase, *J. Biol. Chem. 261:* 11444–11447.
33. Hyslop, P.A., Hinshaw, D.B., Scraufstatter, I.U., Cochrane, C.G., Kunz, S., and Vosbeck, K. (1995) Hydrogen Peroxide as a Potent Bacteriostatic Antibiotic: Implications for Host Defense, *Free Radic. Biol. Med. 19:* 31–37.

34. Imlay, J.A., and Linn, S. (1988) DNA Damage and Oxygen Radical Toxicity, *Science 240:* 1302–1309.
35. Britigan, B.E., Coffman, T.J., and Buettner, G.R. (1990) Spin Trapping Evidence for the Lack of Significant Hydroxyl Radical Production During the Respiration Burst of Human Phagocytes Using a Spin Adduct Resistant to Superoxide-Mediated Destruction, *J. Biol. Chem. 265:* 2650–2656.
36. Cohen, M.S., Britigan, B.E., Hassett, D.J., and Rosen, G.M. (1988) Do Humans Neutrophils Form Hydroxyl Radical? Evaluation of an Unresolved Controversy, *Free Radic. Biol. Med. 5:* 81–88.
37. Samuni, A., Black, C.D., Krishna, C.M., Malech, H.L., Bernstein, E.F., and Russo, A. (1988) Hydroxyl Radical Production by Stimulated Neutrophils Reappraised, *J. Biol. Chem. 263:* 13797–13801.
38. Dunford, H.B. (1987) Free Radicals in Iron-Containing Systems, *Free Radic. Biol. Med. 3:* 405–421.
39. Heinecke, J.W., Li, W., Daehnke, H.L., III, and Goldstein, J.A. (1993) Dityrosine, a Specific Marker of Oxidation, Is Synthesized by the Myeloperoxidase-Hydrogen Peroxide System of Human Neutrophils and Macrophages, *J. Biol. Chem. 268:* 4069–4077.
40. Winterbourn, C.C. (1985) Comparative Reactivities of Various Biological Compounds with Myeloperoxidase-Hydrogen Peroxide-Chloride, and Similarity of the Oxidant to Hypochlorite, *Biochim. Biophys. Acta 840:* 204–210.
41. Schiller, J., Arnhold, J., Grunder, W., and Arnold, K. (1994) The Action of Hypochlorous Acid on Polymeric Components of Cartilage, *Biol. Chem. Hoppe Seyler 375:* 167–172.
42. Schaur, R.J., Jerlich, A., and Stelmaszynska, T. (1998) Hypochlorous Acid as Reactive Oxygen Species, *Curr. Top. Biophys. 22:* 176–185.
43. Albrich, J.M., and Hurst, J.K. (1982) Oxidative Inactivation of *Escherichia coli* by Hypochlorous Acid. Rates and Differentiation of Respiratory from Other Reaction Sites, *FEBS Lett. 144:* 157–161.
44. Albrich, J.M., McCarthy, C.A., and Hurst, J.K. (1981) Biological Reactivity of Hypochlorous Acid: Implications for Microbicidal Mechanisms of Leukocyte Myeloperoxidase, *Proc. Natl. Acad. Sci. USA 78:* 210–214.
45. Nauseef, W.M. (1998) Insights into Myeloperoxidase Biosynthesis from Its Inherited Deficiency, *J. Mol. Med. 76:* 661–668.
46. Panasenko, O.M., Panasenko, O.O., Briviba, K., and Sies, H. (1997) Hypochlorite Destroys Carotenoids in Low Density Lipoproteins Thus Decreasing Their Resistance to Peroxidative Modification, *Biochemistry 62:* 1140–1145.
47. Vogt, W., and Hesse, D. (1994) Oxidants Generated by the Myeloperoxidase-Halide System Activate the Fifth Component of Human Complement, C5, *Immunobiology 192:* 1–9.
48. Weiss, S.J., Klein, R., Slivka, A., and Wei, M. (1982) Chlorination of Taurine by Human Neutrophils. Evidence for Hypochlorous Acid Generation, *J. Clin. Investig. 70:* 598–607.
49. Handelman, G.J., van Kuijk, F.J., Chatterjee, A., and Krinsky, N.I. (1991) Characterization of Products Formed During the Autoxidation of Beta-Carotene, *Free Radic. Biol. Med. 10:* 427–437.
50. Siems, W.G., Sommerburg, O., and van Kuijk, F.J. (1999) Lycopene and Beta-Carotene Decompose More Rapidly than Lutein and Zeaxanthin Upon Exposure to Various Pro-Oxidants *in Vitro*, *Biofactors 10:* 105–113.

51. Sommerburg, O., Langhans, C.D., Arnhold, J., Leichsenring, M., Salerno, C., Crifo, C., Hoffmann, G.F., Debatin, K.M., and Siems, W.G. (2003) Beta-Carotene Cleavage Products After Oxidation Mediated by Hypochlorous Acid—A Model for Neutrophil-Derived Degradation, *Free Radic. Biol. Med. 35:* 1480–1490.
52. Stratton, S.P., Schaefer, W.H., and Liebler, D.C. (1993) Isolation and Identification of Singlet Oxygen Oxidation Products of Beta-Carotene, *Chem. Res. Toxicol. 6:* 542–547.
53. Mordi, R.C., Walton, J.C., Burton, G.W., Hughes, L., Ingold, K.U., and Lindsay, D.A. (1991) Exploratory Study of Beta-Carotene Autoxidation, *Tetrahedron. Lett. 32:* 4203–4206.
54. McClure, T.D., and Liebler, D.C. (1995) A Rapid Method for Profiling the Products of Antioxidant Reactions by Negative Ion Chemical Ionization Mass Spectrometry, *Chem. Res. Toxicol. 8:* 128–135.
55. Marty, C., and Berset, C. (1988) Degradation Products of *trans*-Beta-Carotene Produced During Extrusion Cooking, *J. Food Sci. 53:* 1880–1886.
56. Ouyang, J.M., Daun, H., Chang, S.S., and Ho, C.T. (1980) Formation of Carbonyl Compounds from Beta-Carotene During Palm Oil Deodorization, *J. Food Sci. 45:* 1214–1222.
57. Day, W.C., and Erdman, J.G. (1963) Ionene: A Thermal Degradation Product of Beta-Carotene, *Science 141:* 808.
58. Mader, I. (1964) Beta-Carotene: Thermal Degradation, *Science 144:* 533–534.
59. Schreier, P., Drawert, F., and Bhiwapurkar, S. (1979) Volatile Compounds Formed by Thermal Degradation of Beta-Carotene, *Chem. Mikrobiol. Technol. Lebensm. 6:* 90–91.
60. Onyewu, P.N., Ho, C.T., and Daun, H. (1986) Characterization of Beta-Carotene Thermal Degradation Products in a Model Food System, *J. Am. Oil Chem. Soc. 63:* 1437–1441.
61. Mordi, R.C., Walton, J.C., Burton, G.W., Hughes, L., Ingold, K.U., Lindsay, D.A., and Moffatt, D. (1993) Oxidative Degradation of Beta-Carotene and Beta-Apo-8′-Carotenal, *Tetrahedron. Lett. 49:* 911–928.
62. Gloria, M.B., Grulke, E.A., and Gray, J.I. (1993) Effect of Type of Oxidation on Beta-Carotene Loss and Volatile Products Formation in Model Systems, *Food Chem. 46:* 401–406.
63. Siems, W.G., Sommerburg, O., Schild, L., Augustin, W., Langhans, C.D., and Wiswedel, I. (2002) Beta-Carotene Cleavage Products Induce Oxidative Stress *in Vitro* by Impairing Mitochondrial Respiration, *FASEB J. 16:* 1289–1291.
64. Ferrante, A., and Thong, Y.H. (1978) A Rapid One-Step Procedure for Purification of Mononuclear and Polymorphonuclear Leukocytes from Human Blood Using a Modification of the Hypaque-Ficoll Technique, *J. Immunol. Methods 24:* 389–393.
65. Siems, W.G., Capuozzo, E., Crifo, C., Sommerburg, O., Langhans, C.D., Schlipalius, L., Wiswedel, I., Kraemer, K., and Salerno, C. (2003) Carotenoid Cleavage Products Modify Respiratory Burst and Induce Apoptosis of Human Neutrophils, *Biochim. Biophys. Acta 1639:* 27–33.
66. Steinbeck, M.J., Hegg, G.G., and Karnovsky, M.J. (1991) Arachidonate Activation of the Neutrophil NADPH-Oxidase. Synergistic Effects of Protein Phosphatase Inhibitors Compared with Protein Kinase Activators, *J. Biol. Chem. 266:* 16336–16342.
67. Badwey, J.A., Horn, W., Heyworth, P.G., Robinson, J.M., and Karnovsky, M.L. (1989) Paradoxical Effects of Retinal in Neutrophil Stimulation, *J. Biol. Chem. 264:* 14947–14953.

68. Lochner, J.E., Badwey, J.A., Horn, W., and Karnovsky, M.L. (1986) all-trans-Retinal Stimulates Superoxide Release and Phospholipase C Activity in Neutrophils Without Significantly Blocking Protein Kinase C, *Proc. Natl. Acad. Sci. USA 83:* 7673–7677.
69. Witz, G., Goldstein, B.D., Amoruso, M., Stone, D.S., and Troll, W. (1980) Retinoid Inhibition of Superoxide Anion Radical Production by Human Polymorphonuclear Leukocytes Stimulated with Tumor Promoters, *Biochem. Biophys. Res. Commun. 97:* 883–888.
70. Kensler, T.W., and Trush, M.A. (1981) Inhibition of Phorbol Ester-Stimulated Chemiluminescence in Human Polymorphonuclear Leukocytes by Retinoic Acid and 5,6-Epoxyretinoic Acid, *Cancer Res. 41:* 216–222.
71. Pontremoli, S., Melloni, E., Michetti, M., Sacco, O., Salamino, F., Sparatore, B., and Horecker, B.L. (1986) Biochemical Responses in Activated Human Neutrophils Mediated by Protein Kinase C and a Ca^{2+}-Requiring Proteinase, *J. Biol. Chem. 261:* 8309–8313.
72. Pontremoli, S., Melloni, E., Salamino, F., Sparatore, B., Michetti, M., Sacco, O., and Horecker, B.L. (1986) Phosphorylation of Proteins in Human Neutrophils Activated with Phorbol Myristate Acetate or with Chemotactic Factor, *Arch. Biochem. Biophys. 250:* 23–29.
73. Melloni, E., Pontremoli, S., Salamino, F., Sparatore, B., Michetti, M., Sacco, O., and Horecker, B.L. (1986) ATP Induces the Release of a Neutral Serine Proteinase and Enhances the Production of Superoxide Anion in Membranes from Phorbol Ester-Activated Neutrophils, *J. Biol. Chem. 261:* 11437–11439.
74. Taffet, S.M., Greenfield, A.R., and Haddox, M.K. (1983) Retinal Inhibits TPA Activated, Calcium-Dependent, Phospholipid-Dependent Protein Kinase ("C" Kinase), *Biochem. Biophys. Res. Commun. 114:* 1194–1199.
75. Hirota, K., Hirota, T., Aguilera, G., and Catt, K.J. (1985) Hormone-Induced Redistribution of Calcium-Activated Phospholipid-Dependent Protein Kinase in Pituitary Gonadotrophs, *J. Biol. Chem. 260:* 3243–3246.
76. Parvathenani, L.K., Buescher, E.S., Chacon-Cruz, E., and Beebe, S.J. (1998) Type I cAMP-Dependent Protein Kinase Delays Apoptosis in Human Neutrophils at a Site Upstream of Caspase-3, *J. Biol. Chem. 273:* 6736–6743.
77. Akgul, C., Moulding, D.A., and Edwards, S.W. (2001) Molecular Control of Neutrophil Apoptosis, *FEBS Lett. 487:* 318–322.

Chapter 10

Biological Activity of Lycopene Against Smoke-Induced Lung Lesions: Targeting the IGF-1/IGFBP-3 Signal Transduction Pathway

Xiang-Dong Wang

Nutrition and Cancer Biology Laboratory, Jean Mayer United States Department of Agriculture Human Nutrition Research Center on Aging at Tufts University, Boston, MA 02111

Introduction

The beneficial effects of carotenoid-rich fruits and vegetables on lung cancer risk were noted in many epidemiologic studies. However, the failure of the β-carotene intervention trials to show a benefit against lung carcinogenesis in smokers indicates that carotenoids or phytonutrients other than β-carotene may account for the observed protective effects of fruits and vegetables on lung cancer risk. Recent epidemiologic studies provide supportive evidence that lycopene (tomatoes and related tomato products are the major sources of lycopene) may have a chemopreventive effect against a broad range of epithelial cancers, particularly prostate, breast, colon, and lung cancer (1–5). Lycopene has attracted attention due to its biological and physicochemical properties, especially related to its effects as a natural antioxidant that exhibits a physical quenching rate constant for singlet oxygen almost twice as high as that of β-carotene (6,7). Lycopene may also enhance gap junction communication (8,9), suppress growth factor–stimulated cell proliferation (10,11), and inhibit neoplastic transformation of normal human and mouse cells (12).

Insulin-like growth factors (IGF) are mitogens that play a pivotal role in regulating cell proliferation, differentiation, and apoptosis (13). Disruptions of normal IGF-1 system components leading to hyperproliferation and survival signals were implicated in the development of different tumor types (14). This review focuses on recent evidence that lycopene exerts its protective effects against smoke-induced lung carcinogenesis through upregulating IGFBP-3 as a molecular target, interrupting the signal transduction pathway of IGF-1, inhibiting cell proliferation, and promoting apoptosis. In addition, the issue regarding the alteration of lycopene metabolism in smoke-exposed lung tissue is discussed.

IGF-1/IGF Binding Protein-3 Signaling and Lung Cancer

IGF-1 is the major mediator of the effects of growth hormone and performs a fundamental role in the regulation of cellular proliferation, differentiation, and apoptosis (Fig. 10.1). The action of IGF-1 is mediated predominantly through the IGF-1

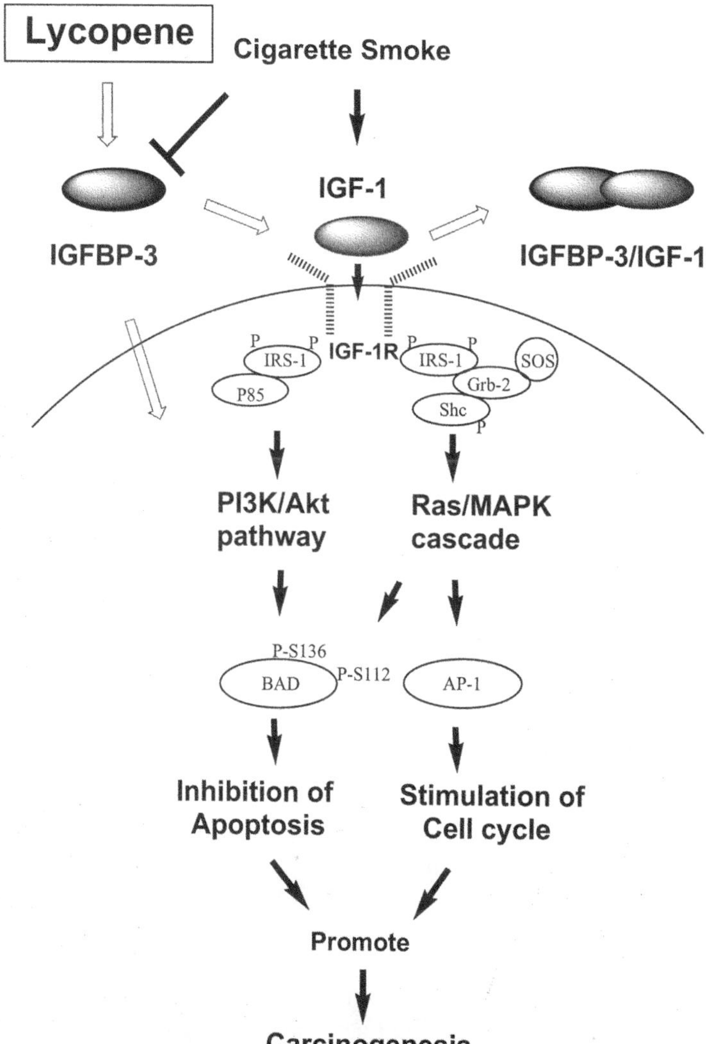

Fig. 10.1. Simplified illustration of proposed pathways of insulin-like growth factor (IGF)-1-dependent and IGF-1-independent IGF binding protein-3 (IGFBP-3). IGFBP-3 regulates bioactivity of IGF-1 by sequestering IGF-1 away from its receptor in the extracellular milieu, thereby inhibiting the mitogenic and antiapoptotic action of IGF-1 and reducing cancer risk. Apart from modulating IGF-1 action, IGFBP-3 may exert intrinsic bioactivity *via* its interaction with other signaling. Our hypothesis is that lycopene or lycopene metabolites exert their protective effects against smoke-induced lung carcinogenesis through upregulating IGFBP-3 as a molecular target, interrupting the signal transduction pathway of IGF-1, downregulating phosphorylation of BAD, promoting apoptosis, and inhibiting cell proliferation, thereby preventing lung carcinogenesis.

receptor (IGF-1R). The downstream pathways of IGF-1R signaling involve the activation of both phosphatidylinositol 3′-kinase (PI3K)/Akt/protein kinase B (PKB) and Ras/Raf/mitogen-activated protein kinase (MAPK) pathways (Fig. 10.1), which regulate various biological processes such as cell cycle, survival, and transformation. The stimulation of IGF-1 to increase cell proliferation can be achieved by regulating activator protein 1 (AP-1) (c-Jun and c-Fos), p21Cip1/Waf1, p27 Kip1, and cyclin D1, whereas the ability of IGF-1 to increase cell survival is mediated by phosphorylating BAD (a member of the BH3-only subfamily of Bcl-2) on serine residues (Ser 136, Ser 112 and Ser 155) (15,16). Ser 136 of BAD is preferentially phosphorylated by the PI3K/Akt/PKB pathway (17,18). Activation of the MAPK pathway phosphorylates BAD on Ser 112 and Ser 155 (19,20). In the presence of phosphorylation, BAD is found in the cytosol, bound to 14-3-3 proteins rather than Bcl-xL, and does not induce cell death, whereas nonphosphorylated BAD is localized to the mitochondria, bound to Bcl-xL, which promotes its proapoptotic effect (15).

IGF binding protein-3 (IGFBP-3), one of the six members of the IGFBP family and a major circulating protein in human plasma (3,21), regulates the bioactivity of IGF-1 by sequestering IGF-1 away from its receptor in the extracellular milieu, thereby inhibiting the mitogenic and antiapoptotic action of IGF-1. The expression of IGFBP-3 is increased by retinoic acid (22), 1,25$(OH)_2$ vitamin D_3 (23), and the p53 tumor suppressor protein (24). IGFBP-3 has both IGF-dependent and -independent antiproliferative and proapoptotic effects as shown by the findings that IGFBP-3 sequesters IGF-1 away from its receptor and that IGFBP-3 suppresses the growth of IGF-1R null fibroblasts (25–27). IGFBP-3 is also a potent inhibitor of both the PI3K/Akt/PKB and MAPK signaling pathways (28). The results from studies in lung cancer cell lines (29) and human populations studies (30–32) strongly support an important role of high levels of IGF-1 and low levels of IGFBP-3 in lung cancer. For example, in a case-control study, the relative risk of lung cancer between the highest and lowest quartiles of IGF-1 serum concentrations was 2.0, whereas increased levels of IGFBP-3 were associated with a 50% reduction [odds ratio (OR) = 0.48] in the relative risk after adjustment for IGF-1 levels (33). In a prospective study, for subjects in the highest vs. the lowest quartile of serum IGFBP-3 levels, the OR for lung cancer was 0.50 after adjustment for IGF-1 levels and smoking (31). Furthermore, it was shown that downregulation of IGFBP-3 is a frequent event in stage I non-small cell lung cancer and correlates with the disease-specific survival probability of patients with stage I non-small cell lung cancer (34). Serum levels of IGFBP-3 are also negatively correlated with the number of cigarettes smoked per day or pack-year history of smoking (35). These results suggest the important role of IGFBP-3 as a tumor suppressor against lung carcinogenesis and other malignancies (34).

Lycopene and Lung Cancer Risk

Beneficial effects of carotenoid-rich fruits and vegetables on lung cancer risk were found in many epidemiologic studies, with strong evidence implicating lycopene as a micronutrient with important health benefits (1,36). Recent studies showed that dietary lycopene was associated with stronger and more significant reductions in lung cancer risk, and this association was slightly stronger than that observed for total fruit and vegetable intake (37,38). In an animal study, Kim *et al.* (39) provided evidence that lycopene decreased the incidence of lung tumors in B6C3F1 mice. Because little is known regarding whether lycopene can inhibit smoke-induced lung preneoplastic lesions through modulation of cell proliferation, and apoptosis, we conducted a study to test the effectiveness of lycopene supplementation against lung preneoplastic lesions in smoke-exposed ferrets, which closely mimic humans in terms of lycopene metabolism and cigarette smoke–induced lung lesions (40). Ferrets in the low-dose lycopene group were supplemented with 1.1 mg lycopene/(kg·d), which is equivalent to an intake of 15 mg/d in humans. This dose of lycopene is slightly higher than the average intake of lycopene (9.4 ± 0.3 mg/d) in U.S. men and women (41). Ferrets in the high-dose lycopene group were supplemented with 4.3 mg lycopene/(kg·d), which is equivalent to 60 mg/d in humans and achievable in a diet enriched with tomato products or supplements. We observed that squamous metaplastic lesions, which precede the appearance of lung cancer, were observed in the lung tissue of six of the six ferrets exposed to smoke alone but in only two of the six ferrets exposed to smoke and supplemented with low-dose lycopene (40). No squamous metaplasia was observed in the lung tissue of ferrets in the control group, the low-dose lycopene group, the high-dose lycopene group, and the high-dose plus smoke exposure group after 9 wk of intervention. Proliferating cellular nuclear antigen (PCNA) expression was increased threefold in the smoke-exposed group, but did not differ between ferrets supplemented with either a low or high dose of lycopene and exposed to smoke, and control ferrets (Fig. 10.2). Further, smoke exposure in ferrets for a 9-wk period significantly decreased apoptosis as indicated by a 74% reduction in cleaved caspase 3 (Fig. 10.2), compared with controls. However, there were no differences in cleaved caspase 3 levels in ferrets supplemented with either a low or high dose of lycopene with or without smoke exposure compared with control, suggesting that lycopene supplementation restored to normal the reduced apoptosis due to smoke exposure. Our study indicates that lycopene supplementation for 9 wk at either a low or high dose prevents smoke-induced squamous metaplasia, cell hyperproliferation, and the reduction of apoptosis in the lungs of smoke-exposed ferrets.

Lycopene and IGF-1/IGFBP-3

A possible mechanism of lycopene action by its interference with the mitogenic pathway of IGF-1 was proposed by Sharoni and his colleagues (10,11). These investigators showed that the growth stimulation by IGF-1 as well as the DNA binding activity of the AP-1 transcription factor were reduced by physiologic concentrations of lycopene in endometrial, mammary (MCF-7), and lung (NCI-H226) cancer cells (10,11). Lycopene treatment was also associated with an increase in membrane-associated IGFBP (11). A recent study in humans also showed that a

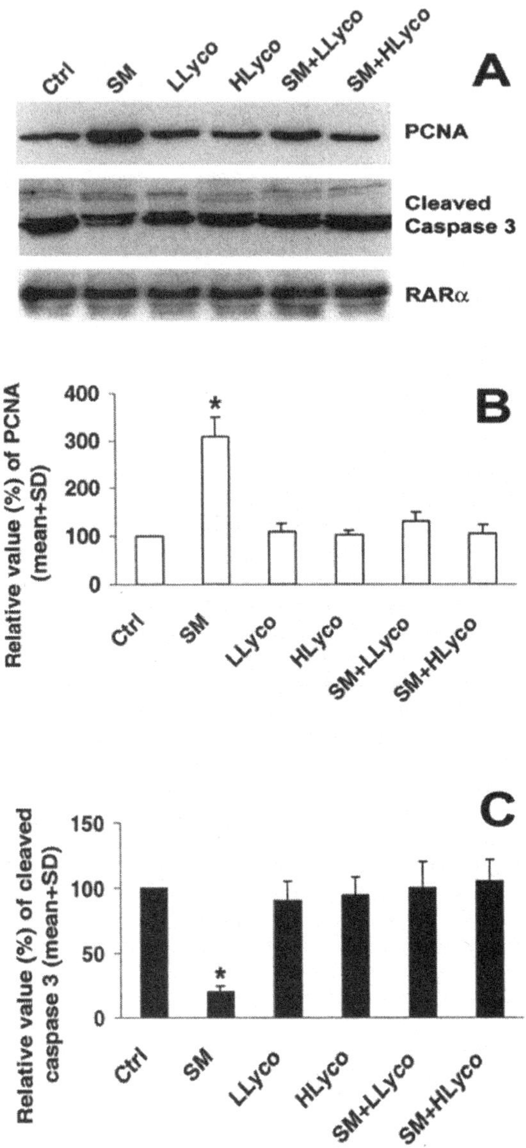

Fig. 10.2. Expressions of proliferating cellular nuclear antigen (PCNA), cleaved Caspase 3, and retinoic acid receptor (RAR)α protein levels in lung tissue from six groups of ferrets (Ctrl, control; SM, smoke-exposed; LLyco, low-dose lycopene supplemented; HLyco, high-dose lycopene supplemented; SM+LLyco, smoke-exposed + low-dose lycopene supplemented; and SM+HLyco, smoke-exposed + high-dose lycopene supplemented). (A) Representative Western blot analysis for PCNA, cleaved Caspase 3, and RARα protein levels. The size of the detected PCNA was 34 kDa. The sizes of the detected cleaved Caspase 3 were 17–19 kDa. The size of the detected RARα was 53 kDa. (B) The intensity of the protein signal of PCNA determined by densitometry (n = 6 ferrets/group) and expressed by the relative values (means ± SD). (C) The intensity of the protein signal of cleaved Caspase 3 determined by densitometry (n = 6 ferrets/group) and expressed by the relative values (means ± SD). *Different from the other groups, $P < 0.05$. *Source:* Reference 40.

higher intake of cooked tomatoes or lycopene was significantly associated with lower circulating levels of IGF-1 (42) and higher levels of IGFBP-3 (43). We hypothesized that cigarette smoke exposure, a strong risk factor for lung cancer, may promote cell proliferation and neoplasia by affecting normal IGF-1 signaling. Because both IGF-1 and IGFBP-3 are produced mainly in the liver and released as circulating proteins in plasma, and a higher ratio of IGF-1 to IGFBP-3 in the circulation is associated with an increased risk for the development of lung cancer and other cancers (breast, prostate and colon), we conducted an *in vivo* study using the ferret model to systemically investigate the effect of lycopene on IGF-1/IGFBP-3 signaling. We observed no significant differences in plasma IGF-1 concentrations in ferrets exposed to smoke alone and those exposed to smoke with or without lycopene supplementation (Table 10.1). In contrast, compared with controls, ferrets exposed to smoke exposure alone had significantly lower plasma IGFBP-3 concentrations (Table 10.1). However, ferrets exposed to smoke and supplemented with lycopene had plasma IGFBP-3 concentrations similar to those supplemented with lycopene alone. Furthermore, the ratio of IGF-1/IGFBP-3 was significantly higher in ferrets exposed to smoke alone than in the control ferrets. However, the ratio of IGF-1/IGFBP-3 was significantly lower in ferrets supplemented with either a low or high dose of lycopene and exposed to smoke than in those exposed to smoke alone. To further examine the effect of lycopene on ferret IGF-1 and IGFBP-3 gene transcription, we cloned ferret cDNA fragments of IGF-1 and IGFBP-3 by reverse transcription-polymerase chain reaction (RT-PCR). Using the ferret IGF-1 and IGFBP-3 cDNA fragments as probes in the RNase protection assay, we compared the mRNA level of IGF-1 and IGFBP-3 in the liver of ferrets with or without lycopene supplementation. The expression of IGFBP-3 was upregulated greatly in the liver of ferrets supplemented with lycopene, whereas that of IGF-1 remained unchanged (Lian, F. and Wang, X.D., unpublished data).

We then examined BAD, which has the potential to be a chemopreventive or therapeutic target because it has a central position between growth factor signaling pathways and apoptosis (20), in the lung tissue of ferrets after 9 wk of treatment. The levels of total BAD and Bcl-xL did not change among the six treatment groups (Fig. 10.3). However, smoke exposure alone substantially increased BAD phosphorylation at both Ser 112 (~4 fold) and Ser 136 (~3.4 fold), compared with con-

TABLE 10.1
Plasma Levels of Insulin-Like Growth Factor 1 (IGF-1) and Insulin-Like Growth Factor Binding Protein 3 (IGFBP-3) in Six Groups of Ferrets After 9 Weeks of Treatment[a,b]

Treatment	IGF-1	IGFBP-3 (ng/mL)	IGF-1/IGFBP-3 Ratio
Control (sham smoke exposure)	754.07 ± 208.30	2.33 ± 0.83^{d}	360.01 ± 134.57^{d}
Low-dose lycopene (15 mg/d)[c]	1022.88 ± 121.83	3.68 ± 0.63^{e}	$282.08 \pm 41.98^{d,e}$
High-dose lycopene (60 mg/d)	928.36 ± 333.93	3.85 ± 1.31^{e}	238.27 ± 46.55^{e}
Smoke exposure	745.02 ± 141.36	1.55 ± 1.15^{f}	706.98 ± 40.65^{f}
Low-dose lycopene (15 mg/d)	929.04 ± 127.71	3.34 ± 0.69^{e}	$281.89 \pm 30.20^{d,e}$
High-dose lycopene (60 mg/d)	993.30 ± 80.25	3.79 ± 0.86^{e}	$275.93 \pm 76.58^{d,e}$

[a]*Source:* Reference (40).
[b]Values are expressed as means ± SD, n = 6 ferrets/group. Means in a column with different superscript letters differ, $P < 0.05$.
[c]Ferrets were given the dosage that was equivalent to (x) mg/d in humans.

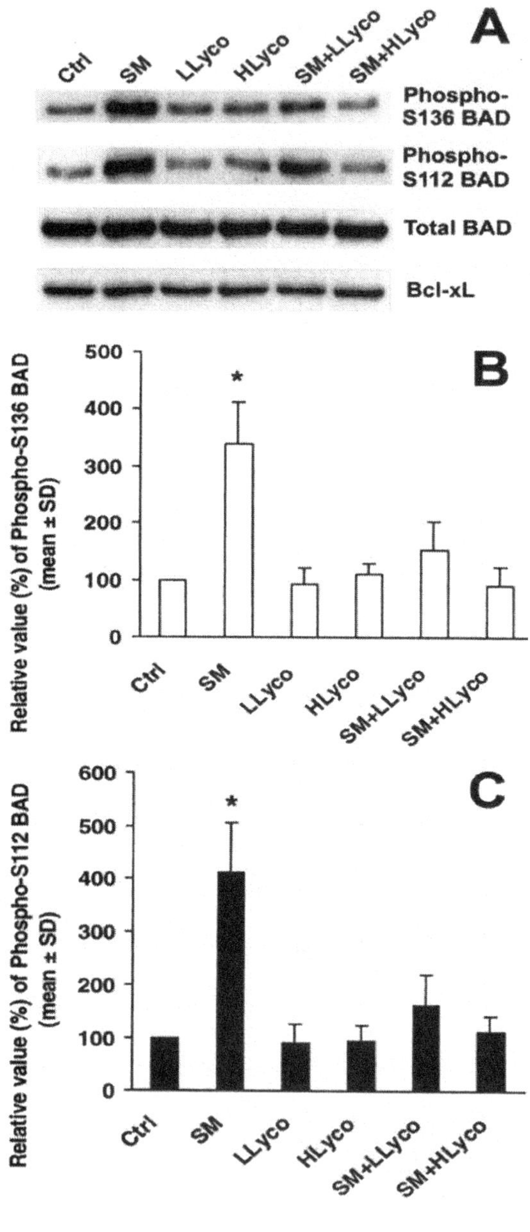

Fig. 10.3. Expressions of phosphorylation of BAD (Ser 136 and Ser 112), total BAD, and Bcl-xL protein levels in lung tissue from six groups of ferrets (Ctrl, control; SM, smoke-exposed; LLyco, low-dose lycopene supplemented; HLyco, high-dose lycopene supplemented; SM+LLyco, smoke-exposed + low-dose lycopene supplemented; and SM+HLyco, smoke-exposed + high-dose lycopene supplemented). (A) Representative Western blot analysis for phosphorylated BAD-Ser 136, phosphorylated BAD-Ser 112, total BAD and Bcl-xL protein levels. The sizes of the detected BAD and phosphorylated BAD were 23 kDa. The size of the detected Bcl-xL was 30 kDa. (B) The intensity of the protein signal of phosphorylated BAD-Ser 136, determined by densitometry (n = 6 ferrets/group) and expressed by the relative values (means ± SD). (C) The intensity of the protein signal of phosphorylated BAD-Ser 112, determined by densitometry (n = 6 ferrets/group) and expressed by the relative values (means ± SD). *Different from the other groups, $P < 0.05$. *Source:* Reference 40.

trols (Fig. 10.3), whereas lycopene supplementation at either a low or high dose prevented the smoke-induced BAD phosphorylation at both Ser 112 and Ser 136. Multiple signaling pathways may be involved in this process because the PI3K appeared to mediate survival factor–induced phosphorylation of BAD Ser 136, whereas MAPK are thought to mediate survival factor–induced phosphorylation of BAD Ser 112. Our observation is in agreement with a previous study that reported that IGFBP-3 can inhibit both PI3K/Akt/PKB and MAPK signaling pathways in non-small cell lung cancer (28). Our results demonstrate the importance of IGFBP-3 in the regulation of smoke-induced lung lesion, proliferation, and apoptosis, suggesting that IGFBP-3 is a molecular target of lycopene for the prevention of lung cancer.

Lycopene Metabolism in Smoke-Exposed Environment

Both the conflicting results of the β-carotene intervention trials in cigarette smokers (which used high doses of β-carotene and increased lung cancer risk) vs. the observational epidemiologic studies showing that diets high in fruits and vegetables containing carotenoids (but at much lower concentrations than in the intervention studies) are associated with a decreased risk for lung cancer, and the conflicting results of lycopene effects on lung carcinogenesis in animal studies (e.g., enhancement of BaP-induced-mutagenesis by lycopene in lung of *LacZ* mice) (44) motivated us to focus our attention on the dosage of lycopene supplementation and interaction of lycopene metabolism with cigarette smoke. One of the important aspects is that humans can absorb and accumulate significant amounts of carotenoids in the tissues and these carotenoids can undergo extensive oxidation into various carotenoid metabolites. Therefore, both the beneficial and adverse effects of carotenoids may be due to their metabolites or decomposition products (45). Previously, we proposed that the harmful effect of β-carotene supplementation in smokers was associated with the pharmacologic doses of β-carotene used in the human intervention studies and the free radical–rich atmosphere in the lungs of cigarette smokers (46–48). With pharmacologic (high) dose β-carotene supplementation, the environment of the lungs of cigarette smokers enhances β-carotene breakdown to produce oxidative by-products, such as β-apo-carotenals and β-carotene-epoxides. These oxidative metabolites of β-carotene may promote lung carcinogenesis by several mechanisms, e.g., enhancement of retinoic acid catabolism (40), downregulation of retinoic acid receptor (RAR)β, which functions as a tumor suppressor, upregulation of protooncogene gene (c-Jun and c-Fos) expression (46,48), and an increase in the binding of metabolites of benzo[a]pyrene to DNA (49). In contrast, we showed that in ferrets *in vivo* (48) and in normal human bronchial epithelial cells *in vitro* (50), low-dose β-carotene treatment, such as would be provided by consuming ~5–9 servings of fruits and vegetables/d, had no detrimental effects, but rather a protective effect against cigarette smoke-induced lung damage and benzo[a]pyrene-reduced RARβ expression. These findings indicated that β-carotene at low dose or its metabolites at a low concentration can act as anticarcinogenic agents.

However, an important question that remains to be answered is whether the anticarcinogenic effect of lycopene is also related to the lycopene dose that is administered *in vivo*. More specifically, would a smaller dose of lycopene provide optimal protection, thus reducing the risk of undesirable metabolic by-products and providing protection against lung carcinogenesis? Lowe *et al.* (51) demonstrated that lycopene and β-carotene protect against oxidative DNA damage (induced by xanthine/xanthine oxidase) in HT29 cells at relatively low concentrations (1–3 μmol/L), but lose this capacity at higher concentrations (4–10 μmol/L). Yeh and Hu (52) demonstrated that lycopene and β-carotene at a high concentration (20 μmol/L) significantly enhanced levels of lipid peroxidation induced by a lipid-soluble radical generator [2,2′-azobis(2,4-dimethylvaleronitrile)]. Therefore, lycopene and β-carotene appear to behave similarly under *in vitro* oxidative conditions. Our results show that lycopene supplementation at both a low and high dose for 9 wk significantly increased the concentrations of lycopene in both plasma and lung tissue of the ferrets (Table 10.2). The concentration of plasma lycopene (range from 226 to 373 nmol/L) in ferrets after lycopene supplementation was similar to the lycopene concentration (range 29–350 nmol/L) reported in humans (53,54). Furthermore, the lycopene concentrations in the lungs of ferrets that were given a low dose of lycopene (equivalent to 15 mg/d in humans) reached 342 nmol/kg, which is within the range of lung lycopene concentration in normal humans (100–500 nmol/kg) (55). We also observed that lycopene concentration in ferrets supplemented with a high dose of lycopene (equivalent to 60 mg/d in humans) increased 3.4-fold in lung tissue and 1.6-fold in plasma, compared with ferrets supplemented with a low dose of lycopene (equivalent to 15 mg/d in humans). The observation that a greater increase in lycopene concentrations occurred in lung tissue than in plasma after lycopene supplementation should be studied further in terms of possible harmful effects of lycopene supplementation. It is important to point out that, in our previous study, when the ferrets were supplemented with β-carotene at a dose of 30 mg/d, the concentration of β-carotene in the lungs of ferrets was 26 μmol/kg lung tissue, which was associated with an enhanced develop-

TABLE 10.2
Plasma and Lung Concentrations of Lycopene in Six Groups of Ferrets After 9 Weeks of Treatment[a,b]

Treatment	Plasma lycopene (nmol/L)	Lung lycopene (nmol/kg)
Control (sham smoke exposure)	ND	ND
Low-dose lycopene (15 mg/d)[c]	226 ± 35[d]	342.2 ± 42.3[d]
High-dose lycopene (60 mg/d)	373 ± 60[e]	1159.2 ± 145[e]
Smoke exposure	ND	ND
Low-dose lycopene (15 mg/d)	142 ± 36[f]	29.2 ± 6.8[f]
High-dose lycopene (60 mg/d)	228 ± 33[d]	100.3 ± 16.5[g]

[a]*Source:* Reference 40.
[b]Values are expressed as means ± SD, n = 6 ferrets/group. Means in a column with different superscript letters differ, $P < 0.05$. ND, not detected.
[c]Ferrets were given the dosage that was equivalent to (x) mg/d in humans.

ment of lung squamous metaplasia induced by cigarette smoke exposure (46). In the present study, the concentration of lycopene in the lungs was only 1.2 μmol/kg lung tissue in ferrets supplemented with lycopene at a dose of 60 mg/d, which caused no harmful effect but rather prevented the development of lung squamous metaplasia and cell proliferation induced by smoke exposure. The different outcomes between the lycopene and β-carotene studies in ferrets may be due to the differences in the levels of carotenoids that accumulated in lung tissue.

In addition, we showed that smoke exposure decreased plasma and lung lycopene concentrations in both the low- and high-dose lycopene groups; the percentage decrease in lycopene concentrations was ~40% in plasma for both the low- and high-dose lycopene groups and 90% in lung tissue for both groups (Table 10.2). Smoke exposure decreased the elevated lycopene concentrations in plasma and lung tissue of ferrets supplemented with lycopene, which is consistent with the data from National Health and Nutrition Examination Survey III (NHANES III), which found that smokers had lower serum levels of lycopene than nonsmokers (56). Very recently, we investigated whether the β-carotene excentric cleaving enzyme, β-carotene-9′,10′-monooxygenase, which was cloned from human and mouse tissues (57), can cleave lycopene at the 9′,10′ double bond to produce apo-10′-lycopenoid. Using HPLC analysis, we demonstrated that a homogenate of COS-1 cells overexpressing ferret carotene-9′,10′-monooxygenase by transient transfection can cleave both β-carotene and lycopene at the 9′,10′ double bond to produce apo-10′-carotenal and apo-10′-lycopenal, respectively (Hu, K.Q., and Wang, X.D., unpublished data). Interestingly, the expression of carotene-9′,10′-monooxygenase in the lungs of ferrets was slightly induced by smoke exposure but greatly induced by lycopene supplementation with or without smoke exposure. Further, the production of apo-10′-lycopenol in the lungs of lycopene-supplemented ferrets was enhanced by smoke exposure, compared with ferrets supplemented with lycopene alone (Liu, C., and Wang, X.D., unpublished data). The importance of these findings warrants further investigation.

Summary

IGFBP-3 is the most abundant IGF binding protein in human serum and was shown to be a growth inhibitory and apoptosis-inducing molecule, *via* IGF-dependent and -independent mechanisms. Lycopene supplementation reversed the decrease in plasma IGFBP-3 concentrations and the increase in the IGF-1/IGFBP-3 ratio induced by smoke exposure. Lycopene supplementation substantially inhibited smoke-induced squamous metaplasia and proliferating cellular nuclear antigen expression in the lungs of ferrets. Furthermore, the elevated phosphorylation of BAD and downregulated apoptosis induced by cigarette smoke in the lungs of ferrets was prevented by lycopene supplementation. Lycopene supplementation increased plasma and lung tissue levels of lycopene, but the elevated levels of lycopene in both the plasma and lungs of ferrets supplemented with lycopene were lowered by smoke exposure. These studies suggest that lycopene may mediate its protective effects against smoke-induced lung lesions in ferrets through upregulating IGFBP-3 as a molecular target, which promotes apoptosis and inhibits cell proliferation, thereby decreasing the risk of lung cancer. The exact mechanism(s) of lycopene targeting of IGFBP-3 warrant further investigation.

Acknowledgments

This material is based upon work supported by the National Institutes of Health Grant CA104932 and the U.S. Department of Agriculture, under agreement No. 1950–51000–064. Any opinions, findings, conclusions, or recommendations expressed in this publication are those of the author(s) and do not necessarily reflect the views of the U.S. Department of Agriculture.

References

1. Giovannucci, E. (1999) Tomatoes, Tomato-Based Products, Lycopene, and Cancer: Review of the Epidemiologic Literature, *J. Natl. Cancer Inst. 91:* 317–331.
2. Giovannucci, E. (2002) A Review of Epidemiologic Studies of Tomatoes, Lycopene, and Prostate Cancer, *Exp. Biol. Med. 227:* 852–859.
3. Giovannucci, E. (1999) Insulin-Like Growth Factor-1 and Binding Protein-3 and Risk of Cancer, *Horm. Res. 51 (Suppl. 3):* 34–41.
4. Clinton, S.K., Emenhiser, C., Schwartz, S.J., Bostwick, D.G., Williams, A.W., Moore, B.J., and Erdman, J.W., Jr. (1996) *cis-trans* Lycopene Isomers, Carotenoids, and Retinol in the Human Prostate, *Cancer Epidemiol. Biomark. Prev. 5:* 823–833.
5. Arab, L., Steck-Scott, S., and Bowen, P. (2001) Participation of Lycopene and Beta-Carotene in Carcinogenesis: Defenders, Aggressors, or Passive Bystanders? *Epidemiol. Rev. 23:* 211–230.
6. Stahl, W., and Sies, H. (1996) Lycopene: A Biologically Important Carotenoid for Humans? *Arch. Biochem. Biophys. 336:* 1–9.
7. Agarwal, S., and Rao, A.V. (2000) Tomato Lycopene and Its Role in Human Health and Chronic Diseases, *Can. Med. Assoc. J. 163:* 739–744.
8. Bertram, J.S. (1999) Carotenoids and Gene Regulation, *Nutr. Rev. 57:* 182–191.
9. Stahl, W., von Laar, J., Martin, H.D., Emmerich, T., and Sies, H. (2000) Stimulation of Gap Junctional Communication: Comparison of Acyclo-Retinoic Acid and Lycopene, *Arch. Biochem. Biophys. 373:* 271–274.
10. Levy, J., Bosin, E., Feldman, B., Giat, Y., Miinster, A., Danilenko, M., and Sharoni, Y. (1995) Lycopene Is a More Potent Inhibitor of Human Cancer Cell Proliferation than Either Alpha-Carotene or Beta-Carotene, *Nutr. Cancer 24:* 257–266.
11. Karas, M., Amir, H., Fishman, D., Danilenko, M., Segal, S., Nahum, A., Koifmann, A., Giat, Y., Levy, J., and Sharoni, Y. (2000) Lycopene Interferes with Cell Cycle Progression and Insulin-Like Growth Factor I Signaling in Mammary Cancer Cells, *Nutr. Cancer 36:* 101–111.
12. Bertram, J.S., and Bortkiewicz, H. (1995) Dietary Carotenoids Inhibit Neoplastic Transformation and Modulate Gene Expression in Mouse and Human Cells, *Am. J. Clin. Nutr. 62:* 1327S–1336S.
13. Yu, H., and Rohan, T. (2000) Role of the Insulin-Like Growth Factor Family in Cancer Development and Progression, *J. Natl. Cancer Inst. 92:* 1472–1489.

14. Jerome, L., Shiry, L., and Leyland-Jones, B. (2003) Deregulation of the IGF Axis in Cancer: Epidemiological Evidence and Potential Therapeutic Interventions, *Endocr. Relat. Cancer 10:* 561–578.
15. Zha, J., Harada, H., Yang, E., Jockel, J., and Korsmeyer, S.J. (1996) Serine Phosphorylation of Death Agonist BAD in Response to Survival Factor Results in Binding to 14-3-3 not Bcl-x(L), *Cell 87:* 619–628.
16. Zhou, X.M., Liu, Y., Payne, G., Lutz, R.J., and Chittenden, T. (2000) Growth Factors Inactivate the Cell Death Promoter BAD by Phosphorylation of Its BH3 Domain on Ser155, *J. Biol. Chem. 275:* 25046–25051.
17. del Peso, L., Gonzalez-Garcia, M., Page, C., Herrera, R., and Nunez, G. (1997) Interleukin-3-Induced Phosphorylation of BAD Through the Protein Kinase Akt, *Science 278:* 687–689.
18. Peruzzi, F., Prisco, M., Dews, M., Salomoni, P., Grassilli, E., Romano, G., Calabretta, B., and Baserga, R. (1999) Multiple Signaling Pathways of the Insulin-Like Growth Factor 1 Receptor in Protection from Apoptosis, *Mol. Cell. Biol. 19:* 7203–7215.
19. Shimamura, A., Ballif, B.A., Richards, S.A., and Blenis, J. (2000) Rsk1 Mediates a MEK-MAP Kinase Cell Survival Signal, *Curr. Biol. 10:* 127–135.
20. Gilmore, A.P., Valentijn, A.J., Wang, P., Ranger, A.M., Bundred, N., O'Hare, M.J., Wakeling, A., Korsmeyer, S.J., and Streuli, C.H. (2002) Activation of BAD by Therapeutic Inhibition of Epidermal Growth Factor Receptor and Transactivation by Insulin-Like Growth Factor Receptor, *J. Biol. Chem. 277:* 27643–27650.
21. Rechler, M. (1997) Growth Inhibition by Insulin-Like Growth Factor (IGF) Binding Protein-3—What's IGF Got to Do with It? *Endocrinology 138:* 2645–2647.
22. Gucev, Z.S., Oh, Y., Kelley, K.M., and Rosenfeld, R.G. (1996) Insulin-Like Growth Factor Binding Protein 3 Mediates Retinoic Acid- and Transforming Growth Factor Beta2-Induced Growth Inhibition in Human Breast Cancer Cells, *Cancer Res. 56:* 1545–1550.
23. Huynh, H., Pollak, M., and Zhang, J.C. (1998) Regulation of Insulin-Like Growth Factor (IGF) 2 and IGF Binding Protein 3 Autocrine Loop in Human PC-3 Prostate Cancer Cells by Vitamin D Metabolite 1,25(OH)2D3 and Its Analog EB1089, *Int. J. Oncol. 13:* 137–143.
24. Buckbinder, L., Talbott, R., Velasco-Miguel, S., Takenaka, I., Faha, B., Seizinger, B.R., and Kley, N. (1995) Induction of the Growth Inhibitor IGF-Binding Protein 3 by p53, *Nature 377:* 646–649.
25. Rajah, R., Valentinis, B., and Cohen, P. (1997) Insulin-Like Growth Factor (IGF)-Binding Protein-3 Induces Apoptosis and Mediates the Effects of Transforming Growth Factor-B1 on Programmed Cell Death Through a p53 and IGF-Independent Mechanism, *J. Biol. Chem. 272:* 12181–12188.
26. Butt, A.J., and Williams, A.C. (2001) IGFBP-3 and Apoptosis—A License to Kill? *Apoptosis 6:* 199–205.
27. Butt, A.J., Fraley, K.A., Firth, S.M., and Baxter, R.C. (2002) IGF-Binding Protein-3-Induced Growth Inhibition and Apoptosis Do Not Require Cell Surface Binding and Nuclear Translocation in Human Breast Cancer Cells, *Endocrinology 143:* 2693–2699.
28. Lee, H.Y., Chun, K.H., Liu, B., Wiehle, S.A., Cristiano, R.J., Hong, W.K., Cohen, P., and Kurie, J.M. (2002) Insulin-Like Growth Factor Binding Protein-3 Inhibits the Growth of Non-Small Cell Lung Cancer, *Cancer Res. 62:* 3530–3537.

29. Hochscheid, R., Jaques, G., and Wegmann, B. (2000) Transfection of Human Insulin-Like Growth Factor-Binding Protein 3 Gene Inhibits Cell Growth and Tumorigenicity: A Cell Culture Model for Lung Cancer, *J. Endocrinol. 166:* 553–563.
30. Yu, H., Spitz, M.R., Mistry, J., Gu, J., Hong, W.K., and Wu, X. (1999) Plasma Levels of Insulin-Like Growth Factor-I and Lung Cancer Risk: A Case-Control Analysis, *J. Natl. Cancer Inst. 91:* 151–156.
31. London, S.J., Yuan, J.M., Travlos, G.S., Gao, Y.T., Wilson, R.E., Ross, R.K., and Yu, M.C. (2002) Insulin-Like Growth Factor 1, IGF-Binding Protein 3, and Lung Cancer Risk in a Prospective Study of Men in China, *J. Natl. Cancer Inst. 94:* 749–754.
32. Wakai, K., Ito, Y., Suzuki, K., Tamakoshi, A., Seki, N., Ando, M., Ozasa, K., Watanabe, Y., Kondo, T., Nishino, Y., and Ohno, Y. (2002) Serum Insulin-Like Growth Factors, Insulin-Like Growth Factor-Binding Protein-3, and Risk of Lung Cancer Death: A Case-Control Study Nested in the Japan Collaborative Cohort (JACC) Study, *Jpn. J. Cancer Res. 93:* 1279–1286.
33. Yu, H., Spitz, M.R., Mistry, J., Gu, J., Hong, W.K., and Wu, X. (1999) Plasma Levels of Insulin-Like Growth Factor-1 and Lung Cancer Risk: A Case-Control Analysis, *J. Natl. Cancer Inst. 91:* 151–156.
34. Chang, Y.S., Kong, G., Sun, S., Liu, D., El-Naggar, A.K., Khuri, F.R., Hong, W.K., Lee, H.Y., and Gong, K. (2002) Clinical Significance of Insulin-Like Growth Factor-Binding Protein-3 Expression in Stage I Non-Small Cell Lung Cancer, *Clin. Cancer Res. 8:* 3796–3802.
35. Kaklamani, V.G., Linos, A., Kaklamani, E., Markaki, I., and Mantzoros, C. (1999) Age, Sex, and Smoking Are Predictors of Circulating Insulin-Like Growth Factor 1 and Insulin-Like Growth Factor-Binding Protein 3, *J. Clin. Oncol. 17:* 813–817.
36. Clinton, S.K. (1998) Lycopene: Chemistry, Biology, and Implications for Human Health and Disease, *Nutr. Rev. 56:* 35–51.
37. Holick, C.N., Michaud, D.S., Stolzenberg-Solomon, R., Mayne, S.T., Pietinen, P., Taylor, P.R., Virtamo, J., and Albanes, D. (2002) Dietary Carotenoids, Serum Beta-Carotene, and Retinol and Risk of Lung Cancer in the Alpha-Tocopherol, Beta-Carotene Cohort Study, *Am. J. Epidemiol. 156:* 536–547.
38. ichaud, D.S., Feskanich, D., Rimm, E.B., Colditz, G.A., Speizer, F.E., Willett, W.C., and Giovannucci, E. (2000) Intake of Specific Carotenoids and Risk of Lung Cancer in 2 Prospective US Cohorts, *Am. J. Clin. Nutr. 72:* 990–997.
39. Kim, D.J., Takasuka, N., Nishino, H., and Tsuda, H. (2000) Chemoprevention of Lung Cancer by Lycopene, *Biofactors 13:* 95–102.
40. Liu, C., Lian, F., Smith, D.E., Russell, R.M., and Wang, X.D. (2003) Lycopene Supplementation Inhibits Lung Squamous Metaplasia and Induces Apoptosis *Via* Up-Regulating Insulin-Like Growth Factor-Binding Protein 3 in Cigarette Smoke-Exposed Ferrets, *Cancer Res. 63:* 3138–3144.
41. Institute of Medicine (2001) *Dietary Reference Intakes: Vitamin A, Vitamin K, Arsenic, Boron, Chromium, Copper, Iodine, Iron, Manganese, Molybdenum, Nickel, Silicon, Vanadium, and Zinc*, National Academy Press, Washington.
42. Mucci, L.A., Tamimi, R., Lagiou, P., Trichopoulou, A., Benetou, V., Spanos, E., and Trichopoulos, D. (2001) Are Dietary Influences on the Risk of Prostate Cancer Mediated Through the Insulin-Like Growth Factor System? *BJU Int. 87:* 814–820.
43. Holmes, M.D., Pollak, M.N., Willett, W.C., and Hankinson, S.E. (2002) Dietary Correlates of Plasma Insulin-Like Growth Factor 1 and Insulin-Like Growth Factor Binding Protein 3 Concentrations, *Cancer Epidemiol. Biomark. Prev. 11:* 852–861.

44. Guttenplan, J.B., Chen, M., Kosinska, W., Thompson, S., Zhao, Z., and Cohen, L.A. (2001) Effects of a Lycopene-Rich Diet on Spontaneous and Benzo[a]pyrene-Induced Mutagenesis in Prostate, Colon and Lungs of the *lac*Z Mouse, *Cancer Lett. 164:* 1–6.
45. Wang, X.D. (2004) Carotenoid Oxidative/Degradative Products and Their Biological Activities, in *Carotenoids in Health and Disease*, Krinsky, N.I., Mayne, S.T., and Sies, H., eds., Marcel Dekker, New York, pp. 313–335.
46. Wang, X.D., Liu, C., Bronson, R.T., Smith, D.E., Krinsky, N.I., and Russell, R.M. (1999) Retinoid Signaling and Activator Protein-1 Expression in Ferrets Given Beta-Carotene Supplements and Exposed to Tobacco Smoke, *J. Natl. Cancer Inst. 91:* 60–66.
47. Wang, X.D., and Russell, R.M. (1999) Procarcinogenic and Anticarcinogenic Effects of Beta-Carotene, *Nutr. Rev. 57:* 263–272.
48. Liu, C., Wang, X.D., Bronson, R.T., Smith, D.E., Krinsky, N.I., and Russell, R.M. (2000) Effects of Physiological Versus Pharmacological Beta-Carotene Supplementation on Cell Proliferation and Histopathological Changes in the Lungs of Cigarette Smoke-Exposed Ferrets, *Carcinogenesis 21:* 2245–2253.
49. Salgo, M.G., Cueto, R., Winston, G.W., and Pryor, W.A. (1999) Beta Carotene and Its Oxidation Products Have Different Effects on Microsome Mediated Binding of Benzo[a]pyrene to DNA, *Free Radic. Biol. Med. 26:* 162–173.
50. Prakash, P., Liu, C., Hu, K.Q., Krinsky, N.I., Russell, R.M., and Wang, X.D. (2004) β-Carotene and β-apo-14′-Carotenoic Acid Prevent the Reduction of Retinoic Acid Receptor β in Benzo[a]pyrene-Treated Normal Human Bronchial Epithelial Cells, *J. Nutr. 134:* 667–673.
51. Lowe, G.M., Booth, L.A., Young, A.J., and Bilton, R.F. (1999) Lycopene and Beta-Carotene Protect Against Oxidative Damage in Ht29 Cells at Low Concentrations but Rapidly Lose This Capacity at Higher Doses, *Free Radic. Res. 30:* 141–151.
52. Yeh, S., and Hu, M. (2000) Antioxidant and Pro-Oxidant Effects of Lycopene in Comparison with Beta-Carotene on Oxidant-Induced Damage in Hs68 Cells, *J. Nutr. Biochem. 11:* 548–554.
53. Lu, Q.Y., Hung, J.C., Heber, D., Go, V.L., Reuter, V.E., Cordon-Cardo, C., Scher, H.I., Marshall, J.R., and Zhang, Z.F. (2001) Inverse Associations Between Plasma Lycopene and Other Carotenoids and Prostate Cancer, *Cancer Epidemiol. Biomark. Prev. 10:* 749–756.
54. Vogt, T.M., Mayne, S.T., Graubard, B.I., Swanson, C.A., Sowell, A.L., Schoenberg, J.B., Swanson, G.M., Greenberg, R.S., Hoover, R.N., Hayes, R.B., and Ziegler, R.G. (2002) Serum Lycopene, Other Serum Carotenoids, and Risk of Prostate Cancer in US Blacks and Whites, *Am. J. Epidemiol. 155:* 1023–1032.
55. Schmitz, H.H., Poor, C.L., Wellman, R.B., and Erdman, J.W., Jr. (1991) Concentrations of Selected Carotenoids and Vitamin A in Human Liver, Kidney and Lung Tissue, *J. Nutr. 121:* 1613–1621.
56. Wei, W., Kim, Y., and Boudreau, N. (2001) Association of Smoking with Serum and Dietary Levels of Antioxidants in Adults: NHANES III, 1988–1994, *Am. J. Public Health 91:* 258–264.
57. Kiefer, C., Hessel, S., Lampert, J.M., Vogt, K., Lederer, M.O., Breithaupt, D.E., and von Lintig, J. (2001) Identification and Characterization of a Mammalian Enzyme Catalyzing the Asymmetric Oxidative Cleavage of Provitamin A, *J. Biol. Chem. 276:* 14110–14116.

Chapter 11

Retinoids and Carotenoids as Cancer Chemopreventive Agents: Role of Upregulated Gap Junctional Communication

Laura M. Hix[a,b], Alex L. Vine[c], Samuel F. Lockwood[b], and John S. Bertram[c]

[a]Department of Cell and Molecular Biology, University of Hawaii at Manoa, Honolulu, HI 96822; [b]Hawaii Biotech, Incorporated, Aiea, HI 96701; and [c]Cancer Research Center of Hawaii, University of Hawaii at Manoa, Honolulu, HI 96813

Cancer Prevention: The Ultimate Goal

Prevention is always better than cure, and this old adage is particularly germane in the case of cancer for which cure, if at all possible, is frequently associated with highly cytotoxic agents and/or invasive procedures. With our growing understanding of the molecular etiology of cancer as being caused by specific mutations in genes associated with cell cycle control, apoptosis, and DNA repair enzymes, it has become apparent that strategies to limit DNA damage and/or increase the probability of DNA repair by inhibiting aberrant proliferation will decrease cancer incidence rates (1). Similarly, our growing understanding of the epidemiology of cancer on a worldwide basis has led to the realization that the dramatic differences in site-specific cancer incidence rates in different geographic locations are primarily a consequence of lifestyle differences rather than genetic differences in the affected populations. In a seminal study commissioned by the U.S. National Cancer Institute, it was estimated that ~70% of human cancer in the U.S. was a consequence of lifestyle and is thus, in theory at least, preventable (2). Unfortunately, the identification of etiological agents does not necessarily convert into effective cancer prevention. A prime example is the very limited success in limiting the use of tobacco products, first identified in the 1940s as responsible for the epidemic of lung cancer incidence in the West and currently considered to cause some 20–30% of cancer in Western countries (3). Other examples include the rapidly rising incidence of obesity in Western populations, which is linked to the increased incidence of many diseases including cancer, particularly of the breast and prostate, and the dramatically increasing rates of melanoma in fair-skinned populations, a phenomenon probably linked to ease of travel to warmer climates with high UV irradiation.

Cancer preventive strategies aimed at removing sources of carcinogen exposure, whether they are chemical as in the case of tobacco, physical, as in the case of UV exposure, or of complex physiology as in the case of diet and obesity, have been termed primary cancer chemoprevention. As exemplified above, such strategies have had mixed success. Others, such as the widespread elimination of asbestos, controls on aflatoxin contamination of foods, and strict limits on exposure to ionizing radiation, have been more successful. However, as is clear from the cancer incidence rates presented in Figures 11.1A and B, incidence rates at many anatomic sites have remained essentially unchanged over ~70 years of observation. Success in primary prevention can be observed in the recently decreasing incidence rates of lung cancer, particularly in men (Fig. 11.1B), which parallel earlier declines in tobacco consumption (3). Also of note are the decreasing rates of stomach cancer in both sexes; this is considered to be an inadvertent consequence of the widespread introduction of refrigeration and canning to preserve meats, leading to the decreased consumption of salt-cured meat products containing carcinogenic nitrosamines.

Secondary cancer prevention relates to inhibiting the consequences of carcinogen exposure. This strategy and the use of retinoids or carotenoids to inhibit the process of carcinogenesis will be the subject of this review. Unfortunately, this strategy has had only limited success in clinical situations either as a consequence of inappropriate intervention, dosage or, in the case of retinoids, unacceptable toxicity. Recently, however, inhibitors of cyclooxygenase-2 (COX-2), the inducible form of this enzyme shown to be elevated in preneoplastic and neoplastic tissues, demonstrated effectiveness in decreasing the incidence of preneoplastic polyps in the colon in individuals suffering from an inheritable form of this disease (4).

Tertiary cancer prevention relies upon the identification and removal of preneoplastic lesions; the success of this approach can be readily observed with the dramatically decreasing rates of uterine cancer resulting from the widespread application of cervical cancer screening (the Pap test) beginning in the 1940s (Fig. 11.1A). At present, this approach is unfortunately limited to tissues that are easily accessed. In 1987, we reviewed the prospects for cancer chemoprevention in some detail with emphasis on the epidemiologic, molecular, and public health aspects of cancer prevention (5). There is currently intense interest in the development of molecular tools for remote detection of aberrant cells.

Retinoids and Carotenoids as Cancer Preventive Agents

Because the physiologically active form of vitamin A, retinoic acid, is derived from carotenoid precursors either directly in the diet, or indirectly through consumption of animal products, there is a natural tendency to group both classes of compounds into a single entity in terms of cancer prevention. However, as will be discussed below, it is now apparent that certain carotenoids that are chemically incapable of being converted to retinoic acid (i.e., the non-provitamin A carotenoids) possess chemopreventive activity in experimental models of cancer and have been associated with lower cancer risk in epidemiologic studies. Thus, wherever possible in this review, these two classes will be discussed separately. Complicating this separation on the basis of chemistry is the knowledge that both agents upregulate expression of the putative tumor suppressor gene, connexin 43 (Cx43); however, this upregulation

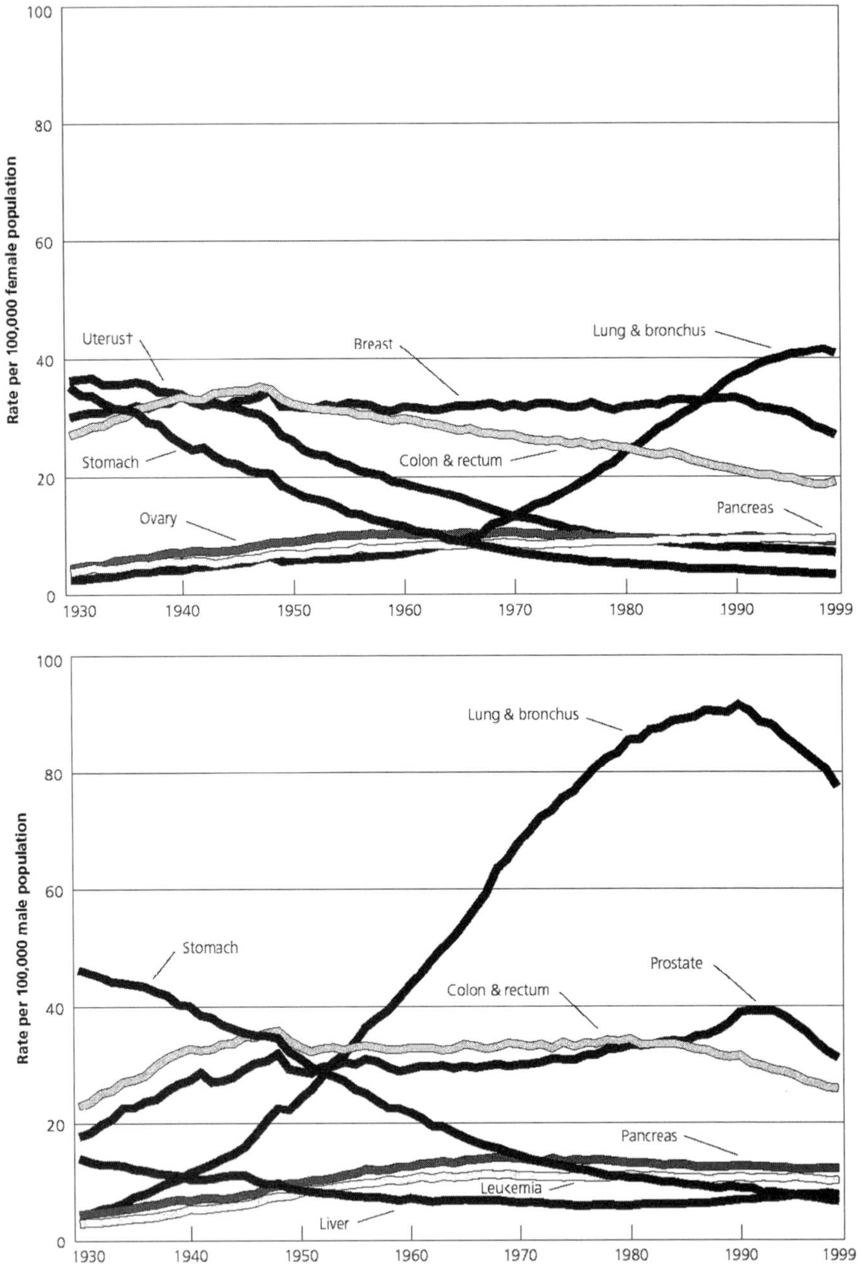

Fig. 11.1. Cancer deaths rates in the USA/100,000 population over the period 1930–1999. Panel A: female rates; B: male death rates. Reproduced by permission American Cancer Society.

appears to involve separate molecular pathways, again emphasizing the need for separate discussion of retinoids and carotenoids as cancer preventive agents. Because upregulation of Cx43 appears central to the ability of both classes of agent to inhibit neoplastic transformation in cell cultures, the role of connexin-mediated cell-cell communication will be discussed first so that actions of retinoids and carotenoids can be viewed in this context.

Proposed Role of Cell-Cell Communication in Suppressing Carcinogenesis

Interest in the role of cell-cell communication and carcinogenesis stemmed from early work conducted in the laboratory of the late Dr. Charles Heidelberger during our development of the C3H/10T1/2 (10T1/2) cell model of *in vitro* cell transformation. In this model, cells are seeded at a low initial density, treated for 24 h with a chemical carcinogen, then allowed to proliferate to form a confluent, growth-inhibited monolayer. In the absence of carcinogen exposure, cells form a monolayer with low rates of proliferation that is stable over many weeks of culture. Carcinogen exposure results in a small proportion of surviving cells, typically <1%, becoming carcinogen-initiated. These cells are latent until ~4 wk after carcinogen exposure; at that point, a few individual cells within a colony of initiated cells begin proliferation to form a focus of morphologically aberrant cells, which progressively invade surrounding normal cells to form a transformed focus (Fig. 11.2). If cells in the focus are cloned and injected into a syngeneic mouse ultimately lethal tumors develop at the site of injection. No such tumors develop if nontransformed cells are similarly injected (6). This model has been widely accepted as a quantitative assay for chemical and physical carcinogens (7).

It was noted early that when the number of cells exposed to the carcinogen was progressively increased, thus increasing the theoretical number of cells at risk for transformation, then paradoxically the number of transformed foci decreased, and foci that developed were at the extreme edge of the cell monolayer (6). Similarly, when the final cell density achieved at confluence was increased by increasing the concentration of serum in the growth medium, transformation frequency was again decreased (8). In the former situation, increased cell seeding density would lead to decreased colony size at confluence, whereas in the latter situation, increased saturation densities would lead to cell crowding, with both situations causing increased cell-cell interactions. The mechanistic significance of these observations was not understood initially; we now know them to be the result of increased cell-cell communication achieved *via* gap junctions.

Retinoids as Cancer Preventive Agents

With the development of the 10T1/2 system as a reproducible, quantitative assay system for neoplastic transformation, it became possible to assess the ability of potential inhibitors of this process to prevent neoplastic transformation. The term chemopre-

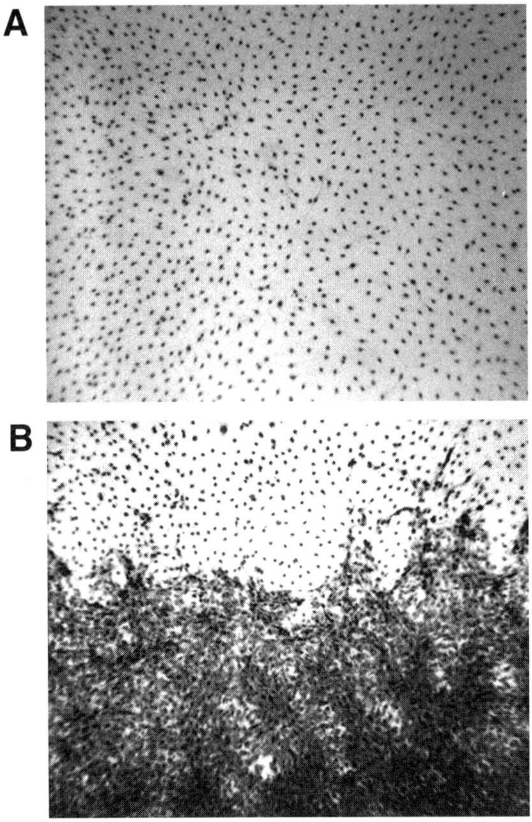

Fig. 11.2. Photomicrographs of 10T1/2 cells stained with Giemsa. Upper panel: cell monolayer showing the highly ordered and low-density organization of cells; lower panel: edge of a transformed focus resulting from exposure to 5 μL of 3-methylcholanthrene (MCA) for 24 h, 5 wk earlier. Magnification: X75.

vention or chemopreventive agents was coined by Dr. Sporn to describe compounds that could inhibit the carcinogenic process. The term retinoids was also introduced by this investigator to encompass natural or synthetic derivatives of vitamin A that could inhibit squamous metaplasia in the vitamin A–deficient hamster trachea (9). Such agents were demonstrated to have activity as chemopreventive agents in whole-animal studies of carcinogenesis. We discovered that retinoids could inhibit in a dose-dependent manner the production of neoplastic foci induced in 10T1/2 cells by exposure to the polycyclic aromatic hydrocarbon methylcholanthrene (MCA). This inhibition occurred when retinoids were added long after exposure to the carcinogen, thus demonstrating that the inhibitory action was not due to interference with carcinogen metabolism, DNA binding, or with subsequent DNA repair processes. Moreover, if retinoids were removed from carcinogen-treated cultures, transformed foci appeared in such cultures at the same frequency as had occurred in replicate cultures not treated with retinoids, but after a delay of 3–4 weeks, precisely the same delay required for production of foci in carcinogen-only treated control cultures (10). It thus became apparent that retinoids were not inhibiting the formation of carcinogen-initiated cells, but were arresting the progression of these cells to cells with a neoplastic phenotype. This effect appeared specific to initiated cells because when retinoids were added to cultures in which transformation had already occurred, as demonstrated microscopically, these transformed cells continued to proliferate to form a macroscopic focus.

Thus, the actions of retinoids in inhibiting transformation in 10T1/2 cultures can be summarized as follows: (i) Inhibition occurs in the postinitiation phase of carcinogenesis and thus does not interfere with carcinogen-induced DNA damage or its processing. (ii) Inhibition is reversible and thus does not involve selective cytotoxicity to initiated cells or the irreversible differentiation of such cells. (iii) Inhibition does not involve activating the immune system because no such effector cells exist in this cloned fibroblast population. (iv) Inhibition does not suppress expression of the transformed phenotype because transformed cells still produce foci in the presence of retinoids. As will be discussed below, inhibition of transformation appears directly related to the ability of retinoids to enhance cell-cell interactions, later discovered to be due to upregulated gap junctional communication (GJC).

Carotenoids as Cancer Preventive Agents

Extensive epidemiologic evidence indicates that a diet rich in fruits and vegetables is protective against cancer at many sites, particularly in the lung. These studies led to a renewed emphasis on characterizing the nutritional and nonnutritional components of such a diet to determine more precisely the nature of the protective factors found in such diets. Again, a consistent finding was that the carotenoid component of these diets most closely correlated with decreased risk, a suggestion further substantiated by analyses of carotenoid concentrations in serum. Of interest, particularly in view of an earlier suggestion that it was the β-carotene component of these diets that conveyed decreased risk, was that protection seemed to be also conveyed by non-provitamin A carotenoids. When we studied the ability of a series of dietary carotenoids to inhibit carcinogen-induced neoplastic transformation in 10T1/2 cells, we discovered that both classes of carotenoid were active, a finding that appeared to rule out conversion to retinoid-like molecules (11). However, both retinoids and carotenoids had remarkably similar activities, i.e., both were found to inhibit in the postinitiation phase of carcinogenesis, both actions were reversible upon removal of the test compound, and both classes of compounds upregulated Cx43 expression in direct proportion to their chemopreventive properties (12,13).

Retinoids Modulate Cell-Cell Interactions

As discussed earlier, the transformation of 10T1/2 cells appears highly dependent upon the inhibition of cell population dynamics in crowded cultures or cultures in which carcinogen-treated cells can form only small colonies. If retinoids were indeed stabilizing initiated cells in a preneoplastic state, we argued that it should be possible to isolate such cells from carcinogen-exposed cultures. To achieve this, we cloned ~160 exposed cells and grew them in the absence or presence of inhibitory concentrations of retinoids. We predicted that between 2 and 4 of these cells would be initiated and would undergo transformation in the absence, but not in the presence of retinoids. We were successful in isolating 2 clones with these properties; in the absence of retinoids, cells underwent transformation that involved the entire cell culture dish, in contrast to the discrete formation of transformed foci in uncloned carcinogen-exposed populations (14).

With our possession of clones of carcinogen-initiated cells, we were able to test these interactions in a highly quantitative fashion. To explore the colony size dependence on transformation, we created a mixture of initiated and parental 10T1/2 cells and seeded these at increasing densities, so that when cells attained confluence, colonies formed by surviving cells would contain from 1000 to ~10 cells. These cultures were then maintained in the absence or presence of retinoids and the incidence of transformed foci monitored after 5 wk of culture as in standard transformation assays. As expected, in the presence of retinoids, no foci appeared, whereas in the absence of retinoids, the transformation frequency (i.e., the number of surviving initiated cells forming transformed foci), was directly proportional to the colony size attained. Thus, ~70% of initiated cells forming large colonies underwent transformation, whereas virtually zero transformation occurred in small colonies. That this phenomenon was not a consequence of the increased numbers of cell divisions necessary to form large colonies was demonstrated by allowing initiated cells to proliferate with repeated reseeding to prevent colony formation (15). Taken together, these results indicated that transformation occurred only when initiated cells were shielded from the influence of normal cells by being surrounded by other initiated cells, i.e., a communication gradient was established that did not penetrate to the center of large initiated cell colonies. As we later discovered, this communication from normal to initiated cells occurred through gap junctions and retinoids and carotenoids, by increasing cell communication between both initiated and normal cells, removing this shielding and allowing signals from normal cells to penetrate large colonies of initiated cells and suppress transformation. Although the chemical identity of this signal remains unknown, we demonstrated that one immediate effect of retinoid-enhanced junctional communication was to dramatically suppress the rate of proliferation in confluent cultures (16). Inappropriate proliferation is a known requirement for carcinogenesis, allowing the fixation of additional mutations and clonal expansion of preneoplastic cells; therefore, suppressed proliferation could well explain the ability of both retinoids and carotenoids to inhibit neoplastic transformation (1,17). This model of gap junctional inhibition of neoplastic transformation is shown in Figure 11.3.

Connexins and Carcinogenesis

At the time of our work with retinoids, GJC was not a widely recognized phenomenon; it was discovered serendipitously by Kanno and Lowenstein with the demon-

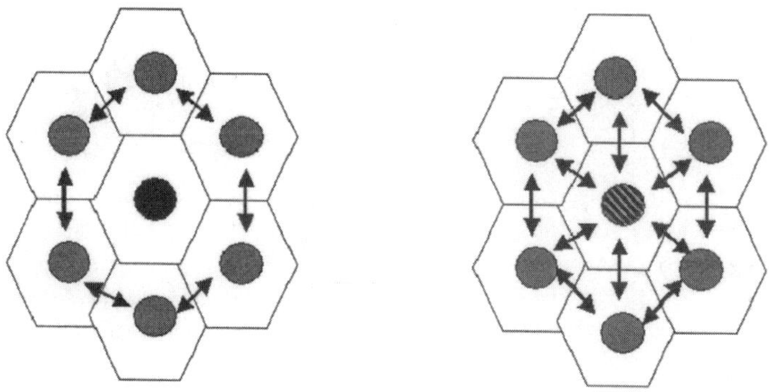

Fig. 11.3. Hypothetical model of growth control by gap junctional communication (GJC). A central carcinogen-initiated cell is surrounded by growth-inhibited normal cells (red nuclei). In the left diagram, the initiated cell is not in communication with normal cells and undergoes inappropriate proliferation (green nucleus). This allows clonal expansion and the progressive accumulation of additional mutations resulting in malignancy. In the right diagram, junctional communication of growth inhibitory signals from surrounding cells is upregulated by retinoids or carotenoids, the initiated cell itself becomes growth arrested, and progression to malignancy is delayed.

stration that cells in culture are electrically coupled. Interest in its role in carcinogenesis was stimulated by the observation that this coupling was disrupted upon malignant transformation of cells in culture (18), a finding that led to the hypothesis of control of proliferation through junctional communication (19). With the advent of molecular biology and the successful cloning of the structural unit of the most widely expressed gap junction protein, Cx43, these observations were extended and confirmed. It is now well established that with the obvious exception of red blood cells, virtually all cell types are in junctional communication with their neighbors, or in the case of leukocytes, can form junctions after activation (20). In addition, it is equally well established that most if not all human solid tumors are deficient in GJC (21) and, as we showed in the human cervix and oral cavity, downregulated expression of Cx43 is an early event in carcinogenesis, with detection occurring *via* dysplasia and leukoplakia, respectively (22). Downregulated expression seems only rarely due to inactivating mutations, although giving rise to several other diseases (23); instead, epigenetic silencing *via* DNA methylation or disrupted trafficking of connexins to the plasma membrane of tumor cells appears frequently responsible for disrupted GJC (24,25). Further supporting the role of GJC in suppressing carcinogenesis are studies in which GJC was inhibited by pharmacologic agents. Tumor promoters are agents that although not intrinsically carcinogenic, enhance the process of carcinogenesis (i.e., they have properties that are opposite to chemopreventive agents). A wide variety of tumor promoters were shown to inhibit GJC both *in vivo* and in cell culture (26). We found that the most widely studied tumor promoter, 12-*O*-tetradecanoylphorbol-13-acetate (TPA), directly antagonizes the ability of retinoids to suppress transformation (9).

Gap Junction Structure. Gap junctions are composed of small water-filled pores that directly connect the cytoplasm of adjacent cells and allow the transfer of water-soluble molecules <1000 Da. Each junction, termed a connexon, is formed by the assembly of 6 connexin proteins donated by one cell with 6 connexin proteins donated by the adjacent cell. These transmembrane proteins each possess 6 highly conserved cysteine residues in their extracellular domains; these are believed to form sulfhydryl bridges with cysteine residues donated by the adjacent cell to stabilize these interactions and form a tight seal, preventing entry of extracellular ions such as calcium into the cytoplasm (27). This organization is depicted diagrammatically in Figure 11.4.

It is now well established that a family of >20 connexins is expressed in mammals with cell and developmental specificity of expression. All connexins are believed to share the same organization in the plasma membrane as shown in Figure 11.4, with major differences between family members being restricted to the length and sequence of the C-terminal cytoplasmic domains. Not all connexins

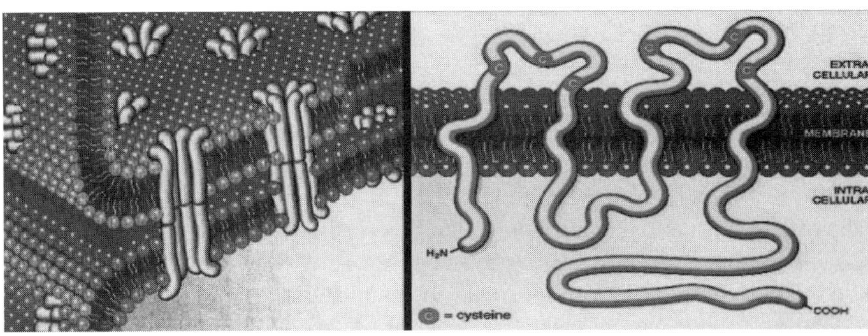

Fig. 11.4. Organization of connexins into the plasma membrane. Left panel: Diagrammatic cross section through an area of cell-cell contact containing gap junctions. Connexin proteins are shown in yellow traversing the phospholipid bilayer in the plasma membrane (blue). Each cell contributes 6 connexins to form a cylinder enclosing a central water-filled pore seen in cross section (foreground). Right panel: A single connexin molecule traverses the plasma membrane 4 times with both N- and C-terminal ends in the cytoplasm. Connexins assemble to form a connexon by forming 3 sulfhydryl bonds between the highly conserved cysteine residues present in each opposing loop (C). Thus, each connexon is bound by 18 sulfhydryl bonds to produce a tight seal blocking the entry of extracellular ions such as Ca^{++}. Reprinted from *Science and Medicine*, with permission.

are compatible, leading to the development of communication compartments. For example, in the heart, Cx43, the major connexin expressed in cardiomyocytes, is located in the intercalated disk and allows the rapid transfer of the contractile stimulus throughout the ventricle or auricle; its absence leads to arrhythmia and death (29). In contrast, connexins 40 and 45 are the major proteins expressed in the conduction fibers, designed for the rapid transit of information from auricle to ventricle (29).

Connexons transport nutrients, waste products, and information. The nature of the relevant signals transmitted between normal and transformed cells that lead to the suppression of transformation is unclear. Candidates include any molecules or ions that are water-soluble and <1000 Da in size; many entities with these characteristics were shown to traverse the gap junction. In a study in which donor cells were labeled with C^{14} glucose, the most abundant labeled molecule traversing gap junctions composed of Cx43, was ATP, a finding that likely reflects simply the high millimolar cytoplasmic levels of this molecule (30). More likely candidates for growth suppression would be the second messenger cAMP, which traverses gap junctions, and inositol triphosphate (IP3) whose transit through the junction is associated with the propagation of Ca^{++} waves (31). The avascular tissues of the lens and cornea appear to rely for survival on junctional communication of nutrients and waste products (32,33). Similarly, gap junctions in the placenta appear necessary for adequate transfer of nutrients (34). However, the supply of metabolic energy seems unlikely to explain the phenomena in cell culture where cells are bathed in a nutrient-rich medium.

Modulation of Connexin Function by Molecular Means. To specifically address the functional consequences of disrupted GJC, several investigators manipulated tumor cells to reestablish junctional communication by the use of molecular constructs to force expression of connexins. Comparable studies utilizing normal cells were compromised by technical problems of transfection and selection, and by the fact that most normal cells in culture are junctionally competent. A consistent finding after forced expression of Cx43 in malignant cells was a decrease in their malignant potential. Using a tetracycline-inducible promoter system to drive Cx43, we showed in a human cervical carcinoma cell line and in a human fibrosarcoma cell line, that compared with noninduced cells, Cx43 expression dramatically decreased anchorage-independent growth and decreased the ability to form tumors in immunocompromised nude mice (Fig. 11.5) (22,35). Using constitutive promoters, others found similar effects in a variety of human and animal malignant cell lines. The studies were reviewed recently (36). However, we have cautioned that when using constitutive promoters to express connexins, the interpretation of results may be problematic because of extensive clonal heterogeneity in cultured cells (25).

As an alternative approach to probe connexin function, a number of mouse strains were created in which different connexin family members were knocked out by homologous recombination. Notable here is the observation that mice in

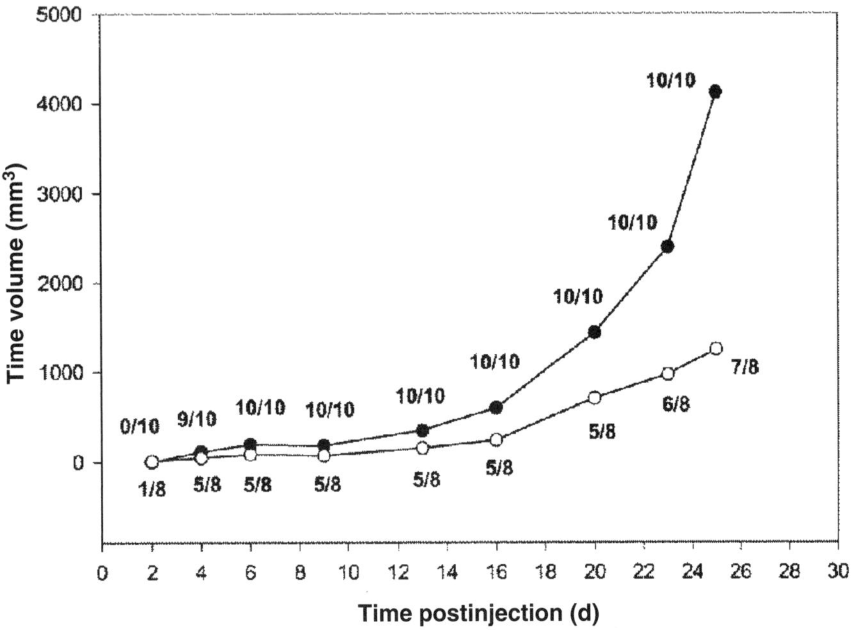

Fig. 11.5. Expression of Cx43 reduces the growth rate of human cervical carcinoma cells in the "nude" mouse. HeLa cells were engineered to express Cx43 under the influence of a bacterial promoter driven by doxycycline. Immunocompromised nude mice were injected subcutaneously with HeLa cells, then randomized to receive doxycycline (0.2 mg/mL in 5% sucrose) in the drinking water, or sucrose alone as controls. Tumor volumes were measured by calipers at the indicated times. (○-○) doxycycline treated; (●-●) sucrose controls. Numbers by each data point represent total number of tumors/number of tumor injections. From (37) with permission.

which connexin 32, a connexin extensively expressed in the liver, is inactivated become more susceptible to liver carcinogenesis (37).

The ability of neoplastic cells to grow in suspension, a phenomenon called anchorage-independent growth, has long been used to distinguish neoplastic cells from their normal counterparts. In normal epithelial and fibroblastic cells, contact with the extracellular matrix leads to the formation of focal contacts and activation of focal adhesion kinase that allow for cell replication after mitogenic stimulus. In the absence of such contacts, normal cells will not proliferate and frequently undergo apoptosis. The lack of this requirement in tumor cells as a result of inappropriate expression of this kinase, presumably reflects their acquired ability to migrate through and proliferate in inappropriate locations (38,39). The ability of forced expression of Cx43 to restore this requirement for cell proliferation was unexpected, but fits into some recently generated clinical data. Here, administration of fairly high doses of supplemental lycopene to patients with prostate cancer for 3 wk before radical prostatectomy, was associated with increased expression of Cx43 and decreased pathological severity of treated vs. control tumors, a finding most easily explained by induction of apoptosis in these tumors (40). If confirmed in a larger group of patients, these studies would indicate that at least in the case of lycopene and prostate cancer, carotenoids could have therapeutic potential. At present, it is unknown how Cx43 expression influences anchorage-independent growth. One possibility is that the plasma membrane–associated connexins act as a focus for the reassembly of cytoskeletal elements known to be disrupted during carcinogenesis. In this context, Cx43 is known to associate with microtubules (41) and with Z0-1, a constituent of epithelial tight junctions that can interact with cytoskeletal elements (42).

The influence of lycopene on proliferation of carcinoma cells is apparently not limited to its ability to modulate Cx43 expression. In studies conducted by others, growth stimulation of MCF7 mammary cancer cells by insulin-like growth factor (IGF)-1 was markedly reduced by physiologic concentrations of lycopene (43). Lycopene treatment markedly reduced the IGF-1 stimulation of tyrosine phosphorylation of insulin receptor substrate 1 and binding capacity of the activator protein (AP)-1 transcription complex, suggesting that effects on proliferation were due to interference in IGF-1 receptor signaling (43). Interactions are likely not limited to cell culture studies because epidemiologic investigations showed a strong inverse correlation between plasma lycopene and circulating IGF-1 (44). IGF-1 is an important mitogen for prostate cells; thus reports of lycopene/IGF-1 interactions in terms of both production of and responses to IGF-1 could be highly significant for the role of lycopene as a chemopreventive and possibly therapeutic agent in prostate cancer.

Mechanisms of Upregulated Cx43 Expression by Retinoids and Carotenoids

There is evidence that both retinoids and carotenoids upregulate expression at both the mRNA and protein levels. Moreover, increased Cx43 mRNA levels are not inhibited by cyclohexamide, indicating direct action, and activity does not involve an increase in mRNA half-life, indicating direct transcriptional activation. However, differences between retinoids and carotenoids do exist. Pharmacologic inhibitors of retinoid nuclear receptors (RAR) inhibited the activity of retinoids at the transcriptional and protein level, whereas they did not inhibit the activity of non-provitamin A carotenoids (Vine *et al.*, unpublished data). Unfortunately, the Cx43 promoter does not contain a known retinoid response element (RARE), and no clear picture has emerged of how this gene is regulated by other factors such as estrogen, stretch, transforming growth factor (TGF)-β, or thyroxin, which also increase its expression in a cell-type specific manner.

Do carotenoids function intact or conversely require metabolic activation? At present, it is unclear whether carotenoids function intact or require metabolic or spontaneous degradation to smaller molecules. However, several lines of evidence suggest that the carotenoids require chemical modification to regulate gene expression. From a pharmacologic perspective, this appears most evident in the case of lycopene, a C-40 polyunsaturated hydrocarbon with no functional groups. Our initial investigations demonstrated that although the retinoids were capable of upregulating Cx43 mRNA within ~6 h of treatment, carotenoids required ~3 times longer to produce the same response (45,46). Although differences in rates of transport of the active molecule to the nucleus could well explain these differences (retinoids, for example, have soluble binding proteins), this delay may also reflect the requirement for production of biologically active molecules. These molecules could be formed as a consequence of cellular metabolism, or the active agent(s) could be produced as a consequence of oxidative modification of the parent carotenoid molecule. If cellular metabolism is involved, then the responsible enzymes must be widely expressed because as discussed above, lycopene was shown to be active in this respect in mouse and human fibroblasts, and in human oral carcinoma cells. There is evidence that the recently discovered β, β-carotene 9′,10′-dioxygenase can cleave lycopene at the 9,10 position (47). The products of such cleavage, presumably C-9 and C-31 aldehydes and alcohols, would still not possess the β-ionone ring required of retinoids for activity, but could still activate other nuclear receptors and lead to transcriptional modulation. Of interest here is our recent discovery that several carotenoids can activate peroxisome proliferator-activated receptor (PPAR)-γ nuclear receptors and that a specific antagonist of PPAR, which inhibits ligand binding, interferes with the ability of carotenoids to upregulate Cx43 (Vine *et al.*, unpublished data). Of additional interest is the observation that PPAR agonists will also induce expression of the β-carotene 15,15′-dioxygenase by activating a PPAR-responsive element within the 5′ promoter element of this gene (48), implying that carotenoids may modulate their own metabolism. A surprising variety of chemical species are capable of activating PPAR; these include fatty acids, prostaglandins, fibrates, and thiazolidinediones, i.e., drugs utilized clinically in obesity and diabetes (49,50). Products of carotenoids derived by chemical or enzymatic oxidation have structural similarities to fatty acids and prostaglandins and thus could activate PPAR-γ.

It seems unlikely, however, particularly in the case of lycopene, that conversion to retinoids is responsible for the actions of carotenoids. Even for carotenoids with this potential, we were unable to demonstrate conversion. In a study utilizing 10T1/2 cells, cultures were treated with C^{14}-labeled β-carotene. Using sensitive HPLC methodology, we were unable to detect any of the expected products of conversion, i.e., retinal, retinaldehyde or retinoic acid (51). In a second study, we inhibited the enzyme-mediated catabolism of retinoic acid with liarozole. Treatment with this agent increased by ~1000-fold the ability of retinoic acid and of 4-oxo-retinoic acid to upregulate Cx43 expression in 10T1/2 cells; however, the activities of β-carotene and canthaxanthin were essentially unchanged. This provided strong evidence that treatment with β-carotene was not producing biologically relevant quantities of retinoic acid, nor was canthaxanthin being symmetrically degraded to the known biologically active retinoid 4-oxo-retinoic acid (52). In a third series of experiments, the ability of canthaxanthin to upregulate the retinoic acid-inducible gene RAR-β was evaluated. Here again, strong upregulated expression of Cx43 was detected, but without modulation of RAR-β expression (45). All of these studies indicated that carotenoids and retinoids had independent pathways of gene regulation. In related studies, several eccentric cleavage products of β-carotene were shown to induce growth arrest of human carcinoma cells *in vitro* by mechanisms not involving interactions with RAR (53). The lack of conversion of canthaxanthin to retinoids was not confirmed by others who presented evidence, utilizing the same 10T1/2 cell line, that conversion of canthaxanthin to 4-oxo retinoic acid was responsible for some or all of its ability to modulate Cx43 expression (54).

The activity of lycopene, if not due to the parent molecule, seems more likely a consequence of the production of biologically active oxidation products. Three such products were described, but none seems to possess the desired potency to fully explain the activity of lycopene on Cx43 expression. A cyclic compound with a 5-membered ring, 2,6-cyclolycopene-1, 5-diol, was described by Paetau *et al.* (55) to be present in human serum after consumption of tomato products. Its formation is likely a result of initial oxidation to an epoxide followed by hydrolysis. When tested in 10T1/2 cells, it was marginally more potent than lycopene itself in regulating Cx43 expression (56). However, it seems unlikely that this derivative is responsible for lycopene's activity because nearly quantitative conversion to this product would be required, yet the concentration in human serum is only ~10% of that of lycopene (55). Moreover, it is not known whether this molecule is produced in cell culture after lycopene treatment. Similar conclusions can be made regarding the biological significance of the second lycopene derivative, 2,7,11-trimethyl-tetradecahexaene-1,14-dial, produced as a consequence of chemical oxidation of lycopene with hydrogen peroxide/osmium tetroxide. Although this compound also upregulates Cx43 expression in 10T1/2 cells, it appears to be less potent than lycopene itself, which suggests that its contribution is minimal (57). Again, it is not known whether this dial is produced *in vivo* or in cell culture. A third molecule, acyclo-retinoic acid, also potentially formed by oxidation of lycopene, was evaluated in 10T1/2 cells. Although possessing activity, this was 500-fold lower than that of lycopene itself as an inducer of Cx43 expression, again implying a minimal contribution to the activity of lycopene in this respect (58).

These studies of lycopene metabolites, although essentially negative in not having discovered a derivative more active than lycopene itself, nevertheless do illustrate the structurally diverse nature of carotenoid derivatives capable of activating this gene. Although potential oxidation products can all be expected to be lipid-phase antioxidants, by virtue of their expected retention of some polyunsaturated conjugated double bonds, as pointed out earlier, antioxidant activity *per se* does not seem to explain activity. Potent antioxidants such as α-tocopherol do not upregulate Cx43 expression, and the *in vitro* antioxidant activity of diverse carotenoids did not correlate with their ability to modulate Cx43 expression (45). Additional studies are ongoing to determine whether activation of PPARγ, a highly promiscuous nuclear receptor, can indeed explain how all these compounds produce a common response.

Bioavailability of Carotenoids

In general, dietary carotenoids are poorly bioavailable due to their highly lipophilic nature and the fact that in many fruits and vegetables, carotenoids are present as crystalline inclusions trapped within the vegetable matrix. This was most clearly demonstrated in the case of lycopene, a carotenoid of growing interest because of epidemiologic evidence linking its consumption, principally in the form of tomato products, with a lower risk of prostate cancer (59). Here, processing the tomato to break down the matrix increased bioavailability several-fold (60). Most clinical studies with carotenoids have relied upon their commercial formulation into "beadlets" in which a microdisbursed solution of carotenoids in vegetable oil is distributed in a water-soluble matrix. At present, carotenoids suitable for clinical application are limited to β-carotene and lycopene.

Studies in experimental animals have further constraints. Most commonly employed laboratory animals absorb carotenoids extremely poorly even when administered in beadlet form, severely limiting the conduct of most carcinogenesis studies that employ mice or rats. Recently, ferrets emerged as an experimental model that absorbs β-carotene efficiently; they were used to probe the unfortunate interactions discovered between high-dose β-carotene and the lungs of active smokers (61). However, ferrets are expensive to purchase and maintain, are not genetically homogeneous as are laboratory rats and mice, and little is known of their susceptibility to carcinogenesis. These studies in cell culture described above were made possible only by our development of tetrahydrofuran (THF) as a solvent that allowed the production of a pseudosolution of carotenoids in cell culture medium, which was nontoxic and allowed for high bioavailability (62). However, THF is not an ideal solvent for animal or clinical studies.

Development of Water-Dispersible Derivatives of Carotenoids

To overcome the poor bioavailability of carotenoids after oral administration and to develop formulations capable of parenteral administration in humans, Hawaii Biotech, Inc. (HBI), located in Aiea, HI, developed a novel carotenoid derivative, the disodium disuccinate derivative of astaxanthin (dAST), Cardax™ (Fig. 11.6). This compound forms a pseudosolution in water at concentrations of up to 8.6 mg/mL (~10 mM). Dispersibility is achieved in aqueous solution secondary to self-assembly of disodium disuccinate astaxanthin monomers into supramolecular complexes. Monomeric solutions of compound can also be achieved by inclusion of ethanol at concentrations up to 50%, thereby disrupting this self-assembly but preserving aqueous solubility (63). Additionally, the formation of an oil:water:lecithin emulsion allows dAST to be formulated at concentrations >50 mg/mL for oral delivery (64,65). These synthetically modified formulations of astaxanthin

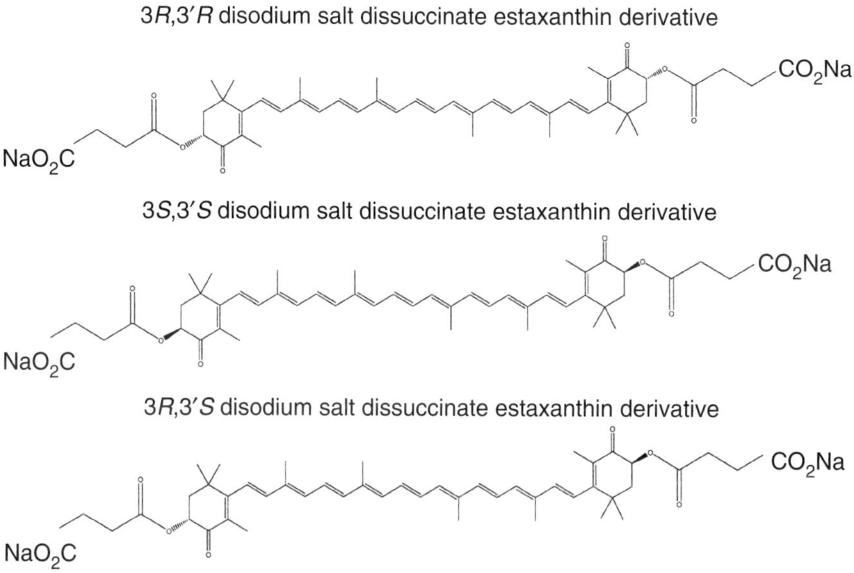

Fig. 11.6. Structures of the three stereoisomers of the disodium salt disuccinate astaxanthin derivatives present in dAST described in the current study (shown as the all-*E* geometric isomers). The racemic mixture of stereoisomers contains (3*S*, 3′*S*)-astaxanthin disuccinate, disodium salt; (3*R*, 3′*R*)-astaxanthin disuccinate, disodium salt; and (3*R*, 3′S; *meso*)-astaxanthin disuccinate, disodium salt in a 1:1:2 ratio.

offer three exciting possibilities: first, they can be rapidly delivered at high concentrations to tissues after either oral or parenteral administration; second, they should be capable of aqueous-phase scavenging of reactive oxygen species (ROS) (e.g., the superoxide anion released by neutrophils) with high efficiency; third, the ester-linkage of the succinate moieties should make this molecule susceptible to serum and cell esterases, leading to the release of free astaxanthin capable of acting as a chain-breaking antioxidant in the lipid environment of lipoprotein or cell membrane. dAST appears to fulfill all these expectations.

dAST is highly bioavailable after oral or intravenous administration. Administration of dAST as an oil:water:lecithin emulsion at a concentration of 50 mg/mL to mice by oral gavage at a dose of 500 mg/kg body weight required a volume of only ~25 μL. This resulted in the rapid appearance of free astaxanthin in serum, liver, and heart, with peak levels achieved ~6 h postadministration. At this time point, mean levels were 1760 nM in liver, 694 nM in heart, whereas plasma levels were ~25% of that found in liver (70). This demonstrated rapid absorption and cleavage to free astaxanthin in these mice together with accumulation in target tissues. Tissue levels were above the reported 50% effective dose of 200 nM for free astaxanthin as a radical scavenger, suggesting that oral administration of this novel carotenoid derivative could potentially achieve significant tissue protection against oxidative stress—in most cases with a single oral dose in emulsion vehicle. dAST was also administered in aqueous solution by intravenous tail vein injection to rats. There was a dose-dependent increase in plasma astaxanthin with a mean peak concentration of 612 nM achieved 24 h after four daily doses of 75 mg/kg (64).

Direct scavenging of the superoxide anion. When incubated with human polymorphonuclear leukocytes stimulated to release superoxide by exposure to phorbol ester, dAST dose dependently decreased the levels of superoxide anion detected by electron paramagnetic resonance spectroscopy utilizing the spin trap DEPMPO. Complete inhibition of the superoxide anion signal was achieved at millimolar concentrations of dAST (65). This *in vitro* test system generates supraphysiologic molar amounts of superoxide anion, levels much higher than those achieved in ischemic tissue *in vivo*. In addition, supramolecular assembly of dAST in aqueous solution of the monomeric compound limits the reactivity of the monomeric compound with aqueous-phase radicals. Therefore, it was postulated that much lower concentrations would be therapeutic in animal models *in vivo*, results that were subsequently achieved at nanomolar concentrations in both rats (64) and dogs (Lockwood, unpublished results).

dAST induces Cx43 expression *in vitro*. Problems associated with the delivery of dietary carotenoids to biological systems were discussed above. To determine whether the disuccinate astaxanthin derivative dAST had activity in addition to its antioxidant properties, we treated 10T1/2 cells with various formulations of dAST. To enhance both solubility and bioavailability, several EtOH/water formulations were tested for aggregation with UV/vis spectroscopy. The "solubility" of the derivatives was significantly enhanced by the use of 1:1 (50% EtOH) and 1:2 (33% EtOH) EtOH/water formulations. These formulations were demonstrated previously to maintain the carotenoid derivatives in monomeric form (65). Induction levels of Cx43, as determined by immunoblotting, were higher with the EtOH formulation at 10^{-5} M than for formulations in sterile water alone, demonstrating enhanced biological availability using EtOH as a co-solvent, as suggested by the previous physicochemical studies. The mixture of stereoisomers of dAST in pure aqueous formulation was able to upregulate Cx43 expression with equivalent or greater potency than that previously observed for other carotenoids in THF (12,45). Importantly, treated cells were found to assemble Cx43 into immunoreactive plaques in regions of cell-cell contact, consistent with the formation of gap junctions. This was confirmed by functional studies, utilizing a dye-microinjection technique, which demonstrated that treated cells were more extensively coupled than cells treated with solvent alone (67). Other modified carotenoids are being synthesized and tested (68). They too appear to have enhanced bioavailability and show promise in a variety of clinical conditions involving ROS and/or inappropriate proliferation.

Conclusions

Both retinoids and carotenoids inhibit experimentally induced carcinogenesis in the postinitiation phase, consistent with an ability to arrest the progression of preneoplastic cells to fully transform cells. In both cases, activity is tightly correlated with their abilities to upregulate expression of Cx43 and thus increase gap junctional communication. This action is consistent with studies of human neoplasia and preneoplasia, in which downregulated expression of Cx43 is an early event, and with studies utilizing forced expression of Cx43 to study its function in which expression was associated with decreased indices of neoplasia. Retinoids appear to act *via* interactions with retinoic acid nuclear receptors, whereas there is evidence that carotenoids may activate PPAR receptors. To overcome problems in bioavailability of carotenoids, a novel astaxanthin derivative was synthesized; it is water dispersible, inactivates aqueous-phase ROS, and can upregulate Cx43 expression when delivered in an aqueous vehicle.

References

1. Bertram, J.S. (2001) The Molecular Biology of Cancer, *Mol. Asp. Med. 21*: 167–223.
2. Doll, R., and Peto, R. (1981) *The Causes of Cancer*, Oxford University Press, Oxford.
3. Vineis, P., Alavanja, M., Buffler, P., Fontham, E., Franceschi, S., Gao, Y.T., Gupta, P.C., Hackshaw, A., Matos, E., Samet, J.M., Sitas, F., Smith, J., Stayner, L., Straif, K., Thun, M.J., Wichmann, H.E., Wu, A.H., Zaridze, D., Peto, R., and Doll, D. (2004) Tobacco and Cancer: Recent Epidemiological Evidence, *J. Natl. Cancer Inst. 96:* 99–106.
4. Marnett, L.J., and DuBois, R.N. (2002) COX-2: A Target for Colon Cancer Prevention, *Annu. Rev. Pharmacol. Toxicol. 42:* 55–80.
5. Bertram, J.S., Kolonel, L.N., and Meyskens, F.L. (1987) Rationale and Strategies for Chemoprevention of Cancer in Humans, *Cancer Res. 47:* 3012–3031.
6. Reznikoff, C.A., Bertram, J.S., Brankow, D.W., and Heidelberger, C. (1973) Quantitative and Qualitative Studies of Chemical Transformation of Cloned C3H Mouse Embryo Cells Sensitive to Postconfluence Inhibition of Cell Division, *Cancer Res. 33:* 2339–2349.
7. Heidelberger, C.H., Freeman, A.W., Pienta, R.J. Sivak, A., Bertram, J.S., Casto, B.C., Dunkel, V.C., Francis, M.W., Kakunage, T., Little, J.B., and Schechtman, L.M. (1983) Cell Transformation by Chemical Agents: A Review and Analysis of the Literature, *Mutat. Res. 114:* 283–385.
8. Bertram, J.S., Bertram, B.B., and Janik, P. (1982) Inhibition of Neoplastic Cell Growth by Quiescent Cells Is Mediated by Serum Concentration and cAMP Phosphodiesterase Inhibitors, *J. Supramol. Struct. Cell. Biochem. 18:* 515–538.
9. Sporn, M.B., and Roberts, A.B. (1983) Role of Retinoids in Differentiation and Carcinogenesis, *Cancer Res. 43:* 3034–3040.
10. Merriman, R., and Bertram, J.S. (1979) Reversible Inhibition by Retinoids of 3-Methylcholanthrene-Induced Neoplastic Transformation in C3H10T1/2 Cells, *Cancer Res. 39:* 1661–1666.
11. Bertram, J.S., Pung, A., Churley, M., Kappock, T.J.I., Wilkins, L.R., and Cooney, R.V. (1991) Diverse Carotenoids Protect Against Chemically Induced Neoplastic Transformation, *Carcinogenesis 12:* 671–678.
12. Zhang, L.-X., Cooney, R.V., and Bertram, J.S. (1991) Carotenoids Enhance Gap Junctional Communication and Inhibit Lipid Peroxidation in C3H/10T1/2 Cells: Relationship to Their Cancer Chemopreventive Action, *Carcinogenesis 12:* 2109–2114.

13. Hossain, M.Z., Wilkens, L.R., Mehta, P.P., Loewenstein, W.R., and Bertram, J.S. (1989) Enhancement of Gap Junctional Communication by Retinoids Correlates with Their Ability to Inhibit Neoplastic Transformation, *Carcinogenesis 10:* 1743–1748.
14. Mordan, L.J., Bergin, L.M., Budnick, J.L., Meegan, R., and Bertram, J.S. (1982) Isolation of Methylcholanthrene-Initiated C3H/10T1/2 Cells by Inhibiting Neoplastic Progression with Retinyl Acetate, *Carcinogenesis 3:* 279–285.
15. Mordan, L.J., Martner, J.E., and Bertram, J.S. (1983) Quantitative Neoplastic Transformation of C3H/10T1/2 Fibroblasts: Dependence upon the Size of the Initiated Cell Colony at Confluence, *Cancer Res. 43:* 4062–4067.
16. Mordan, L.J., and Bertram, J.S. (1983) Retinoid Effects on Cell-Cell Interactions and Growth Characteristics of Normal and Carcinogen-Treated C3H/10T1/2 Cells, *Cancer Res. 43:* 567–571.
17. Ames, B.N., Shigenaga, M.K., and Gold, L.S. (1993) DNA Lesions, Inducible DNA Repair, and Cell Division: Three Key Factors in Mutagenesis and Carcinogenesis, *Environ. Health Perspect. 101 (Suppl. 5):* 35–44.
18. Loewenstein, W.R., and Kanno, Y. (1966) Intercellular Communication and the Control of Tissue Growth: Lack of Communication Between Cancer Cells, *Nature 209:* 1248–1249.
19. Loewenstein, W.R. (1979) Junctional Intercellular Communication and the Control of Growth, *Biochim. Biophys. Acta 560:* 1–65.
20. Simon, A.M., and Goodenough, D.A. (1998) Diverse Functions of Vertebrate Gap Junctions, *Trends Cell Biol. 8:* 477–483.
21. Neveu, M., and Bertram, J. S. (2000) Gap Junctions and Neoplasia, in *Gap Junctions*, Hertzberg, E.L., and Bittar, E.E., eds., JAI Press, Greenwich, CT, pp. 221–262.
22. King, T.J., Fukushima, L.H., Hieber, A.D., Shimabukuro, K.A., Sakr, W.A., and Bertram, J.S. (2000) Reduced Levels of Connexin43 in Cervical Dysplasia: Inducible Expression in a Cervical Carcinoma Cell Line Decreases Neoplastic Potential with Implications for Tumor Progression, *Carcinogenesis 21:* 1097–1109.
23. Krutovskikh, V., and Yamasaki, H. (2000) Connexin Gene Mutations in Human Genetic Diseases, *Mutat. Res. 462:* 197–207.
24. Tan, L.W., Bianco, T., and Dobrovic, A. (2002) Variable Promoter Region CpG Island Methylation of the Putative Tumor Suppressor Gene Connexin 26 in Breast Cancer, *Carcinogenesis 23:* 231–236.
25. King, T.J., Fukushima, L.H., Donlon, T.A., Hieber, A.D., Shimabukuro, K.A., and Bertram, J.S. (2000) Correlation Between Growth Control, Neoplastic Potential and Endogenous Connexin43 Expression in HeLa Cell Lines: Implications for Tumor Progression, *Carcinogenesis 21:* 311–315.
26. Trosko, J.E., and Chang, C.C. (2001) Mechanism of Up-Regulated Gap Junctional Intercellular Communication During Chemoprevention and Chemotherapy of Cancer, *Mutat. Res. Fund. Mol. Mech. Mutagen. 480:* 219–229.
27. Yeager, M. (1998) Structure of Cardiac Gap Junction Intercellular Channels, *J. Struct. Biol. 121:* 231–245.
28. Gutstein, D.E., Morley, G.E., Tamaddon, H., Vaidya, D., Schneider, M.D., Chen, J., Chien, K.R., Stuhlmann, H., and Fishman, G.I. (2001) Conduction Slowing and Sudden Arrhythmic Death in Mice with Cardiac-Restricted Inactivation of Connexin43, *Circulation Res. 88:* 333–339.
29. Gros, D.B., and Jongsma, H.J. (1996) Connexins in Mammalian Heart Function, *BioEssays 18:* 719–730.

30. Goldberg, G.S., Lampe, P.D., and Nicholson, B.J. (1999) Selective Transfer of Endogenous Metabolites Through Gap Junctions Composed of Different Connexins, *Nat. Cell Biol. 1:* 457–459.
31. Alexander, D.B., and Goldberg, G.S. (2003) Transfer of Biologically Important Molecules Between Cells Through Gap Junction Channels, *Curr. Med. Chem. 10:* 2045–2058.
32. Zampighi, G.A., Eskandari, S., and Kreman, M. (2000) Epithelial Organization of the Mammalian Lens, *Exp. Eye Res. 71:* 415–435.
33. White, T.W., Goodenough, D.A., and Paul, D.L. (1998) Targeted Ablation of Connexin50 in Mice Results in Microphthalmia and Zonular Pulverulent Cataracts, *J. Cell Biol. 143:* 815–825.
34. Gabriel, H.D., Jung, D., Butzler, C., Temme, A., Traub, O., Winterhager, E., and Willecke, K. (1998) Transplacental Uptake of Glucose Is Decreased in Embryonic Lethal Connexin26-Deficient Mice, *J. Cell Biol. 140:* 1453–1461.
35. King, T.J., Fukushima, L.H., Yasui, Y., Lampe, P.D., and Bertram, J.S. (2002) Inducible Expression of the Gap Junction Protein Connexin43 Decreases the Neoplastic Potential of HT-1080 Human Fibrosarcoma Cells *In Vitro* and *In Vivo*, *Mol. Carcinog. 35:* 29–41.
36. Vine, A.L., and Bertram, J.S. (2002) Cancer Chemoprevention by Connexins, *Cancer Metastasis Rev. 21:* 199–216.
37. King, T.J., and Lampe, P.D. (2004) Mice Deficient for the Gap Junction Protein Connexin32 Exhibit Increased Radiation-Induced Tumorigenesis Associated with Elevated Mitogen-Activated Protein Kinase (p44/Erk1, p42/Erk2) Activation, *Carcinogenesis 25:* 669–680.
38. Hecker, T.P., and Gladson, C.L. (2003) Focal Adhesion Kinase in Cancer, *Front. Biosci. 8:* S705–S714.
39. Hauck, C.R., Hsia, D.A., and Schlaepfer, D.D. (2002) The Focal Adhesion Kinase—A Regulator of Cell Migration and Invasion, *IUBMB Life 53:* 115–119.
40. Kucuk, O., Sarkar, F.H., Sakr, W., Djuric, Z., Pollak, M.N., Khachik, F., Li, Y.W., Banerjee, M., Grignon, D., Bertram, J.S., Crissman, J.D., Pontes, E.J., and Wood, D.P., Jr. (2001) Phase II Randomized Clinical Trial of Lycopene Supplementation Before Radical Prostatectomy, *Cancer Epidemiol. Biomark. Prev. 10:* 861–868.
41. Giepmans, B.N.G., Verlaan, I., Hengeveld, T., Janssen, H., Calafat, J., Falk, M., and Moolenaar, W. (2001) Gap Junction Protein Connexin-43 Interacts Directly with Microtubules, *Curr. Biol. 11:* 1364–1368.
42. Giepmans, B.N.G., and Moolenaar, W.H. (1998) The Gap Junction Protein Connexin43 Interacts with the Second PDZ Domain of the Zona Occludens-1 Protein, *Curr. Biol. 8:* 931–934.
43. Karas, M., Amir, H., Fishman, D., Danilenko, M., Segal, S., Nahum, A., Koifmann, A., Giat, Y., Levy, J., and Sharoni, Y. (2000) Lycopene Interferes with Cell Cycle Progression and Insulin-Like Growth Factor I Signaling in Mammary Cancer Cells, *Nutr. Cancer 36:* 101–111.
44. Mucci, L.A., Tamimi, R., Lagiou, P., Trichopoulou, A., Benetou, V., Spanos, E., and Trichopoulos, D. (2001) Are Dietary Influences on the Risk of Prostate Cancer Mediated Through the Insulin-Like Growth Factor System? *BJU. Int. 87:* 814–820.
45. Zhang, L.-X., Cooney, R.V., and Bertram, J.S. (1992) Carotenoids Up-Regulate Connexin43 Gene Expression Independent of Their Pro-Vitamin A or Antioxidant Properties, *Cancer Res. 52:* 5707–5712.

46. Rogers, M., Berestecky, J.M., Hossain, M.Z., Guo, H.M., Kadle, R., Nicholson, B.J., and Bertram, J.S. (1990) Retinoid-Enhanced Gap Junctional Communication Is Achieved by Increased Levels of Connexin 43 mRNA and Protein, *Mol. Carcinog. 3:* 335–343.
47. Kiefer, C., Hessel, S., Lampert, J.M., Vogt, K., Lederer, M.O., and Breithaupt, D.E., and von Lintig, J. (2001) Identification and Characterization of a Mammalian Enzyme Catalyzing the Asymmetric Oxidative Cleavage of Provitamin A, *J. Biol. Chem. 276:* 14110–14116.
48. Zomer, A.W.M., Van der Burg, B., Jansen, G.A., Wanders, R.J.A., Poll-The, B.T., and Van der Saag, P.T. (2000) Pristanic Acid and Phytanic Acid: Naturally Occurring Ligands for the Nuclear Receptor Peroxisome Proliferator-Activated Receptor A, *J. Lipid Res. 41:* 1801–1807.
49. Berger, J., and Moller, D.E. (2002) The Mechanisms of Action of PPARs, *Annu. Rev. Med. 53:* 409–435.
50. Bocos, C., Göttlicher, M., Gearing, K., Banner, C., Enmark, E., Teboul, M., Crickmore, A., and Gustafsson, J.A. (1995) Fatty Acid Activation of Peroxisome Proliferator-Activated Receptor (PPAR), *J. Steroid Biochem. Mol. Biol. 53:* 467–473.
51. Rundhaug, J.E., Pung, A., Read, C.M., and Bertram, J.S. (1988) Uptake and Metabolism of β-Carotene and Retinal by C3H/10T1/2 Cells, *Carcinogenesis 9:* 1541–1545.
52. Acevedo, P., and Bertram, J.S. (1995) Liarozole Potentiates the Cancer Chemopreventive Activity of and the Up-Regulation of Gap Junctional Communication and Connexin43 Expression by Retinoic Acid and β-Carotene in 10t1/2 Cells, *Carcinogenesis 16:* 2215–2222.
53. Tibaduiza, E.C., Fleet, J.C., Russell, R.M., and Krinsky, N.I. (2002) Excentric Cleavage Products of β-Carotene Inhibit Estrogen Receptor Positive and Negative Breast Tumor Cell Growth *In Vitro* and Inhibit Activator Protein-1-Mediated Transcriptional Activation, *J. Nutr. 132*, 1368–1375.
54. Stahl, W., Hanusch, M., and Sies, H. (1996) 4-Oxo-Retinoic Acid Is Generated from Its Precursor Canthaxanthin and Enhances Gap Junctional Communication in 10T1/2 Cells, *Adv. Exp. Med. Biol. 387:* 121–128.
55. Paetau, I., Khachik, F., Brown, E.D., Beecher, G.R., Kramer, T.R., Chittams, J., and Clevidence, B.A (1998) Chronic Ingestion of Lycopene-Rich Tomato Juice or Lycopene Supplements Significantly Increases Plasma Concentrations of Lycopene and Related Tomato Carotenoids in Humans, *Am. J. Clin. Nutr. 68:* 1187–1195.
56. Bertram, J.S., King, T.J., Fukushima, L., and Khachik, F. (2000) Enhanced Activity of an Oxidation Product of Lycopene Found in Tomato Products and Human Serum Relevant to Cancer Prevention, in *Antioxidant and Redox Regulation of Genes*, Sen, K.S., Sies, H., and Baeuerle, P.A., eds., Academic Press, San Diego.
57. Aust, O., Ale-Agha, N., Zhang, L., Wollersen, H., Sies, H., and Stahl, W. (2003) Lycopene Oxidation Product Enhances Gap Junctional Communication, *Food Chem. Toxicol. 41:* 1399–1407.
58. Stahl, W., Von Laar, J., Martin, H.D., Emmerich, T., and Sies, H. (2000) Stimulation of Gap Junctional Communication: Comparison of *acyclo*-Retinoic Acid and Lycopene, *Arch. Biochem. Biophys. 373:* 271–274.
59. Gann, P.H., Ma, J., Giovannucci, E., Willett, W., Sacks, F.M., Hennekens, C.H., and Stampfer, M.J. (1999) Lower Prostate Cancer Risk in Men with Elevated Plasma Lycopene Levels: Results of a Prospective Analysis, *Cancer Res. 59:* 1225–1230.

60. Gärtner, C., Stahl, W., and Sies, H. (1997) Lycopene Is More Bioavailable from Tomato Paste than from Fresh Tomatoes, *Am. J. Clin. Nutr. 66:* 116–122.
61. Wang, X.D., Liu, C., Bronson, R.T., Smith, D.E., Krinsky, N.I., and Russell, R.M. (1999) Retinoid Signaling and Activator Protein-1 Expression in Ferrets Given β-Carotene Supplements and Exposed to Tobacco Smoke, *J. Natl. Cancer Inst. 91:* 60–66.
62. Cooney, R.V., Kappock, T.J., Pung, A., and Bertram, J.S. (1993) Solubilization, Cellular Uptake, and Activity of β-Carotene and Other Carotenoids as Inhibitors of Neoplastic Transformation in Cultured Cells, *Methods Enzymol. 214:* 55–68.
63. Zsila, F., Simonyi, M., and Lockwood, S.F. (2003) Interaction of the Disodium Disuccinate Derivative of Meso-Astaxanthin with Human Serum Albumin: From Chiral Complexation to Self-Assembly, *Bioorg. Med. Chem. Lett. 13:* 4093–4100.
64. Gross, G.J., and Lockwood, S.F. (2004) Cardioprotection and Myocardial Salvage by a Disodium Disuccinate Astaxanthin Derivative, *Life Sci. 75:* 215–224.
65. Cardounel, A.J., Dumitrescu, C., Zweier, J.L., and Lockwood, S.F. (2003) Direct Superoxide Anion Scavenging by a Disodium Disuccinate Astaxanthin Derivative: Relative Efficacy of Individual Stereoisomers Versus the Statistical Mixture of Stereoisomers by Electron Paramagnetic Resonance Imaging, *Biochem. Biophys. Res. Commun. 307:* 704–712.
66. Showalter, L.A., Weinman, S.A, Osterlie, M., and Lockwood, S.F. (2004) Plasma Appearance and Tissue Accumulation of Non-Esterified, Free Astaxanthin in C57Bl/6 Mice After Oral Dosing of a Disodium Disuccinate Diester of Astaxanthin (Heptax™), *Comp. Biochem. Physiol. Part C 137:* 227–236.
67. Hix, L.M., Lockwood, S.F., and Bertram, J.S. (2004) Upregulation of Connexin 43 Protein Expression and Increased Gap Junctional Communication by Water Soluble Disodium Disuccinate Astaxanthin Derivatives, *Cancer Lett. 211:* 25–37.
68. Foss, B.J., Sliwka, H-R., Partali, V., Cardounel, A.J., Zweier, J.L., and Lockwood, S.F. (2004) Direct Superoxide Anion Scavenging by a Highly Water-Dispersible Carotenoid Phospholipid Evaluated by Electron Paramagnetic Resonance (EPR) Spectroscopy, *Bioorg. Med. Chem. Lett. 14:* 2807–2812.

Chapter 12

Lycopene and Risk of Cardiovascular Disease

Lauren Petr and John W. Erdman, Jr.

Division of Nutritional Sciences, University of Illinois, Urbana, IL 61820

Introduction

In 2001, cardiovascular disease (CVD) affected ~64 million Americans and accounted for ~40% of total deaths nationwide (1). Medical costs associated with this disease are estimated to reach $368 billion for the year 2004. Although the cause of CVD is multifaceted, it is theorized that oxidative stress may induce development of this disease (2) and that oxidation of lipoproteins, mainly LDL (Ox-LDL), may initiate the process (3,4). As a component in atherosclerotic plaques, the Ox-LDL particle itself may propagate atherogenic effects such as dysregulation of vessel dilation, recruitment and activation of monocytes, and promotion of monocyte adhesion to the cell wall (5). Alterations to the lipoprotein during oxidation result in the formation of surface tags that are recognized by circulating macrophages. Due to the increase in surface receptors during oxidative stress, uptake of Ox-LDL by macrophages is augmented, promoting foam cell development. Although the mechanism has not been fully elucidated, research suggests that oxidative stress can additionally result in injury to the arterial wall and trigger an inflammatory response. Through mechanisms involved in inflammation, damage may occur to the endothelium, rendering it impaired and allowing for unregulated access of particles into the subendothelium (6). Accumulation of particles, particularly foam cells, in this locale initiates fatty streak formation, indicative of the early stages of atherosclerosis (7). Due to the purported role of oxidation in atherosclerosis, it was hypothesized that prevention of oxidation *via* antioxidants may be a key step in battling CVD (5).

Fruit and Vegetable Intake and Risk of CVD

Fruits and vegetables provide an abundant source of natural antioxidants. Evidence from epidemiology trials suggests that a diet rich in fruits and vegetables is protective against chronic diseases such as CVD and cancer (8,9). In a study evaluating the relation between CVD and fruit and vegetable consumption, it was found that individuals who consumed the highest quintile of these foods often had a 20% diminution in incidence [95% confidence interval (CI), 0.69–0.93] compared with individuals in the lowest quintile (10). Further, the authors suggested that with each additional daily serving of fruits or vegetables, there was a concomitant 4%

decrease in risk ($P = 0.01$). A high consumption of fruits and vegetables elevates serum concentrations of numerous antioxidants in the body, including carotenoids (11). Therefore, it is conceivable that carotenoids, fat-soluble antioxidants naturally found in plant foods, play a role in the protective effect against CVD.

Carotenoids and CVD

Carotenoids are typically found in plants, fungi, insects, birds, and crustaceans and can be obtained by animals only *via* the diet. Their primary functions in plants are to absorb light and protect from excess oxidation during photosynthesis; however, they also provide coloration for many species. All carotenoids share the common structural feature of a polyisoprenoid chain composed of multiple conjugated bonds. It is this polyene structure that allows for the quenching of singlet oxygen and other free radical compounds (12). Species such as $O_2^{\bullet -}$, $NO^{\bullet}$, and $^{\bullet}OH$, as well as singlet oxygen are thought to have detrimental effects on cells by initiating oxidation (13); it is theorized that carotenoids act as antioxidants to inhibit oxidative damage of DNA, lipids, and proteins. By dissipating excess energy obtained from the radical throughout the conjugated structure, carotenoids are able to return to their ground state, prepared for another cycle of oxidation (14). Numerous epidemiologic studies describe inverse correlations between blood levels of carotenoids and CVD (11,15,16).

There are >600 carotenoids found in nature; however, relatively few exist in significant amounts in the human blood. Major carotenoid concentrations in blood include lycopene (20–40%), β-carotene (15–30%), β-cryptoxanthin (13–20%), lutein (10–20%), α-carotene (5–10%), and zeaxanthin (1–5%) (16). Lycopene, a non-provitamin A carotenoid, which exhibits the distinctive red pigment coloring in mature tomatoes, watermelon, and pink grapefruit, has received attention recently regarding protection against CVD (11,17). This acyclic carotenoid, found in nature primarily in the all-*trans* configuration, is more commonly isolated in human tissues as *cis* isomers (18). The reasons for this phenomenon may include the notion that *cis* isomers of lycopene are more bioavailable than isomers existing in the *trans* configuration. It was suggested that increased bioavailability of *cis* isomers may result from enhanced solubility into micelles and a decreased tendency to form aggregates (18). Re *at al.* (19) demonstrated that isolated *trans*-lycopene underwent isomerization, forming *cis* isomers once the pH fell below 2.0 during a 3-h incubation with human gastric acid. In contrast, lycopene in tomato purée appeared to be more resistant to isomerization, signifying that the food matrix may be a determinant in isomerization, possibly by sequestering lycopene inside the food matrix (19).

The bioavailability of lycopene from foods appears to be enhanced during heating and processing, indicating that tomato sauces, pastes, and even catsup are more potent lycopene sources than raw tomatoes. The maturity of fruits and vegetables also appears to affect lycopene concentrations; as a tomato ripens, the deeper red color reflects a higher concentration of lycopene (20). As a lipid-soluble compound, the presence of dietary fat during the meal is essential for absorption. Once ingested, lycopene is absorbed by the intestines and transported in the blood by lipoproteins, mainly LDL (21). The main storage sites of lycopene in the human body include liver, spleen, adipose tissue, and adrenals (22); ~80% of lycopene consumed in a typical American diet (23) is supplied from tomato products (Table 12.1).

Observations/ Epidemiology

Much of the evidence supporting lycopene as a protective compound against CVD originated from favorable epidemiology and observational studies (Table 12.2). Results from the European multicenter case-control study on Antioxidants, Myocardial Infarction and Breast Cancer (EURAMIC) revealed that lycopene was the sole carotenoid to remain independently protective against myocardial infarctions (MI) once adjustments were made for typical confounding factors ($P = 0.005$) (17). Subjects for this study were recruited from participating hospitals in 10 European countries and were men <70 y old who had survived a first-time MI. Cases were matched to controls on the basis of age. Samples of subcutaneous adipose tissue were collected *via* needle aspiration and analyzed for carotenoids as well as polyunsaturated fatty acid (PUFA) concentrations. In addition to the inverse relationship with MI occurrence, lycopene concentrations were also significantly negatively correlated with body mass index ($P < 0.05$, $r = -0.24$). Interestingly, the beneficial effects of lycopene were more pronounced as adipose concentrations of PUFA increased. Cardiovascular protective effects of lycopene were significant in individuals with >16.1% fat as PUFA [odds ratio (OR) = 0.38, 95% CI 0.21–0.71]. Due to the increased susceptibility of oxidative damage to

TABLE 12.1
Concentrations of Lycopene in Foods[a]

Food	Amount	Lycopene (mg)
Tomato paste, canned	1 cup	75.4
Tomato puree, canned	1 cup	54.4
Pasta (spaghetti) sauce	1 cup	40.0
Tomato sauce, canned	1 cup	37.1
Tomato soup, canned and prepared	1 cup	25.6
Vegetable juice, canned	1 cup	23.4
Tomato juice, canned	1 cup	22.0
Watermelon, raw	1 wedge	13.0
Tomatoes, stewed and canned	1 cup	10.3
Tomatoes, raw and ripe	1 cup	4.6
Catsup	1 Tbsp	2.6
Grapefruit, raw, pink and red	1/2 fruit	1.7
Salsa, ready-to-serve	1 Tbsp	1.7

[a]*Source:* Reference 49.

TABLE 12.2
Lycopene and CVD: Summary of Epidemiology and Observational Studies[a]

Study	Subjects	n/Length	Endpoints	Outcomes
EURAMIC study (17)	Men < 70 y old; 1st time MI survivors vs. matched	662 cases 717 controls	Adipose lycopene and adipose PUFA	Lycopene levels were inversely correlated with MI events and the effects were more pronounced as adipose PUFA concentrations increased.
Linköping-Vilinus coronary disease risk assessment study (24)	Lithuanian and Swedish men aged 50 y	109 Lithuania 101 Swedish	Resistance to LDL oxidation *ex vivo* and plasma lycopene	In a population with high CVD, Lithuanian men had decreased oxidative resistance and decreased plasma lycopene.
KIHD study (25,11)	Men aged 46–64 y	1028 subjects/ 4-y follow-up	Serum lycopene, IMT-CCA and occurrence of coronary event	Individuals that experienced coronary events had 39% lower serum lycopene and IMT-CCA was inversely proportional to serum lycopene.
ASAP study (25)	Smoking and nonsmoking men and postmenopausal women aged 45–69 y	520 subjects with elevated serum cholesterol	Plasma lycopene and IMT-CCA	Men with low plasma lycopene had 18% higher IMT-CCA, but women had non-significant differences.
Women's Health Study (26)	Women health professionals >45 y old and postmeno-pausal	483 case-control pairs/7.2-y mean follow-up	Plasma lycopene and occurrence of coronary event	Inverse relationship of plasma lycopene and coronary events (34%).
Women's Health Study (27)	Women health professionals >45 y old and postmeno-pausal	483 subjects/7.2-y mean follow-up	Intake of dietary lycopene and tomato-based foods and occurrence of coronary events	A nonsignificant inverse trend for lycopene and CVD, but a significant 32% decrease with tomato-based food intake and CVD

[a]Abbreviations: ASAP, Antioxidant Supplementation in Atherosclerosis Prevention study; CVD, cardiovascular disease; EURAMIC, European study on Antioxidants, Myocardial Infarction and Breast Cancer; IMT-CCA, intima-media thickness of the common carotid artery; KIHD, Kuopio Ischemic Heart Disease study; LDL, low density lipoprotein; MI, myocardial infarction; PUFA, polyunsaturated fatty acids.

PUFA, individuals with elevated concentrations of PUFA may be more likely to reap the antioxidant benefits of lycopene.

The Linköping-Vilinus coronary disease risk assessment study measured plasma concentrations of fat-soluble antioxidants and susceptibility of LDL particles to *ex vivo* oxidation in 50-y-old Lithuanian and Swedish men to investigate possible explanations for an increased incidence of heart disease evident among the Lithuanian population. Results revealed that Lithuanian men had lower resistance to *ex vivo* LDL oxidation than Swedish men. The lag phase was 67.6 and 79.5 min for Lithuanian and Swedish men, respectively ($P < 0.001$) (24). Lower plasma levels of β-carotene (377 vs. 510 nmol/L, $P < 0.01$) and lycopene (327 vs. 615 nmol/L, $P < 0.001$) were also seen in the Lithuanian men, suggesting that low carotenoid status may be a risk factor for heart disease secondary to amplified susceptibility to LDL oxidation.

The Kuopio Ischemic Heart Disease (KIHD) study assessed middle-aged men (aged 46–64 y) living in eastern Finland by measuring serum antioxidant levels as well as intima-media thickness of the common carotid artery (IMT-CCA), a marker of early atherosclerosis, during a 5-y follow-up. Serum levels of lycopene were significantly (39%) lower in men who suffered from a coronary event compared with controls ($P = 0.003$) (25) and the severity of IMT was shown to be inversely proportional to serum lycopene levels ($r = -0.22$, $P < 0.001$) (11).

Although the majority of the research was conducted among male populations, some studies were conducted among women because it is possible that gender differences in response to lycopene and CVD may exist. The Antioxidant Supplementation in the Atherosclerosis Prevention (ASAP) study was designed to investigate the effects of vitamin E and C supplementation on atherosclerotic progression in both smoking and nonsmoking men and postmenopausal women of the Finnish population. Subjects (aged 45–69 y), despite serum cholesterol concentrations >5 mmol/L, were relatively healthy. Measurements included plasma carotenoid concentrations and IMT-CCA. In men, depressed plasma lycopene levels were associated with a significant 18% ($P = 0.003$) increase in IMT compared with male counterparts with lycopene concentrations above the median (25). Plasma lycopene concentrations tended to be higher and IMT lower in women ($P = 0.81$). Interestingly, mean lycopene levels were significantly higher in women (0.17 ± 0.11 μmol/L) than in men (0.14 ± 0.12 μmol/L) ($P = 0.007$). Sesso *et al.* (26) examined plasma lycopene concentrations in middle-aged and elderly women participating in the Women's Health Study who were initially free of CVD and cancer and living in the United States. Analysis for plasma lycopene concentrations was completed on 483 case-control pairs matched for age, smoking status, and follow-up time. After adjusting for plasma total cholesterol, there was a significant inverse association between CVD and plasma lycopene. The relative risks (RR) for increasing quartiles of plasma lycopene were 1.00, 0.78 (0.55–1.11), 0.56 (0.39–0.82) and 0.62 (0.43–0.90) with 95% CI. Comparing the upper and lower halves of plasma lycopene concentrations, a 34% (RR 0.66, CI 0.47–0.95) reduction in risk occurred in individuals with higher lycopene concentrations. Once angina cases were excluded, women in the upper three quartiles of plasma lycopene displayed a 50% greater decline in CVD than those in the lowest quartile (26).

As mentioned above, the majority of lycopene in the diet is supplied by tomato products; therefore, several studies investigated the relation between tomato product consumption, in addition to blood lycopene concentrations, and CVD risk. A prospective study conducted among women participating in the Women's Health Study examined whether dietary lycopene and/or tomato-based products were associated with a decreased risk of CVD (27). Analysis was completed on women ≥45 y old that were free of CVD and cancer. Lycopene intakes were calculated on the basis of food-frequency questionnaires, and subjects were asked to self-report typical heart disease risk factors. The participants were followed for ~7.2 y and monitored for CVD events including MI, revascularization, stroke, and cardiovascular death. The correlation between reported lycopene intakes and corresponding plasma levels was determined on a subpopulation of 483 women ($r = 0.14$, $P = 0.004$). Lycopene intakes were distributed into quintiles; individuals with lycopene consumption in the highest quintile were more likely to be younger, nonsmoking, exercise more frequently, and use hormone replacement therapy. A nonsignificant trend was detected for a decreasing risk of CVD as lycopene consumption increased ($P = 0.30$). Although the association of lycopene and CVD was not significant, there was a significant correlation between the intake of tomato-based foods and CVD. The average consumption of tomato-based foods was 4.32 ± 3.23 servings/wk. Women consuming 7–10 servings/wk of tomato-based foods had a 32% lower risk for CVD (95% CI 0.49–0.96) compared with women consuming ≤1.5 servings/wk (27).

Not all research supports a significant link between lycopene and CVD. An investigation examining the relation between calcified plaques present in the abdominal aorta and serum carotenoids was conducted among men and women ≥55 y old participating in the Rotterdam study (28). Subjects displaying moderate-to-severe atherosclerosis, based on severity of plaque formation, were matched with controls based on age and sex. Analysis of serum carotenoids revealed that serum concentrations of lycopene, α- and β-carotene, and β-cryptoxanthin appeared to be depressed in patients with atherosclerosis compared with their matched controls; however, after adjusting for age and sex, these associations were not significant. Interestingly, lycopene was the only carotenoid to show a trend between highest plaque formation and lowest serum concentrations ($P = 0.13$). A prospective, case-control study conducted among subjects free of CVD and participating in the Physicians' Health study assessed the relation of serum carotenoids and MI occurrence on a subgroup of 531 case-control pairs matched for age and smoking status. Elevated serum carotenoid levels, lycopene included, were not associated with any significant decrease in risk for MI (29). In addition, the Atherosclerosis Risk in Communities (ARIC) study revealed only a nonsignificant trend between low serum lycopene concentrations and IMT-CCA in individuals ≥55 y old.

Interestingly, negative correlations were significant for serum lutein + zeaxanthin and β-cryptoxanthin levels (30). Despite the fact that epidemiologic studies reveal confounding results, evidence suggests a connection between lycopene concentrations and CVD.

Tomato and Lycopene Supplementation and Biomarkers of CVD

Human supplementation studies investigating lycopene and biomarkers of CVD risk factors have been inconclusive. Due to the antioxidant properties of lycopene, major endpoints evaluated typically include measurements of oxidative stress. Oxidative stress was defined by the Institute of Medicine as "an imbalance between the production of various reactive species and the ability of the organism's natural protective mechanisms to cope with these reactive compounds and prevent adverse effects" (31). Reactive species are thought to attack lipids, proteins, and DNA within the body, producing oxidized compounds such as F_2-isoprostanes, protein carbonyls, and purine or pyrimidine metabolites. To determine whether a dietary compound acts as an antioxidant, evidence of decreased oxidation *in vivo* is necessary. Nelson *et al.* (32) evaluated differences in oxidative stress in subjects administered various treatment combinations of antioxidants. Subjects were given differing concentrations of antioxidants based upon assignment into one of four intervention groups: antioxidant capsule, antioxidant tablet, high carotenoid diet, or placebo. For the carotenoid intervention, the tablet group received 4.0 mg/d of both β-carotene and lutein/zeaxanthin; the capsule group received 2.4, 6.0, 0.5, and 100 mg/d of β-carotene, lutein/zeaxanthin, lycopene, and spinach leaf powder, respectively. For the diet intervention group, instruction was given to subjects to consume four foods daily from a specified list yielding an average daily intake of 6 mg lutein, 0.6 mg zeaxanthin, 10 mg lycopene, and 11 mg β-carotene. Analysis of total oxidative stress included measurements of urinary isoprostanes and total alkenals and serum lipid peroxides. All groups except the placebo group significantly raised their serum lycopene levels from baseline; however, only the diet group had significantly higher concentrations compared with the placebo group postintervention (P = 0.038). Serum lipid peroxides were unaffected, but urinary isoprostanes and total alkenals were significantly lowered in both the capsule and diet intervention groups (isoprostanes P = 0.009, P = 0.001; total alkenals P = 0.015, P = 0.006) (32). Interestingly, these were the only groups that received lycopene during the intervention.

Agarwal and Rao (33) investigated the effect of lycopene consumption from tomato-based foods on *ex vivo* LDL oxidation in humans. Lycopene sources included tomato juice, spaghetti sauce, and tomato oleoresin, a natural extract of tomato lipids. Each food was supplemented for 1 wk and resulted in a twofold increase in serum lycopene compared with baseline. Although serum cholesterol levels did not change, markers of both lipid peroxidation and LDL oxidation were significantly decreased. A study conducted by Visoli *et al.* (34) provided additional support to the above findings. Baseline measurements for plasma carotenoid concentrations, antioxidant capacity, susceptibility of LDL to *ex vivo* oxidation, and urinary isoprostanes were taken in 12 healthy women after they followed a low-carotenoid diet for 1 wk. The women were given instructions to supplement the low-carotenoid diet with tomato products for 3 wk to yield ~8 mg of lycopene intake/d. After supplementation, plasma lycopene concentrations significantly increased ($P < 0.001$) and LDL susceptibility was attenuated ($P < 0.001$). Although antioxidant capacity was not affected, urinary isoprostane secretion was reduced by 53% ($P < 0.05$), suggesting a reduction in lipid peroxidation.

Bub *et al.* (35) conducted a human trial to demonstrate the antioxidant effects of a short-term dietary intervention of carotenoid-containing foods in healthy, nonsmoking men aged 27–40 y consuming a low-carotenoid diet. Subjects adhered to that diet for 2 wk to determine baseline levels of plasma carotenoids, lipid peroxidation, and LDL susceptibility to oxidation. Afterward, subjects were given in order a series of supplemented foods including tomato juice, carrot juice, and a liquid preparation of spinach powder. Individual foods were consumed for 2-wk periods before switching to the next food. A washout period separated each intervention period. Carotenoid concentrations were markedly different for each food source. Lycopene was the dominant carotenoid present in the tomato juice (40 mg/d); carrot juice was more concentrated in α- (15.7 mg/d) and β-carotene (22.3 mg/d), and the spinach powder contained lutein (11.3 mg/d) and β-carotene (3.1 mg/d). Analysis was completed on plasma and lipoprotein carotenoid concentrations, lipid peroxidation, and *ex vivo* LDL oxidation. Carotenoid levels increased significantly in lipoproteins for all groups after supplementation. The tomato juice significantly reduced lipid peroxidation by 12% ($P < 0.05$) and *ex vivo* LDL oxidation by 18% ($P < 0.001$). Lycopene was the only carotenoid negatively correlated with LDL oxidation and lipid peroxidation ($r = -0.816$, $P < 0.004$). Neither the carrot juice nor the spinach supplementations affected LDL oxidation or lipid peroxidation (35).

Consumption of a lycopene-free diet was shown to reduce blood lycopene levels and increase oxidative products, particularly Ox-LDL particles (36). However, a major limitation in associating oxidative products with disease is the unanswered question concerning whether the oxidation products are a reflection of the development of disease or products of the disease itself (31). Studies also measure resistance to LDL oxidation utilizing an *ex vivo* situation. This is not an adequate method to be used alone when determining antioxidant activity, and more research is required in the development of methods to quantify *in vivo* LDL oxidation.

Synergistic Effects of Carotenoids in Foods and Supplements

Recently, results from a study conducted in our laboratory suggested that the whole tomato may be more effective than lycopene alone in preventing prostate carcinogenesis in an animal model (37). It is possible that an analogous situation may be occurring for lycopene and cardiovascular disease as well. As described earlier in this review, one study concluded that although plasma lycopene concentrations were not associated with a reduced risk of CVD, consumption of tomato-based products was significantly inversely correlated (27). In addition to lycopene, other components present in the tomato such as vitamins C and E, flavonoids, and other carotenoids may be necessary for the protective effect (20).

Antioxidants are typically found in foods in combinations and may be working concurrently to produce additive or synergistic protective effects. Therefore, consumption of an individual antioxidant may not be effective itself. To address this hypothesis, Fuhrman *et al.* (38) investigated the protective effect of lycopene against LDL oxidation in conjunction with other compounds. Isolated human LDL particles were incubated with either lycopene or tomato oleoresin at increasing concentrations, and susceptibility to copper-induced oxidation was measured. Oxidation was attenuated by 22% for lycopene and by 90% for tomato oleoresin when given at equal concentrations based on lycopene. When incubated in the presence of vitamin E, lycopene appeared to act synergistically against LDL oxidation by exceeding the anticipated additive value of resistance of the two individual compounds. At 5 μmol/L lycopene, the addition of 10 μmol/L of vitamin E suppressed LDL oxidation by 45% more than the expected value. This study provides support that the tomato may be more effective in reducing oxidation vs. lycopene alone.

A possible explanation for synergistic effects could be due to structural differences among antioxidants. Lycopene was shown to have the strongest antioxidant capacity *in vitro* followed by β-carotene, whereas lutein was least efficient (39). These results imply that the number of conjugated bonds and polarity of the compound may influence antioxidant efficacy. Compared with β-carotene and lutein, lycopene has the greatest number of conjugated bonds in which excess energy extracted from the free radicals can resonate. As a hydrocarbon, lycopene is thought to be sequestered in the hydrophobic center of lipoproteins, unlike the more polar xanthophylls lutein and zeaxanthin, which tend to orient in the membrane. The presence of polar carotenoids in the membrane may increase the rigidity of the lipoprotein and restrict movement of the carotenoid, both of which reduce antioxidant effectiveness. Hydrocarbons located in the lipid core allow for fluidity of the membrane to enhance free radical permeation and freedom of movement of the carotenoid (13). Stahl *et al.* (39) showed that lycopene and lutein together provided the most efficient carotenoid mixture in resisting lipid peroxidation in multilamellar liposomes. Perhaps the most effective combinations of antioxidants are those in which the compounds differ in polarities.

Mechanisms of Action

Most of the studies investigating lycopene and CVD risk were designed to measure the effects of lycopene on one or more markers of oxidation. However, there are several other suggestions as to the mechanism of action.

Cholesterol Synthesis Inhibition. It was proposed that carotenoids may work by inhibiting cholesterol synthesis through the regulation of 3-hydroxy-3-methylglutaryl coenzyme A (HMGCoA) reductase. Fuhrman *et al.* (40) illustrated a dose-dependent response inhibition of macrophage cholesterol synthesis with both lycopene and β-carotene in the human J-774A.1 macrophage cell line. At a concentration of 10 μmol/L, inhibition was 73 vs. 63% for lycopene and β-carotene, respectively (40). In addition, an increase in degradation of LDL particles (34 and 25% for lycopene and β-carotene, respectively) suggested enhancement of the macrophage LDL receptor activity. A feeding trial was carried out in six healthy men who received 60 mg lycopene/d (LycoRed) for 6 mo. Results showed a significant 14% ($P < 0.02$) reduction in plasma LDL cholesterol with no alterations in HDL cholesterol.

Inflammation. Several studies suggested an anti-inflammatory response of lycopene. Yaping *et al.* (41) investigated the effect of lycopene on inflammation using oil-induced ear edema in male Kingming mice, a common screening method for determining anti-inflammatory activity. Lycopene, given as tomato oleoresin, was most effective against inflammation at a dose of 0.5 g/kg body weight ($P < 0.05$). Interestingly, the inhibitory effects that occurred with a dose of 0.1 g lycopene/kg body weight were similar to those in the control group using the anti-inflammatory drug amoxicillin at the same dosage.

A cross-sectional study conducted in 85 nuns aged 77–99 y demonstrated that elevated levels of C-reactive protein (CRP), an acute inflammatory phase protein, were linked to significant decreases in plasma lycopene ($P = 0.03$), α-carotene ($P = 0.02$), β-carotene ($P = 0.02$), as well as total carotenoids ($P = 0.01$) (42). CRP has received much attention for its potential use as a biomarker for atherosclerosis (43) and results from this study support an association between carotenoids and reduced oxidative-induced inflammatory responses thought to be involved in CVD development.

Immune Function. A study carried out by Porrini *et al.* (44) investigated the effects of spinach plus tomato purée consumption in nine healthy women. Subjects consumed 150 g of spinach plus 25 g of tomato purée each day providing 9.0 mg lutein, 0.6 mg zeaxanthin, 4.3 mg β-carotene, and 7.0 mg lycopene; the women consumed the intervention diet for 3 wk. Carotenoid concentrations in lymphocytes were determined and the Comet assay was performed to analyze effects on DNA resistance to *ex vivo* oxidative stress. The intervention significantly increased both lutein and lycopene levels in lymphocytes ($P < 0.01$ and < 0.05, respectively), and the resistance of DNA to oxidative stress was significantly increased ($P < 0.01$).

Endothelial Function/Gap Junction Communication. As described earlier, Ox-LDL may play a role in inflammation-induced injury to the endothelium. Martin *et al.* (45) tested carotenoid effects on vascular endothelium using human aortic endothelial cell cultures. Preincubation of lycopene resulted in a 13% decreased expression of the vascular cell adhesion molecule, a molecule present on activated endothelial tissues that aids in the recruitment of leukocytes. There has also been considerable interest in the increase in gap junction communication (GJC) with carotenoids, especially lycopene, as it relates to cancer (46). Optimal communication between endothelial cells in the artery walls is also desirable; this enhancement of GJC by lycopene may help maintain an intact and healthy endothelial surface of arteries. More research is warranted in this area.

Conclusions

In contrast to the evidence that supports an association between tomatoes and lycopene with a reduced risk of prostate cancer, research regarding the relation of lycopene and CVD risk is at an earlier stage of investigation. More studies have to be conducted with appropriate animal models of both carotenoid absorption and disease development to determine whether lycopene, other tomato components, or whole tomato reduces CVD development.

In conclusion, the function of lycopene as an antioxidant as well as other potential mechanisms of action suggests biological plausibility for a role for lycopene in preventing CVD, and this possible relation is supported by epidemiology and tomato and lycopene supplementation trials. However, more research is warranted because it is possible that lycopene may be only a biomarker for tomato consumption and lycopene itself is not effective. In addition, synergistic effects between lycopene and other components present in tomatoes may be necessary for the protective action. Other carotenoids such as phytoene and phytofluene and even oxidized metabolites of lycopene have been suggested to have a bioactive role (47,48).

References

1. American Heart Association (2003) *Heart Disease and Stroke Statistics—2004 Update*, American Heart Association, Dallas, TX.
2. Nelson, J.L., Bernstein, P.S., Schmidt, M.C., Von Tress, M.S., and Askew, E.W. (2003) Dietary Modification and Moderate Antioxidant Supplementation Differentially Affect Serum Carotenoids, Antioxidant Levels and Markers of Oxidative Stress in Older Humans, *J. Nutr. 133:* 3117–3123.
3. Reaven, P.D., Khouw, A., Beltz, W.F., Parthasarathy, S., and Witztum, J.L. (1993) Effect of Dietary Antioxidant Combinations in Humans: Protection of LDL by Vitamin E but Not by β-Carotene, *Arterioscler. Thromb. 13:* 590–600.
4. Maxwell, S., and Greig, L. (2001) Anti-Oxidants—A Protective Role in Cardiovascular Disease? *Expert Opin. Pharmacother. 2:* 1737–1750.
5. Berliner, J.A., and Heinecke, J.W. (1996) The Role of Oxidized Lipoproteins in Atherogenesis, *Free Radic. Biol. Med. 20:* 707–727.
6. Griendling, K.K., and Fitzgerald, G.A. (2003) Oxidative Stress and Cardiovascular Injury, Part I: Basic Mechanisms and *in Vivo* Monitoring of ROS, *Circulation 108:* 1912–1916.

7. Fernandopulle, R.J. (1993) Atherosclerosis, in *Pathophysiology of Heart Disease: A Collaborative Project of Students and Faculty*, Lilly, L.S., ed., Williams and Wilkins, Baltimore, pp. 84–97.
8. Kerver, J.M., Yang, E.J., Bianchi, L., and Song, W.O. (2003) Dietary Patterns Associated with Risk Factors for Cardiovascular Disease in Healthy US Adults, *Am. J. Clin. Nutr. 78:* 1103–1110.
9. Verlangieri, A.J., Kapeghian, J.C., el-Dean, S., and Bush, M. (1985) Fruit and Vegetable Consumption and Cardiovascular Mortality, *Med. Hypotheses 16:* 7–15.
10. Joshipura, K.J., Hu, F.B., Manson, J.E., Stampfer, M.J., Rimm, E.B., Speizer, F.E., Colditz, G., Ascherio, A., Rosner, B., Spiegelman, D., and Willett, W.C. (2001) The Effect of Fruit and Vegetable Intake on Risk for Coronary Heart Disease, *Ann. Intern. Med. 134:* 1106–1114.
11. Rissanen, T., Voutilainen, S., Nyyssonen, K., Salonen, R., Kaplan, G.A., and Salonen, J.T. (2003) Serum Lycopene Concentrations and Carotid Atherosclerosis: The Kuopio Ischaemic Heart Disease Risk Factor Study, *Am. J. Clin. Nutr. 77:* 133–138.
12. Clinton, S.K. (1998) Lycopene: Chemistry, Biology, and Implications for Human Health and Disease, *Nutr. Rev. 56:* 35–51.
13. Cantrell, A., McGarvey, D.J., Truscott, T.G., Rancan, F., and Bohm, F. (2003) Singlet Oxygen Quenching by Dietary Carotenoids in a Model Membrane Environment, *Arch. Biochem. Biophys. 412:* 47–54.
14. Stahl, W., and Sies, H. (2003) Antioxidant Activity of Carotenoids, *Mol. Aspects Med. 24:* 345–351.
15. Kohlmeier, L., and Hastings, S.B. (1995) Epidemiologic Evidence of a Role of Carotenoids in Cardiovascular Disease Prevention, *Am. J. Clin. Nutr. 62 (Suppl.):* 1370S–1376S.
16. Kritchevsky, S.B. (1999) β-Carotene, Carotenoids and the Prevention of Coronary Heart Disease, *J. Nutr. 129:* 5–8.
17. Kohlmeier, L., Kark, J.D., Gomez-Gracia, E., Martin, B.C., Steck, S.E., Kardinaal, A.F.M., Ringstad, J., Thamm, M., Masaev, V., Riemersma, R., Martin-Moreno, J.M., Huttunen, J.K., and Kok, F.J. (1997) Lycopene and Myocardial Infarction Risk in the EURAMIC Study, *Am. J. Epidemiol. 146:* 618–626.
18. Boileau, T.W.-M., Boileau, A.C., and Erdman, J.E., Jr. (2002) Bioavailability of All-*trans* and *cis*-Isomers of Lycopene, *Exp. Biol. Med. 227:* 914–919.
19. Re, R., Fraser, P.D., Long, M., Bramley, P.M., and Rice-Evans, C. (2001) Isomerization of Lycopene in the Gastric Milieu, *Biochem. Biophys. Res. Commun. 281:* 579–581.
20. Willcox, J.K., Catignani, G.L., and Lazarus, S. (2003) Tomatoes and Cardiovascular Health, *Crit. Rev. Food Sci. Nutr. 43:* 1–18.
21. Bramley, P.M. (2000) Is Lycopene Beneficial to Human Health? *Phytochemistry 54:* 233–236.
22. Zaripheh, S., Boileau, T.-W.-M., Lila, M.A., and Erdman, Jr., J.W. (2003) [^{14}C]-Lycopene and [^{14}C]-Labeled Polar Products Are Differentially Distributed in Tissues of F344 Rats Prefed Lycopene, *J. Nutr. 133:* 4189–4195.
23. Clinton, S.K. (1998) Lycopene: Chemistry, Biology, and Implications for Human Health and Disease, *Nutr. Rev. 56:* 35–51.
24. Kristenson, M., Zieden, B., Kucinskiene, Z., Elinder, L.S., Bergdahl, B., Elwing, B., Abaravicius, A., Razinkoviene, L., Calkauskas, H., and Olsson, A.G. (1997) Antioxidant State and Mortality from Coronary Heart Disease in Lithuanian and Swedish Men: Concomitant Cross Sectional Study of Men Aged 50, *Br. Med. J. 314:* 629–633.

25. Rissanen, T., Voutilainen, S., Nyyssonen, K., and Salonen, J.T. (2002) Lycopene, Atherosclerosis, and Coronary Heart Disease, *Exp. Biol. Med. 227:* 900–907.
26. Sesso, H.D., Buring, J.E., Norkus, E.P., and Gaziano, J.M. (2004) Plasma Lycopene, Other Carotenoids, and Retinol and the Risk of Cardiovascular Disease in Women, *Am. J. Clin. Nutr. 79:* 47–53.
27. Sesso, H.D., Liu, S., Gaziano, J.M., and Buring, J.E. (2003) Dietary Lycopene, Tomato-Based Food Products and Cardiovascular Disease in Women, *J. Nutr. 133:* 2336–2341.
28. Klipstein-Grobusch, K., Launer, L.J., Geleijnse, J.M., Boeing, H., Hofman, A., and Witteman, J.C. (2000) Serum Carotenoids and Atherosclerosis, The Rotterdam Study, *Atherosclerosis 148:* 49–56.
29. Hak, A.E., Stampfer, M.J., Campos, H., Sesso, H.D., Gaziano, J.M., Willett, W., and Ma, J. (2003) Plasma Carotenoids and Tocopherols and Risk of Myocardial Infarction in a Low-Risk Population of US Male Physicians, *Circulation 108:* 802–807.
30. Iribarren, C., Folsom, A.R., Jacobs, D.R., Jr., Gross, M.D., Belcher, J.D., and Eckfeldt, J.H. (1997) Association of Serum Vitamin Levels, LDL Susceptibility to Oxidation, and Autoantibodies Against MDA-LDL with Carotid Atherosclerosis, A Case-Control Study, The ARIC Study Investigators. Atherosclerosis Risk in Communities, *Arterioscler. Thromb. Vasc. Biol. 17:* 1171–1177.
31. Institute of Medicine (2000) *Dietary Reference Intakes for Vitamin C, Vitamin E, Selenium, and Carotenoids*, Food and Nutrition Board, Institute of Medicine, Washington.
32. Nelson, J.L., Bernstein, P.S., Schmidt, M.C., Von Tress, M.S., and Askew, E.W. (2003) Dietary Modification and Moderate Antioxidant Supplementation Differentially Affect Serum Carotenoids, Antioxidant Levels and Markers of Oxidative Stress in Older Humans, *J. Nutr. 133:* 3117–3123.
33. Agarwal, S., and Rao, A.V. (1998) Tomato Lycopene and Low Density Lipoprotein Oxidation: a Human Dietary Intervention Study, *Lipids 33:* 981–984.
34. Visioli, F., Riso, P., Grande, S., Galli, C., and Porrini, M. (2003) Protective Activity of Tomato Products on *In Vivo* Markers of Lipid Oxidation, *Eur. J. Nutr. 42:* 201–206.
35. Bub, A., Watzl, B., Abrahamse, L., Delincee, H., Adam, S., Wever, J., Muller, H., and Rechkemmer, G. (2000) Moderate Intervention with Carotenoid-Rich Vegetable Products Reduces Lipid Peroxidation in Men, *J. Nutr. 130:* 2200–2206.
36. Gomez-Aracena, J., Bogers, R., Van't Veer, P., Gomez-Gracia, E., Garcia-Rodriguez, A., Wedel, H., and Fernandez-Crehuet Navajas, J. (2003) Vegetable Consumption and Carotenoids in Plasma and Adipose Tissue in Malaga, Spain, *Int. J. Vitam. Nutr. Res. 73:* 24–31.
37. Boileau, T.W., Liao, Z., Kim, S., Lemeshow, S., Erdman, J.W., Jr., and Clinton, S.K. (2003) Prostate Carcinogenesis in *N*-Methyl-*N*-nitrosourea (NMU)-Testosterone-Treated Rats Fed Tomato Powder, Lycopene, or Energy-Restricted Diets, *J. Natl. Cancer Inst. 95:* 1578–1586.
38. Fuhrman, B., Volkova, N., Rosenblat, M., and Aviram, M. (2000) Lycopene Synergistically Inhibits LDL Oxidation in Combination with Vitamin E, Glabridin, Rosmarinic Acid, Carnosic Acid or Garlic, *Antioxid. Redox Signal. 2:* 491–506.
39. Stahl, W., Junghans, A., de Boer, B., Driomina, E.S., Briviba, K., and Sies, H. (1998) Carotenoid Mixtures Protect Multilamellar Liposomes Against Oxidative Damage: Synergistic Effects of Lycopene and Lutein, *FEBS Lett. 427:* 305–308.
40. Fuhrman, B., Elis, A., and Aviram, M. (1997) Hypocholesterolemic Effect of Lycopene and Beta-Carotene Is Related to Suppression of Cholesterol Synthesis and Augmentation

of LDL Receptor Activity in Macrophages, *Biochem. Biophys. Res. Commun. 233:* 658–662.
41. Yaping, Z., Wenli, Y., Weile, H., and Ying, Y. (2003) Anti-Inflammatory and Anticoagulant Activities of Lycopene in Mice, *Nutr. Res. 23:* 1591–1595.
42. Boosalis, M.G., Snowdon, D.A., Tully, C.L., and Gross, M.D. (1996) Acute Phase Response and Plasma Carotenoid Concentrations in Older Women: Findings from the Nun Study, *Nutrition 12:* 475–478.
43. Ridker, P.M., Hennekens, C.H., Buring, J.E., and Rifai, N. (2000) C-Reactive Protein and Other Markers of Inflammation in the Prediction of Cardiovascular Disease in Women, *N. Engl. J. Med. 342:* 836–843.
44. Porrini, M., Riso, P., and Oriani, G. (2002) Spinach and Tomato Consumption Increases Lymphocyte DNA Resistance to Oxidative Stress but This Is Not Related to Cell Carotenoid Concentrations, *Eur. J. Nutr. 41:* 95–100.
45. Martin, K.R., Wu, D., and Meydani, M. (2000) The Effect of Carotenoids on the Expression of Cell Surface Adhesion Molecules and Binding of Monocytes to Human Aortic Endothelial Cells, *Atherosclerosis 150:* 265–274.
46. Rao, A.V., and Agarwal, S. (2000) Role of Antioxidant Lycopene in Cancer and Heart Disease, *J. Am. Coll. Nutr. 19:* 563–569.
47. Khachik, F., Carvalho, L., Bernstein, P.S., Muir, G.J., Zhao, D.-Y., and Katz, M.B. (2002) Chemistry Distribution and Metabolism of Tomato Carotenoids and Their Impact on Human Health, *Exp. Biol. Med. 227:* 845–851.
48. Rao, A.V. (2002) Lycopene, Tomatoes, and the Prevention of Coronary Heart Disease, *Exp. Biol. Med. 227:* 908–913.
49. U.S. Department of Agriculture, Agriculture Research Services (2003) USDA National Nutrition Database for Standard Reference, http://www.nal.usda.gov/fnic/foodcomp.

Chapter 13

Effect of Feeding and Then Depleting a High Fruit and Vegetable Diet on Oxidizability in Human Serum

Kyung-Jin Yeum[a], Giancarlo Aldini[b], Elizabeth J. Johnson[a], Robert M. Russell[a], and Norman I. Krinsky[a,c]

[a]The Jean Mayer USDA-Human Nutrition Research Center on Aging, Tufts University, Boston, MA; [b]Istituto Chimico Farmaceutico Tossicologico, University of Milan, Viale Abruzzi 42–20131 Milan, Italy; and the [c]Department of Biochemistry, School of Medicine, Tufts University, Boston, MA 02111

Introduction

The effects of diets high in fruits and vegetables in preventing the development of various chronic diseases are well documented (1,2). To a large extent, these effects were attributed to the antioxidants present in such diets (3), inasmuch as oxidative stress was suggested to lead to carcinogenesis, aging, coronary vascular disease, and inflammation (4). However, it has not been possible to attribute the protective effects of fruits and vegetables to a single nutrient or even to a family of compounds. For example, early epidemiologic evidence strongly suggested that diets rich in fruits and vegetables containing β-carotene might be protective with respect to lung cancer (5). However, subsequent intervention trials proved this hypothesis to be false when tested in smokers or asbestos workers (6–8).

One of the key issues in determining whether dietary components such as those found in fruits and vegetables alter the oxidative stress status in humans is the selection of appropriate biomarkers. For many years, determination of thiobarbituric acid-reactive substances (TBARS) such as malondialdehyde (MDA) was assumed to be a valid measure of lipid peroxidation, but we now know that this is a somewhat nonspecific biomarker. Nevertheless, changes in MDA levels were used to evaluate the effects of added nutrients such as carotenoids in cases in which oxidative stress might arise. Other investigators examined the effects of diets high in fruits and vegetables on biomarkers of oxidative stress. Even dehydrated fruit and vegetable extracts were used to demonstrate that they cause a decrease in lipid peroxide levels (9). Dixon and her associates (10,11) gave women carotenoid-deficient diets and observed an increase in plasma MDA levels. This effect could be reversed when the diets were supplemented with a mixture of carotenoids, strongly supporting the idea that dietary carotenoids can serve to decrease oxidative stress in humans.

We recently developed a new technique with which to evaluate oxidizability in both the hydrophilic and lipophilic compartments of human plasma/serum

(12,13). In this study, we compared this technique with the measurement of MDA levels in subjects supplemented with a high fruit and vegetable diet for a total of 8 wk; these biomarkers were then followed during a 4-wk period in which the subjects consumed a very low fruit and vegetable diet. In addition, we analyzed the lipid-soluble antioxidants, which include carotenoids, during these periods to see whether there was any correlation between their levels and those of the lipid oxidation biomarkers.

Subjects and Methods

Subjects. Healthy adults (n = 4; >60 y old) were recruited from the New England region. All study participants were in good health as determined by a medical history questionnaire, physical examination, and normal results for clinical laboratory tests. All of the study participants fulfilled the following eligibility criteria: (i) no history of cardiovascular, hepatic, gastrointestinal, or renal disease; (ii) no alcoholism or heavy alcohol use; (iii) no antibiotic or supplemental vitamin and/or mineral use for > 8 wk before the start of the study; and (iv) no smoking. The study protocol was approved by the Human Investigation Review Committee of Tufts University and the New England Medical Center, and written informed consent was obtained from each study participant.

The four healthy elderly adults lived at the Metabolic Research Unit (MRU) at the Human Nutrition Research Center and consumed high fruit and vegetable diets for 2 wk (~10 servings/d), followed by a 6-wk free-living period during which subjects were advised to consume a high fruit and vegetable diet. Then the subjects returned to the MRU and consumed a low fruit and vegetable diet (2 servings of light colored fruit and vegetables/d) for 4 wk.

All meals consumed during the residency period were prepared under the supervision of a dietitian at the MRU. With the exception of water, no other foods or beverages were allowed during that period. The diet consumed throughout the residency periods was a 2-d rotating menu diet based on foods that are commonly consumed by Americans (14). When averaged over the 2 d, the percentages of energy from protein, fat, and carbohydrate were 15, 31, and 54%, respectively. To control for potential variation in the carotenoid content of the vegetables, a single lot of each of the frozen vegetables was purchased for each study participant, protected from light, and stored at –20°C until the time of consumption.

In Table 13.1, the carotenoid-containing foods of the high fruit and vegetable diets are listed. After cooking, this diet contained a mean of 13 mg carotenoids/d (lutein, 2.71 mg; lutein isomer, 0.21 mg, zeaxanthin, 0.28 mg; cryptoxanthin, 1.04 mg; α-carotene, 1.13 mg; *cis*-β-carotene, 0.56 mg; *trans*-β-carotene, 3.73 mg; *cis*-lycopene, 0.57 mg; *trans*-lycopene, 2.90 mg). The low fruit and vegetable diets provided <1 mg/d total carotenoids. The carotenoid concentrations of the diet were determined by using individual aliquots of replicate meals with the modified official method of analysis by the Association of Official Analytical Chemists (15).

TABLE 13.1
Carotenoid-Containing Foods in the High Fruit and Vegetable 2-d Rotating Diet

Food		Amount (g)
Day 1	Pumpkin, canned	35
	Orange, raw	100
	Cantaloupe, raw	200
	Peaches, canned	100
	Green pepper, raw	30
	Green beans, canned	100
	Mangos, raw	100
Day 2	Apricot, canned	200
	Yams, canned	200
	Oranges, raw	200
	Tomato sauce, canned	50

Blood samples were collected from fasting subjects 3 times/wk during both MRU residential periods. Changes in carotenoid concentrations and total antioxidant performance using a lipophilic radical initiator and lipophilic probe in serum were determined.

Chemicals. All-*trans*-β-carotene (type II), α-carotene, and lycopene were purchased from Sigma Chemical (St. Louis, MO). Lutein was purchased from Kemin Industries (Des Moines, IA). Cryptoxanthin and echinenone were gifts from Hoffmann-La Roche (Nutley, NJ). The fatty acid analog 4,4-difluoro-5-(4-phenyl-1,3-butadienyl)-4-bora-3a,4a-diaza-s-indacene-3-undecanoic acid (BODIPY 581/591) was purchased from Molecular Probes (Eugene, OR). The radical initiators 2,2′-azobis(4-methoxy-2,4-dimethylvaleronitrile) (MeO-AMVN) and 2,2′-azobis(2,4-dimethylvaleronitrile) (AMVN) were gifts from Wako Chemicals (Richmond, VA). All HPLC solvents were obtained from J.T. Baker Chemical (Philipsburg, NJ) and passed through a 0.45-μm membrane filter before use.

Carotenoid Analysis in Serum. Serum antioxidants were measured by a HPLC system as previously described with minor modifications (16). Briefly, 200 μL serum was extracted with 2 mL of chloroform:methanol, 2:1, followed by 3 mL of hexane extraction. Samples were dried under nitrogen and resuspended in 150 μL ethanol:methyl-*tert*-butyl ether (2:1) and 50 μL of this mixture was injected onto the HPLC.

The HPLC system consisted of a series 410 LC pump (Perkin-Elmer, Norwalk, CT), a Waters 717 plus autosampler (Millipore, Milford, MA), a C30 carotenoid column (3 μm, 150 × 4.6 mm, YMC, Wilmington, NC), an HPLC column temperature controller (model 7950; column heater/chiller, Jones Chromatography, Lakewood, CO), a Waters 994 programmable photodiode array detector, and a Waters 840 digital 350 data station. The Waters 994 programmable photodiode array detector was set at

450 nm for carotenoids. The HPLC mobile phase was methanol:methyl-*tert*-butyl ether:water (83:15:2, by vol, 1.5% ammonium acetate in the water, solvent A) and methanol:methyl-*tert*-butyl ether:water (8:90:2, by vol, 1.0% ammonium acetate in the water, solvent B). The gradient procedure, at a flow rate of 1 mL/min (16°C), was as follows: (i) 100% solvent A was used for 2 min followed by a 6-min linear gradient to 70% solvent A; (ii) 3-min hold followed by a 10-min linear gradient to 45% solvent A; (iii) a 2-min hold, then a 10-min linear gradient to 5% solvent A; (iv) a 4-min hold, then a 2-min linear gradient back to 100% solvent A.

Using this method, lutein, zeaxanthin, cryptoxanthin, α-carotene, 13-*cis*-β-carotene, all-*trans*-β-carotene, 9-*cis*-β-carotene, and *cis*- and *trans*-lycopenes were adequately separated. Carotenoids were quantified by determining peak areas in the HPLC chromatograms calibrated against known amounts of standards. The amounts were corrected for extraction and handling losses by monitoring the recovery of the internal standard, echinenone. The lower limit of detection was 0.2 pmol for carotenoids. Our HPLC analysis for carotenoids provides coefficients of variation as 4% for intraassay (n = 25) and 4% for interassay (n = 9). The recovery of the internal standard averaged 97%. The accuracy, determined by the recovery of added β-carotene to a serum sample, averaged 95%.

Oxidizability of Serum. Blood samples from fasting subjects were collected in evacuated containers and kept on ice. Samples were protected from light and centrifuged for 20 min (800 × *g*, 4°C) within 30 min of collection. Aliquots of serum were stored at –70°C and analyzed within 3 y. The oxidation of serum was achieved by incubating serum:PBS (1:5) at 37°C with the lipophilic radical initiator, MeO-AMVN. The oxidation of serum was monitored by the appearance of the green fluorescent oxidation product of BODIPY 581/591, (λex = 500 nm, λem = 520 nm) as reported earlier (9). The fluorescence measurements were carried out using a Perkin-Elmer spectrofluorometer (model 650-10s).

Measurement of Lipid Peroxidation. Lipid peroxidation was assessed by the measurement of MDA using an HPLC system (17). The optimized method involved a protein precipitation step using trichloroacetic acid (TCA), acid hydrolysis and formation of an MDA thiobarbituric acid (TBA) adduct. Briefly, serum or serum, which was incubated with 5 mM of AMVN at 42°C for 2 h was treated with butylated hydroxytoluene (5% in EtOH), followed by the addition of TCA (10% wt/vol). The mixture was heated for 20 min at 95°C. After cooling in ice, the mixture was centrifuged at 1000 × *g* for 10 min at 4°C. KOH (3.065 M) was added to the supernatant, followed by the addition of TBA (0.4% wt/vol, in acetate buffer, pH 3.5). The mixture was heated for 45 min at 95°C. After cooling, an aliquot was filtered (0.45 μm) and injected onto an HPLC system with a Pecosphere-3 C18 column (83 × 4.6 mm). The HPLC column temperature controller (model 7950; column heater/chiller, Jones Chromatography) was set at 20°C. The scanning fluorescence detector (Waters 474; Millipore) was set at Ex 515 nm and Em 553 nm. The HPLC mobile phase was 20 mM potassium phosphate buffer:acetonitrile (80:20, vol/vol) and the flow rate was set at 0.8 mL/min. The lower limit of detection was 0.2 pmol for $(MDA\text{-}TBA)_2$ adduct.

Statistical Analysis. Results are expressed as means ± SEM; the significance of differences was determined by Student's *t* test or ANOVA using the STATVIEW II program (Abacus Concepts, Berkeley, CA).

Results

Table 13.2 shows the changes in serum lutein, β-carotene, and lycopene concentrations during the MRU residency periods in the study participants. Compared with baseline, serum lutein concentrations increased significantly to 165 and 209% at wk 1 and 2, respectively, after consumption of the high fruit and vegetable diets, and decreased significantly to 65, 42, 39, and 34% at wk 1, 2, 3, and 4, respectively, after consumption of the low fruit and vegetable diets. Serum β-carotene concentrations increased significantly to 181, and 204% at wk 1 and 2, respectively, with the high fruit and vegetable diets, and decreased to 77, 51, 43, and 35% at wk 1, 2, 3, and 4, respectively, with the low fruit and vegetable diet. However, due to the large variability in response among the subjects, only the 4-wk concentration was significantly different. Serum lycopene concentrations increased significantly to 169% at wk 1 of the high fruit and vegetable diet and the levels tended to remain elevated at wk 2 of high fruit and vegetable diet. With consumption of the low fruit and vegetable diet, serum lycopene levels decreased to 72, 47, 42, and 39% at wk 1, 2, 3, and 4, respectively.

Serum total carotenoid concentrations were significantly increased within 3 d ($P < 0.05$) of high fruit and vegetable diet consumption (13 mg carotenoids/d) as shown in Figure 13.1. The serum oxidizability was significantly decreased within 3

TABLE 13.2
Serum Carotenoid Responses to the High and Low Fruit and Vegetable Diets[a]

		Serum carotenoids (μg/dL)		
Dietary intervention	Time (wk)	Lutein	β-Carotene	Lycopene
High fruit and vegetable (13 mg/d carotenoids)	Baseline	9.85 ± 1.10	42.63 ± 12.46	14.77 ± 2.89
	1	16.20 ± 0.82*	77.38 ± 19.83	25.05 ± 4.04*
	2	20.59 ± 1.84*	87.31 ± 19.49	24.00 ± 4.77
Low fruit and vegetable (<1 mg/d carotenoids)	Baseline	20.78 ± 4.30	80.04 ± 17.62	19.85 ± 4.98
	1	13.58 ± 2.23*	61.77 ± 20.02	14.32 ± 2.97
	2	8.78 ± 0.92*	40.97 ± 13.54	9.37 ± 1.66*
	3	8.14 ± 1.32*	34.63 ± 12.29	8.35 ± 0.79*
	4	7.00 ± 0.98*	28.47 ± 8.60*	7.83 ± 0.99*

[a]Values are means ± SE; *significantly different from baseline of each intervention period. $P < 0.05$.

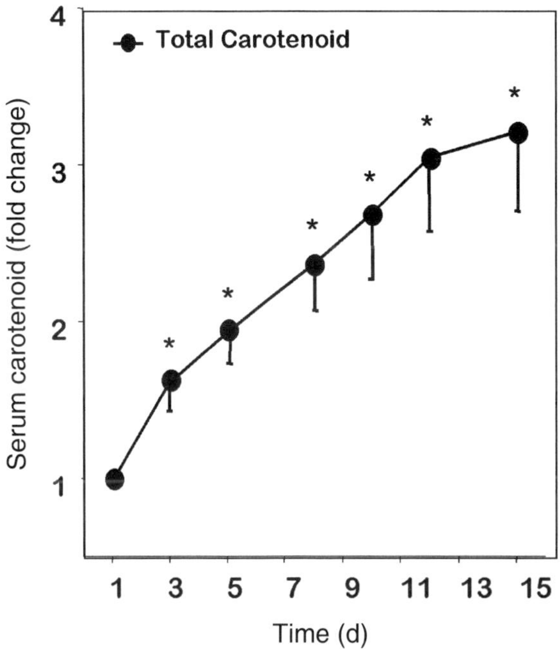

Fig. 13.1. Serum carotenoid response to the high fruit and vegetable diet. Values are means ± SEM, n = 4. Asterisks indicate significant difference from baseline, **P* < 0.05.

d ($P < 0.05$), and continued to decrease over the 2-wk period of low fruit and vegetable diet consumption (Fig. 13.2).

Serum total carotenoid levels were significantly decreased ($P < 0.0005$) within 1 wk of ingesting a low fruit and vegetable diet (<1 mg carotenoids/d; Fig. 13.3). To evaluate the susceptibility against oxidative stress, both endogenous MDA and free radical generated MDA were determined. The latter were generated by adding AMVN to plasma, thus inducing MDA formation. Although there was a trend for an increase in the ratio of AMVN-induced MDA/endogenous MDA, no significant difference was found. On the other hand, serum oxidizability was increased by the ingestion of the low fruit and vegetable diet and the difference was significant at 4 wk (Fig. 13.4).

Figure 13.5 shows the correlation between carotenoid levels and oxidizability in serum of two subjects measured 3 times/wk for 4 wk in response to consumption of the low fruit and vegetable diet. There was an inverse linear relationship between the oxidizability and serum β-carotene ($r^2 = 0.799$, $P < 0.0001$) , lycopene ($r^2 = 0.802$, $P < 0.0001$), lutein ($r^2 = 0.724$, $P < 0.0001$), and cryptoxanthin ($r^2 = 0.688$, $P < 0.0001$).

Discussion

Our data clearly indicate that the level of serum carotenoids can be readily altered by either increasing or decreasing the number of servings of fruits and vegetables

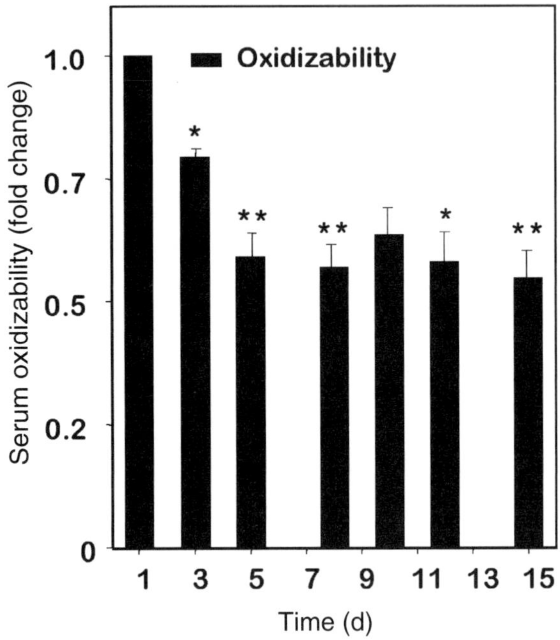

Fig. 13.2. Changes of serum oxidizability in response to the high fruit and vegetable diet. Values are means ± SEM, n = 4. Asterisks indicate significant difference from baseline, $*P < 0.05$, $**P < 0.005$.

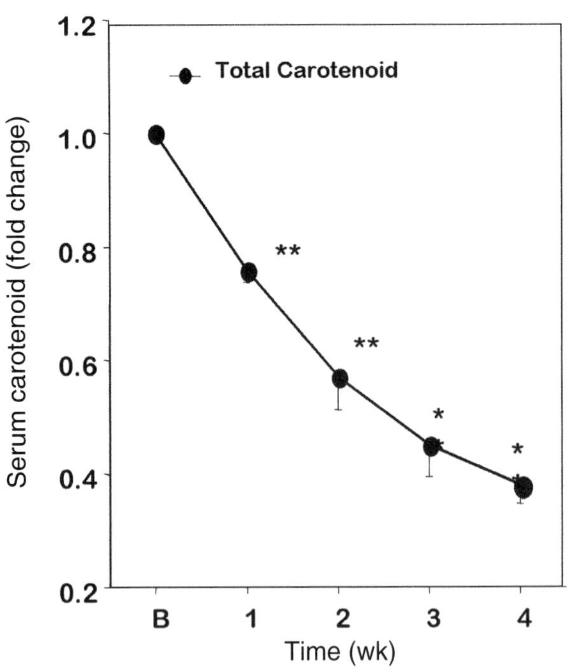

Fig. 13.3. Serum carotenoid response to the low fruit and vegetable diet. Values are means ± SEM, n = 4. Asterisks indicate significant difference from the start of the low-carotenoid period, $*P < 0.05$, $**P < 0.005$.

in the diet. In our subjects who consumed a high fruit and vegetable diet with an intake of 13 mg total carotenoids/d, a significant increase in serum carotenoids was observed by d 3 of this diet, and the increase continued for the entire 2-wk feeding period (Fig. 13.1). Other investigators also reported an increase in plasma/serum carotenoid levels when subjects consumed high fruit and vegetable intakes (18,19), but there are few reports indicating a significant increase in plasma/serum carotenoids by d 3. Accompanying this increase in serum carotenoids is the observation of a significant decrease in serum oxidizability by d 3 of high fruit and vegetable diet intake (Fig. 13.2). Again, other workers observed decreases in parameters of oxidizability such as plasma TBARS and LDL oxidation when carotenoid-rich vegetable products were added to the diet (20). However, these observations,

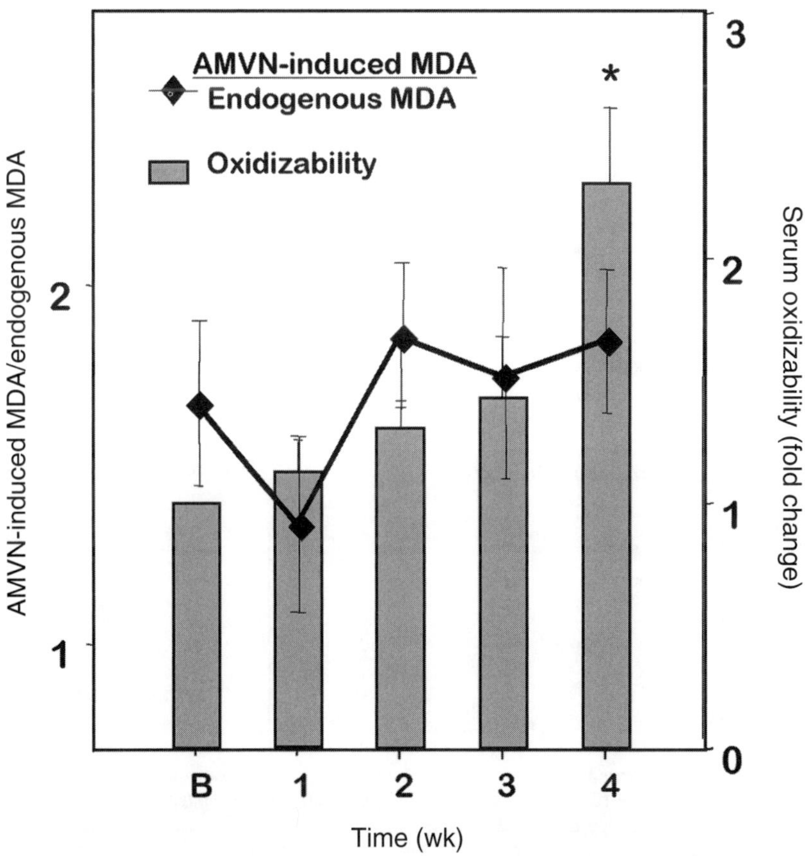

Fig. 13.4. Changes of serum oxidizability and AMVN induced/endogenous MDA in response to the high fruit and vegetable diet. Values are means ± SEM, n = 4. Asterisks indicate significant difference from the start of the low-carotenoid period, $^*P < 0.05$.

as well as ours, are associative in nature and do not prove that the changes in oxidizability are directly related to changes in the plasma carotenoid levels.

When these subjects consumed a low fruit and vegetable diet (<1 mg carotenoid/d), plasma carotenoids decreased significantly within 1 wk and continued decreasing over the 4-wk depletion period (Fig. 13.3). We used two biomarkers of oxidative stress to determine whether carotenoid deletion was associated with any changes in lipid oxidizability. Although we used a HPLC method to determine MDA, we found a very large variance in our subjects. We also measured AMVN-induced MDA formation and reported the ratio of AMVN-induced MDA to endogenous MDA formation. These results, shown in Figure 13.4, indicated that there was no significant change in the ratio over the 4-wk depletion period. It is possible that depletion of fruits and vegetables might have increased either AMVN-induced MDA or endogenous MDA alone. For example, Dixon *et al.* (11)

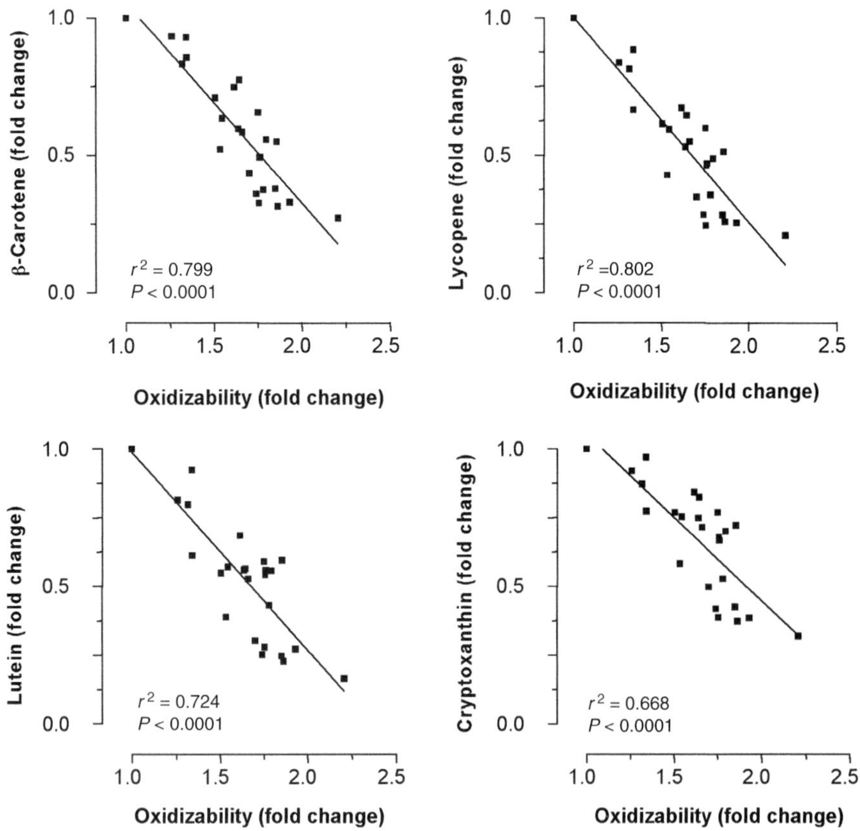

Fig. 13.5. Correlation between carotenoid concentrations and oxidizability in serum in response to the low fruit and vegetable diet.

reported an increase in MDA-TBARS in a group of women who consumed a carotenoid-deficient diet for 60 d. However, when serum oxidizability was determined by the BODIPY 581/591 method (12), a significant increase was observed by wk 4 of depletion (Fig. 13.4).

The change to a low fruit and vegetable diet involved dietary changes in nutrients other than carotenoids. For example, vitamin C consumption was also reduced during this period. However, the diet change did not markedly affect vitamin E consumption. For two of the subjects, we analyzed lipid antioxidants and oxidizability every 3 d during the 4-wk depletion period. When we compared the changes in individual carotenoids with the change in oxidizability (Fig. 13.5), a remarkable inverse linear relationship was observed for β-carotene, lycopene, lutein, and cryptoxanthin ($P < 0.0001$). We conclude that carotenoids either are the direct determinants of oxidizability in human serum, or that they are serving as a marker for those dietary factors that would determine serum oxidizability (21). Thus, the oxidizability in serum can be modified by diet and is related to the carotenoid content of serum.

Acknowledgments

The authors thank the study participants and staffs at the Metabolic Research Unit at the Jean Mayer USDA-Human Nutrition Research Center on Aging at Tufts University. This research was supported in part by the U.S. Department of Agriculture, under agreement number 1950-51000-048-01A. The contents of this publication do not necessarily reflect the views or policies of the U.S. Department of Agriculture, nor does mention of trade names, commercial products, or organizations imply endorsement by the U.S. Government.

References

1. Ziegler, R.G. (1991) Vegetables, Fruits, and Carotenoids and the Risk of Cancer, *Am. J. Clin. Nutr. 53:* 251S–259S.
2. Liu, S., Lee, I.-M., Ajani, U., Cole, S.R., Buring, J.E., and Manson, J.E. (2001) Intake of Vegetables Rich in Carotenoids and Risk of Coronary Heart Disease in Men: The Physicians' Health Study, *Int. J. Epidemiol. 30:* 130–135.
3. Ames, B.N., Shigenaga, M.K., and Hagen, T.M. (1993) Oxidants, Antioxidants, and the Degenerative Diseases of Aging, *Proc. Natl. Acad. Sci. USA 90:* 7915–7922.
4. Sies, H. (1986) Biochemistry of Oxidative Stress, *Angew Chem. Int. Ed. 25:* 1058–1071.
5. Peto, R., Doll, R.J., Buckley, J.D., and Sporn, M.B. (1981) Can Dietary β-Carotene Materially Reduce Human Cancer Rates? *Nature 290:* 201–208.
6. The Alpha-Tocopherol-Beta-Carotene Cancer Prevention Study Group (1994) The Effect of Vitamin E and Beta Carotene on the Incidence of Lung Cancer and Other Cancers in Male Smokers, *N. Engl. J. Med. 330:* 1029–1035.
7. Omenn, G.S., Goodman, G.E., Thornquist, M.D., Balmes, J., Cullen, M.R., Glass, A., Keogh, J.P., Meyskens, F.L., Jr., Valanis, B., Williams, J.H., Jr., Barnhart, S., and Hammar, S. (1996) Effects of a Combination of Beta Carotene and Vitamin A on Lung Cancer and Cardiovascular Disease, *N. Engl. J. Med. 334:* 1150–1155.
8. Hennekens, C.H., Buring, J.E., Manson, J.E., Stampfer, M., Rosner, B., Cook, N.R., Belanger, C., LaMotte, F., Gaziano, J.M., Ridker, P.M., Willett, W., and Peto, R. (1996)

Lack of Effect of Long-Term Supplementation with Beta Carotene on the Incidence of Malignant Neoplasms and Cardiovascular Disease, *N. Engl. J. Med. 334:* 1145–1149.

9. Wise, J.A., Morin, R.J., Sanderson, R., and Blum, K. (1996) Changes in Plasma Carotenoid, Alpha-Tocopherol, and Lipid Peroxide Levels in Response to Supplementation with Concentrated Fruit and Vegetable Extracts: A Pilot Study, *Curr. Ther. Res. 57:* 445–461.
10. Dixon, Z.R., Burri, B.J., Clifford, A., Frankel, E.N., Schneeman, B.O., Parks, E., Keim, N.L., Barbieri, T., Wu, M.-M., Fong, A.K.H., Kretsch, M.J., Sowell, A.L., and Erdman, J.W., Jr. (1994) Effects of a Carotene-Deficient Diet on Measures of Oxidative Susceptibility and Superoxide Dismutase Activity in Adult Women, *Free Radic. Biol. Med. 17:* 537–544.
11. Dixon, Z.R., Shie, F.S., Warden, B.A., Burri, B.J., and Neidlinger, T.R. (1998) The Effect of a Low Carotenoid Diet on Malondialdehyde-Thiobarbituric Acid (MDA-TBA) Concentrations in Women: A Placebo-Controlled Double-Blind Study, *J. Am. Coll. Nutr. 17:* 54–58.
12. Aldini, G., Yeum, K.-J., Russell, R.M., and Krinsky, N.I. (2001) A Method to Measure the Oxidizability of the Aqueous and Lipid Compartments of Plasma, *Free Radic. Biol. Med. 31:* 1043–1050.
13. Aldini, G., Yeum, K.-J., Russell, R.M., and Krinsky, N.I. (2003) A Selective Assay to Measure Antioxidant Capacity in Both the Aqueous and Lipid Compartments of Plasma, *Nutr. Sci. 6:* 12–19.
14. Block, G., Dresser, C.M., Hartman, A.M., and Carroll, M.D. (1985) Nutrient Sources in the American Diet: Quantitative Data from the NHANES II Survey. I. Vitamins and Minerals, *Am. J. Epidemiol. 122:* 13–26.
15. Deutsch, M.J. (1984) Vitamins and Other Nutrients, in *Official Methods of Analysis of the Association of Official Analytical Chemists*, Williams, S., ed., AOAC International, Arlington, VA, pp. 830–836.
16. Yeum, K.-J., Booth, S., Sadowski, J., Lin, C., Tang, G., Krinsky, N.I., and Russell, R.M. (1996) Human Plasma Carotenoid Response to the Ingestion of Controlled Diets High in Fruits and Vegetables, *Am. J. Clin. Nutr. 64:* 594–602.
17. Templar, J., Kon, S.P., Milligan, T.P., Newman, D.J., and Raftery, M.J. (1999) Increased Plasma Malondialdehyde Levels in Glomerular Disease as Determined by a Fully Validated HPLC Method, *Nephrol. Dial. Transplant. 14:* 946–951.
18. Broekmans, W.M., Klöpping-Ketelaars, I.A., Schuurman, C.R., Verhagen, H., van den Berg, H., Kok, F.J., and van Poppel, G. (2000) Fruits and Vegetables Increase Plasma Carotenoids and Vitamins and Decrease Homocysteine in Humans, *J. Nutr. 130:* 1578–1583.
19. Smith-Warner, S.A., Elmer, P.J., Tharp, T.M., Fosdick, L., Randall, B., Gross, M., Wood, J., and Potter, J.D. (2000) Increasing Vegetable and Fruit Intake: Randomized Intervention and Monitoring in an at-Risk Population, *Cancer Epidemiol. Biomark. Prev. 9:* 307–317.
20. Bub, A., Watzl, B., Abrahamse, L., Delincée, H., Adam, S., Wever, J., Müller, H., and Rechkemmer, G. (2000) Moderate Intervention with Carotenoid-Rich Vegetable Products Reduces Lipid Peroxidation in Men, *J. Nutr. 130:* 2200–2206.
21. Yeum, K.-J., Aldini, G., Chung, H.-Y., Krinsky, N.I., and Russell, R.M. (2003) The Activities of Antioxidant Nutrients in Human Plasma Depend on the Localization of the Attacking Radical Species, *J. Nutr. 133:* 2688–2691.

Chapter 14

Mitochondria as Novel Targets for Proapoptotic Synthetic Retinoids

Numsen Hail, Jr., and Reuben Lotan

Department of Thoracic/Head and Neck Medical Oncology, The University of Texas M.D. Anderson Cancer Center, Houston, TX 77030

Introduction

Natural and synthetic analogs of vitamin A are collectively classified as retinoids. Certain retinoids have shown promise in the prevention and/or treatment of cancer. The results obtained from preclinical studies in animal carcinogenesis models (1,2) and human clinical trials (3,4) suggest that certain retinoids may impede the process of carcinogenesis (3,5). Indeed, chemoprevention with retinoids can promote the partial regression of premalignant lesions when these agents are given at high dosage (6). Ultimately, the modulation of carcinogenesis by retinoids is believed to lower the risk of developing invasive or clinically significant disease (5).

The mechanisms through which retinoids accomplish their cancer preventive and therapeutic effects *in vivo* are not fully understood. Retinoids are thought to act *via* nuclear receptor–mediated transactivation of target genes. There are two types of nuclear retinoid receptors: retinoic acid receptors (RAR) and retinoid X receptors (RXR), which belong to the steroid hormone receptor superfamily (6,7). There are three subtypes of RAR and RXR encoded by different genes designated α, β, and γ. The RAR bind the natural ligands all-*trans* retinoic acid (ATRA, Fig. 14.1) and 9-*cis* retinoic acid (9-*cis* RA), whereas the RXR bind only 9-*cis* RA. The receptors form RXR-RAR heterodimers or RXR-RXR homodimers that bind DNA sequences called response elements (RARE and RXRE, respectively). These receptors become active transcription factors upon ligand binding, and regulate the transactivation of RARE- or RXRE-containing target genes (6,7).

The synthetic retinoid *N*-(4-hydroxyphenyl)retinamide (4HPR or fenretinide) does not bind with high affinity to RAR because it lacks a carboxyl functional group that is believed to be required for this activity (8). However, 4HPR is able to enhance the transcriptional activity of these receptors in intact cells (8–10). The synthetic retinoid 6-[3-(1-adamantyl)-4-hydroxyphenyl]-2-naphthalene carboxylic acid (CD437 or AHPN) exhibits a degree of selectivity for binding to RARγ and RARβ. CD437 transactivates the transcription of RARE reporter gene constructs *via* these receptors, albeit with an ~10-fold lower potency than ATRA (11,12). The

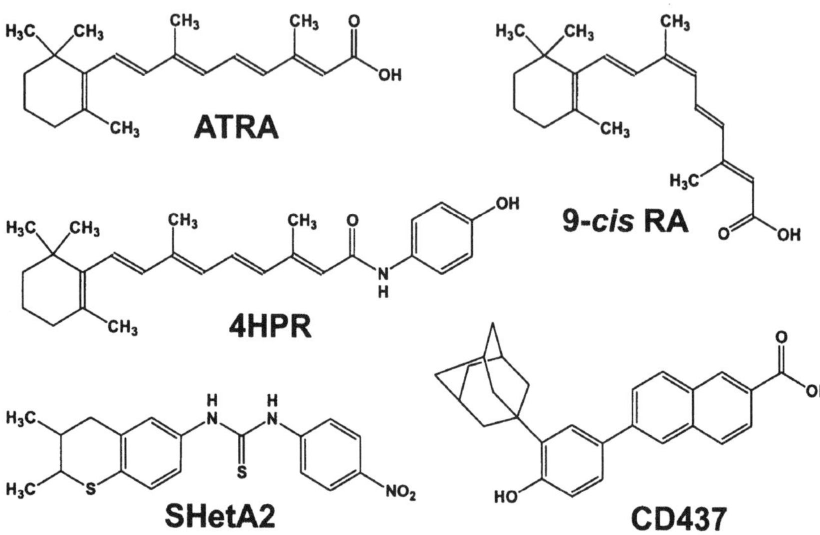

Fig. 14.1. The chemical structures of all-*trans* retinoic acid (ATRA), 9-*cis* retinoic acid (9-*cis* RA), *N*-(4-hydroxyphenyl)retinamide (4HPR), 6-[3-(1-adamantyl)-4-hydroxyphenyl]-2-naphthalene carboxylic acid (CD437), and [(4-nitrophenyl)amino] [2,2,4,4-tetramethyl thiochroman-6-yl)amino] methane-1-thione (SHetA2).

synthetic retinoid [(4-nitrophenyl)amino][(2,2,4,4-tetramethyl thiochroman-6-yl)amino] methane-1-thione (SHetA2), like 4HPR, lacks a carboxyl functional group, and did not initiate retinoid receptor transcriptional activation at concentrations as high as 10 μM in cells transfected with the reporter constructs of individual retinoid receptors (13). In addition, 4HPR (10,14–16), CD437 (17–21), and SHetA2 (13) apparently induce apoptosis in a nuclear receptor–independent fashion in various tumor cell types because several retinoid receptor antagonists were ineffective in blocking cell death.

Work conducted in our laboratory over the past 10 years has focused on evaluating the effects of 4HPR, CD437, and, more recently, SHetA2, in human carcinoma cells and, in certain instances, in normal human epithelial cell strains with an emphasis on the mechanisms associated with the induction of apoptosis in the carcinoma cells. We observed that 4HPR (10,15,16,22), CD437 (18,19,21), and SHetA2 (13) are far more potent than ATRA and 9-*cis* RA in inducing the marked intranucleosomal DNA fragmentation that is characteristic of apoptosis in carcinoma cells derived from lung, head and neck, prostate, and skin as measured by the terminal deoxynucleotidyl transferase dUTP nick-end labeling technique. In addition, we also observed selectivity in CD437-induced apoptosis in malignant epithelial cells relative to their normal counterparts (18,23). Most importantly, all of these synthetic retinoids appear to target the mitochondria to trigger apoptosis in transformed cells *in vitro*, which may be important for their anticancer potential.

Common Mechanistic Attributes of Proapoptotic Synthetic Retinoids

It is generally accepted that apoptosis is the mechanism utilized by metazoans to eliminate redundant or potentially deleterious cells. As such, apoptosis can be viewed as an essential process for regulating tissue homeostasis and maintaining genomic integrity. Many human diseases can be directly linked to the dysregulation of apoptosis. For example, excessive apoptosis can lead to Alzheimer's disease or acquired immunodeficiency syndrome. Conversely, the loss of apoptosis regulation can lead to autoimmune disorders and cancer (24). The cellular machinery associated with apoptosis is highly conserved, with many similarities existing between phylogenetically divergent species. The process of apoptosis is relatively linear. It is triggered by an initiation phase that is remarkably varied, depending on cell type and apoptotic stimuli (e.g., oxidative stress, DNA damage, ion fluctuations, and cytokines). This is followed by an effector phase that is associated with the systematic activation of catabolic proteases and hydrolases that ultimately participate in the cleavage of proteins and DNA characteristic of the degradation phase (25–28). In the following subsections, we will discuss the common mechanistic aspects associated with the apoptogenic activity of 4HPR, CD437, and SHetA2, with an emphasis on correlating these aspects with the initiation, effector, or degradation phases of apoptosis.

Reactive Oxygen Species Generation

Reactive oxygen species (ROS) can promote divergent cellular effects depending on the extent of their production and the enzymatic or nonenzymatic mechanisms available for their dismutation in a given cell type. Thus, ROS can serve as mitogenic stimuli, senescence promoters, or cell death mediators (29). The mitochondria are the primary site of ROS production in most cells (29–32); under certain conditions, elevated mitochondrial ROS generation can serve as an apoptogenic signal (29,30,33).

Within 15 min after exposure to 4HPR, the oxidation of 2′,7′-dichlorofluorescin to 2′,7′-dichlorofluorescein increased in a linear fashion in various tumor cell types (10,16,34,35). The increase in 2′,7′-dichlorofluorescein fluorescence intensity is most likely the result of enhanced hydroperoxide generation in these cells. The prooxidant and proapoptotic properties of 4HPR can be blocked by several antioxidants including L-ascorbic acid (16), pyrrolidine dithiocarbamate (36), or butylated hydroxyanisole (BHA) (10), suggesting that the cytotoxicity of 4HPR is linked to oxidative stress in certain cancer cell types.

Several lines of evidence suggest that 4HPR-induced ROS generation arises from the mitochondria in cancer cells. There has been speculation that 4HPR can be converted to a radical species in the mitochondria (37), and classical inhibitors of mitochondrial electron transport (e.g., rotenone, myxothiazol, and cyanide) diminish ROS generation promoted by 4HPR in skin (38) and cervical (36) cancer cells. A study by Maurer *et al.* (39) reported that neuroblastoma cells cultured under hypoxic conditions became resistant to the cytotoxic effects of 4HPR. In this scenario, the low oxygen conditions should favor decreased ROS production by reducing the amount of oxygen in the intracellular environment capable of undergoing nonenzymatic one-electron reduction (40). Moreover, hypoxic conditions can inhibit respiration in isolated mitochondria (41) and intact cells (42). Finally, in respiration-deficient skin cancer cells depleted of mitochondrial DNA (ρ^0) by chronic exposure to ethidium bromide, the prooxidant and apoptogenic effects of 4HPR were markedly suppressed, implying that the mitochondria, specifically mitochondrial respiration, was the primary target of 4HPR in skin cancer cells (38).

CD437 promoted the oxidation of dihydroethidium to ethidium, presumably *via* enhanced superoxide generation, in myeloma (17) and skin cancer cells (43). In these cells, the enhanced superoxide generation was believed to be associated predominately with the induction of the mitochondrial permeability transition. Interestingly, L-ascorbic acid inhibited CD437-induced apoptosis in skin cancer cells (43), and α-tocopherol acetate provided a similar protective effect in lymphocytic leukemia cells exposed to CD437 (44). These findings suggest that free radical generation is intimately associated with triggering CD437-induced cell death in certain cell types. Moreover, L-ascorbic acid also blocked the CD437-mediated inhibition of oxygen consumption by skin cancer cells, implying that the mitochondria were both the source and the target of oxidative stress in these cells (43).

SHetA2 increased the generation of intracellular ROS in head and neck carcinoma cells within 60 min after treatment. A linear and time-dependent increase in 2′,7′-dichlorofluorescein fluorescence intensity was observed up to 240 min in this assay. Under the same conditions, ATRA promoted only a slight, but discernible and reproducible, increase in ROS generation, whereas 4HPR promoted ROS generation more effectively than SHetA2 in head and neck carcinoma cells. The increase of ROS by SHetA2 treatment was inhibited by co-treatment with the antioxidant BHA. Furthermore, BHA was able to suppress the induction of apoptosis by SHetA2, suggesting that ROS generation was important for apoptosis induction in head and neck carcinoma cells (13).

Induction of the Mitochondrial Permeability Transition

The intrinsic (i.e., type 2 or mitochondria-dependent) effector pathway of apoptosis relies solely on the permeabilization of mitochondrial membranes to release the apoptogenic mitochondrial proteins [e.g., cytochrome *c* (45), endonuclease G (46,47), Smac/DIABLO (48), Omi/HtrA2 (49), apoptosis-inducing factor (AIF) (50), and its homologue AIF-homologous mitochondrion-associated inducer of death (AMID) (51) required for caspase activation and/or apoptosis. The induction of the mitochondrial permeability transition (MPT) is a process associated with the opening of nonspecific proteinaceous pores that increase the permeability of an otherwise intrinsically impermeable inner mitochondrial membrane, resulting in the collapse of the mitochondrial inner transmembrane potential ($\Delta\Psi_m$) (52–54). The MPT is a rate-limiting and self-amplifying process that is regulated at various levels by several mitochondrial proteins. Many of these proteins are believed to constitute the permeability transition pore complex (PTPC) (55,56). Normally, proteins in the outer and inner mitochondrial membranes that constitute the PTPC are predictably in close proximity to each other and in a closed or low conductance conformation. Numerous pathologic stimuli (e.g., ROS and Ca^{2+}) as well as various chemical agents can cause the PTPC to adopt an open conformation (55,56). Once this occurs, water and solutes can infiltrate the mitochondrial matrix (57). This results in colloidal osmotic swelling of the mitochondrial matrix and permeabilization of the outer mitochondrial membrane, presumably due to physical rupture of the outer membrane (27). Once the outer membrane fragments, the apoptogenic mitochondrial proteins are released to the cytosol to participate in the degradation phase of apoptosis. The loss of cytochrome *c* from the mitochondria and/or the induction of MPT would be expected to result in bioenergetic catastrophe in respiring cells (25,30,33). Consequently, the events associated with the permeabilization of mitochondrial membranes can be considered the decisive regulatory point in cell death signaling (53,55).

Exposure to 4HPR caused a time-dependent loss of $\Delta\Psi_m$, indicative of MPT in skin (16,38) and cervical (36) cancer cells. This process was measured in intact cells by the decreased retention of the lipophilic cationic fluorescent dyes rhodamine 123 or 3,3′-dihexyloxacarbocyanine iodide [$DiOC_6(3)$] alone, or concurrently with dihydroethidium to measure superoxide production (38). As observed with ROS generation, the antioxidant L-ascorbic acid (16) or pyrrolidine dithiocarbamate (36) suppressed the induction of the MPT in skin or cervical cancer cells, respectively. The MPT antagonists cyclosporin A and bongkrekic acid also inhibited the induction of MPT in 4HPR-treated skin cancer cells, partially diminishing 4HPR-induced ROS generation (16). Furthermore, 4HPR exposure did not trigger MPT in ρ^0 skin cancer cells, suggesting that MPT was required for apoptosis induction (38).

An elegant study by Marchetti *et al.* (17) was the first to illustrate that CD437-induced apoptosis required the rapid loss of $\Delta\Psi_m$ and MPT induction in myeloma cells. This study also illustrated that short-term exposure to CD437 could promote mitochondrial swelling indicative of MPT in isolated mouse liver mitochondria respiring on succinate (17). On a molar basis, CD437 was by far the most potent MPT inducer of the three synthetic retinoids discussed here. In addition to myeloma cells (17), we observed similar biochemical characteristics of CD437-induced MPT in skin cancer cells (43). In both cell types, MPT induction could be modulated by various inhibitors of MPT (17,43).

We compared the biochemical changes associated with CD437-induced MPT in COLO 16 skin cancer cells with the morphological changes in the mitochondria from these cells to investigate the MPT phenomenon further. The dissipation of

$\Delta\Psi_m$, represented by decreased retention of the cationic probe $DiOC_6(3)$, and enhanced superoxide production, represented by the oxidation of dihydroethidium to ethidium, can be monitored concurrently in intact cells by flow cytometry, and serve as a surrogate indicator of MPT (43). The control cells in Figure 14.2A were gated because we assumed that they exhibited high $DiOC_6(3)$ fluorescence and low ethidium fluorescence. Therefore, most of the cell population is situated in the lower right quadrant of the panel. A 4-h exposure to 1 µM CD437 shifted ~73% of the cell population to the upper left quadrant of the panel in Figure 14.2B. These cells exhibited low $DiOC_6(3)$ fluorescence intensity and high ethidium fluorescence intensity, suggesting that the mitochondria in these cells had undergone MPT. An electron micrograph of the mitochondria in a control COLO 16 cell (Fig. 14.2C) revealed that these organelles were electron dense, and the matrix exhibited a condensed conformation. A 4-h exposure to 1 µM CD437 promoted marked swelling of the mitochondria as illustrated by the electron micrograph in Figure 14.2D. The cristae were absent and the disruption of outer membrane integrity was evident, illustrating that CD437 was acutely mitochondriotoxic in COLO 16 cells. The biochemical characteristics of CD437-induced MPT, as well as apoptosis induction, were absent in ρ^0 skin cancer cells. This would suggest that CD437 directly inhibited bioenergetics to trigger MPT and apoptosis in the parental cells (43).

Exposure to SHetA2 diminished $\Delta\Psi_m$ and enhanced the release of cytochrome *c* from the mitochondria of head and neck carcinoma cells. Both of these effects were prevented by the MPT inhibitor cyclosporin A, which also decreased SHetA2-induced apoptosis. Moreover, another MPT inhibitor, trifluoperazine (17), also suppressed the induction of apoptosis by SHetA2 treatment (13). The ability of an antioxidant, cyclosporin A, or trifluoperazine to inhibit SHetA2-induced apoptosis suggests that this synthetic retinoid, like 4HPR and CD437, interferes with mitochondrial function, perhaps mitochondrial respiration, to engage apoptosis. These results demonstrate that SHetA2 is a potent apoptosis inducer that acts on the mitochondria in head and neck carcinoma cells.

Caspase-3-Like Protease Activation

A central component of the degradation phase of apoptosis is the proteolytic system comprised of a family of cysteine proteases called caspases. Caspases are constitutively present in most cells, and reside in the cytosol as a single-chain proenzyme. Over a dozen caspases have been identified in mammalian cells. Approximately two thirds of these were suggested to function in apoptosis (58,59). There are two types of caspases, upstream caspases called initiator caspases (e.g., caspases 8, 9, and 10), and their downstream targets known as effector or executioner caspases (e.g., caspases 3, 6, and 7) (59). Caspases selectively cleave a restricted set of target proteins, usually at one, or at most a few positions in the primary sequence [always after an aspartate (Asp) residue] (59,60). Cleavage of spe-

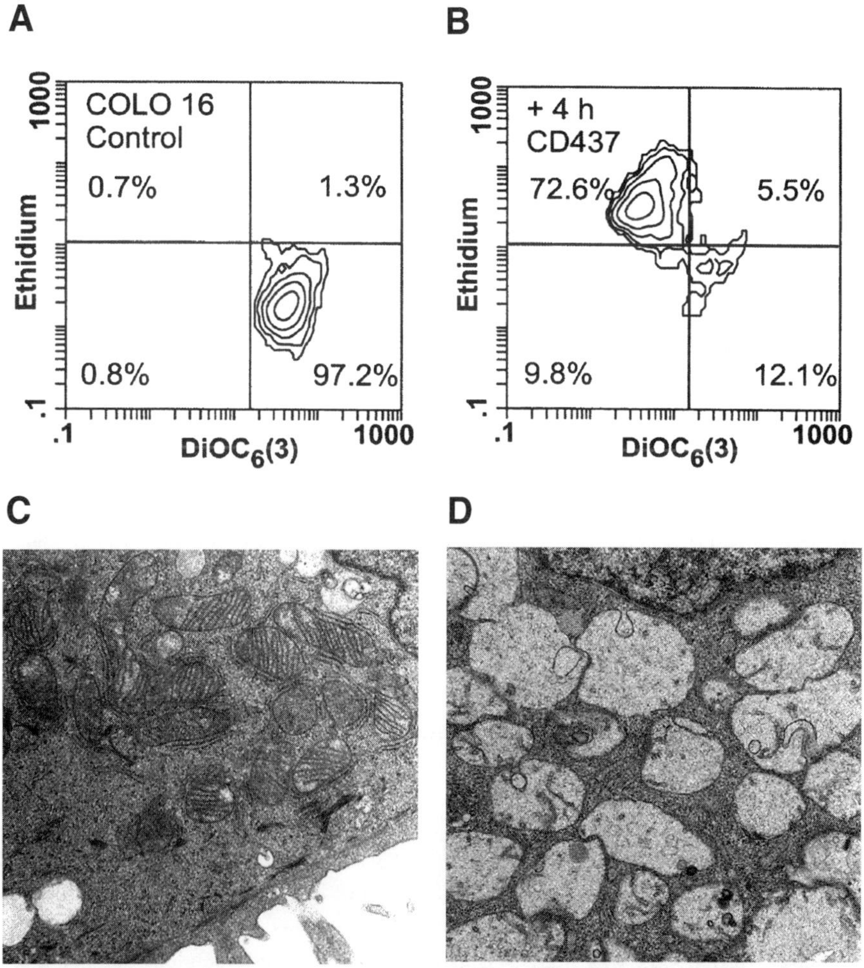

Fig. 14.2. Comparison of the biochemical and morphological characteristics of CD437-induced mitochondrial permeability transition (MPT) in COLO 16 skin cancer cells. MPT induction was evaluated in COLO 16 cells after a 4-h exposure to the vehicle control (dimethyl sulfoxide, DMSO) (A), or 1 μM CD437 (B). MPT induction was measured by flow cytometry after concurrent staining with $DiOC_6(3)$ and dihydroethidium as described in (38). Electron micrographs of the mitochondria of a COLO 16 cell treated for 4 h with the vehicle control (DMSO) (C), or 1 μM CD437 (D). The mitochondria were examined using a JEOL 1010 transmission electron microscope (JEOL USA, Peabody, MA) and images were acquired using an Advanced Microscopy Techniques imaging system (Advanced Microscopy Techniques, Danvers, MA). The electron microscopy procedures were performed as described in (84). The arrows in C and D indicate mitochondria, and the scale bar = 500 nm.

cific cellular proteins by caspases was implicated in the removal of endogenous apoptosis inhibitors, morphological changes, and DNA fragmentation. For example, cleavage of inhibitory regulators of apoptosis like Bcl-2 or Bcl-X_L not only inactivates their inhibitory function, but also produces a protein fragment that promotes apoptosis (61). Cleavage of nuclear lamins and cytoskeletal proteins such as fodrin and gesolin is required for the acquisition of the apoptotic phenotype (58), and the cleavage of the inhibitor of the caspase-activated DNAse releases this DNAse, leading to DNA fragmentation (62).

Exposure to 4HPR promoted the cleavage of the zymogen of caspase-3 to yield the active form of caspase-3 in head and neck (34), lung (34), and cervical (36) cancer cells. Exposure to 4HPR also promoted the cleavage of a fluorogenic caspase substrate (DEVD-rhodamine) in skin cancer cells, which was presumably mediated *via* the DEVD-dependent activity of caspase-3 (38). The kinetics of this process revealed that caspase-3 activity was not profoundly evident until ~8 h after treatment with 4HPR, which was ~6 h after the dissipation of $\Delta\Psi_m$ had commenced in these cells. Interestingly, the caspase-3–like protease activity declined after a 12-h exposure of skin cancer cells to 4HPR (38). These results suggest that 4HPR-induced caspase activity was contingent on MPT induction.

In lung (63) and prostate (64) carcinoma cells, exposure to CD437 promoted the proteolytic activation of caspase-3. The activation of caspase-3 and apoptosis induction in lung cancer cells exposed to CD437 could be inhibited when these cells were pretreated with the caspase inhibitors Z-VAD-FMK and Z-DEVD-FMK (63). A 4-h exposure to CD437 promoted a marked increase in the fluorescence intensity of a fluorogenic caspase substrate (DEVD-rhodamine) in skin cancer cells, suggesting the activation of caspase-3 (43). Pretreating the skin cancer cells with cyclosporin A blocked caspase-3 activity, and inhibited DNA fragmentation in these cells, which would indicate that MPT was responsible for initiating caspase activation (43).

Exposure to SHetA2 promoted the activation of capases-3 in head and neck cells as illustrated by both a Western blot and a fluorescent substrate assay (13). Caspase-3 processing was increased efficiently in a time-dependent manner after SHetA2 treatment. The level of procaspase-3 decreased after treatment with SHetA2, indicating the activation of caspase-3; indeed, an analysis of caspase-3 activity by spectrofluorimetry showed an increase in the cleavage of a synthetic fluorogenic substrate. The caspase-3 inhibitor, Ac-DEVD-CHO, inhibited SHetA2-induced caspase-3 activation and apoptosis in head and neck cells. The ability of cyclosporin A and trifluoperazine to block SHetA2-induced apoptosis would suggest that caspase activation resulted from the induction of the MPT (13).

Conclusions and Perspectives

Mitochondria constitute 15–50% of the total cytoplasmic volume in most cells, and they participate in more metabolic functions, especially those involved in cellular energy production, than do any other organelles (65). The mitochondria have also emerged as important regulators of apoptosis (27,55). The endosymbiont hypothesis of apoptosis evolution proposes that certain proteins, situated in the primitive endosymbiot/host interface, would have played a strategic role in the establishment of endosymbiosis, giving rise to mitochondria. Eucaryotic cells would evolve to exploit the bioenergetic windfall associated with endosymbiosis and, with the caveat of nuclear control (e.g., p53 and pro-apoptotic Bcl-2 family members), would also be provided with an intrinsic or mitochondrial-mediated mechanism for cell elimination through the release of soluble apoptogenic mitochondrial proteins (66). Bioenergetic processes can produce excessive ROS under certain conditions that are potentially deleterious to mitochondrial and other cellular functions. Thus, it was also suggested that the intrinsic pathway of apoptosis, regulated by MPT induction, serves a teleolonomic function in metazoans by deleting aberrant ROS-producing cells to facilitate the maintenance of tissue homeostasis (30). Given that the process of carcinogenesis is associated with the systematic dysregulation of apoptosis, presumably due to the loss of nuclear control mechanisms (e.g., p53, pro-apoptotic Bcl-2 family members, and caspases) (55), targeting the mitochondria and MPT induction in cancerous or precancerous cells to promote apoptosis would seem to be an effective and direct means of abating or preventing cancer.

A mitochondrial-mediated approach to cancer cell elimination was proposed during the early part of the 20th century by Warburg on the basis of metabolic studies demonstrating abnormally high rates of aerobic glycolysis in tumor cells (67). Warburg believed that mitochondrial bioenergetics was compromised in these cells, and treatments causing nonspecific mitochondrial damage would be more detrimental to cancer cells, compared with their normal counterparts, because of their already diminished respiratory function. Many studies described defects in cancer cell mitochondria, including a reduced capacity to oxidize NADH-linked substrates, reduced expression and/or activity of oxidative phosphorylation enzymes, and mitochondrial DNA mutations. Yet, these alterations do not seem to fit a general pattern (68). Recently, several reports showed that certain carcinoma cell types can accumulate lipophilic cationic agents to a higher degree than normal epithelial cells, apparently due to higher $\Delta\Psi_m$ in the tumor cells. This activity can conceivably be exploited to selectively trigger MPT-induced apoptosis in tumor cells (69–71). To date, it is still unclear what factors contribute to this phenomenon. However, it was suggested that the higher $\Delta\Psi_m$ exhibited by tumor cells may be due to differences in substrate oxidation and phosphorylation of ADP (65), glycolysis (65), and/or mitochondrial mass (70), which may reflect transformation-related characteristics of carcinoma cell mitochondria (69,71,72). The role of the mitochondria in the cancer cell metabolic phenotype is probably highly dependent on cell type as well as the mechanism of transformation.

We observed that the apoptogenic effects of *N*-(4-hydroxyphenyl)retinamide (38), CD437 (43), capsaicin (73), resiniferatoxin (73), and deguelin (74) are conspicuously diminished in ρ^0 skin cancer cells that are functionally deficient in mitochondrial respiration. Furthermore, several studies demonstrated that ρ^0 cells are more resistant than their parental counterparts to the apoptogenic effects of various cancer chemotherapeutic agents (75,76), ceramide (77), and extrinsic mediators of apoptosis-like tumor necrosis factor (78), and TRAIL (79), implying that the disruption of mitochondrial respiration was associated with cell death. The inability of the ρ^0 mitochondria to conduct bioenergetic processes would appear to correspond to the mitochondrial alterations observed in various malignant tumor cells *ex vivo* (67,80–83). This would imply that certain aspects of mitochondrial function [e.g., electron transport, which is inhibited by *N*-(4-hydroxyphenyl)retinamide, CD437, capsaicin, resiniferatoxin, and deguelin] are realistic targets for cancer chemoprevention or therapy. This does not seem to be a broad conceptual leap considering that several novel cancer chemotherapeutic agents are also considered mitochondriotoxic (55). Continued examination of the mechanisms of synthetic retinoid–induced apoptosis should provide further clues to support the notion that the mitochondria are important targets for both cancer chemoprevention and therapy.

Acknowledgments

We thank Karen Ramirez in the Department of Immunology at the University of Texas M. D. Anderson Cancer Center for her assistance in the acquisition of the flow cytometry data presented in Figure 2, and Kenneth Dunner, Jr. in the Department of Cancer Biology at the University of Texas M. D. Anderson Cancer Center for his assistance with the acquisition of the electron micrographs presented in Figure 2.

References

1. McCormick, D.L., Bagg, B.J., and Hultin, T.A. (1987) Comparative Activity of Dietary or Topical Exposure to Three Retinoids in the Promotion of Skin Tumor Induction in Mice, *Cancer Res. 47:* 5989–5993.
2. McCormick, D.L., and Moon, R.C. (1986) Antipromotional Activity of Dietary *N*-(4-Hydroxyphenyl)retinamide in Two-Stage Skin Tumorigenesis in CD-1 and SENCAR Mice, *Cancer Lett. 31:* 133–138.
3. Hong, W.K., and Sporn, M.B. (1997) Recent Advances in Chemoprevention of Cancer, *Science 278:* 1073–1077.
4. Kakizoe, T. (2003) Chemoprevention of Cancer—Focusing on Clinical Trials, *Jpn. J. Clin. Oncol. 33:* 421–442.
5. Sporn, M.B. (1976) Approaches to Prevention of Epithelial Cancer During the Preneoplastic Period, *Cancer Res. 36:* 2699–2702.
6. Lotan, R. (1996) Retinoids in Cancer Prevention, *FASEB J. 10:* 1031–1039.
7. Chambon, P. (1996) A Decade of Molecular Biology of Retinoic Acid Receptors, *FASEB J. 10:* 940–954.
8. Formelli, F., Barua, A.B., and Olson, J.A. (1996) Bioactivities of *N*-(4-Hydroxyphenyl)-retinamide and Retinoyl β-Glucuronide, *FASEB J. 10:* 1014–1024.
9. Fanjul, A.N., Delia, D., Pierotti, M.A., Rideout, D., Qiu, J., and Pfahl, M. (1996) 4-Hydroxyphenyl Retinamide Is a Highly Selective Activator of Retinoid Receptors, *J. Biol. Chem. 271:* 22441–22446.

10. Sun, S.-Y., Yue, P., and Lotan, R. (1999) Induction of Apoptosis by *N*-(4-Hydroxyphenyl)retinamide and Its Association with Reactive Oxygen Species, Nuclear Retinoic Acid Receptors, and Apoptosis Related Genes in Human Prostate Carcinoma Cells, *Mol. Pharmacol. 55:* 403–410.
11. Shao, Z.-M., Dawson, M.I., Li, X.S., Rishi, A.K., Sheikh, M.S., Han, Q.X., Ordonez, J.V., Shroot, B., and Fontana, J.A. (1995) p53 Independent G_0/G_1 Arrest and Apoptosis Induced by a Novel Retinoid in Human Breast Cancer Cells, *Oncogene 11:* 493–504.
12. Bernard, B.A., Bernardon, J.-M., Delescluse, C., Martin, B., Lenoir, M.-C., Mangnan, J., Charpentier, B., Pilgrim, W.R., Reichert, U., and Shroot, B. (1992) Identification of Synthetic Retinoids with Selectivity for Human Nuclear Retinoic Acid Receptor γ, *Biochem. Biophys. Res. Commun. 186:* 977–983.
13. Chun, K.H., Benbrook, D.M., Berlin, K.D., Hong, W.K., and Lotan, R. (2003) The Synthetic Heteroarotinoid SHetA2 Induces Apoptosis in Squamous Carcinoma Cells Through a Receptor-Independent and Mitochondria-Dependent Pathway, *Cancer Res. 63:* 3826–3832.
14. Delia, D., Aiello, A., Lombardi, L., Pelicci, P.G., Grignani, F., Jr., Grignani, F., Formelli, F., and Menard, S. (1993) *N*-(4-Hydroxyphenyl)retinamide Induces Apoptosis of Malignant Hemopoietic Cell Lines Including Those Unresponsive to Retinoic Acid, *Cancer Res. 53:* 6036–6041.
15. Oridate, N., Lotan, D., Xu, X.C., Hong, W.K., and Lotan, R. (1996) Differential Induction of Apoptosis by All-*trans*-Retinoic Acid and *N*-(4-Hydroxyphenyl)retinamide in Human Head and Neck Squamous Cell Carcinoma Cell Lines, *Clin. Cancer Res. 2:* 855–863.
16. Hail, N., Jr., and Lotan, R. (2000) Mitochondrial Permeability Transition Is a Central Coordinating Event in *N*-(4-Hydroxyphenyl)retinamide-Induced Apoptosis, *Cancer Epidemiol. Biomark. Prev. 9:* 1293–1301.
17. Marchetti, P., Zamzami, N., Joseph, B., Schraen-Maschke, S., Méreau-Richard, C., Costantini, P., Métivier, D., Susin, S.A., Kroemer, G., and Fromstecher, P. (1999) The Novel Retinoid 6-[3-(1-Adamantyl)-4-hydroxyphenyl]-2-naphtalene Carboxylic Acid Can Trigger Apoptosis Through a Mitochondrial Pathway Independent of the Nucleus, *Cancer Res. 59:* 6257–6266.
18. Hail, N., Jr., and Lotan, R. (2001) The Synthetic Retinoid CD437 Promotes Rapid Apoptosis in Malignant Human Epidermal Keratinocytes and G_1 Arrest in Their Normal Counterparts, *J. Cell. Physiol. 186:* 24–34.
19. Sun, S.-Y., Yue, P., Shroot, B., Hong, W.K., and Lotan, R. (1997) Induction of Apoptosis in Human Non-Small Cell Lung Carcinoma Cells by the Novel Synthetic Retinoid CD437, *J. Cell. Physiol. 173:* 279–284.
20. Sun, S.-Y., Yue, P., Dawson, M.I., Shroot, B., Michel, S., Lamph, W.W., Heyman, R.A., Teng, M., Chandraratna, R.A.S., Shudo, K., Hong, W.K., and Lotan, R. (1997) Differential Effects of Synthetic Nuclear Retinoid Receptor-Sensitive Retinoids on the Growth of Human Non-Small Cell Lung Carcinoma Cells, *Cancer Res. 57:* 4931–4939.
21. Oridate, N., Higuchi, M., Suzuki, S., Shroot, B., Hong, W.K., and Lotan, R. (1997) Rapid Induction of Apoptosis in Human C33A Cervical Carcinoma Cells by the Synthetic Retinoid 6-[3-(1-Adamantyl)hydroxyphenyl]-2-naphtalene Carboxylic Acid (CD437), *Int. J. Cancer 70:* 484–487.
22. Sun, S.Y., Li, W., and Yue, P. (1999) Mediation of *N*-(4-Hydoxyphenyl)retinamide-Induced Apoptosis in Human Cancer Cells by Different Mechanisms, *Cancer Res. 59:* 2493–2498.

23. Sun, S.Y., Yue, P., Chen, X., Hong, W.K., and Lotan, R. (2002) The Synthetic Retinoid CD437 Selectively Induces Apoptosis in Human Lung Cancer Cells While Sparing Normal Human Lung Epithelial Cells, *Cancer Res. 62:* 2430–2436.
24. Mignotte, B., and Vayssière, J.-L. (1998) Mitochondria and Apoptosis, *Eur. J. Biochem. 252:* 1–15.
25. Kroemer, G., Petit, P., Zamzami, N., and Vayssière, J.-L. (1995) The Biochemistry of Programmed Cell Death, *FASEB J. 9:* 1277–1287.
26. Raff, M. (1998) Cell Suicide for Beginners, *Nature 396:* 119–122.
27. Green, D.R., and Reed, J.C. (1998) Mitochondria and Apoptosis, *Science 281:* 1309–1312.
28. Ashkenazi, A., and Dixit, V.M. (1998) Death Receptors: Signaling and Modulation, *Science 281:* 1305–1308.
29. Fleury, C., Mignotte, B., and Vayssière, J.-L. (2002) Mitochondrial Reactive Species in Cell Death Signaling, *Biochimie 84:* 131–141.
30. Skulachev, V.P. (1999) Mitochondrial Physiology and Pathology: Concepts of Programmed Cell Death of Organelles, Cells, and Organisms, *Mol. Aspects Med. 20:* 139–184.
31. Turrens, J.F. (1997) Superoxide Production by the Mitochondrial Respiratory Chain, *Biosci. Rep. 17:* 3–7.
32. Chance, B., Sies, H., and Boveris, A. (1979) Hydroperoxide Metabolism in Mammalian Organs, *Physiol. Rev. 59:* 527–605.
33. Skulachev, V.P. (1996) Why Are Mitochondria Involved in Apoptosis? *FEBS Lett. 397:* 7–10.
34. Sun, S.Y., Yue, P., Kelloff, G.J., Steele, V.E., Lippman, S.M., Hong, W.K., and Lotan, R. (2001) Identification of Retinamides That Are More Potent than *N*-(4-Hydroxyphenyl)retinamide in Inhibiting Growth and Inducing Apoptosis of Human Head and Neck and Lung Cancer Cells, *Cancer Epidemiol. Biomark. Prev. 10:* 595–601.
35. Oridate, N., Suzuki, S., Higuchi, M., Mitchell, M.F., Hong, W.K., and Lotan, R. (1997) Involvement of Reactive Oxygen Species in *N*-(4-Hydroxyphenyl)retinamide-Induced Apoptosis in Cervical Carcinoma Cells, *J. Natl. Cancer Inst. 89:* 1191–1198.
36. Suzuki, S., Higuchi, M., Proske, R.J., Oridate, N., Hong, W.K., and Lotan, R. (1999) Implication of Mitochondria-Derived Reactive Oxygen Species, Cytochrome C and Caspase-3 in *N*-(4-Hydroxyphenyl)retinamide-Induced Apoptosis in Cervical Carcinoma Cells, *Oncogene 18:* 6380–6387.
37. Waliszewski, P., Skwarek, R., Jeromin, L., and Manikoski, H. (1999) On the Mitochondrial Aspect of Reactive Oxygen Species Action in External Magnetic Fields, *J. Photochem. Photobiol. 52:* 137–140.
38. Hail, N., Jr., and Lotan, R. (2001) Mitochondrial Respiration Is Uniquely Associated with the Prooxidant and Apoptotic Effects of *N*-(4-Hydroxyphenyl)retinamide, *J. Biol. Chem. 276:* 45614–45621.
39. Maurer, B., Metelitsa, L., Seeger, R., Cabot, M., and Reynolds, C. (1999) Increase of Ceramide and Induction of Mixed Apoptosis/Necrosis by *N*-(4-Hydroxyphenyl)retinamide in Neuroblastoma Cell Lines, *J. Natl. Cancer Inst. 91:* 1138–1146.
40. Skulachev, V.P. (1996) Role of Uncoupled and Non-Coupled Oxidations in Maintenance of Safely Low Levels of Oxygen and Its One-Electron Reductants, *Q. Rev. Biophys. 29:* 169–202.
41. Gnaiger, E., Méndez, G., and Hand, S. (2000) High Phosphorylation Efficiency and Depression of Uncoupled Respiration in Mitochondria Under Hypoxia, *Proc. Natl. Acad. Sci. USA 97:* 11080–11085.

42. Chandel, N.S., Budinger, G.R.S., Choe, S.H., and Schumacker, P.T. (1997) Cellular Respiration During Hypoxia, *J. Biol. Chem. 272:* 18808–18816.
43. Hail, N., Jr., Youssef, E.M., and Lotan, R. (2001) Evidence Supporting a Role for Mitochondrial Respiration in Apoptosis Induction by the Synthetic Retinoid CD437, *Cancer Res. 61:* 6698–6702.
44. Zang, Y., Beard, R.L., Chandraratna, R.A., and Kang, J.X. (2001) Evidence of a Lysosomal Pathway for Apoptosis Induced by the Synthetic Retinoid CD437 in Human Leukemia HL-60 Cells, *Cell Death Differ. 8:* 477–485.
45. Liu, X., Kim, C.N., Yang, J., Jemmerson, R., and Wang, X. (1996) Induction of Apoptotic Program in Cell-Free Extracts: Requirement for dATP and Cytochrome *c*, *Cell 86:* 147–157.
46. Li, L.Y., Lou, X., and Wang, X. (2001) Endonuclease G is an Apoptotic DNAse When Released from Mitochondria, *Nature 412:* 95–99.
47. van Loo, G., Schotte, P., van Gurp, M., Demol, H., Hoorelbeke, B., Gevaert, K., Rodriguez, I., Ruiz-Carrillo, A., Vandekerckhove, J., Declercq, W., Beyaert, R., and Vandenabeele, P. (2001) Endonuclease G: A Mitochondrial Protein Released in Apoptosis and Involved in Caspase-Independent DNA Degradation, *Cell Death Differ. 8:* 1136–1142.
48. Chai, J., Du, C., Wu, J.W., Kyin, S., Wang, X., and Shi, Y. (2000) Structural and Biochemical Basis of Apoptotic Activation by Smac/DIABLO, *Nature 406:* 855–862.
49. van Loo, G., van Gurp, M., Depuydt, B., Srinivasula, S.M., Rodriguez, I., Alnemri, E.S., Gevaert, K., Vandekerckhove, J., Declercq, W., and Vandenabeele, P. (2002) The Serine Protease Omi/HtrA2 Is Released from Mitochondria During Apoptosis. Omi/HtrA2 Interacts with Caspase-Inhibitor XIAP and Induces Enhanced Caspase Activity, *Cell Death Differ. 9:* 20–27.
50. Susin, S.A., Lorenzo, H.K., Zamzami, N., Marzo, I., Snow, B.E., Brothers, G.M., Magion, J., Jacotot, E., Costantini, P., Loeffler, M., Larochette, N., Goodlett, D.R., Abersold, R., Siderovski, D.P., Penninger, J.M., and Kroemer, G. (1999) Molecular Characterization of Mitochondrial Apoptosis-Inducing Factor, *Nature 397:* 441–446.
51. Wu, M., Xu, L.-G., Li, X., Zhai, Z., and Shu, H.-B. (2002) AMID, an Apoptosis-Inducing Factor-Homologus Mitochondrion-Associated Protein, Induces Caspase-Independent Apoptosis, *J. Biol. Chem. 277:* 25617–25623.
52. Crompton, M. (1999) The Mitochondrial Permeability Transition Pore and Its Role in Cell Death, *Biochem. J. 341:* 233–249.
53. Green, D., and Kroemer, G. (1998) The Central Executioners of Apoptosis: Caspases or Mitochondria, *Trends Cell Biol. 8:* 267–271.
54. Kroemer, G., and Reed, J.C. (2000) Mitochondrial Control of Cell Death, *Nat. Med. 6:* 513–519.
55. Costantini, P., Jocotot, E., Decaudin, D., and Kroemer, G. (2000) Mitochondrion as a Novel Target of Anticancer Chemotherapy, *J. Natl. Cancer Inst. 92:* 1042–1053.
56. Zoratti, M., and Szabò, I. (1995) The Mitochondrial Permeability Transition, *Biochim. Biophys. Acta 1241:* 139–176.
57. Bernardi, P. (1992) Modulation of the Mitochondrial Cyclosporin A-Sensitive Permeability Transition Pore by the Proton Electrochemical Gradient. Evidence That the Pore Can Be Opened by Membrane Depolarization, *J. Biol. Chem. 267:* 8834–8839.
58. Hengartner, M.O. (2000) The Biochemistry of Apoptosis, *Nature 407:* 770–776.
59. Thornberry, N.A., and Lazebnik, Y. (1998) Caspases: Enemies Within, *Science 281:* 1312–1316.

60. Cohen, G.M. (1997) Caspases: The Executioners of Apoptosis, *Biochem. J. 326:* 1–16.
61. Reed, J.C., Jurgensmeier, J.M., and Matsuyama, S. (1998) Bcl-2 Family Proteins and Mitochondria, *Biochim. Biophys. Acta 1366:* 127–137.
62. Sakahira, H., Enari, M., and Nagata, S. (1998) Cleavage of CAD Inhibitor in CAD Activation and DNA Degradation During Apoptosis, *Nature 391:* 96–99.
63. Sun, S.Y., Yue, P., Shroot, B., Hong, W.K., and Lotan, R. (1999) Implication of c-Myc in Apoptosis Induced by the Retinoid CD437 in Human Lung Carcinoma Cells, *Oncogene 18:* 3894–3901.
64. Sun, S.Y., Yue, P., and Lotan, R. (2000) Implication of Multiple Mechanisms in Apoptosis Induced by the Synthetic Retinoid CD437 in Human Prostate Carcinoma Cells, *Oncogene 19:* 4513–4522.
65. Modica-Napolitano, J.S., Steele, G.D., and Chen, L.B. (1989) Aberrant Mitochondria in Two Human Colon Carcinoma Cell Lines, *Cancer Res. 49:* 3369–3373.
66. Kroemer, G. (1997) Mitochondrial Implication in Apoptosis. Towards an Endosymbiont Hypothesis of Apoptosis Evolution, *Cell Death Differ. 4:* 443–456.
67. Warburg, O. (1956) On the Origin of Cancer Cells, *Science 123:* 309–314.
68. Hockenbery, D.M. (2002) A Mitochondrial Achilles' Heel in Cancer, *Cancer Cell 2:* 1–2.
69. Fantin, V.R., Bernardi, M.J., Scorrano, L., Korsmeyer, S.J., and Leder, P. (2002) A Novel Mitochondriotoxic Small Molecule That Selectively Inhibits Tumor Cell Growth, *Cancer Cell 2:* 29–42.
70. Davis, S., Weiss, M.J., Wong, J.R., Lampidis, T.J., and Chen, L.B. (1985) Mitochondrial and Plasma Membrane Potentials Cause Unusual Accumulation and Retention of Rhodamine 123 by Human Breast Adenocarcinoma-Derived MCF-7 Cells, *J. Biol. Chem. 260:* 13844–13850.
71. Modica-Napolitano, J.S., Koya, K., Weisberg, E., Brunelli, B.T., Li, Y., and Chen, L.B. (1996) Selective Damage to Carcinoma Mitochondria by the Rhodacyanine MKT-077, *Cancer Res. 56:* 544–550.
72. Nadakavukaren, K.K., Nadakavukaren, J.J., and Chen, L.B. (1985) Increased Rhodamine 123 Uptake by Carcinoma Cells, *Cancer Res. 45:* 6093–6099.
73. Hail, N., Jr., and Lotan, R. (2002) Examining the Role of Mitochondrial Respiration in Vanilloid-Induced Apoptosis, *J. Natl. Cancer Inst. 94:* 1281–1292.
74. Hail, N., Jr., and Lotan, R. (2004) Apoptosis Induction by the Natural Product Cancer Chemopreventive Agent Deguelin Is Mediated Through the Inhibition of Mitochondrial Respiration, *Apoptosis 9:* 437–447.
75. Singh, K.K., Russell, J., Sigala, B., Zhang, Y., Williams, J., and Keshav, K.F. (1999) Mitochondrial DNA Determines the Cellular Response to Cancer Therapeutic Agents, *Oncogene 18:* 6641–6646.
76. Joshi, B., Li, L., Taffe, B.G., Zhu, Z., Wahl, S., Tian, H., Ben-Josef, E., Taylor, J.D., Porter, A.T., and Tang, D.G. (1999) Apoptosis Induction by a Novel Anti-Prostate Cancer Compound, BMD188 (a Fatty Acid Containing Hydroxamic Acid), Requires the Mitochondrial Respiratory Chain, *Cancer Res. 59:* 4343–4355.
77. Poppe, M., Reimertz, C., Munstermann, G., Kogel, D., and Prehn, J.H. (2002) Ceramide-Induced Apoptosis of D283 Medulloblastoma Cells Requires Mitochondrial Respiratory Chain Activity but Occurs Independently of Caspases and Is Not Sensitive to Bcl-xL Overexpression, *J. Neurochem. 82:* 482–494.

78. Higuchi, M., Aggarwal, B.B., and Yeh, E.T. (1997) Activation of CPP32-Like Protease in Tumor Necrosis Factor-Induced Apoptosis Is Dependent on Mitochondrial Function, *J. Clin. Investig. 99:* 1751–1758.
79. Kim, J.-Y., Kim, Y.-H., Chang, I., Kim, S., Pak, Y.K., Oh, B.-H., Yagita, H., Jung, Y.K., Oh, Y.J., and Lee, M.-S. (2002) Resistance of Mitochondrial DNA-Deficient Cells to TRAIL: Role of Bax in TRAIL-Induced Apoptosis, *Oncogene 21:* 3139–3148.
80. Dang, C.V., and Semenza, G.L. (1999) Oncogenic Alterations of Metabolism, *Trends Biochem. Sci. 24:* 68–72.
81. Cavalli, L.R., and Liang, B.C. (1998) Mutagenesis, Tumorigenicity, and Apoptosis: Are the Mitochondria Involved? *Mutat. Res. 398:* 19–26.
82. John, A.P. (2001) Dysfunctional Mitochondria, Not Oxygen Insufficiency, Cause Cancer Cells to Produce Inordinate Amounts of Lactic Acid: The Impact of This on the Treatment of Cancer, *Med. Hypotheses 57:* 429–431.
83. Mathupala, S.P., Rempel, A., and Pedersen, P.L. (1995) Glucose Catabolism in Cancer Cells, *J. Biol. Chem. 270:* 16918–16925.
84. Hail, N., Jr., Konopleva, M., Sporn, M., Lotan, R., and Andreeff, M. (2004) Evidence Supporting a Role for Calcium in Apoptosis Induction by the Synthetic Triterpenoid 2-Cyano-3,12-dioxooleana-1,9-dien-28-oic Acid (CDDO), *J. Biol. Chem. 279:* 11179–11187.

Chapter 15

Molecular Analysis of the Vitamin A Biosynthetic Pathway

Johannes von Lintig

Institute of Biology I, Department of Animal Physiology and Neurobiology, University of Freiburg, D-79104 Freiburg, Germany

Introduction

We are all familiar with carotenoids as the yellow-to-red coloring of fruits, flowers, and vegetables. These colored compounds, C_{40} isoprenoids, are synthesized in plants, certain fungi, and bacteria. Their characteristic chemical and physical properties are responsible for their light absorption as well as for the inactivation of free radicals [for a recent review see (1)]. Among the various classes of pigments found in nature, the diverse family of carotenoids is the most widespread, with important functions beyond those in carotenoid-producing organisms. Some animals use dietary carotenoids for coloration; well-known examples include the feathers of flamingos and the red color of salmon. Carotenoids not only color the world around us; they are also being investigated intensively at present for their potential to prevent chronic disease and vitamin A deficiency. Because of their antioxidative properties, beneficial effects were reported for carotenoids in the reduction of the risk of coronary heart diseases, certain kinds of cancer, and age-related macular degeneration (AMD) [reviewed in (2)]. Most importantly, certain carotenoids are the precursors (provitamins) for the formation of vitamin A in animals.

In humans, vitamin A deficiency (VAD) in milder forms leads to night blindness, whereas more severe progression results in corneal malformations, e.g., xerophthalmia. In addition to visual defects, this deficiency affects the immune system, leads to infertility, or causes malformations during embryogenesis. The molecular basis for these diverse effects is found in the dual role exerted by vitamin A derivatives in animal physiology. In the entire animal kingdom, 11-*cis* retinal or closely related compounds such as 11-*cis*-3-hydroxy-retinal serve as the chromophores of the visual pigments (rhodopsin) (3,4). Light activation of these G protein–coupled receptors is the first step in phototransduction, the process by which light is converted into a photoreceptor's electrical response. In addition to being essential for vision, the vitamin A derivative retinoic acid (RA) is a major signal controlling a wide range of biological processes in vertebrates. RA is the ligand of two classes of nuclear receptors, the retinoic acid receptors (RAR) and the retinoid X receptors (RXR) (5–8). The active receptor complex is an RAR/RXR heterodimer that binds DNA regulatory sequences and regulates gene transcription in response to RA binding. RXR is not only the heterodimer partner of the RAR receptor but also an obligate partner for other nuclear receptors [for a recent review, see (9)]. The pleiotropic effects of vitamin A are explained by the discovery that the RA-responsive target genes are involved in a panoply of biological processes as diverse as pattern formation during embryonic development, cell differentiation, and control of certain metabolic activities.

Because animals cannot synthesize vitamin A *de novo*, they rely on a continuous supply of dietary precursors. This demand can be met either by preformed vitamin A or by carotenoids with provitamin A activity. Today, we know that all naturally occurring vitamin A in the food chain derives from provitamin A conversion and that the world's population relies mainly on carotenoids from staple food sources to satisfy their demand for vitamin A (10). Despite the importance of provitamin A metabolism, its molecular details have remained elusive for a long time. This chapter will focus on recent advances in this field of research. The use of genetically well-defined model organisms led to the identification of respective genes, and loss-of-function analyses provided first insights into the basic principles of this metabolism. After this breakthrough, old questions in vitamin A research could finally begin to be addressed definitively on the molecular level, contributing to a better understanding, e.g., of the tissue specificity of provitamin A conversion and the regulation of vitamin A homeostasis, all with substantial effect on animal physiology and human health.

Identification of β,β-Carotene-15,15′-Oxygenase, the Key Enzyme in Vitamin A Formation

In 1930, Moore (11) provided the first evidence that a carotenoid is the precursor of vitamin A by describing β-carotene conversion in the small intestine of mammals. For this reaction, a central cleavage mechanism at the C-15,C-15′ double bond for the conversion of β-carotene to vitamin A was proposed soon thereafter by Karrer (12). Then Goodman and Huang (13) and Olson and Hyaishi (14) characterized the respective enzymatic activity in cell-free homogenates from rat small intestine. The β-carotene-cleavage enzyme depended on molecular oxygen; thus the enzyme was termed β,β-carotene-15,15′-oxygenase (BCO). It was reported to be soluble, to have a slightly alkaline pH-optimum, and to be inhibited by ferrous iron chelators and by sulfhydryl-binding compounds, indicating that it contains a ferrous iron cofactor (15). Subsequently, this enzyme was also characterized in different mammalian species (16), and substrate specificity was determined for different β-carotene stereoisomers (17). Recent investigation of the mode of action of BCO provided strong evidence that oxidative cleavage at the central (15,15′) double bond is catalyzed in a monooxygenase mechanism *via* a transient carotene epoxide (18).

In 2000, two research groups independently succeeded in cloning the key enzyme in vitamin A formation (19–21). The approach by von Lintig and Vogt relied on sequence homology to the plant carotenoid–cleaving enzyme VP14, which catalyzes 9-*cis* epoxycarotenoid cleavage in the biosynthetic pathway of the plant growth factor abscisic acid (22). By employing an expression cloning strategy in an *Escherichia coli* strain genetically engineered to produce all of the enzymes required to synthesize β-carotene *de novo*, they identified a β,β-carotene-15,15′-oxygenase from the fruit fly *Drosophila melanogaster*. The enzymatic properties of the purified recombinant carotene oxygenase revealed that it catalyzed exclusively the centric cleavage of β-carotene (C_{40}) to yield retinal (C_{20}) and that it depended on ferrous iron as cofactor (19). Direct genetic evidence that this enzyme catalyzes the key step in vitamin A formation was provided by mutant analysis. Among the various available *Drosophila* mutants affected in visual performance, the *ninaB* mutant lacks the visual chromophore, when raised on standard media with carotenoids as the sole source for vitamin A formation. The *ninaB* mutation was cytologically mapped in the *Drosophila* genome on chromosome 3 at position 87E-F (23), coinciding with the physical location of the *Drosophila bco* gene. By analyzing the molecular basis of the blindness of *ninaB* mutants, von Lintig and colleagues (24) showed that this phenotype is caused by mutations in the *bco/ninaB* gene, thus unequivocally demonstrating that BCO/NinaB actually catalyzes vitamin A synthesis *in vivo*.

Confirmation that this type of enzyme generally catalyzes the first step in vitamin A metabolism in metazoans was provided by Wyss and colleagues (20,25) by cloning a *bco* from chicken. Their approach relied on partial protein purification and determination of peptide sequences; they then used this information to synthesize corresponding degenerate oligonucleotide primers for polymerase chain reaction to generate a partial cDNA and screen a cDNA library derived from chicken small intestine. Amino acid sequence comparison between the *Drosophila* and chicken BCOs showed an overall similarity with several highly conserved regions and a significant similarity to some domains of the plant carotenoid oxygenase VP14 (21).

By sequence similarity to the thus far identified genes from *Drosophila* and chicken, their counterparts from mouse and human were then identified and functionally characterized in several laboratories (26–30). It was shown that these mammalian homologues catalyze exclusively the centric oxidative cleavage of β-carotene to yield retinal. Expression of the murine BCO in various carotenoid-accumulating *E. coli* strains revealed the cleavage of carotenoid substrates such as β- and α-carotene, but also lycopene, resulting in the last case in the formation of acyclic retinoids (26). The purified recombinant BCO, however, catalyzed only the cleavage of carotenoid substrates with at least one unsubstituted β-ionone ring, such as β-carotene and β-cryptoxanthin, and there was no significant cleavage of lycopene or zeaxanthin (29). The K_m values for β-carotene were estimated to be in the range of 1–10 μM for BCOs from the different animal species (19,26,27,29). BCO exhibits a slightly alkaline pH-optimum, and enzymatic activity is sensitive to chelating agents such as *o*-phenanthroline and α,α′-bipyridyl, indicating that it depends on ferrous iron (19,29). Thus, the purified recombinant BCOs share biochemical properties that were already described for the native BCOs.

Purification of the recombinant BCOs as fusion proteins by affinity chromatography was achieved without the addition of detergents. This characteristic and the predicted amino acid sequences of the various BCOs indicate that we are dealing with hydrophilic, non-membrane-bound proteins. Indeed, a cytosolic localization of the native BCO was recently demonstrated for its human representative (29). Therefore, *in vitro* tests for enzymatic activity must be conducted in the presence of detergents to mimic the interaction between the enzyme and its insoluble substrate. *In vivo*, however, the cytosolic localization of BCO may require specific binding proteins to deliver the carotenoid substrate as well as to pick up the retinoid product because both are highly lipophilic compounds. On the product side, three different types of cellular retinoid-binding proteins (CRBP 1–3) were characterized in mice [(31) and references therein]. However, no direct protein-protein interaction between a recombinant murine BCO-glutathione-*S*-transferase fusion protein and CRBP could be detected in pull-down experiments (27). Even though these results argue against a tight protein-protein interaction of CRBP with BCO, it seems likely that CRBP may facilitate β-carotene cleavage by binding retinal. In mouse testis homogenates, an L-lactate dehydrogenase C was identified and shown to interact specifically with BCO. At present, the exact physiologic role of this type of alcohol dehydrogenase is not known, and there is no experimental evidence that this enzyme catalyzes either the oxidation or reduction of aldehydes such as retinal. Hence, it remains to be elucidated whether BCO may interact in a tissue-specific manner with a certain subset of proteins involved in retinoid metabolism. Such proteins might control the metabolic flow of the primary cleavage product retinal, either to retinol formation for vitamin A transport and storage, or in the direction of RA formation for retinoid-signaling.

In summary, this recent research led to the molecular identification of β,β-carotene-15,15′-oxygenases from various metazoan species. The recombinant enzymes share common biochemical properties with the native BCO from tissue homogenates. On the basis of its structural and biochemical properties, BCO from animals belongs to an ancient family of carotenoid-modifying enzymes first described in plants [Fig. 15.1; for a review, (32)].

Two Additional bco*-Homologous Genes,* rpe65 *and* bco2*, in Vertebrates*

In the entire *Drosophila* genome only one *bco* gene was found; it is encoded by the *ninaB* gene (24). In vertebrates, however, in addition to *bco,* two additional genes, *rpe65* and *bco2*, encode putative *bco/ninaB* homologues (19,30,33). On the level of the deduced amino acid sequence RPE65, BCO2 and BCO share ~40% overall sequence identity (30). Additionally, six histidine residues within the amino acid sequence are conserved in their positions as well as several highly conserved regions that can be considered as protein family motifs (26,30). In humans, *bco*, *rpe65*, and *bco2* map genomically to the chromosomal positions 16q21, 1q31, and 11q23, respectively.

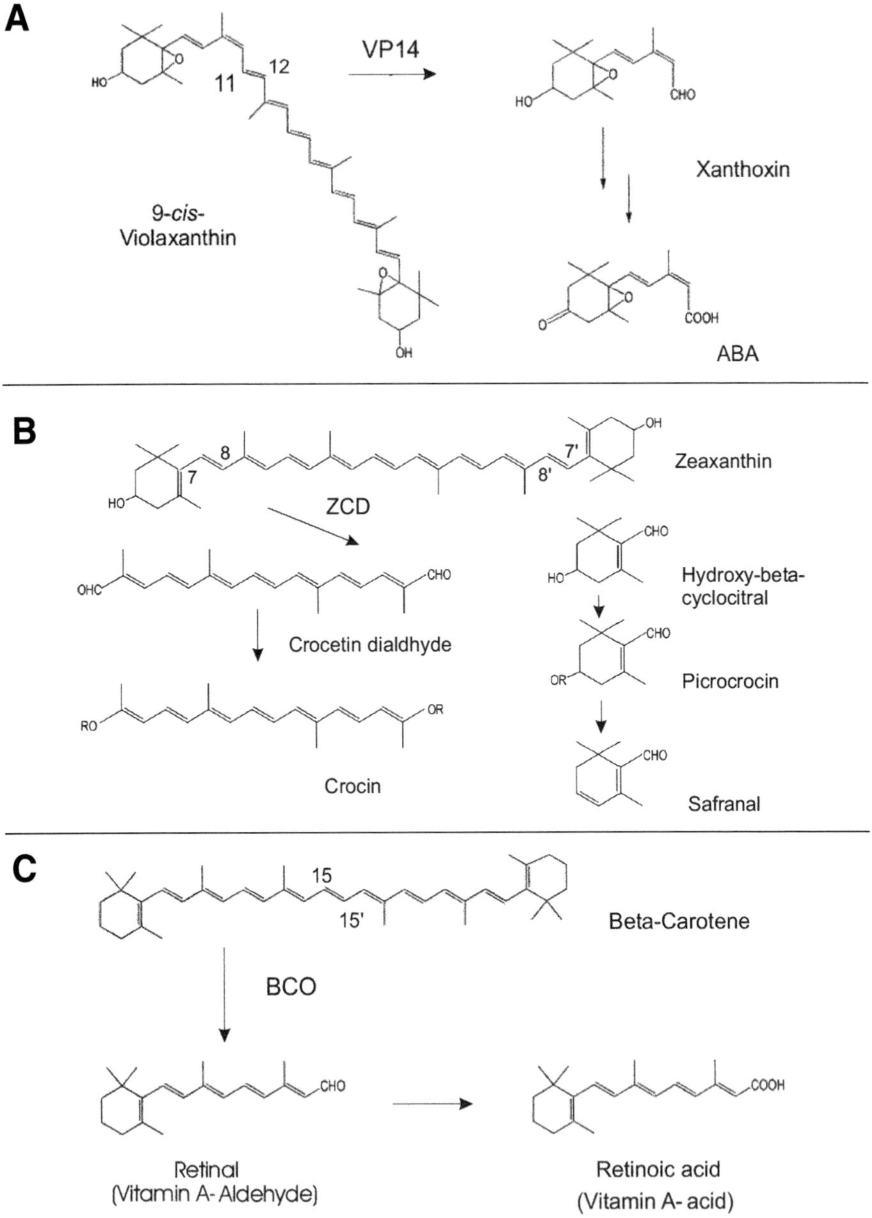

Fig. 15.1. Proposed pathways for apo-carotenoid biosynthesis in plants and animals [for recent review, see (32) and references cited therein]. (A) Abscisic acid biosynthesis. (B) Zeaxanthin oxidation in saffron styles. (C) Vitamin A biosynthesis. Abbreviations: VP14, 9-*cis* violaxanthin-oxygenase; BCO, β,β-carotene-15,15-oxygenase; ZCD, zeaxanthin-cleavage oxygenase.

RPE65 was first described as an abundant protein of the retinal pigment epithelium (RPE) with a molecular mass of 65 kDa (34), but a relation to carotenoid/retinoid metabolism was not established initially. After the cloning and sequencing of *bco/ninaB*, it became evident that RPE65 belongs to the gene family of putative carotenoid-modifying enzymes. However, expression of the bovine or murine RPE65 in a β-carotene–accumulating *E. coli* strain did not result in the formation of apocarotenoid cleavage products, indicating that the RPE65 function does not include β-carotene cleaving activity (19). Most interestingly, mutations in *rpe65* are associated with specific forms of blindness in humans, such as Leber's congenital disease, autosomal recessive retinitis pigmentosa, and rod-cone dystrophy (35,36). A direct involvement of RPE65 in the retinoid (visual) cycle of the eyes was demonstrated by the analysis of *rpe65*-deficient mice. These mice lack rhodopsin, despite the presence of the opsin apoprotein in the rod outer segments. In their eyes, 11-*cis* retinal is not detectable; instead, there is an accumulation of all-*trans*-retinyl esters (36), intermediates of the retinoid cycle in the RPE. All-*trans*-retinyl esters are the substrate for the proposed isomerohydrolase, which may catalyze the key step in the regeneration of the visual pigments in a combined isomerization and ester-hydrolase reaction (37). Even though RPE65 seems to be essential for this reaction, the recombinant RPE65 lacks isomerohydrolase activity *in vitro*. Recent biochemical studies suggest that RPE65 binds stereospecifically all-*trans*-retinyl ester and stimulates the intrinsic isomerohydrolase activity of RPE membranes (38,39). Thus, it was proposed that RPE65 is an all-*trans*-retinyl ester-binding protein. Because RPE65 shares overall sequence similarities to BCO and BCO2, which are both enzymatically active proteins, it remains to be further elucidated whether this representative of the *bco*-gene family is simply a retinoid-binding protein or an enzymatically active component of the carotenoid/retinoid metabolism in the eyes.

The murine *bco2* gene encodes a protein of 532 amino acid residues that shares ~40% overall sequence identity with both the murine BCO and RPE65. After cloning its full-length cDNA, expression of BCO2 in a β-carotene–accumulating *E. coli* strain revealed that it catalyzes the formation of β-10′-carotenal and β-ionone (30). The resulting apocarotenoids were identified from their chemical properties, spectral characteristics, and mass spectroscopy. Thus, BCO2 is a β,β-carotene-9′,10′-oxygenase specifically catalyzing the oxidative cleavage at one site of its symmetric substrate β-carotene at the C-9′,C-10′ double bond. The existence of this type of carotene oxygenase in other vertebrates was established by cloning this gene from humans and a lower vertebrate, the zebrafish *Danio rerio*. The molecular identification and functional characterization of BCO2 in several vertebrate species provides strong evidence that in addition to centric (C-15,C-15′), an additional eccentric (C-9′,C-10′) cleavage pathway for β-carotene exists in vertebrates. The existence of such a cleavage pathway for carotenoids was proposed earlier in several studies. Glover (40) provided evidence that an eccentric cleavage of β-carotene results in the formation of retinoids in mammals. In addition, the existence of such a cleavage reaction was demonstrated in cell-free homogenates (41). Furthermore, it was shown that long-chain β-apocarotenoids (>C20) as primary cleavage products from this reaction can be shortened to retinoic acid (C20). For this process a stepwise shortening reaction was proposed comparable to the β-oxidation of fatty acids (42,43) Thus, the molecular cloning of this second, eccentrically cleaving carotene oxygenase, BCO2, indicates that indeed two different pathways for the formation of biologically active retinoids exist in vertebrates.

In summary, in vertebrates, a small gene family of *bco*-homologues exists. On the level of the deduced amino acid sequences, all three representatives in a given vertebrate genome possess several common structural features. Furthermore, sequence comparison revealed that the three different representatives from vertebrates all have a higher degree of similarity to each other than to the protein encoded by *Drosophila bco/ninaB*. With the emerging number of sequences for carotenoid-cleaving enzymes from animals, and also from plants, now becoming available in the public database, this information can be used to define their catalytic domains and identify their active sites.

Molecular Analyses of the Vitamin A Biosynthetic Pathway

In the past few years, a large number of different components of vitamin A metabolism were identified molecularly [for a review, see (44)]. By reverse genetics, animal models with mutations in these genes were established. This strategy proved to be extremely powerful to learn more about individual aspects of the complex effects of this vitamin, e.g., it could be demonstrated that a coordinated expression of RA synthesizing and catabolizing enzymes is crucial to fine-tune RA-signaling in the embryo. Furthermore, natural mutations in the genes necessary for the metabolism of vitamin A (visual cycle) in the eyes recently emerged as an important class of genetic defects responsible for a wide range of retinal dystrophies and dysfunctions in humans [for a recent review, (45)], e.g., as already discussed above, mutations in the BCO-homologous gene RPE65 are responsible for inherited blinding diseases.

The recent molecular cloning of BCO and BCO2 provided molecular tools to analyze their functional role in animal retinoid metabolism in more functional detail. The following highlights recent research dealing with different aspects of provitamin A metabolism such as the effect on the formation of biologically active retinoids, provitamin A transport, and body distribution as well as its tissue specificity and its regulation.

Tissue-Specific Expression of BCO

In vertebrates, most of the vitamin A is already synthesized in epithelial cells of the intestinal mucosa by the conversion of provitamin A carotenoids and then transported to the liver for storage. Upon cloning BCO, its tissue-specific expression patterns were analyzed in several vertebrate species (25–30). In chickens, the tissue-specific expression patterns of *bco* were analyzed by a combination of Northern blot and *in situ* hybridization experiments. Its mRNA was localized mainly in liver, in duodenal microvilli, as well as in tubular structures of the lung and the kidney (25). In mice, *bco* mRNA was detectable in small intestine and liver but also in kidney, testes, uterine tissues, skin, and skeletal muscle (26,27,30). Analyses of *bco* mRNA expression in humans revealed a comparable picture (29). Although Yan and colleagues (28) reported that *bco* is preferentially expressed in the retinal pigment epithelium (RPE) of the human eye and only at much lower levels in other tissues, more recent results dealing with *bco* expression in the eye showed only low mRNA levels in the RPE of humans and monkeys (46). In sum, the surprising result of all these current investigations dealing with the tissue specificity of *bco* expression is that its steady-state mRNA levels are quite high in peripheral nondigestive tissues. Testes, for example, require retinoids for spermatogenesis, and vitamin A is needed for retinoid signaling in almost all tissues. Thus, *bco* expression in peripheral tissues indicates that in addition to an external vitamin A supply *via* the circulation, provitamin A may affect retinoid metabolism in a tissue-specific manner in various cell types and tissues.

Provitamin A as an Essential Precursor for the Retinoic Acid Signaling Pathway in Zebrafish Embryos

It is generally assumed that retinoid-dependent physiologic processes in vertebrates can be maintained by a dietary supply of preformed vitamin A. The finding that *bco* is expressed in a variety of different tissues raises the question of whether provitamin A is an essential precursor for the formation of biologically active retinoids in vertebrate physiology. Direct evidence for such a role comes from analysis of BCO function in embryos of the zebrafish (*Danio rerio*) (47). In a multitude of recent studies, this fish has proven to be a valuable model organism for the analysis of complex molecular processes in vertebrate biology. As in all vertebrate embryos, impairments in retinoid metabolism result in severe embryonic malformations due to an interference with retinoid-signaling events during development (48,49). Lampert and colleagues (47) addressed the question of whether BCO is required for embryonic development in the fish. First, they demonstrated that in the zebrafish embryo, *bco* is expressed in clearly defined spatial compartments and translated into protein. HPLC analyses for lipophilic compounds revealed the existence of both provitamin A (β-carotene) and vitamin A (all-*trans*-retinal) in the egg yolk. Thus, retinoids as well as β-carotene exist as precursors for the synthesis of biologically active retinoids such as RA and 11-*cis* retinal for use in zebrafish development. To test whether there is an actual requirement for the BCO-mediated β-carotene conversion to retinal, they performed loss-of-function studies. A targeted gene knock-down of BCO led to abnormalities of the craniofacial skeleton, hindbrain, and eyes, which are all impairments well known from VAD vertebrate embryos (47). Indeed, analyses of changes in the expression of marker genes revealed that several RA-dependent processes were severely impaired. The defects could be rescued by supplementation of RA or could also be elicited in wild-type embryos treated with Citral, an inhibitor of RA generation. These results suggest that provitamin A conversion is an essential upstream step in the pathway for retinoic acid synthesis in several specific developmental processes in the fish (Fig. 15.2).

The use of the nontoxic provitamin for RA signaling may provide an additional control mechanism to finely balance retinoid levels at the cellular level in local tissue environments. Interestingly, embryonic expression of BCO was also reported in mice, suggesting that this developmental role of provitamin A–dependent retinoic acid signaling might also be evolutionarily conserved in higher vertebrates (36). But it remains to be elucidated whether there is an essential role of BCO in mammalian embryos with their continuous maternal vitamin A supply *via* the placenta or whether this is a special characteristic of egg-laying lower vertebrates. Even though this question has not yet been answered, the demonstrated role of BCO in zebrafish as well as finding *bco* expression in several distinct cell types in embryonic and adult vertebrates provide strong evidence that a local, tissue-specific provitamin A conversion takes place, which may directly influence retinoid-dependent physiologic processes.

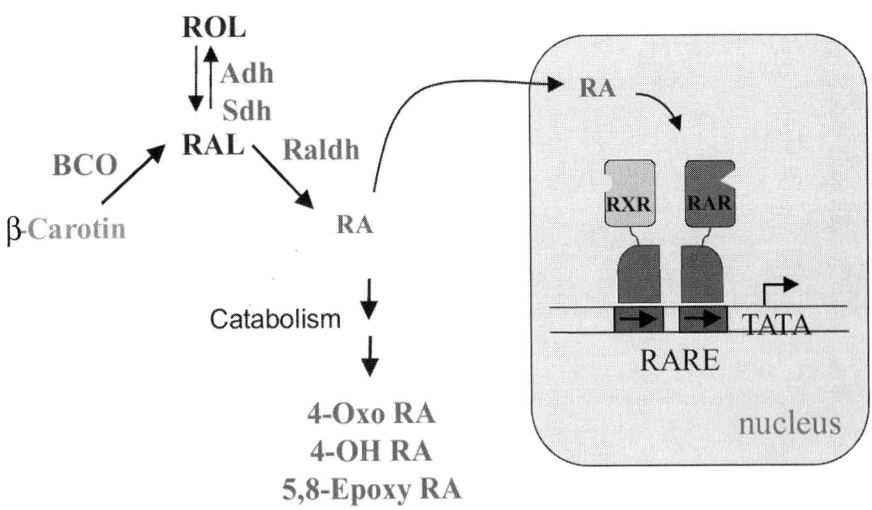

Fig. 15.2. Proposed pathway for provitamin A–dependent retinoic acid formation in the zebrafish embryo. Abbreviations: ROL, retinol; RA, retinoic acid; Adh, alcohol dehydrogenase; Sdh, short chain reductase; Raldh, retinal dehydrogenase; BCO, β,β-carotene-15,15′-oxygenase; BCO2, β,β-carotene-9′,10′-oxygenase; RARE, retinoid receptor responsive element; RAR, retinoic acid receptor; RXR, retinoid X receptor; 4-OXO RA, 4-oxo-retinoic acid; 4-OH RA, 4-hydroxy retinoic acid; 5,8-Epoxy RA, 5,8-epoxy-retinoic acid.

A Putative Role of Class B Scavenger Receptors in the Cellular Uptake of Provitamin A

Recent analyses of *bco* function questioned how the provitamin A is transported within the body and delivered to different target tissues. Carotenoids are highly lipophilic molecules, suggesting that as for other lipids, specific binding/transport proteins may exist. First insights into carotenoid transport were again provided by the analysis of a *Drosophila* mutant. In the blind mutant, *ninaD*, the carotenoid content was significantly altered compared with wild-type flies and was ineffective in mediating visual pigment synthesis. Molecular analyses revealed that this phenotype is caused by a defect in the uptake and body distribution of dietary carotenoids (50). The *ninaD* gene encodes a cell-surface receptor with significant sequence similarity to the mammalian class B scavenger receptor, SR-BI. SR-BI plays a key role in HDL-metabolism in mammals (51) by mediating the bidirectional flux of cholesterol between lipoproteins and target cells. In *ninaD* flies, a nonsense mutation is found in the gene encoding this receptor, which abolished its function. Direct functional evidence for a role of the *ninaD* receptor in cellular uptake of carotenoids was provided by gene rescue, using P-element–mediated transformation of flies with a wild-type *ninaD* allele. Heat shock–induced expression of the wild-type allele in the genetic background of *ninaD* flies restored carotenoid uptake and visual pigment synthesis (50). This provided genetic and functional evidence that lipoprotein-bound carotenoids are distributed to target tissues within the body by a protein-mediated transport process involving this type of scavenger receptor (Fig. 15.3). The existence of homologous receptors in vertebrates indicates that class B scavenger receptors may play a more general role in the cellular uptake of carotenoids from the circulating lipoprotein classes. A better understanding of the molecular mechanisms contributing to the body distribution of carotenoids is of interest not only for retinoid metabolism but also because carotenoids are important in a variety of physiologic processes. For example, lutein and zeaxanthin accumulate as macular pigments in human eyes. The beneficial effects of carotenoids, due mainly to their antioxidative properties, were discussed in the context of several diseases. Thus, the results coming from the analyses of the *ninaD* mutant promise to elucidate new aspects of class B scavenger receptor functions in further research.

Regulation of the Vitamin A Biosynthetic Pathway

Unlike vitamin A itself, high-dose supplementation of β-carotene in humans causes no hypervitaminosis, indicating that β-carotene cleavage to vitamin A is tightly regulated. Several investigations with animal models showed that the vitamin A status of the individual affects BCO enzymatic activity (52,53). Recent analyses provided evidence that BCO regulation in the small intestine is mediated on the transcriptional level, possibly *via* a negative feedback regulation mechanism involving RA and its nuclear receptors (54). More detailed analyses of the regula-

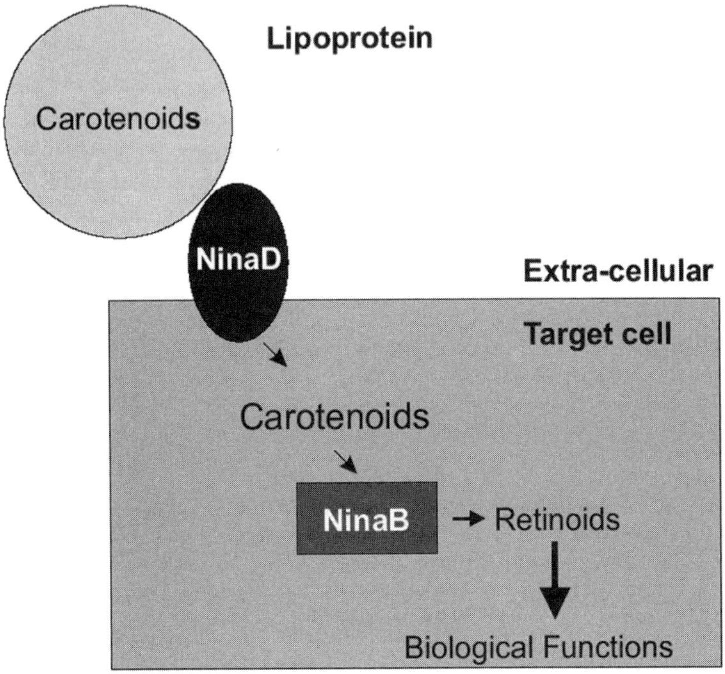

Fig. 15.3. Schematic overview of the vitamin A biosynthetic pathway in *Drosophila*. The cellular uptake of carotenoids is mediated by the class II scavenger receptor *ninaD* from circulating lipophorins of the hemolymph. For retinoid synthesis in *Drosophila*, carotenoids are converted by the BCO enzyme encoded by the *ninaB* gene.

tion of the murine *bco* gene by Boulanger and colleagues (55) provided strong evidence that the murine *bco* promoter contains a peroxisome proliferator response element (PPRE) as a key regulatory switch and is regulated by peroxisome proliferator activated receptor γ (PPARγ). PPAR constitute a subfamily of the steroid hormone superfamily [for a recent review, see (56)]. Most of the known naturally occurring ligands of PPAR are diet-derived fatty acids and their metabolites (57). PPARγ activates genes involved in anabolic pathways, particularly in adipose tissues, and is required for placental, cardiac, and adipose tissue development (58). RXR is the obligate heterodimeric partner of the PPAR transcription factors. The promoter analysis showed that *Bco* expression is positively regulated by both the PPAR/RXR heterodimer and the RXR/RXR homodimer, implying that the expression of the key enzyme for vitamin A synthesis can be upregulated by 9-*cis* RA (55).

A role of retinoid signaling in the positive regulation of retinoid-metabolizing enzymes on the transcriptional level was also demonstrated for the lecithin:retinol acyltransferase, LRAT (59). Interestingly, the cellular retinol-binding protein II (*crbp2*) gene is the only other gene in carotenoid and retinoid metabolism known at present to contain a PPRE. As reported for *Bco*, this gene is also upregulated by ligands of PPAR and by 9-*cis* RA (60). *crbp2* is expressed in large amounts in the small intestine of adult mammals, the major site of vitamin A synthesis. As described above, *crbp2* may act downstream of BCO in binding retinal, the primary cleavage product of provitamin A conversion. LRAT catalyzes the synthesis of retinyl esters, the storage form for vitamin A in the liver. The expression of LRAT is induced by feeding RA, but the details concerning its gene promoter responsive elements are missing at present.

Common mechanisms in the regulation of the genes involved in the vitamin A biosynthetic pathway may contribute to vitamin A homeostasis, and the involvement of PPAR may interlink vitamin A formation to the regulation of overall lipid metabolism. However, some recent controversial results from studies on the *Bco* promoter (positive regulation by 9-*cis* RA) and the analysis of BCO enzymatic activity in the gut (negative regulation by RA) warrant further clarification. It may turn out that the latter is the result of indirect effects due to the influence of RA on additional genes involved in this process.

In summary, a better understanding of the regulation of factors dealing with the bioavailability of carotenoids, their subsequent conversion to vitamin A, and the regulatory factors interlinking this process to lipid metabolism as a whole is certainly of future importance. This interest inherently arises from the fact that carotenoids in staple foods are the major source of satisfaction for the world population's vitamin A demand.

Centric vs. Eccentric Cleavage: The Role of BCO2 in Retinoid Metabolism

Although the role of BCO in the vitamin A biosynthetic pathway has been well established, the role of the second putative carotene oxygenase, BCO2, remains somewhat elusive. There has long been a controversy over centric vs. eccentric cleavage of β-carotene in the synthesis of vitamin A. Evidence that eccentric cleavage of carotenoids also occurs was provided by several investigations. Napoli and Race (61), for example, showed that, in addition to the formation of RA from retinal as the initial product of symmetric β-carotene cleavage, RA is formed directly from β-carotene in cell-free homogenates. As outlined above, it was shown that long-chain apocarotenoids (>C20) are shortened to RA in a stepwise process that is most probably mechanistically related to the β-oxidation of fatty acids (42,43). Thus, BCO2 may catalyze the first step in an alternative pathway for RA formation. Further evidence for the existence of additional yet uncharacterized pathways for RA generation was provided by analyses of transgenic mouse embryos. By directly measuring RA generation *via* a retinoic acid responsive reporter transgene, it could be shown that in addition to the three known retinal dehydrogenases (Raldh1, 2, and 3), which catalyze the final step in RA-generation, additional RA-generating systems exist in the heart and in the spinal cord (61,62).

Kiefer and colleagues investigated the expression pattern of the eccentric carotenoid-cleaving enzyme in mice (30). Here, the *bco2* gene had an expression pattern comparable to that of *bco*. The mRNA expression of both types of carotene-oxygenases in the same tissues, e.g., small intestine and liver, confirms biochemical investigations and explains the observation in biochemical experiments that both centric and eccentric cleavage activities are present in cell-free homogenates of the same tissue. Additionally, low-abundance steady-state mRNA levels of *bco2* were present in spleen, brain, lung, and heart in mice (30). Specific expression of *bco2* was also found in the developing heart of zebrafish embryos and in the fetal heart of humans upon analyzing a commercial multitissue RNA panel (A. Isken and J. von Lintig, unpublished results). These results, and the existence of an alternative pathway for RA generation in the heart, may indicate that BCO2 plays a particular role for the development of the cardiovascular system.

Biological activities of β-apocarotenoids different from retinoids were also reported by various studies in animals (e.g., 63). *In vitro*, in addition to β-carotene cleavage, BCO2 catalyzes the oxidative cleavage of lycopene (30). Favorable effects of lycopene, e.g., on certain kinds of cancers, have been repeatedly reported (64,65). Thus, in addition to being a putative precursor for RA formation, in the case of β-carotene cleavage, it may be speculated that apocarotenoids different from retinoids may represent biologically active substances.

To summarize, much further work is required for a full understanding of the precise physiologic functions of BCO2. These analyses must include a detailed biochemical investigation of its enzymatic properties, substrate specificity, and subcellular localization and of the fate of the primary cleavage products of the reaction. Furthermore, animal models with mutations in this gene must be established to unequivocally elucidate the physiologic role of this enzyme.

Conclusions

The molecular identification of the different metazoan carotene oxygenases has established the existence of an ancient family of nonheme iron oxygenases in animals. With these enzymes, animals have access to and can modulate their retinoids as required for biological processes as diverse as vision, cell differentiation, and development. With the increasing number of carotene oxygenases in the database, sequence comparison can be used to predict common structural features and to identify functional domains and active site residues. The identification of proteins involved in the transport of carotenoids in insects demonstrated that this process is protein mediated, as described earlier for other lipids. The identification of these genes provides a starting point with which to characterize analogous genes in mammals. The advanced state of knowledge about the molecular components of the vitamin A biosynthetic pathway gained in the past few years will surely help in the worldwide public fight against VAD and will open new avenues of research, dealing with biochemical, physiologic, developmental, and medical aspects of carotenoids and their numerous derivatives.

References

1. Demmig-Adams, B., and Adams, W.W., 3rd (2002) Antioxidants in Photosynthesis and Human Nutrition, *Science 298:* 2149–2153.
2. Mayne, S.T. (2003) Antioxidant Nutrients and Chronic Disease: Use of Biomarkers of Exposure and Oxidative Stress Status in Epidemiologic Research, *J. Nutr. 133:* 933–940.
3. Wald, G. (1960) The Molecular Basis of Visual Excitation, *Nature 219:* 800–807.
4. Vogt, K. (1984) Is the Fly Visual Pigment a Rhodopsin? *Z. Naturforsch. 39c:* 196–197.
5. Giguere, V., Ong, E.S., Segui, P., and Evans, R.M. (1987) Identification of a Receptor for the Morphogen Retinoic Acid, *Nature 330:* 624–629.
6. Petkovich, M., Brand, N.J., Krust, A., and Chambon, P. (1987) A Human Retinoic Acid Receptor Which Belongs to the Family of Nuclear Receptors, *Nature 330:* 444–450.
7. Mangelsdorf, D.J., and Evans, R.M. (1995) The RXR Heterodimers and Orphan Receptors, *Cell 83:* 841–850.
8. Chambon, P. (1996) A Decade of Molecular Biology of Retinoic Acid Receptors, *FASEB J. 9:* 940–954.
9. Chawla, A., Repa, J.J., Evans, R.M., and Mangelsdorf, D.J. (2001) Nuclear Receptors and Lipid Physiology: Opening the X-Files, *Science 294:* 1866–1870.
10. Underwood, B.A. (2004) Vitamin A Deficiency Disorders: International Efforts to Control a Preventable "Pox," *J. Nutr. 134:* 231–236.
11. Moore, T. (1930) Vitamin A and Carotene. VI. The Conversion of Carotene to Vitamin A *In Vivo*, *Biochem. J. 24:* 692–702.
12. Karrer, P., Helfenstein, A., Wehrli, H., and Wettstein, A. (1930) Über die Konstitution des Lycopins und Carotins, *Helv. Chim. Acta 13:* 1084.
13. Goodman, D.S., and Huang, H.S. (1965) Biosynthesis of Vitamin A with Rat Intestinal Enzyme, *Science 149:* 879–880.
14. Olson, J.A., and Hayaishi, O. (1965) The Enzymatic Cleavage of Beta-Carotene into Vitamin A by Soluble Enzymes of Rat Liver and Intestine, *Proc. Natl. Acad. Sci. USA 54:* 1364–1370.
15. Goodman, D.S., Huang, H.S., and Shiratori, T. (1966) Mechanism of the Biosynthesis of Vitamin A from Beta-Carotene, *J. Biol. Chem. 241:* 1929–1932.
16. Fidge, N.H., Smith, F.R., and Goodman, D.S. (1969) Vitamin A and Carotenoids. The Enzymic Conversion of Beta-Carotene into Retinal in Hog Intestinal Mucosa, *Biochem. J. 114:* 689–694.
17. Nagao, A., and Olson, J.A. (1994) Enzymatic Formation of 9-*cis*, 13-*cis*, and All-*trans*-Retinals from Isomers of Beta-Carotene, *FASEB J. 12:* 968–973.
18. Leuenberger, M.G., Engeloch-Jarret, C., and Woggon, W.D. (2001) The Reaction Mechanism of the Enzyme-Catalyzed Central Cleavage of β-Carotene to Retinal, *Angew Chem. Int. Ed. Engl. 40:* 2613–2617.
19. von Lintig, J., and Vogt, K. (2000) Filling the Gap in Vitamin A Research. Molecular Identification of an Enzyme Cleaving Beta-Carotene to Retinal, *J. Biol. Chem. 275:* 11915–11920.
20. Wyss, A., Wirtz, G., Woggon, W.D., Brugger, R., Wyss, M., Friedlein, A., Bachmann, H., and Hunziker, W. (2000) Cloning and Expression of β,β-carotene 15,15′-dioxygenase, *Biochem. Biophys. Res. Commun. 271:* 334–336.
21. von Lintig, J., and Wyss, A. (2001) Molecular Analysis of Vitamin A Formation: Cloning and Characterization of β-15,15′-Dioxygenases, *Arch. Biochem. Biophys. 385:* 47–52.

22. Schwartz, S.H., Tan, B.C., Gage, D.A., Zeevaart, J.A., and McCarty, D.R. (1997) Specific Oxidative Cleavage of Carotenoids by VP14 of Maize, *Science 276:* 1872–1874.
23. Stephenson, R.S., O'Tousa, J., Scavarda, N.J., Randall, L.L., and Pak, W.L. (1983) in *The Biology of Photoreception,* Cosens, D.J., and Vince-Price, D., eds., Cambridge University Press, Cambridge, pp. 477–501.
24. von Lintig, J., Dreher, A., Kiefer, C., Wernet, M.F., and Vogt, K. (2001) Analysis of the Blind Drosophila Mutant ninaB Identifies the Gene Encoding the Key Enzyme for Vitamin A Formation *In Vivo*, *Proc. Natl. Acad. Sci. USA 98:* 1130–1135.
25. Wyss, A., Wirtz, G.M., Woggon, W.D., Brugger, R., Wyss, M., Friedlein, A., Riss, G., Bachmann, H., and Hunziker, W. (2001) Expression Pattern and Localization of Beta,Beta-Carotene 15,15′-Dioxygenase in Different Tissues, *Biochem. J. 354:* 521–529.
26. Redmond, T.M., Gentleman, S., Duncan, T., Yu, S., Wiggert, B., Gantt, E., and Cunningham, F.X., Jr. (2001) Identification, Expression, and Substrate Specificity of a Mammalian Beta-Carotene 15,15′-Dioxygenase, *J. Biol. Chem. 276:* 6560–6565.
27. Paik, J., During, A., Harrison, E.H., Mendelsohn, C.L., Lai, K., and Blaner, W.S. (2001) Expression and Characterization of a Murine Enzyme Able to Cleave Beta-Carotene: The Formation of Retinoids, *J. Biol. Chem. 276:* 32160–32168.
28. Yan, W., Jang, G.F., Haeseleer, F., Esumi, N., Chang, J., Kerrigan, M., Campochiaro, M., Campochiaro, P., Palczewski, K., and Zack, D.J. (2001) Cloning and Characterization of a Human Beta,Beta-Carotene-15,15′-Dioxygenase That Is Highly Expressed in the Retinal Pigment Epithelium, *Genomics 72:* 193–202.
29. Lindqvist, A., and Andersson, S. (2002) Biochemical Properties of Purified Recombinant Human Beta-Carotene 15,15′-Monooxygenase, *J. Biol. Chem. 277:* 23942–23948.
30. Kiefer, C., Hessel, S., Lampert, J.M., Vogt, K., Lederer, M.O., Breithaupt, D.E., and von Lintig, J. (2001) Identification and Characterization of a Mammalian Enzyme Catalyzing the Asymmetric Oxidative Cleavage of Provitamin A, *J. Biol. Chem. 276:* 14110–14116.
31. Vogel, S., Mendelsohn, C.L., Mertz, J.R., Piantedosi, R., Waldburger, C., Gottesman, M.E., and Blaner, W.S. (2001) Characterization of a New Member of the Fatty Acid-Binding Protein Family That Binds All-*trans*-Retinol, *J. Biol. Chem. 276:* 1353–1360.
32. Giuliano, G., Al-Babili, S., and von Lintig, J. (2003) Carotenoid Oxygenases: Cleave It or Leave It, *Trends Plant Sci. 8:* 145–149.
33. Redmond, T.M., Yu, S., Lee, E., Bok, D., Hamasaki, D., Chen, N., Goletz, P., Ma, J.X., Crouch, R.K., and Pfeifer, K. (1998) Rpe65 Is Necessary for Production of 11-*cis*-Vitamin A in the Retinal Visual Cycle, *Nat. Genet. 20:* 344–351.
34. Hamel, C.P., Tsilou, E., Pfeffer, B.A., Hooks, J.J., Detrick, B., and Redmond, T.M. (1993) Molecular Cloning and Expression of RPE65, A Novel Retinal Pigment Epithelium-Specific Microsomal Protein That Is Post-Transcriptionally Regulated *In Vitro*, *J. Biol. Chem. 268:* 15751–15757.
35. Marlhens, F., Bareil, C., Griffoin, J.M., Zrenner, E., Amalric, P., Eliaou, C., Liu, S.Y., Harris, E., Redmond, T.M., Arnaud, B., Claustres, M., and Hamel, C.P. (1997) Mutations in RPE65 Cause Leber's Congenital Amaurosis, *Nat. Genet. 17:* 139–141.
36. Gu, S.M., Thompson, D.A., Srikumari, C.R., Lorenz, B., Finckh, U., Nicoletti, A., Murthy, K.R., Rathmann, M., Kumaramanickavel, G., Denton, M.J., and Gal, A. (1997) Mutations in RPE65 Cause Autosomal Recessive Childhood-Onset Severe Retinal Dystrophy, *Nat. Genet. 17:* 194–197.
37. Bernstein, P.S., Law, W.C., and Rando, R.R. (1987) Isomerization of All-*trans*-Retinoids to 11-*cis* Retinoids *In Vitro*, *Proc. Natl. Acad. Sci. USA 84:* 1849–1853.

38. Mata, N.L., Moghrabi, W.N., Lee, J.S., Bui, T.V., Radu, R.A., Horwitz, J., and Travis, G.H. (2004) Rpe65 Is a Retinyl Ester Binding Protein That Presents Insoluble Substrate to the Isomerase in Retinal Pigment Epithelial Cells, *J. Biol. Chem. 279:* 635–643.
39. Gollapalli, D.R., Maiti, P., and Rando, R.R. (2003) RPE65 Operates in the Vertebrate Visual Cycle by Stereospecifically Binding All-*trans*-Retinyl Esters, *Biochemistry 42:* 11824–11830.
40. Glover, J., and Redfearn, E.R. (1954) The Mechanism of the Transformation of β-Carotene into Vitamin A *In Vivo*, *Biochem. J. 58:* 15.
41. Wang, X.D., Tang, G.W., Fox, J.G., Krinsky, N.I., and Russell, R.M. (1991) Enzymatic Conversion of Beta-Carotene into Beta-Apo-Carotenals and Retinoids by Human, Monkey, Ferret, and Rat Tissues, *Arch. Biochem. Biophys. 285:* 8–16.
42. Sharma, R.V., Mathur, S.N., and Ganguly, J. (1976) Studies on the Relative Biopotencies and Intestinal Absorption of Different Apo-Beta-Carotenoids in Rats and Chickens, *Biochem. J. 158:* 377–383.
43. Wang, X.D., Russell, R.M., Liu, C., Stickel, F., Smith, D.E., and Krinsky, N.I. (1996) Beta-Oxidation in Rabbit Liver *In Vitro* and in the Perfused Ferret Liver Contributes to Retinoic Acid Biosynthesis from Beta-Apocarotenoic Acids, *J. Biol. Chem. 271:* 26490–26498.
44. Clagett-Dame, M., and DeLuca, H.F. (2002) The Role of Vitamin A in Mammalian Reproduction and Embryonic Development, *Annu. Rev. Nutr. 22:* 347–381.
45. Thompson, D.A., and Gal, A. (2003) Genetic Defects in Vitamin A Metabolism of the Retinal Pigment Epithelium, *Dev. Ophthalmol. 37:* 141–154.
46. Bhatti, R.A., Yu, S., Boulanger, A., Fariss, R.N., Guo, Y., Bernstein, S.L., Gentleman, S., and Redmond, T.M. (2003) Expression of Beta-Carotene 15,15′ Monooxygenase in Retina and RPE Choroid, *Investig. Ophthalmol. Vis. Sci. 44:* 44–49.
47. Lampert, J.M., Holzschuh, J., Hessel, S., Driever, W., Vogt, K., and von Lintig, J. (2003) Provitamin A Conversion *Via* the Beta,Beta-Carotene-15,15′-Oxygenase Is Essential for Pattern Formation and Differentiation During Zebrafish Embryogenesis, *Development 130:* 2173–2186.
48. Begemann, G., Schilling, T.F., Rauch, G.-J., Geisler, R., and Ingham, P.W. (2001) The Zebrafish Neckless Mutation Reveals a Requirement for raldh2 in Mesodermal Signals That Pattern the Hindbrain, *Development 128:* 3081–3094.
49. Grandel, H., Lun, K., Rauch, G.J., Rhinn, M., Piotrowski, T., Houart, C., Sordino, P., Kuchler, A.M., Schulte-Merker, S., Geisler, R., Holder, N., Wilson, S.W., and Brand, M. (2002) Retinoic Acid Signalling in the Zebrafish Embryo Is Necessary During Pre-Segmentation Stages to Pattern the Anterior-Posterior Axis of the CNS and to Induce a Pectoral Fin Bud, *Development 129:* 2851–2865.
50. Kiefer, C., Sumser, E., Wernet, M.F., and Von Lintig, J. (2002) A Class B Scavenger Receptor Mediates the Cellular Uptake of Carotenoids in *Drosophila*, *Proc. Natl. Acad. Sci. USA 99:* 10581–10586.
51. Rigotti, A., Trigatti, B.L., Penman, M., Rayburn, H., Herz, J., and Krieger, M. (1997) A Targeted Mutation in the Murine Gene Encoding the High Density Lipoprotein (HDL) Receptor Scavenger Receptor Class B Type I Reveals Its Key Role in HDL Metabolism, *Proc. Natl. Acad. Sci. USA 94:* 12610–12615.
52. van Vliet, T., van Vlissingen, M.F., van Schaik, F., and van den Berg, H. (1996) β-Carotene Absorption and Cleavage in Rats Is Affected by the Vitamin A Concentration of the Diet, *J. Nutr. 126:* 499–508.

53. Parvin, S.G., and Sivakumar, B. (2000) Nutritional Status Affects Intestinal Carotene Cleavage Activity and Carotene Conversion to Vitamin A in Rats, *J. Nutr. 130:* 573–577.
54. Bachmann, H., Desbarats, A., Pattison, P., Sedgewick, M., Riss, G., Wyss, A., Cardinault, N., Duszka, C., Goralczyk, R., and Grolier, P. (2002) Feedback Regulation of β,β-Carotene 15,15′-Monooxygenase by Retinoic Acid in Rats and Chickens, *J. Nutr. 132:* 3616–3622.
55. Boulanger, A., McLemore, P., Copeland, N.G., Gilbert, D.J., Jenkins, N.A., Yu, S.S., Gentleman, S., and Redmond, T.M. (2003) Identification of Beta-Carotene 15,15′-Monooxygenase as a Peroxisome Proliferator-Activated Receptor Target Gene, *FASEB J. 17:* 1304–1306.
56. Desvergne, B., and Wahli, W. (1999) Peroxisome Proliferator-Activated Receptors: Nuclear Control of Metabolism, *Endocr. Rev. 20:* 649–688.
57. Peters, J.M., Aoyama, T., Cattley, R.C., Nobumitsu, U., Hashimoto, T., and Gonzalez, F.J. (1998) Role of Peroxisome Proliferator-Activated Receptor Alpha in Altered Cell Cycle Regulation in Mouse Liver, *Carcinogenesis 19:* 1989–1994.
58. Barak, Y., Nelson, M.C., Ong, E.S., Jones, Y.Z., Ruiz-Lozano, P., Chien, K.R., Koder, A., and Evans, R.M. (1999) PPAR Gamma Is Required for Placental, Cardiac, and Adipose Tissue Development, *Mol. Cell 4:* 585–595.
59. Randolph, R.K., and Ross, A.C. (1991) Vitamin A Status Regulates Hepatic Lecithin: Retinol Acyltransferase Activity in Rats, *J. Biol. Chem. 266:* 16453–16457.
60. Suruga, K., Mochizuki, K., Kitagawa, M., Goda, T., Horie, N., Takeishi, K., and Takase, S. (1999) Transcriptional Regulation of Cellular Retinol-Binding Protein, Type II Gene Expression in Small Intestine by Dietary Fat, *Arch. Biochem. Biophys. 362:* 159–166.
61. Napoli, J.L., and Race, K.R. (1988) Biogenesis of Retinoic Acid from Beta-Carotene. Differences Between the Metabolism of Beta-Carotene and Retinal, *J. Biol. Chem. 263:* 17372–17377.
62. Mic, F.A., Haselbeck, R.J., Cuenca, A.E., and Duester, G. (2002) Novel Retinoic Acid Generating Activities in the Neural Tube and Heart Identified by Conditional Rescue of Raldh2 Null Mutant Mice, *Development 129:* 2271–2282.
63. Niederreither, K., Vermot, J., Fraulob, V., Chambon, P., and Dolle, P. (2002) Retinaldehyde Dehydrogenase 2 (RALDH2)-Independent Patterns of Retinoic Acid Synthesis in the Mouse Embryo, *Proc. Natl. Acad. Sci. USA 99:* 16111–16116.
64. Tibaduiza, E.C., Fleet, J.C., Russell, R.M., and Krinsky, N.I. (2002) Excentric Cleavage Products of β-Carotene Inhibit Estrogen Receptor Positive and Negative Breast Tumor Cell Growth *In Vitro* and Inhibit Activator Protein-1-Mediated Transcriptional Activation, *J. Nutr. 132:* 1368–1375.
65. Clinton, S.K. (1998) Lycopene: Chemistry, Biology, and Implications for Human Health and Disease, *Nutr. Rev. 56:* 35–51.

Chapter 16

Regulation of Transcription by Antioxidant Carotenoids[1]

Yoav Sharoni[a], Riad Agbaria[b], Hadar Amir[a], Anat Ben-Dor[a], Noga Dubi[a], Yudit Giat[a], Keren Hirsh[a], Gaby Izumchenko[a], Marina Khanin[a], Elena Kirilov[a], Amit Nahum[a], Michael Steiner[a], Yossi Walfisch[a], Shlomo Walfisch[c], Michael Danilenko[a], and Joseph Levy[a]

[a]Department of Clinical Biochemistry, [b]Department of Pharmacology, and [c]The Colorectal Unit, Faculty of Health Sciences, Ben-Gurion University of the Negev and Soroka Medical Center of Kupat Holim, Beer-Sheva, Israel

Introduction

Carotenoids have been implicated as important dietary phytonutrients having cancer preventive activity (1). Much of the evidence, particularly in terms of cancer prevention, is derived from observational studies of dietary carotenoid intake; thus, the findings must be interpreted with caution. In such studies, it is not clear whether an association between diet and disease is due to the specific carotenoid, other micronutients present in the specific diet, or the combined effect of several of these active ingredients. Many studies evaluated the relation between carotenoid intake and cancer (2,3). The best evidence for an inverse association exists for lung, colon, breast, and prostate cancer. The epidemiologic data are reinforced by studies showing the inhibitory effect of carotenoids on tumor growth in animal models *in vivo* (4–8). Additional support for the effect of carotenoids was found with diverse cancer cells *in vitro* (9–12). Particularly, we demonstrated that lycopene inhibits mammary, endometrial, lung, and leukemic cancer cell growth in a dose-dependent manner [50% inhibitory concentration (IC_{50}) = ~2 μM] (10,12).

Mechanism of Cancer Cell Growth Inhibition at the Protein Expression Level

The mechanisms underlying the anticancer activity of carotenoids may involve changes in pathways leading to cell growth or cell death. These include hormone and growth factor signaling, regulatory mechanisms of cell cycle progression, cell differentiation, and apoptosis. Examples of carotenoid effects on some of these

[1]Reprinted from *Molecular Aspects of Medicine 24*, Sharoni, Y., Agbaria, R., Amir, H., Ben-Dor, A., Bobilev, I., Doubi, N., Giat, Y., Hirsh, K., Izumchenko, G., Khanin, M., Kirilov, E., Krimer, R., Nahum, A., Steiner, M., Walfisch, Y., Walfisch, S., Zango, G., Danilenko, M., and Levy, J, Modulation of Transcriptional Activity by Antioxidant Carotenoids, 371–384, Copyright (2003), with permission from Elsevier.

pathways will be described below with emphasis on the changes in protein expression associated with these effects.

Gap Junctional Communication. One of the earliest discoveries related to carotenoids and modulation of protein level was made by Bertram and colleagues (13,14) who found that carotenoids increased gap junctional communication between cells and induced the synthesis of connexin43, a component of the gap junction structure. This effect was achieved independently of provitamin A or the antioxidant properties of the carotenoids. Loss of gap junctional communication may be important for malignant transformation, and its restoration may reverse the malignant process (15).

Growth Factor Signaling. Growth factors, either in the blood or as part of the autocrine or paracrine loops, are important for cancer cell growth. Recently, insulin-like growth factor (IGF)-I was implicated as a major cancer risk factor. It was reported that high blood levels of IGF-I, existing years before malignancy detection, predicted an increase in risk for breast, prostate, colorectal, and lung cancer (16–19). Accordingly, two possible strategies might be used to reduce IGF-related cancer risk: (i) reduction in IGF-I blood levels, and (ii) interference with IGF-I activity in the cancer cell. Preliminary results of our studies on the former strategy suggest that tomato phytonutrients lower IGF-I blood levels. In addition, we showed that lycopene inhibits the mitogenic action of IGF-I in human cancer cells. In mammary cancer cells, lycopene treatment markedly reduced IGF-I stimulation of both tyrosine phosphorylation of insulin receptor substrate-1 and DNA binding capacity of the activator protein (AP)-1 transcription factor (20). These effects were not associated with changes in the number or affinity of IGF-I receptors, but rather with an increase in membrane-associated IGF-binding proteins (IGFBP). This finding can explain the suppression of IGF-I-signaling by lycopene, on the basis of our earlier studies (21–23) showing that membrane-associated IGFBP-3 inhibits IGF-I receptor signaling in an IGF-dependent manner.

Cell Cycle Progression. Growth factors have a major effect in promoting cell cycle progression, primarily during the G_1 phase. We showed that lycopene treatment of MCF-7 mammary cancer cells slowed down IGF-I–stimulated cell cycle progression (20), which was not accompanied by either apoptotic or necrotic cell death. Lycopene-induced delay in progression through the G_1 and S phases was also observed in other cancer cell lines (leukemic, endometrial, lung, and prostate) tested in our laboratory [(12) and unpublished data]. A similar effect of another carotenoid, α-carotene, was demonstrated in GOTO human neuroblastoma cells (24). Similarly, β-carotene induced a cell-cycle delay in the G_1 phase in normal human fibroblasts (25). Cell cycle transition through a late G_1 checkpoint is governed by a mechanism known as the "pRb pathway" (26). The central element in this pathway, retinoblastoma protein (pRb), is a tumor suppressor that prevents premature G_1/S transition *via* physical interaction with transcription factors of the E2F family. The activity of pRb is regulated by an assembly of cyclins, cyclin-dependent kinases (Cdk) and Cdk inhibitors. Phosphorylation of pRb by Cdk results in the release of E2F, which leads to the synthesis of various cell growth–related proteins. Cdk activity is modulated in both a positive and a negative manner by cyclins and Cdk inhibitors, respectively. It is well documented that growth factors affect the cell cycle apparatus primarily during the G_1 phase, and that the D-type cyclins are the main elements acting as growth factor sensors (27). Moreover, cyclin D1 is known as an oncogene and was found to be overexpressed in many breast cancer cell lines as well as in primary tumors (28). In a recent study (29) we demonstrated that cancer cells arrested by serum deprivation in the presence of lycopene are incapable of returning to the cell cycle after serum readdition. This inhibition correlated with a reduction in cyclin D1 protein levels that resulted in inhibition of both cdk4 and cdk2 kinase activity and in hypophosphorylation of pRb. Inhibition of cdk4 was related directly to the lower amount of cyclin D1-cdk4 complexes, whereas inhibition of cdk2 action was related to the retention of p27 molecules in cyclin E-cdk2 complexes due to the reduction in cyclin D1 level.

Differentiation-Related Proteins. Induction of differentiation to mature cells with distinct functions similar to nonmalignant cells was proposed as an alternative to cytotoxic chemotherapy and may be useful for chronic chemoprevention. Differentiation therapy was quite effective in treating acute promyelocytic leukemia and is currently under investigation for treatment of solid tumors. Differentiation inducers that are presently under laboratory and clinical investigation include vitamin D and its analogs, retinoids, polyamine inhibitors, and others. We showed that lycopene alone induces differentiation of HL-60 promyelocytic leukemia cells (12). A similar effect was described also for other carotenoids such as β-carotene and lutein (30,31). The differentiation effect of lycopene was associated with elevated expression of several differentiation-related proteins, such as cell surface antigen (CD14), oxygen burst oxidase (12), and chemotactic peptide receptors.

The mechanism of the differentiating activity of lycopene and its ability to synergize with $1,25(OH)_2D_3$ in this effect (12) is largely unclear. However, in a similar study, we showed recently that the differentiation-enhancing effect of another phytonutrient, carnosic acid from rosemary, was associated with induction of multiple differentiation-related proteins, such as Cdk inhibitor, p21Cip1, early growth response gene-1 (EGR-1), and Cdk5 and its activator protein, p35Nck5a (32,33). Most importantly, carnosic acid and its combinations with $1,25(OH)_2D_3$ and retinoic acid transcriptionally activated the expression of nuclear hormone receptors, such as the vitamin D_3 receptor (VDR), retinoic acid receptor (RARα) and retinoid X receptor (RXRa) (32). This may represent a molecular basis for synergy between phytonutrients and differentiation inducers. The possibility that lycopene as well as other carotenoids and/or their derivatives may affect nuclear signaling pathways is an attractive suggestion that should be studied.

Synergy Between Carotenoids and Other Phytonutrients. The epidemiologic and laboratory studies reviewed above suggested a cancer-preventive activity for carotenoids and led to collaborative large intervention studies with synthetic β-carotene. However, the use of a single plant-derived compound in human prevention studies was not successful, revealing either no beneficial effect (34) or even a negative effect (35,36). These results led to the hypothesis that a single micronutrient cannot replace the power of the concerted action of multiple compounds derived from a diet rich in fruits and vegetables.

To support the hypothesis that a concerted action of several micronutrients is responsible for the anticancer activity of diet enriched with fruits and vegetables, it has to be shown that plant-derived constituents, such as carotenoids, have the ability to synergize with other phytonutrients. Therefore, we have been studying the effects of combinations of various micronutrients or their metabolites on cancer cell proliferation and differentiation. These micronutrients include carotenoids (β-carotene, lycopene, phytoene, phytofluene and astaxanthin); a polyphenolic antioxidant (carnosic acid from rosemary); an organosulfur compound (allicin from garlic); the active metabolite of vitamin D (1,25-dihydroxyvitamin D_3); the metabolite of β-carotene and vitamin A (retinoic acid); and a synthetic derivative of lycopene (*acyclo*-retinoic acid). We found that various combinations of these compounds produce a synergistic or additive inhibition of cancer cell proliferation. For example, as discussed above, the combination of low concentrations of lycopene with 1,25-dihydroxyvitamin D_3 [$1,25(OH)_2D_3$] synergistically inhibited proliferation and induced differentiation in HL-60 leukemic cells (12). In addition, Pastori and colleagues found that the simultaneous addition of lycopene and another vitamin, α-tocopherol, at physiologic physiologic concentrations, resulted in a strong synergistic inhibition of prostate carcinoma cell proliferation (11).

Carotenoids and Transcription

As discussed above, carotenoids modulate the basic mechanisms of cell proliferation, growth factor signaling, gap junctional intercellular communication (GJIC), and produce changes in the expression of many proteins participating in these processes, for example, connexins, cyclins, cyclin-dependent kinases, and their inhibitors. Therefore, the question that arises is by what mechanisms do carotenoids affect so many diverse cellular pathways? The changes in the expression of multiple proteins suggest that the initial effect of carotenoids involves modulation of transcription. This may be due to either direct interaction of the carotenoid molecules or their derivatives with transcription factors, e.g., with ligand-activated nuclear receptors or indirect modification of transcriptional activity, e.g., *via* changes in the status of cellular redox, which affects redox-sensitive transcription systems, such as AP-1, nuclear factor (NF)κB and antioxidant response element (ARE).

The idea that carotenoid derivatives can activate nuclear receptors is not new; until recently, however, it was limited to retinoic acid, which is produced from β-carotene and other vitamin A precursors. Emerging evidence now suggests that derivatives of other carotenoids or the carotenoids themselves may also modulate the activity of transcription factors. For example, the synergistic inhibition of cancer cell proliferation by lycopene in combination with $1,25(OH)_2D_3$ or retinoic acid (12), the ligands of two members of the nuclear receptor superfamily, suggests that lycopene or one of its derivatives may also interact with members of this family of receptors.

Retinoid Receptors. Retinoic acid is the parent compound of ligands known as retinoids. It exerts multiple effects on cell proliferation and differentiation through two classes of nuclear receptors, RAR and RXR. All-*trans* retinoic acid binds only to RAR, whereas its isomer, 9-*cis* retinoic acid, binds to both RAR and RXR. Upon DNA binding, RXR/RAR heterodimers regulate gene expression of retinoic acid target genes in a ligand-dependent manner. RXR is also able to form heterodimers with other members of the nuclear hormone receptors superfamily, such as the thyroid hormone receptor, the vitamin D receptor, the peroxisome proliferator-activated receptor (PPAR), and possibly other receptors with unknown ligands designated orphan receptors.

The possibility that a lycopene derivative mediates the inhibitory action of this carotenoid on cell growth is suggested above. To test the hypothesis that such a derivative acts as a ligand for nuclear receptors and mediates the anticancer activity of lycopene, Ben-Dor *et al.* (37) analyzed the effect of a hypothetical oxidation product of lycopene, *acyclo*-retinoic acid (38), on cancer cell growth and the transactivation of retinoic acid–regulated reporter gene. *acyclo*-Retinoic acid transactivated a reporter gene containing the retinoic acid response element (RARE) with a ~100-fold lower potency than retinoic acid (37). Lycopene exhibited only very modest activity in this system. In contrast to the transactivation data, *acyclo*-retinoic acid, retinoic acid, and lycopene inhibited MCF-7 cell growth and slowed down cell cycle progression from the G_1 to the S phase with a similar potency. Furthermore, the two retinoids decreased serum-stimulated cyclin D1 expression. On the other hand, they had dissimilar effects on the level of p21. In the presence of *acyclo*-retinoic acid, the level of p21, established during serum stimulation, was comparable to that of control cells, whereas in retinoic acid–treated cells, the p21 level was much lower, suggesting that the effects of *acyclo*-retinoic acids are not entirely mediated by the RAR. Moreover, a comparable potency of *acyclo*-retinoic acid and lycopene in inhibition of cell growth suggests that *acyclo*-retinoic acid is unlikely to be the active metabolite of this carotenoid. A similar conclusion was made by Stahl *et al.* (38) who found that retinoic acid was much more potent than *acyclo*-retinoic acid in transactivation of the retinoic acid responsive promoter of RAR-β2 and that lycopene and retinoic acid are more active than *acyclo*-retinoic acid in the activation of GJIC. Muto *et al.* (39,40) synthesized a series of acyclic retinoids and found that some acyclic retinoids caused transactivation of an RAR reporter gene, comparable to that achieved by retinoic acid. One of these acyclic retinoids was recently shown to inhibit cell cycle progression, which was associated with a reduction in cyclin D1 level and an increase in the level of p21 (41), similar to the results obtained with *acyclo*-retinoic acid (37). It is interesting to note that the acyclic retinoids described by Muto and colleagues may be potential derivatives of phytoene and phytofluene, two carotenoids present in tomatoes that were found in our laboratory to inhibit the proliferation of various cancer cells (unpublished data).

The anticancer activity of carotenoid derivatives is not necessarily mediated by activation of the retinoid receptors. For example, several cleavage products of β-carotene strongly inhibited AP-1 transcriptional activity (42). Lycopene was also shown to inhibit AP-1 activation (20). We hypothesize that carotenoids or their oxidized derivatives interact with a network of transcription factors that are activated by different ligands at low affinity and specificity. The activation of several transcription factor systems by different compounds may lead to the synergistic inhibition of cell growth. In addition to the retinoid receptors and AP-1, other candidate transcription systems that may participate in this network are the PPAR (43–45), the ARE (46,47), the xenobiotic receptor (48,49), NFκB (50), and yet unidentified orphan receptors.

Peroxisome Proliferator-Activated Receptors. These nuclear hormone receptors have a key role in the differentiation of adipocytes, although recently their role in cancer cell growth inhibition and differentiation was also demonstrated. One of the PPAR subtypes, PPARγ, is expressed at significant levels in human primary and metastatic breast adenocarcinomas (51) and liposarcomas (52). Colon cancer in humans was shown to be associated with loss-of-function mutations in PPARγ (53). Ligand activation of PPARγ in cultured breast cancer cells causes extensive lipid accumulation, changes in breast epithelial gene expression associated with a more differentiated, less malignant state, and a reduction in growth rate and clonogenic capacity of the cells (51). These data suggest that the PPARγ transcription system can induce terminal differentiation of malignant breast epithelial cells and thus may provide a novel therapy for human breast cancer. Human prostate cancer cells were found to express PPARγ at prominent levels, whereas normal prostate tissues demonstrated a very low expression (44). Activation of this receptor with specific ligands, such as troglitazone, exerted an inhibitory effect on the growth of prostate cancer cell lines (45). In prostate cancer patients with no metastatic disease, troglitazone treatment prevented an increase in prostate-specific antigen level that was evident in the untreated patient group (45). These data suggest that PPARγ may serve as a biological modifier in human prostate cancer; therefore, its therapeutic potential in this disease should be further investigated.

The presence of PPARγ receptors in various cancer cells and their activation by fatty acids, prostaglandins and related hydrophobic agents in the micromolar range make these ligand-dependent transcription factors an interesting target for carotenoid derivatives. Takahashi *et al.* (54) found that the isoprenols farnesol and geranylgeraniol transactivated PPAR reporter gene at a high concentration (100 μM). Moreover, these isoprenols upregulated the expression of lipid metabolic target genes of PPAR. However, different carotenoids at 100 μM, a concentration scarcely achievable in aqueous solutions, did not have any significant effect. We recently compared the relative efficacy of several carotenoids found in tomatoes in transactivation of PPAR response element (PPARE). Preliminary results indicated that lycopene, phytoene, phytofluene, and β-carotene transactivate PPARE in MCF-7 cells cotransfected with PPARγ. However, it is not clear whether activation of the PPAR system contributes to the inhibition of cancer cell growth by carotenoids.

The Antioxidant Response Element. Induction of phase II enzymes, which conjugate reactive electrophiles and act as indirect antioxidants, appears to be an effective means for achieving protection against a variety of carcinogens in animals and humans. Transcriptional control of the expression of these enzymes is mediated, at least in part, through ARE found in the regulatory regions of their genes. The transcription factor Nrf2, which binds to ARE, appears to be essential for the induction of phase II enzymes, such as glutathione *S*-transferase (GST), NAD(P)H:quinone oxidoreductase (NQO1) (55), as well as the thiol-containing reducing factor, thioredoxin (56). Constitutive hepatic and gastric activities of GST and NQO1 were decreased by 50–80% in Nrf2-deficient mice compared with wild-type mice (55). Several studies showed that antioxidants present in the diet, such as terpenoids, phenolic flavonoids (e.g., green tea polyphenols and epigallocatechin-3-gallate), and isothiocyanates may work as anticancer agents by activating this transcription system (57,58).

Gradelet *et al.* (59) showed that some carotenoids are capable of inducing phase II metabolizing enzymes, *p*-nitrophenol-UDP-glucuronosyl transferase and NQO1 in rats. In this study, male rats were fed diets containing different carotenoids for 15 d; canthaxanthin and astaxanthin, but not lutein and lycopene, were active in the induction of these enzymes. In another study, Bhuvaneswari *et al.* (60) showed an association between the reduction in the incidence of dimethyl benz[a]antracene (DMBA)– hamster buccal pouch tumors by lycopene and a concomitant rise in the level of glutathione (GSH), the phase II enzyme GST and enzymes of the GSH redox cycle. Based on these results, these authors suggested that the lycopene-induced increase in the levels of GSH and the phase II enzyme GST in the buccal pouch mucosa inactivated carcinogens by forming conjugates that were less toxic and readily excreted from the body.

To date, no studies have considered the direct activation of the antioxidant response element by carotenoids. However, preliminary results of our experiments on this subject show that in transiently transfected mammary cancer and hepatocarcinoma cells, lycopene transactivates the expression of the reporter gene luciferase fused with ARE sequences present in NQO1 and in γ-glutamylcysteine synthetase (the rate-limiting enzyme in GSH synthesis).

Xenobiotic and Other Orphan Nuclear Receptors. Orphan receptors are structurally related to nuclear hormone receptors but lack known physiologic ligands. One family of orphan nuclear receptors called xenobiotic receptors makes up part of the defense mechanism against foreign lipophilic chemicals (xenobiotics). This family of receptors includes the steroid and xenobiotic receptor/pregnane X receptor (SXR/PXR), constitutive androstane receptor (CAR) (48,49), and the aryl hydrocarbon receptor (AhR) (61). These receptors respond to a wide variety of drugs, environmental pollutants, carcinogens, and dietary and endogenous compounds; they regulate the expression of cytochrome P_{450} (CYP) enzymes, conjugating enzymes, and transporters involved in the metabolism and elimination of xenobiotics (49).

Animal studies showed that in addition to the phase II xenobiotic metabolizing enzymes described above (59), some carotenoids are capable of inducing CYP enzymes, the constituents of the phase I detoxification pathway. Canthaxanthin and astaxanthin induced liver CYP1A1 and CYP1A2 in rats and similar effects were observed with β-apo-8′-carotenal. β-Carotene, lutein, and lycopene were not active (59,62–64). In the mouse liver, canthaxanthin induced weak effects, whereas the other carotenoids did not stimulate CYP1A1 activity at all (65). The mechanism underlying CYP enzyme induction by carotenoids is not fully understood, but there is evidence that the AhR-dependent pathway is involved. However, carotenoids did not directly bind to this receptor (66). Of several carotenoids tested in rats (β-carotene, bixin, lycopene, lutein, canthaxanthin, and astaxanthin), only bixin, canthaxanthin, and astaxanthin were capable of inducing the activity of CYP1A1 in liver, lung, and kidney and CYP1A2 in liver and lung (67). In another study, the administration of lycopene to rats at doses ranging from 0.001 to 0.1 g/kg was shown to induce the liver CYP types 1A1/2, 2B1/2 and 3A in a dose-dependent manner (68). The observation that these enzymatic activities were induced at very low lycopene plasma levels led the authors to suggest that modulation of drug-metabolizing enzymes by carotenoids might be relevant to humans (68). Indeed, the activity of CYP1A2 in humans was shown to be correlated with plasma levels of micronutrients (69), and it was proposed that about one third of the variation in enzyme activity is related to dietary factors. Plasma lutein levels were negatively associated with CYP1A2 activity, whereas lycopene levels were positively correlated with the enzyme activity.

The direct effect of carotenoids on a xenobiotic receptor system was tested in an *in vitro* transcription system (R. Rühl and F.J. Schweigert, personal communication, 2002). They found that in transiently transfected HepG2 hepatoma cells, β-carotene can transactivate the PXR reporter gene in a manner comparable to rifampicin. Furthermore, an upregulation of CYP3A4 and CYP3A5 was obtained in these cells, pointing to a potential effect of the carotenoid on the metabolism of xenobiotics.

It has become clear that orphan nuclear receptors represent a unique and pivotal resource with which to uncover new regulatory systems that affect both health and human diseases. The discussion above suggests that at least one group of this receptor family, the xenobiotic receptors, are affected by carotenoids. Thus, it is possible that other, yet unknown, types of orphan receptors are involved in the cellular action of carotenoids.

Conclusions

In studies using cell and animal models, carotenoids were shown to influence diverse molecular and cellular processes, which can form the basis for the beneficial effects of carotenoids on human health and disease prevention. However, despite the promising results, it is difficult at present to relate available experimental data directly to human pathophysiology. One problem is that many results were obtained in studies using carotenoid levels that are much higher than those achievable in human blood. On the other hand, the evidence presented suggests a synergistic action of low concentrations of various carotenoids and other micronutrients. A growing body of experimental data indicates that this synergy may be based on the ability of different dietary compounds to modulate a network of transcription systems. The concerted action of multiple micronutrients accomplished by activation of transcription and probably by other mechanisms can explain the beneficial effect of diets rich in fruits and vegetables. An important question that remains open is whether the changes described in various cellular pathways are due to direct effects of the carotenoid molecules or are mediated by their derivatives. Although some information on this issue is presented in this chapter, additional studies are warranted to identify and characterize these putative active carotenoid derivatives.

Acknowledgments

Studies from the authors' laboratory were supported in part by the Israel Cancer Association; by the Chief Scientist, Israel Ministry of Health; by the Israel Science Foundation founded by the Israel Academy of Science and Humanities; by the European Community (Project No. FAIR CT 97–3100); by LycoRed Natural Products Industries, Beer-Sheva, Israel, and by the S. Daniel Abraham International Center for Health and Nutrition, Ben-Gurion University of the Negev.

References

1. van Poppel, G. (1993) Carotenoids and Cancer: An Update with Emphasis on Human Intervention Studies, *Eur. J. Cancer 29A:* 1335–1344.
2. van Poppel, G., and Goldbohm, R.A. (1995) Epidemiologic Evidence for Beta-Carotene and Cancer Prevention, *Am. J. Clin. Nutr. 62:* 1393S–1402S.
3. Giovannucci, E. (1999) Tomatoes, Tomato-Based Products, Lycopene, and Cancer: Review of the Epidemiologic Literature, *J. Natl. Cancer Inst. 91:* 317–331.
4. Nagasawa, H., Mitamura, T., Sakamoto, S., and Yamamoto, K. (1995) Effect of Lycopene on Spontaneous Mammary Tumor Development in SHN Virgin Mice, *Anticancer Res. 15:* 1173–1178.

5. Sharoni, Y., Giron, E., Rise, M., and Levy, J. (1997) Effects of Lycopene Enriched Tomato Oleoresin on 7,12-Dimethyl-benz[a]anthracene-Induced Rat Mammary Tumors, *Cancer Detect. Prev. 21:* 118–123.
6. Narisawa, T., Fukaura, Y., Hasebe, M., Ito, M., Aizawa, R., Murakoshi, M., Uemura, S., Khachik, F., and Nishino, H. (1996) Inhibitory Effects of Natural Carotenoids, Alpha-Carotene, Beta-Carotene, Lycopene and Lutein, on Colonic Aberrant Crypt Foci Formation in Rats, *Cancer Lett. 107:* 137–142.
7. Kim, J.M., Araki, S., Kim, D.J., Park, C.B., Takasuka, N., Baba-Toriyama, H., Ota, T., Nir, Z., Khachik, F., Shimidzu, N., Tanaka, Y., Osawa, T., Uraji, T., Murakoshi, M., Nishino, H., and Tsuda, H. (1998) Chemopreventive Effects of Carotenoids and Curcumins on Mouse Colon Carcinogenesis After 1,2-Dimethylhydrazine Initiation, *Carcinogenesis 19:* 81–85.
8. Okajima, E., Tsutsumi, M., Ozono, S., Akai, H., Denda, A., Nishino, H., Oshima, S., Sakamoto, H., and Konishi, Y. (1998) Inhibitory Effect of Tomato Juice on Rat Urinary Bladder Carcinogenesis After *N*-Butyl-*N*-(4-hydroxybutyl)nitrosamine Initiation, *Jpn. J. Cancer Res. 89:* 22–26.
9. Wang, C.-J., and Lin, J.-K. (1989) Inhibitory Effects of Carotenoids and Retinoids on the *In Vitro* Growth of Rat C-6 Glioma Cells, *Proc. Natl. Sci. Counc. B ROC 13:* 176–183.
10. Levy, J., Bosin, E., Feldman, B., Giat, Y., Miinster, A., Danilenko, M., and Sharoni, Y. (1995) Lycopene Is a More Potent Inhibitor of Human Cancer Cell Proliferation than Either α-Carotene or β-Carotene, *Nutr. Cancer 24:* 257–267.
11. Pastori, M., Pfander, H., Boscoboinik, D., and Azzi, A. (1998) Lycopene in Association with Alpha-Tocopherol Inhibits at Physiological Concentrations Proliferation of Prostate Carcinoma Cells, *Biochem. Biophys. Res. Commun. 250:* 582–585.
12. Amir, H., Karas, M., Giat, J., Danilenko, M., Levy, R., Yermiahu, T., Levy, J., and Sharoni, Y. (1999) Lycopene and 1,25-Dihydroxyvitamin-D3 Cooperate in the Inhibition of Cell Cycle Progression and Induction of Differentiation in HL-60 Leukemic Cells, *Nutr. Cancer 33:* 105–112.
13. Zhang, L.-X., Cooney, R.V., and Bertram, J. (1991) Carotenoids Enhance Gap Junctional Communication and Inhibit Lipid Peroxidation in C3H/10T1/2 Cells: Relationship to Their Cancer Chemopreventive Action, *Carcinogenesis 12:* 2109–2114.
14. Zhang, L.X., Cooney, R.V., and Bertram, J.S. (1992) Carotenoids Up-Regulate Connexin-43 Gene Expression Independent of Their Provitamin-A or Antioxidant Properties, *Cancer Res. 52:* 5707–5712.
15. Hotz-Wagenblatt, A., and Shalloway, D. (1993) Gap Junctional Communication and Neoplastic Transformation, *Crit. Rev. Oncog. 4:* 541–558.
16. Mantzoros, C.S., Tzonou, A., Signorello, L.B., Stampfer, M., Trichopoulos, D., and Adami, H.O. (1997) Insulin-Like Growth Factor 1 in Relation to Prostate Cancer and Benign Prostatic Hyperplasia, *Br. J. Cancer 76:* 1115–1118.
17. Hankinson, S.E., Willett, W.C., Colditz, G.A., Hunter, D.J., Michaud, D.S., Deroo, B., Rosner, B., Speizer, F.E., and Pollak, M. (1998) Circulating Concentrations of Insulin-Like Growth Factor I and Risk of Breast Cancer, *Lancet 351:* 1393–1396.
18. Ma, J., Pollak, M.N., Giovannucci, E., Chan, J.M., Tao, Y., Hennekens, C.H., and Stampfer, M.J. (1999) Prospective Study of Colorectal Cancer Risk in Men and Plasma Levels of Insulin-Like Growth Factor (IGF)-I and IGF-Binding Protein-3, *J. Natl. Cancer Inst. 91:* 620–625.

19. Yu, H., Spitz, M.R., Mistry, J., Gu, J., Hong, W.K., and Wu, X. (1999) Plasma Levels of Insulin-Like Growth Factor-I and Lung Cancer Risk: A Case-Control Analysis, *J. Natl. Cancer Inst. 91:* 151–156.
20. Karas, M., Amir, H., Fishman, D., Danilenko, M., Segal, S., Nahum, A., Koifmann, A., Giat, Y., Levy, J., and Sharoni, Y. (2000) Lycopene Interferes with Cell Cycle Progression and Insulin-Like Growth Factor I Signaling in Mammary Cancer Cells, *Nutr. Cancer 36:* 101–111.
21. Karas, M., Danilenko, M., Fishman, D., LeRoith, D., Levy, J., and Sharoni, Y. (1997) Membrane-Associated IGFBP-3 Inhibits Insulin-Like Growth Factor-I (IGF-I)-Induced IGF-I Receptor Signaling in Ishikawa Endometrial Cancer Cells, *J. Biol. Chem. 272:* 16514–16520.
22. Karas, M., Kleinman, D., Danilenko, M., Roberts, C.T., Jr., LeRoith, D., Levy, J., and Sharoni, Y. (1995) Components of the IGF System Mediate the Opposing Effects of Tamoxifen on Endometrial and Breast Cancer Cell Growth, *Prog. Growth Factor Res. 6:* 513–520.
23. Kleinman, D., Karas, M., Danilenko, M., Arbeli, A., Roberts, C.T., Jr., LeRoith, D., Levy, J., and Sharoni, Y. (1996) Stimulation of Endometrial Cancer Cell Growth by Tamoxifen Is Associated with Increased IGF-I Induced Tyrosine Phosphorylation and Reduction in IGF Binding Proteins, *Endocrinology 137:* 1089–1095.
24. Murakoshi, M., Takayasu, J., Kimura, O., Kohmura, E., Nishino, H., Okuzumi, J., Sakai, T., Sugimoto, T., Imanishi, J., and Iwasaki, R. (1989) Inhibitory Effects of α-Carotene on Proliferation of the Human Neuroblastoma Cell Line GOTO, *J. Natl. Cancer Inst. 81:* 1649–1652.
25. Stivala, L.A., Savio, M., Quarta, S., Scotti, C., Cazzalini, O., Rossi, L., Scovassi, I.A., Pizzala, R., Melli, R., Bianchi, L., Vavvivi, V., and Prosperi, E. (2000) The Antiproliferative Effect of Beta-Carotene Requires p21waf1/cip1 in Normal Human Fibroblasts, *Eur. J. Biochem. 267:* 2290–2296.
26. Bartek, J., Bartkova, J., and Lukas, J. (1997) The Retinoblastoma Protein Pathway in Cell Cycle Control and Cancer, *Exp. Cell Res. 237:* 1–6.
27. Sherr, C.J. (1995) D-Type Cyclins, *Trends Biochem. Sci. 20:* 187–190.
28. Buckley, M.F., Sweeney, K.J., Hamilton, J.A., Sini, R.L., Manning, D.L., Nicholson, R.I., deFazio, A., Watts, C.K., Musgrove, E.A., and Sutherland, R.L. (1993) Expression and Amplification of Cyclin Genes in Human Breast Cancer, *Oncogene 8:* 2127–2133.
29. Nahum, A., Hirsch, K., Danilenko, M., Watts, C.K., Prall, O.W., Levy, J., and Sharoni, Y. (2001) Lycopene Inhibition of Cell Cycle Progression in Breast and Endometrial Cancer Cells Is Associated with Reduction in Cyclin D Levels and Retention of p27(Kip1) in the Cyclin E-cdk2 Complexes, *Oncogene 20:* 3428–3436.
30. Liu, Y., Chang, R.L., Cui, X.X., Newmark, H.L., and Conney, A.H. (1997) Synergistic Effects of Curcumin on All-*trans* Retinoic Acid- and 1α,25-Dihydroxyvitamin D_3-Induced Differentiation in Human Promyelocytic Leukemia HL-60 Cells, *Oncology Res. 9:* 19–29.
31. Sokoloski, J.A., Hodnick, W.F., Mayne, S.T., Cinquina, C., Kim, C.S., and Sartorelli, A.C. (1997) Induction of the Differentiation of HL-60 Promyelocytic Leukemia Cells by Vitamin E and Other Antioxidants in Combination with Low Levels of Vitamin D-3: Possible Relationship to NF-kappa B, *Leukemia 11:* 1546–1553.
32. Danilenko, M., Wang, X., and Studzinski, G.P. (2001) Carnosic Acid and Promotion of Monocytic Differentiation of HL60-G Cells Initiated by Other Agents, *J. Natl. Cancer Inst. 93:* 1224–1233.

33. Steiner, M., Priel, I., Giat, J., Levy, J., Sharoni, Y., and Danilenko, M. (2001) Carnosic Acid Inhibits Proliferation and Augments Differentiation of Human Leukemic Cells Induced by 1,25-Dihydroxyvitamin D_3 and Retinoic Acid, *Nutr. Cancer 41:* 135–144.
34. Hennekens, C.H., Buring, J.E., Manson, J.E., Stampfer, M., Rosner, B., Cook, N.R., Belanger, C., LaMotte, F., Gaziano, J.M., Ridker, P.M., Willett, W., and Peto, R. (1996) Lack of Effect of Long-Term Supplementation with Beta Carotene on the Incidence of Malignant Neoplasms and Cardiovascular Disease, *N. Engl. J. Med. 334:* 1145–1149.
35. The Alpha-Tocopherol, Beta Carotene Cancer Prevention Study Group (1994) Effect of Vitamin E and Beta Carotene on the Incidence of Lung Cancer and Other Cancers in Male Smokers, *N. Engl. J. Med. 330:* 1029–1035.
36. Omenn, G.S., Goodman, G.E., Thornquist, M.D., Balmes, J., Cullen, M.R., Glass, A., Keogh, J.P., Meyskens, F.L., Valanis, B., Williams, J.H., Jr., Barnhart, S., and Hammar, S. (1996) Effects of a Combination of Beta Carotene and Vitamin A on Lung Cancer and Cardiovascular Disease, *N. Engl. J. Med. 334:*1150–1155.
37. Ben-Dor, A., Nahum, A., Danilenko, M., Giat, Y., Stahl, W., Martin, H.D., Emmerich, T., Noy, N., Levy, J., and Sharoni, Y. (2001) Effects of *acyclo*-Retinoic Acid and Lycopene on Activation of the Retinoic Acid Receptor and Proliferation of Mammary Cancer Cells, *Arch. Biochem. Biophys. 391:* 295–302.
38. Stahl, W., von Laar, J., Martin, H.D., Emmerich, T., and Sies, H. (2000) Stimulation of Gap Junctional Communication: Comparison of *acyclo*-Retinoic Acid and Lycopene, *Arch. Biochem. Biophys. 373:* 271–274.
39. Muto, Y., Moriwaki, H., and Omori, M. (1981) *In Vitro* Binding Affinity of Novel Synthetic Polyprenoids (Polyprenoic Acids) to Cellular Retinoid-Binding Proteins, *Gann 72:* 974–977.
40. Araki, H., Shidoji, Y., Yamada, Y., Moriwaki, H., and Muto, Y. (1995) Retinoid Agonist Activities of Synthetic Geranyl Geranoic Acid Derivatives, *Biochem. Biophys. Res. Commun. 209:* 66–72.
41. Suzui, M., Masuda, M., Lim, J.T., Albanese, C., Pestell, R.G., and Weinstein, I.B. (2002) Growth Inhibition of Human Hepatoma Cells by Acyclic Retinoid Is Associated with Induction of p21(CIP1) and Inhibition of Expression of Cyclin D1, *Cancer Res. 62:* 3997–4006.
42. Tibaduiza, E.C., Fleet, J.C., Russell, R.M., and Krinsky, N.I. (2002) Excentric Cleavage Products of β-Carotene Inhibit Estrogen Receptor Positive and Negative Breast Tumor Cell Growth *In Vitro* and Inhibit Activator Protein-1-Mediated Transcriptional Activation, *J. Nutr. 132:* 1368–1375.
43. Butler, R., Mitchell, S.H., Tindall, D.J., and Young, C.Y. (2000) Nonapoptotic Cell Death Associated with S-Phase Arrest of Prostate Cancer Cells *Via* the Peroxisome Proliferator-Activated Receptor Gamma Ligand, 15-Deoxy-delta12,14-prostaglandin J_2, *Cell. Growth Differ. 11:* 49–61.
44. Kubota, T., Koshizuka, K., Williamson, E.A., Asou, H., Said, J.W., Holden, S., Miyoshi, I., and Koeffler, H.P. (1998) Ligand for Peroxisome Proliferator-Activated Receptor Gamma (Troglitazone) Has Potent Antitumor Effect Against Human Prostate Cancer Both *In Vitro* and *In Vivo*, *Cancer Res. 58:* 3344–3352.
45. Mueller, E., Smith, M., Sarraf, P., Kroll, T., Aiyer, A., Kaufman, D.S., Oh, W., Demetri, G., Figg, W.D., Zhou, X.P., Eng, C., Spiegelman, B.M., and Kantoff, P.W. (2000) Effects of Ligand Activation of Peroxisome Proliferator-Activated Receptor Gamma in Human Prostate Cancer, *Proc. Natl. Acad. Sci. USA 97:* 10990–10995.

46. Jaiswal, A.K. (2000) Regulation of Genes Encoding NAD(P)H:Quinone Oxidoreductases, *Free Radic. Biol. Med. 29:* 254–262.
47. Venugopal, R., and Jaiswal, A.K. (1996) Nrf1 and Nrf2 Positively and c-Fos and Fra1 Negatively Regulate the Human Antioxidant Response Element-Mediated Expression of NAD(P)H:Quinone Oxidoreductase1 Gene, *Proc. Natl. Acad. Sci. USA 93:* 14960–14965.
48. Xie, W., Barwick, J.L., Simon, C.M., Pierce, A.M., Safe, S., Blumberg, B., Guzelian, P.S., and Evans, R.M. (2000) Reciprocal Activation of Xenobiotic Response Genes by Nuclear Receptors SXR/PXR and CAR, *Genes Dev. 14:* 3014–3023.
49. Xie, W., and Evans, R.M. (2001) Orphan Nuclear Receptors: The Exotics of Xenobiotics, *J. Biol. Chem. 276:* 37739–37742.
50. Seo, J.Y., Kim, H., Seo, J.T., and Kim, K.H. (2002) Oxidative Stress Induced Cytokine Production in Isolated Rat Pancreatic Acinar Cells: Effects of Small-Molecule Antioxidants, *Pharmacology 64:* 63–70.
51. Mueller, E., Sarraf, P., Tontonoz, P., Evans, R.M., Martin, K.J., Zhang, M., Fletcher, C., Singer, S., and Spiegelman, B.M. (1998) Terminal Differentiation of Human Breast Cancer Through PPAR Gamma, *Mol. Cell 1:* 465–470.
52. Demetri, G.D., Fletcher, C.D., Mueller, E., Sarraf, P., Naujoks, R., Campbell, N., Spiegelman, B.M., and Singer, S. (1999) Induction of Solid Tumor Differentiation by the Peroxisome Proliferator-Activated Receptor-Gamma Ligand Troglitazone in Patients with Liposarcoma, *Proc. Natl. Acad. Sci. USA 96:* 3951–3956.
53. Sarraf, P., Mueller, E., Smith, W.M., Wright, H.M., Kum, J.B., Aaltonen, L.A., de la Chapelle, A., Spiegelman, B.M., and Eng, C. (1999) Loss-of-Function Mutations in PPAR Gamma Associated with Human Colon Cancer, *Mol. Cell 3:* 799–804.
54. Takahashi, N., Kawada, T., Goto, T., Yamamoto, T., Taimatsu, A., Matsui, N., Kimura, K., Saito, M., Hosokawa, M., Miyashita, K., and Fushiki, T. (2002) Dual Action of Isoprenols from Herbal Medicines on Both PPARγ and PPARα in 3T3-L1 Adipocytes and HepG2 Hepatocytes, *FEBS Lett. 514:* 315–322.
55. Ramos-Gomez, M., Kwak, M.K., Dolan, P.M., Itoh, K., Yamamoto, M., Talalay, P., and Kensler, T.W. (2001) Sensitivity to Carcinogenesis Is Increased and Chemoprotective Efficacy of Enzyme Inducers Is Lost in Nrf2 Transcription Factor-Deficient Mice, *Proc. Natl. Acad. Sci. USA 98:* 3410–3415.
56. Kim, Y.C., Masutani, H., Yamaguchi, Y., Itoh, K., Yamamoto, M., and Yodoi, J. (2001) Hemin-Induced Activation of the Thioredoxin Gene by Nrf2. A Differential Regulation of the Antioxidant Responsive Element by a Switch of Its Binding Factors, *J. Biol. Chem. 276:* 18399–18406.
57. Kwak, M.K., Egner, P.A., Dolan, P.M., Ramos-Gomez, M., Groopman, J.D., Itoh, K., Yamamoto, M., and Kensler, T.W. (2001) Role of Phase 2 Enzyme Induction in Chemoprotection by Dithiolethiones, *Mutat. Res. 480–481:* 305–315.
58. Kong, A.N., Owuor, E., Yu, R., Hebbar, V., Chen, C., Hu, R., and Mandlekar, S. (2001) Induction of Xenobiotic Enzymes by the Map Kinase Pathway and the Antioxidant or Electrophile Response Element (ARE/EpRE), *Drug Metab. Rev. 33:* 255–271.
59. Gradelet, S., Astorg, P., Leclerc, J., Chevalier, J., Vernevaut, M.F., and Siess, M.H. (1996) Effects of Canthaxanthin, Astaxanthin, Lycopene and Lutein on Liver Xenobiotic-Metabolizing Enzymes in the Rat, *Xenobiotica 26:* 49–63.
60. Bhuvaneswari, V., Velmurugan, B., Balasenthil, S., Ramachandran, C.R., and Nagini, S. (2001) Chemopreventive Efficacy of Lycopene on 7,12-Dimethylbenz[a]anthracene-Induced Hamster Buccal Pouch Carcinogenesis, *Fitoterapia 72:* 865–874.

61. Denison, M.S., and Nagy, S.R. (2003) Activation of the Aryl Hydrocarbon Receptor by Structurally Diverse Exogenous and Endogenous Chemicals, *Annu. Rev. Pharmacol. Toxicol. 43:* 309–334.
62. Astorg, P., Berges, R., and Suschetet, M. (1994) Induction of Gamma GT- and GST-P Positive Foci in the Liver of Rats Treated with 2-Nitropropane or Propane 2-Nitronate, *Cancer Lett. 79:* 101–106.
63. Astorg, P., Gradelet, S., Leclerc, J., Canivenc, M.C., and Siess, M.H. (1994) Effects of Beta-Carotene and Canthaxanthin on Liver Xenobiotic-Metabolizing Enzymes in the Rat, *Food Chem. Toxicol. 32:* 735–742.
64. Gradelet, S., Leclerc, J., Siess, M.H., and Astorg, P.O. (1996) β-Apo-8′-carotenal, But Not β-Carotene, Is a Strong Inducer of Liver Cytochromes P4501A1 and 1A2 in Rat, *Xenobiotica 26:* 909–919.
65. Astorg, P., Gradelet, S., Leclerc, J., and Siess, M.H. (1997) Effects of Provitamin A or Non-Provitamin Carotenoids on Liver Xenobiotic-Metabolizing Enzymes in Mice, *Nutr. Cancer 27:* 245–249.
66. Gradelet, S., Astorg, P., Pineau, T., Canivenc, M.C., Siess, M.H., Leclerc, J., and Lesca, P. (1997) Ah Receptor-Dependent CYP1A Induction by Two Carotenoids, Canthaxanthin and β-Apo-8′-carotenal, with No Affinity for the TCDD Binding Site, *Biochem. Pharmacol. 54:* 307–315.
67. Jewell, C., and O'Brien, N.M. (1999) Effect of Dietary Supplementation with Carotenoids on Xenobiotic Metabolizing Enzymes in the Liver, Lung, Kidney and Small Intestine of the Rat, *Br. J. Nutr. 81:* 235–242.
68. Breinholt, V., Lauridsen, S.T., Daneshvar, B., and Jakobsen, J. (2000) Dose-Response Effects of Lycopene on Selected Drug-Metabolizing and Antioxidant Enzymes in the Rat, *Cancer Lett. 154:* 201–210.
69. Le Marchand, L., Franke, A.A., Custer, L., Wilkens, L.R., and Cooney, R.V. (1997) Lifestyle and Nutritional Correlates of Cytochrome CPY1A2 Activity: Inverse Associations with Plasma Lutein and Alpha-Tocopherol, *Pharmacogenetics 7:* 11–19.

Chapter 17

Vitamin A in Health and Disease in Developing Countries

Machteld van Lieshout[a] and Clive E. West[b,*]

[a]Department of Nutrition, North-West University, Potchefstroom Campus, Potchefstroom, South Africa and [b]Department of Human Nutrition, Wageningen University, Wageningen, The Netherlands and the Department of Gastroenterology and Hepatology, University Medical Center St. Radboud, Nijmegen, The Netherlands

Introduction

Worldwide, 150 million children <5 y old (32%) are malnourished, i.e., they have an insufficient energy and protein intake (1) and 243 million adults, mainly women of child-bearing age, are severely undernourished, i.e., they have a BMI <17 kg/m^2 (2). As in the proverb "Ein Unglück selten allein," malnutrition seldom comes alone. Of the 10 million children dying every year, ~56% die from infectious diseases and in >50% of these cases, the underlying cause of death is malnutrition (3). The causes of malnutrition in turn are rooted in poverty. In the 1990s, UNICEF developed a conceptual framework showing that the causes of malnutrition are multisectoral and multilevel. At the root, at the societal level, there are basic causes related to the quantity and quality of actual resources and their management, and the potential resources, which are sometimes blocked by political, cultural, religious, economic, or social systems. At the household or family level, there are underlying causes of malnutrition such as insufficient access to food, inadequate maternal and child-care practices, poor water/sanitation, and inadequate health services. At the individual level, there are immediate causes of malnutrition such as disease, inadequate dietary intake, and the interaction between these two factors (4). Here we will focus on these immediate causes and their link with health and disease. We will do so using examples from vitamin A nutrition and metabolism, but many of the principles could be applied also to other nutrients.

Vitamin A Deficiency

"Hidden hunger" refers to the lack of micronutrients in the diet of many people in developing countries; it is much less visible than hunger caused by a plain lack of food and is an even larger problem than the protein and energy malnutrition mentioned above. Over 2 billion women and children worldwide suffer from micronu-

*In memory of C.E. West, who died August 27, 2004.

trient malnutrition such as deficiencies in vitamin A, folate, iodine, iron, and zinc. Vitamin and mineral deficiencies are estimated to cost some countries the equivalent of >5% of their gross national product in lost lives, physical and mental disability, and lost productivity (4). Worldwide, ~127 million children <5 y old and >7.2 million women of child-bearing age suffer from vitamin A deficiency (serum retinol concentration <0.70 μmol/L) (5). When looking at the immediate causes only, a nutrient deficiency can be explained as an imbalance between need and intake of a nutrient, or more specifically by inadequate ingestion, absorption, or utilization and increased excretion or requirement (6). This imbalance often results in people entering a vicious circle of malnutrition and infection (7) whereby malnutrition increases susceptibility to infections, and prolonged periods of infection aggravate malnutrition. Here, we have visualized the various elements that play a role in this balance (see Fig. 17.1). In the text, we will explain step-by-step all of the terms used in this figure. Differences during health and disease will be exemplified. This breakdown of nutrient intake into its various components is important because it helps to clarify at which levels interventions to improve vitamin A status should be targeted to have maximum effect. At the end of this chapter, we will briefly summarize successful intervention strategies employed over the past decade and possible interventions to be employed in the future to reach the goal of eliminating vitamin A deficiency. An adequate vitamin A status will enable vitamin A to perform its role in improving health and combating disease as discussed in this chapter.

Vitamin A Intake: Food Consumption, Nutrient Content, and Bioefficacy

Food and Pharmanutrient Consumption

Vitamin A intake is determined by many factors, but the contribution of each factor is poorly understood (8). Sources of vitamin A are unmodified foods rich in preformed vitamin A (retinol) and provitamin A carotenoids that have to be converted into retinol in the body before they can perform their functions as vitamin A; modified foods such as fortified and enriched foods; and pharmanutrients. The term "pharmanutrient" was introduced to describe a pharmaceutical preparation of a nutrient because the definition of "supplement" extends beyond pharmanutrient to food provided in addition to the normal diet. Breast milk and animal products such as liver, milk, butter, and eggs are good sources of retinol. Plant foods do not contain retinol but some are good sources of provitamin A carotenoids. Plant foods such as red palm oil, dark-green leafy vegetables, yellow and orange fruits, and red/orange roots and tubers are rich in the provitamin A carotenoids, mainly β-carotene, α-carotene, and β-cryptoxanthin. Margarine and sugar fortified with retinyl palmitate are examples of modified foods that have contributed to the elimination of vitamin A deficiency in some parts of the world. In the near future, biofortified plant foods, such as orange-fleshed sweet potato, maize, rice, and wheat fortified with β-carotene through plant breeding or

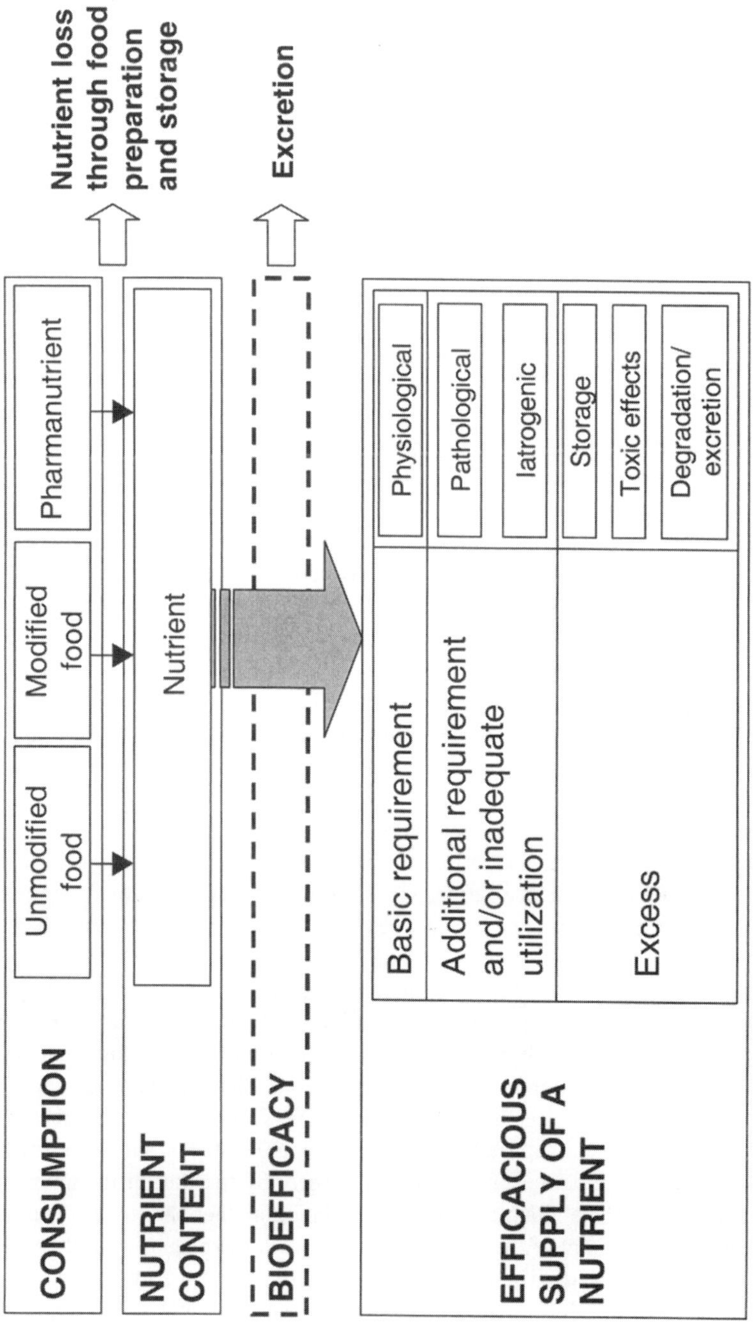

Fig. 17.1. Balance between the supply of nutrients and requirements. See text for definitions and examples.

genetic engineering, may become important sources of vitamin A for population groups now consuming inadequate amounts of vitamin A (9). The type of foods consumed by various population groups may sometimes be culturally or religiously determined, but for the majority of people in developing countries, this choice is unfortunately determined economically. Foods rich in preformed vitamin A, such as animal products, are usually more expensive than plant foods. As a consequence, plant foods are the major source of vitamin A in the diet of many people living in developing countries. Many diseases also have an effect on food consumption patterns. Bacterial, viral, and parasitic infections common in developing countries, for example, pneumonia, HIV/AIDS, and *Ascaris lumbricoides*, cause appetite loss or even anorexia (10,11).

Vitamin A Content

The next apparent factor that affects overall vitamin A intake is the actual vitamin A content of the food consumed. For plant foods, a variety of factors along the food chain affect the final provitamin A content. At the food production end of the food chain, aspects such as cultivar, maturity, and duration of ripening determine the provitamin A carotenoid content (12). Stored at the right temperature, the carotenoid content on a dry weight basis of some fruits, vegetables, and roots, but generally not of leafy vegetables, can increase substantially postharvest (13). During storage, particularly at or above ambient temperatures but also in freezers, considerable loss of carotenoids may occur (14,15). In many developing countries, drying of fruits and vegetables is a common preservation technique. Sun-drying was found to decrease the carotenoid content of various fruits and vegetables substantially, 38–79% (16); for paprika, the losses are >80% (15). During solar-drying, when food is protected from direct sunlight, losses of carotenoids are generally lower than during sun-drying (13) but not all studies have found this (17). Food processing and preparation may affect the carotenoid content positively or negatively. For example, Edwards and colleagues (18) raw-chopped, boiled-mashed, or puréed carrots from the same batch and found that there was a 7 and 28% decrease in the amount of β-carotene on a dry weight basis compared with the raw-chopped carrots during mashing and puréeing, respectively. Various carotenoids respond differently to food processing. Khachik and colleagues (19) found that most carotenoids that play an important role in the human diet are stable during heat treatments such as steaming, boiling, and microwaving, whereas considerable amounts of epoxycarotenoids, such as violaxanthin, are lost. For animal foods and breast milk, the retinol content is determined largely by the diet of the animal or mother, respectively. The vitamin A status of lactating mothers is strongly correlated with the retinol content of their breast milk and with the plasma retinol concentration of their infants (20). The effect of various diseases on the retinol content of breast milk are not well documented, but it can be assumed that diseases that lower serum retinol concentrations also lower retinol concentrations in breast milk.

Similarly, supplementation of lactating mothers increases their vitamin A status, the retinol concentration of the breast milk, and the serum retinol concentration of the infant (21–23). The retinol content of food can also be affected during storage; for example, direct sunlight destroys retinol in milk.

Bioavailability, Bioconversion, and Bioefficacy

In addition to food consumption and nutrient content, a number of factors influence the actual fraction of an ingested nutrient, e.g., provitamin A carotenoids, that is absorbed and converted to the active form of that nutrient, e.g., retinol, in the body, i.e., the bioefficacy (24). If a compound is not converted to another compound before it becomes active as a nutrient, bioefficacy and bioavailability are in fact identical, then, because they both refer to the fraction of an ingested nutrient that is available for utilization in normal functions and for storage (25). Bioconversion refers to the fraction of an ingested nutrient that is converted to the active form of that nutrient. It has long been assumed that the bioavailability of retinol was rather high and constant. From the scarce literature available on this topic, Blomhoff and colleagues estimated absorption of retinol in rats and humans to be 50–75% for physiologic doses and even lower doses when sufficient fat was consumed concomitantly (26,27). In two studies from our laboratory, using the CarRetPIE method, a stable isotope technique based on β-***car***otene and ***ret***inol in serum reaching a ***p***lateau of ***i***sotopic ***e***nrichment during prolonged intake of multiple low doses of β-carotene and retinol, each specifically labeled with 10 ^{13}C-atoms, we found in some children that the bioavailability of β-carotene in oil was higher than the bioavailability of retinol in oil (24,28). We therefore hypothesize that a number of the factors that affect the bioefficacy and bioavailability of β-carotene also affect the bioavailability of retinol. These factors were listed in the acronym SLAMENGHI (29,30) where S) denotes the species of carotenoids, L) the molecular linkage, A) the amount of carotenoids or retinol consumed in a meal, M) the matrix in which the carotenoids or retinol are imbedded, E) the effectors of absorption of carotenoids or retinol and bioconversion of provitamin A carotenoids, N) the nutrient status of the host, G) the genetic factors, H) the host-related factors, and I) the mathematical interactions among factors. Over the past decade a number of other reviews have been written on all (29–36) or a selection (37–40) of factors known to affect the bioavailability, bioconversion, and bioefficacy of carotenoids. The first five factors, SLAME, refer to properties of the nutrients and the food and diet consumed, whereas the next three factors, NGH, refer to properties of the host and may thus be affected during certain conditions or diseases. Therefore, we will focus here on these three factors (see Table 17.1).

Efficacious Supply of Vitamin A

The product of food consumption, nutrient content, and bioefficacy determines the "efficacious supply of a nutrient," by which we mean the amount of nutrient, in

TABLE 17.1
Effect of Certain Conditions or Diseases on Factors Known to Affect the Bioefficacy of Carotenoids and the Bioavailability of Retinol

Factor and condition/disease	Possible route of effect
Nutrient status of the host	
Vitamin A status	Inverse correlation between vitamin A status and bioefficacy of β-carotene probably through feedback mechanism of intestinal β-carotene 15,15′-dioxygenase activity (41,42).
Zinc status	Positive correlation between zinc status and bioefficacy of β-carotene either through effect of zinc on absorption of β-carotene (43) or through role of zinc in enzymes involved in the conversion of β-carotene (44) or of zinc in the synthesis of retinol-binding protein.
Protein status	Low protein status reduces intestinal β-carotene 15,15′-dioxygenase activity (45). Sufficient protein status also required for synthesis of retinol-binding protein and to ensure normal carotenoid and retinol metabolism.
Genetic factors	Some people are unable to convert β-carotene to retinol, likely the result of a genetic defect (46). Others may suffer from inherited fat malabsorption, which may result in low bioefficacy of carotenoids and low bioavailability of retinol.
Host-related factors	
Gender	An effect of gender on bioefficacy is often suggested because serum response after oral administration of β-carotene is sometimes higher in women than in men (47). However, this effect may be due to the effect of female hormones on lipid and lipoprotein metabolism (48).
Age	Bioefficacy may decline with age through the increasing burden of atrophic gastritis, which is common in ~20% of elderly (32), but reduced bioefficacy with age was not found in all studies (49).
Pregnancy/lactation	The effects of pregnancy and lactation on carotenoid bioefficacy and vitamin A bioavailability are not well studied. An ongoing study in lactating women in Indonesia by Firmansah and West using the CarRetPIE method will soon elucidate this topic.
Gastrointestinal disorders	Gastrointestinal infections such as *Helicobacter pylori* and parasites, such as *Giardia lamblia, Ascaris lumbricoides,* and hookworm, can cause maldigestion, malabsorption, and excessive loss of gut epithelium. Whether the effect of parasites is through changes in mucosa (50), decreased fat absorption (51–53), because the parasites consume the nutrients (54), or through a combination of these mechanisms is still poorly understood. Pathogens that cause diarrhea and/or fat malabsorption most certainly lower bioefficacy and bioavailability, but these effects have not yet been quantified.
Infections	Many generalized infections induce malnutrition through infection-induced anorexia (55), refusal to take food because of stomatitis, mouth ulcerations, *Candida albicans* or dysphagia (56), increased utilization (57), increased excretion in urine (58,59), redistribution of nutrients (60), loss of nutrients through diarrhea and/or vomiting. As described above, this impaired nutrient status subsequently can alter carotenoid bioefficacy and retinol bioavailability either positively or negatively.

this case vitamin A, that is capable of producing all of the effects of that nutrient in the body. This "efficacious supply of a nutrient" can be applied to meet basic requirements, additional requirements, and/or increased utilization, storage and/or excretion (see lower half of Fig. 17.1).

Vitamin A Requirements

Vitamin A plays an important role in many processes in the body, as witnessed by the fact that retinoid receptors have been found in almost all types of human cells of human beings. Many processes in which vitamin A is involved are described elsewhere in this book and in books by Krinsky and colleagues (61) and Blomhoff (62). Some diseases common in developing countries result in increased vitamin A requirements. If these effects are specifically through factors known to affect the bioefficacy of carotenoids or the bioavailability of vitamin A, they are mentioned in Table 17.1. However, if a disease affects vitamin A requirements, it is discussed here. The exact causes of the increased requirements are sometimes unknown, in which case, proven beneficial effects of increased efficacious supply of vitamin A, primarily through supplementation, may be used as circumstantial evidence for increased requirements.

Pregnancy and Lactation. The transfer of vitamin A to the fetus and breast milk increases maternal vitamin A requirements during pregnancy and lactation (63). In Nepal, night-blind pregnant women were five times more likely to die from infections than were women who were not night-blind (64). Vitamin A may improve maternal immunity and therefore decrease the risk of bacterial and viral infections. Supplementation with vitamin A (retinol and β-carotene) was shown to reduce self-reported prevalence of diarrhea, symptoms of tuberculosis, and incidence of prolonged labor by approximately 10–30% in pregnant women with night blindness (65). Supplementation with vitamin A and iron during pregnancy was shown to improve maternal hematological status even more than iron supplementation alone (66).

Infant and Child Morbidity and Mortality. A third of all child deaths occur in the first 28 d of life (3). Maternal supplementation with vitamin A during pregnancy does not seem to lower neonatal mortality or mortality in the first 6 mo of life (67) nor morbidity during y 1 of life (68). Maternal supplementation with vitamin A during lactation was shown to increase the retinol concentration of breast milk and infant serum (69) (see also the section on vitamin A content), but studies on its effect on morbidity or mortality in infants are scarce. A study by Filteau and colleagues (70) showed that it is unlikely that beneficial effects of supplementation during lactation are obtained through an increased concentration of immune factors in breast milk. The next best thing would appear to be supplementing the infants themselves; because they are born with very low vitamin A stores, their requirements during the first 6 mo of life are very high. Part of their efficacious supply of vitamin A is used to build up their liver vitamin A stores. However, studies on supplementation of infants have been inconclusive to date. Supplementation of infants within 48 h of birth was shown to reduce mortality by 22–64% (71,72), whereas studies in which supplements were given 3 wk postpartum did not find a reduction in mortality (73,74). These latter findings are in contrast to the impressive findings in the early 1990s of a meta-analysis of eight large trials, which established that vitamin A supplementation of 6- to 59-mo-old children living in areas in which vitamin A deficiency is endemic can reduce their risk of dying by an average of 23% (75). This effect is mainly *via* a reduction in deaths attributed to diarrhea and measles. Deaths attributed to acute respiratory diseases, such as pneumonia, are not reduced. Subsequent studies have confirmed this (76,77) and unless children are clinically vitamin A deficient, adverse effects may even arise from vitamin A supplementation during non-measles pneumonia (78–80). These seemingly paradoxical findings are discussed extensively elsewhere (81); therefore these acute respiratory infections are not further discussed here.

Diarrhea. More than one fifth of all child deaths can be attributed to diarrhea (3), 88% of which are due to the ingestion of unsafe water, inadequate availability of water for hygiene, and lack of access to sanitation (82). Exclusive breast-feeding during the first 6 mo of life can significantly reduce these risks (83,84). Nonexclusive breast-feeding during the next 6 mo of life can even prevent some deaths due to diarrhea (85). Malnutrition in general (86) and vitamin A deficiency in particular (87) were found to increase the risk of developing diarrhea two- to threefold. The duration of diarrheal episodes is also two- to threefold longer in malnourished individuals (86). These effects are probably due to depressed immunity during malnutrition and vitamin A deficiency. The diarrhea in turn predisposes individuals to malnutrition (86) and vitamin A deficiency (87). This latter effect may be explained, in part, by an increased urinary loss of vitamin A during acute diarrhea, especially that due to rotavirus and in the presence of fever (59). Therefore, vitamin A supplementation is currently advised as a safe, low-cost, preventive intervention for diarrhea (88). Clinical trials failed to clearly demonstrate a therapeutic benefit of vitamin A supplementation on the current diarrhea episode (76,89), although vitamin A supplementation during diarrhea may still increase liver stores of vitamin A and thus possibly provide a beneficial effect for subsequent episodes of diarrhea and other infections. We hypothesize that the more clearly and unequivocally observed beneficial effects of therapeutic zinc supplementation during diarrhea (90) may be due to the fact that the bioavailability of zinc is less affected by diarrhea than is the bioavailability of vitamin A. The efficacious supply of vitamin A in the above-mentioned clinical trials was probably lower than the minimum dose required for vitamin A to have an effect. Therefore, food fortification, providing multiple lower doses of vitamin A, may well be more effective in reducing diarrhea-related morbidity and mortality than vitamin A supplementation.

Measles. Mass campaigning rapidly reduced measles incidence to near zero in the Americas (91) and in Southern Africa in the second half of the 1990s (92). Similar successes were reported recently for West Africa (93). The goal of >50% reduction in measles mortality by 2005 seems achievable, which will also considerably decrease the number of infants and children disabled as a consequence of complications of measles, such as chronic lung disease, malnutrition, blindness due to vitamin A deficiency, deafness, and recurrent infections (94). It was recently estimated that 1% of all child deaths can be attributed to measles (3); most of these deaths occur in Africa, especially in high-risk groups such as young infants, children who are immunocompromised (including those with HIV or AIDS), migrants, refugees, or those with severe malnutrition. The 1% of child deaths due to measles may not seem that high, but is more than that seen with other vaccine-preventable diseases. Although there is no specific therapy for measles, the adverse consequences can be reduced by treatment with vitamin A (88), by providing good medical care, and through the treatment of complications. In 1932, Ellison reported the therapeutic effect of vitamin A in the treatment of measles (95). Later, a meta-analysis of four hospital-based studies showed that vitamin A therapy reduced measles-related mortality by an impressive 67% (96). Morbidity was reduced to a similar extent. Another meta-analysis showed that prophylactic vitamin A supplementation reduced measles-related mortality by 36% (75). Vitamin A supplementation seems to reduce the immunosuppressive effects of measles, thereby reducing the infectious complications such as pneumonia and diarrhea. This negative spiral of measles/infection and malnutrition in general (97) and vitamin A deficiency in particular (98) was summarized by Scrimshaw and colleagues in a monograph in 1968 (7). A more recent review suggested a less important role of general poor nutritional status as a risk factor for measles (99), whereas others have confirmed that vitamin A deficiency is a risk factor for measles (100).

Malaria. Of all child deaths, 9% can be attributed to malaria (3). The majority of these deaths occur in Sub-Saharan Africa. In addition to children, mainly pregnant women are at high risk for malaria. It has long been thought that this was because these two groups often have a poor nutritional status, which would increase susceptibility to infection by the malaria parasite, of which *Plasmodium falciparum* is the most deadly species that infects humans. In the 1970s, however, the hypothesis was launched that malnourished children were less susceptible to malaria infection, morbidity, and mortality. Subsequent animal studies appeared to support this hypothesis. A recent review of these studies and of more recent data from studies in humans led to the conclusion that malnutrition exacerbates malaria and increases the risk of mortality (101). This conclusion was confirmed by a community-based cluster survey among Kenyan children aged 2–36 mo, who were asymptomatic for malaria or anemia, which showed that children with malaria and those who were stunted suffered from more severe anemia and had higher serum concentrations of C-reactive protein and serum transferrin receptor than would be expected from the combined effect of the two working independently (102). As with other infectious diseases discussed elsewhere in this chapter, malaria in turn strongly affects nutritional status and may thus affect the bioavailability of nutrients, thereby further worsening the nutritional status and increasing the susceptibility to malaria of infected individuals. A recent double-blind, placebo-controlled trial in Papua, New Guinea, showed that vitamin A supplementation reduced the frequency of *P. falciparum* episodes by 30% among preschool children (103). A cross-sectional study on the dietary intake of β-carotene–rich red palm oil and the severity of malaria found a weak negative association, which was ascribed to the relatively adequate vitamin A status of these Nigerian preschool children (104).

HIV/AIDS. Of all child deaths, 3% can be attributed to AIDS (3). At the end of 2003, ~2.5 million children <15 y old and 37 million adults worldwide were living with HIV/AIDS, and 3 million of all children and adults with HIV/AIDS died in 2003 (105). The majority of HIV-infected people live in Sub-Saharan Africa (26.6 million) as did the majority of those who died from AIDS (2.3 million). There are two related theories linking micronutrient status in general and vitamin A status in particular with the pathogenesis of HIV/AIDS: (i) the free radical theory, in which reactive oxygen intermediates are generated during the killing of microorganisms by activated macrophages and neutrophils, and these intermediates are balanced by the antioxidant defense system; and (ii) the nutritional immunologic theory, in which micronutrients, such as vitamins A and E and zinc are involved in normal immune function, such as the growth and function of T and B cells, antibody responses and in which vitamin A is also involved in maintenance of the mucosal epithelia. During HIV/AIDS, food consumption may be decreased due to loss of appetite (55), aversion to food, dysphagia, nausea, and vomiting (56). Fat malabsorption is common in all stages of infection. This and the increased prevalence of diarrhea are likely to lower the incorporation of β-carotene and vitamin A into mixed micelles. Changes in the metabolism in liver, pancreas, and intestinal wall may affect the production of enzymes and other factors involved in the breakdown of the food matrix, and thus in the release of β-carotene, which may reduce the absorption of β-carotene. Production of β-carotene-15,15′-mono- and dioxygenase, enzymes responsible for the bioconversion of β-carotene, may also be affected. Altogether, the bioefficacy of carotenoids and the bioavailability of vitamin A will be much lower in people living with HIV/AIDS than in healthy individuals. However, the extent of the reduction has not yet been quantified. Increased urinary excretion of vitamin A due to infection and diarrhea (59) will further contribute to the higher vitamin A requirements of people living with HIV/AIDS. Although one study reported a reduction in the risk of transmission in preterm births (106), it is now firmly believed that vitamin A supplementation cannot be recommended as an intervention to decrease mother-to-child-transmission of HIV (107). In general, vitamin A supplementation may be of some benefit to HIV-infected children and pregnant women in developing countries (108–110), but it is now generally believed that supplementation with multiple micronutrients is required to improve the health of people living with HIV/AIDS (111–113).

Tuberculosis. It is estimated that 2 million people died from tuberculosis in 2002. This estimate includes people living with HIV/AIDS, indicating that a large number of these deaths occurred in Africa. In absolute figures, the incidence of tuberculosis is highest in South-East Asia. A case-control study of the nutritional status of patients with active tuberculosis and their uninfected neighbors in Indonesia showed that significantly more patients were anemic and more had low plasma concentrations of retinol and zinc (114). As discussed above, these factors affect vitamin A requirements through their effects on bioefficacy of carotenoids and bioavailability of vitamin A. Contrary to a study in South African children in which some negative effects of two consecutive doses of vitamin A supplementation on clinical response to antituberculosis treatment were observed (115), a subsequent double-blind, placebo-controlled intervention study in Indonesia, in which newly diagnosed tuberculosis patients received daily vitamin A and zinc or placebo for 6 mo, in addition to standard antituberculosis treatment, showed that especially during the first 2 mo, this supplementation was of great benefit for improving the effectiveness of the antituberculosis treatment and for increasing plasma retinol concentrations (116). Before the emphasis on chemotherapy for the treatment of tuberculosis, cod-liver oil was long used (117,118) to compensate for the decreased dietary intake and bioefficacy and the increased demand for and urinary losses of vitamin A during tuberculosis.

Fate of Vitamin A in Excess of Requirements

Nutrients in excess of the requirements can be stored, excreted, or in some cases, these nutrients may have toxic effects. The toxic (119) and teratogenic (120) effects of high doses of vitamin A, *via* disturbed differentiation, are well documented. β-Carotene does not exert such effects. In healthy individuals, the liver contains ~80–90% of the total body stores of vitamin A, most of which is in the form of retinyl esters. Factors responsible for excretion of large amounts of nutrients were described above.

Possible Intervention Strategies

A two-pronged approach is required to control micronutrient malnutrition: increasing the efficacious supply of vitamin A and reducing the body's vitamin A demand, e.g., by controlling infection. Some examples of the latter are immunization and improved personal hygiene, which were shown to improve vitamin A status effectively (2,88), but development of successful public health programs is a long-term process. Many examples of improving the efficacious supply of vitamin A were mentioned above. Food-based approaches include increasing the use of available foods such as through the promotion of small indigenous fish rich in vitamin A (121). Food-based approaches also include what is generally known as "food fortification" but more appropriately as nutrification that encompasses fortification, enrichment, restoration, and addition of nutrients to substitute foods (e.g., adding retinyl palmitate to margarine to give it a vitamin A content similar to that of butter). In North America and many European countries, vitamin A deficiency was kept at bay by the introduction in the late 1930s of vitamin A fortification, especially of margarine. Similar successes were obtained in Guatemala with the fortification of sugar (122) and in other countries with other food vehicles. Biofortification is also a form of nutrification. It includes not only increasing the nutrient content of foods *via* conventional breeding or transgenic methods, but also altering the content of effectors to improve bioefficacy or bioavailability of nutrients (9). In 2003, Harvest Plus, a CGIAR-initiative, funded in part by the Bill & Melinda Gates Foundation, was launched with the aim of breeding nutrient-rich staple food crop varieties, demonstrating their effect on human nutrition and distributing them to those most at risk of being micronutrient deficient. In addition to food-based approaches, pharmanutrient approaches such as massive dosing and targeted supplementation programs have proven effective, but sometimes expensive tools in reducing vitamin A deficiency.

Conclusions

Considering the many functions of vitamin A in disease, as discussed here, it is not surprising that the U.S. Institute of Medicine warned that its most recent conversion factors for the vitamin A activity of β-carotene may differ for malnourished individuals or for those with a certain disease/condition (123). This would support findings from Asia (124–126) that, instead of 12 μg, as suggested by the IOM, 21 μg of β-carotene in a mixed diet has the same vitamin A activity as 1 μg retinol (127).

With collective efforts at international, national and community levels, ending malnutrition is both a credible and achievable goal (2).

Acknowledgments

We thank Alida Melse-Boonstra at the Department of Human Nutrition of Wageningen University for her advice on the manuscript in general and Figure 17.1 in particular.

References

1. UNICEF (2001) *Progress Since the World Summit for Children: A Statistical Review*, pp. 4–5, Oxford University Press, Oxford.
2. ACC/SCN Commission on the Nutrition Challenges of the 21st Century (2000) *Ending Malnutrition by 2020: An Agenda for Change in the Millenium*, pp. i–xvi, 71–77, ACC/SCN, Geneva.
3. Black, R.E., Morris, S.S., and Bryce, J. (2003) Where and Why Are 10 Million Children Dying Every Year? *Lancet 361:* 2226–2234.

4. UNICEF (1998) *The State of the World's Children 1998*, pp. 7–90, Oxford University Press, Oxford.
5. West, K.P. (2002) Extent of Vitamin A Deficiency Among Preschool Children and Women of Reproductive Age, *J. Nutr. 132:* 2857S–2866S.
6. Herbert, V. (1973) The Five Possible Causes of All Nutrient Deficiency: Illustrated by Deficiencies of Vitamin B 12, *Am. J. Clin. Nutr. 26:* 77–86.
7. Scrimshaw, N.S., Taylor, C.E., and Gordon, J.E. (1968) *Interactions of Nutrition and Infection*, WHO, Geneva.
8. West, C.E., and Castenmiller, J.J.M. (1998) Quantification of the "SLAMENGHI" Factors for Carotenoid Bioavailability and Bioconversion, *Int. J. Vitam. Nutr. Res. 68:* 371–377.
9. Bouis, H.E., Chassy, B.A., and Ochanda, J.O. (2003) Genetically Modified Food Crops and Their Contribution to Human Nutrition and Food Quality, *Trends Food Sci. Technol. 14:* 191–209.
10. Hadju, V., Stephenson, L.S., Abadi, K., Mohammed, H.O., Bowman, D.D., and Parker, R.S. (1996) Improvements in Appetite and Growth in Helminth-Infected Schoolboys Three and Seven Weeks After a Single Dose of Pyrantel Pamoate, *Parasitology 113:* 497–504.
11. Chang, H.R., Dulloo, A.G., and Bistrian, B.R. (1998) Role of Cytokines in AIDS Wasting, *Nutrition 14:* 853–863.
12. Rodriguez-Amaya, D.B. (1999) *A Guide to Carotenoid Analysis in Foods*, pp. 1–13, ILSI, Washington.
13. Rodriguez-Amaya, D.B. (1997) *Carotenoids and Food Preparation: The Retention of Provitamin A Carotenoids in Prepared, Processed, and Stored Foods*, pp. 27–36, 39–52, OMNI/USAID, Washington.
14. Fish, W.W., and Davis, A.R. (2003) The Effects of Frozen Storage Conditions on Lycopene Stability in Watermelon Tissue, *J. Agric. Food Chem. 51:* 3582–3585.
15. Topuz, A., and Ozdemir, F. (2003) Influences of Gamma-Irradiation and Storage on the Carotenoids of Sun-Dried and Dehydrated Paprika, *J. Agric. Food Chem. 51:* 4972–4977.
16. Hedren, E., Mulokozi, G., and Svanberg, U. (2002) *In Vitro* Accessibility of Carotenes from Green Leafy Vegetables Cooked with Sunflower Oil or Red Palm Oil, *Int. J. Food Sci. Nutr. 53:* 445–453.
17. Pott, I., Marx, M., Neidhart, S., Muhlbauer, W., and Carle, R. (2003) Quantitative Determination of β-Carotene Stereoisomers in Fresh, Dried, and Solar-Dried Mangoes (*Mangifera indica* L.), *J. Agric. Food Chem. 51:* 4527–4531.
18. Edwards, A.J., Nguyen, C.H., You, C.S., Swanson, J.E., Emenhiser, C., and Parker, R.S. (2002) α- and β-Carotene from a Commercial Puree Are More Bioavailable to Humans than from Boiled-Mashed Carrots, as Determined Using an Extrinsic Stable Isotope Reference Method, *J. Nutr. 132:* 159–167.
19. Khachik, F., Goli, M.B., Beecher, G.R., Holden, J.M., Lusby, W.R., Tenorio, M.D., and Barrera, M.R. (1992) Effect of Food Preparation on Qualitative and Quantitative Distribution of Major Carotenoid Constituents of Tomatoes and Several Green Vegetables, *J. Agric. Food Chem. 40:* 390–398.
20. Dijkhuizen, M.A., Wieringa, F.T., West, C.E., Muherdiyantiningsih, and Muhilal (2001) Concurrent Micronutrient Deficiencies in Lactating Mothers and Their Infants in Indonesia, *Am. J. Clin. Nutr. 73:* 786–791.

21. Allen, L.H. (1994) Maternal Micronutrient Malnutrition: Effects on Breast Milk and Infant Nutrition, and Priorities for Intervention, *SCN News 11:* 21–24.
22. Canfield, L.M., Taren, D.L., Kaminsky, R.G., and Mahal, Z. (1999) Short-Term β-Carotene Supplementation of Lactating Mothers Consuming Diets Low in Vitamin A, *J. Nutr. Biochem. 10:* 532–538.
23. Canfield, L.M., Kaminsky, R.G., Taren, D.L., Shaw, E., and Sander, J.K. (2001) Red Palm Oil in the Maternal Diet Increases Provitamin A Carotenoids in Breastmilk and Serum of the Mother-Infant Dyad, *Eur. J. Nutr. 40:* 30–38.
24. van Lieshout, M., West, C.E., Muhilal, Permaesih, D., Wang, Y., Xu, X., van Breemen, R.B., Creemers, A.F.L., Verhoeven, M.A., and Lugtenburg, J. (2001) Bioefficacy of β-Carotene Dissolved in Oil Studied in Children in Indonesia, *Am. J. Clin. Nutr. 73:* 949–958.
25. Jackson, M.J. (1997) The Assessment of Bioavailability of Micronutrients: Introduction, *Eur. J. Clin. Nutr. 51 (Suppl. 1):* S1–S2.
26. Blomhoff, R., Green, M.H., and Norum, K.R. (1992) Vitamin A: Physiological and Biochemical Processing, *Annu. Rev. Nutr. 12:* 37–57.
27. Blomhoff, R., Green, M.H., Green, J.B., Berg, T., and Norum, K.R. (1991) Vitamin A Metabolism: New Perspectives on Absorption, Transport, and Storage, *Physiol. Rev. 71:* 951–990.
28. van Lieshout, M., West, C.E., Wang, Y., van Breemen, R.B., Permaesih, D., Muhilal, Verhoeven, M.A., Creemers, A.F.L., and Lugtenburg, J. (2001) in *Bioavailability and Bioefficacy of β-Carotene Measured Using ^{13}C-Labeled β-Carotene and Retinol: Studies in Indonesian Children*, van Lieshout, M., ed., pp. 49–72, Wageningen University, Wageningen.
29. de Pee, S., and West, C.E. (1996) Dietary Carotenoids and Their Role in Combating Vitamin A Deficiency: A Review of the Literature, *Eur. J. Clin. Nutr. 50 (Suppl. 3):* S38–S53.
30. Castenmiller, J.J.M., and West, C.E. (1998) Bioavailability and Bioconversion of Carotenoids, *Annu. Rev. Nutr. 18:* 19–38.
31. Borel, P. (2003) Factors Affecting Intestinal Absorption of Highly Lipophilic Food Microconstituents (Fat-Soluble Vitamins, Carotenoids and Phytosterols), *Clin. Chem. Lab. Med. 41:* 979–994.
32. Yeum, K.J., and Russell, R.M. (2002) Carotenoid Bioavailability and Bioconversion, *Annu. Rev. Nutr. 22:* 483–504.
33. Tanumihardjo, S.A. (2002) Factors Influencing the Conversion of Carotenoids to Retinol: Bioavailability to Bioconversion to Bioefficacy, *Int. J. Vitam. Nutr. Res. 72:* 40–45.
34. Olson, J.A. (1999) Bioavailability of Carotenoids, *Arch. Latinoam. Nutr. 49:* 21S–25S.
35. Parker, R.S. (1996) Absorption, Metabolism, and Transport of Carotenoids, *FASEB J. 10:* 542–551.
36. Williams, A.W., Boileau, T.W., and Erdman, J.W. (1998) Factors Influencing the Uptake and Absorption of Carotenoids, *Proc. Soc. Exp. Biol. Med. 218:* 106–108.
37. Ribaya-Mercado, J.D. (2002) Influence of Dietary Fat on β-Carotene Absorption and Bioconversion into Vitamin A, *Nutr. Rev. 60:* 104–110.
38. van het Hof, K.H., West, C.E., Weststrate, J.A., and Hautvast, J.G.A.J. (2000) Dietary Factors That Affect the Bioavailability of Carotenoids, *J. Nutr. 130:* 503–506.
39. Solomons, N.W., and Bulux, J. (1993) Effects of Nutritional Status on Carotene Uptake and Bioconversion, *Ann. N.Y. Acad. Sci. 691:* 96–109.

40. van den Berg, H. (1999) Carotenoid Interactions, *Nutr. Rev. 57:* 1–10.
41. Ribaya-Mercado, J.D., Solon, F.S., Solon, M.A., Cabal-Barza, M.A., Perfecto, C.S., Tang, G., Solon, J.A., Fjeld, C.R., and Russell, R.M. (2000) Bioconversion of Plant Carotenoids to Vitamin A in Filipino School-Aged Children Varies Inversely with Vitamin A Status, *Am. J. Clin. Nutr. 72:* 455–465.
42. Villard, L., and Bates, C.J. (1986) Carotene Dioxygenase (EC 1.13.11.21) Activity in Rat Intestine: Effects of Vitamin A Deficiency and of Pregnancy, *Br. J. Nutr. 56:* 115–122.
43. Noh, S.K., and Koo, S.I. (2003) Low Zinc Intake Decreases the Lymphatic Output of Retinol in Rats Infused Intraduodenally with β-Carotene, *J. Nutr. Biochem. 14:* 147–153.
44. Dijkhuizen, M.A., Wieringa, F.T., West, C.E., and Muhilal (2001) in *Vitamin A, Iron and Zinc Deficiency in Indonesia. Micronutrient Interactions and Effects of Supplementation*, Dijkhuizen, M.A., and Wieringa, F.T., eds., pp. 151–165, Wageningen University, Wageningen.
45. Gronowska, S.A., and Wolf, G. (1970) Effect of Dietary Protein on the Enzyme from Rat and Human Intestine Which Converts β-Carotene to Retinal, *J. Nutr. 100:* 300–308.
46. Blomstrand, R., and Werner, B. (1967) Studies on the Intestinal Absorption of Radioactive β-Carotene and Vitamin A in Man. Conversion of β-Carotene into Vitamin A, *Scand. J. Clin. Lab. Investig. 19:* 339–345.
47. Johnson, E.J., Krasinski, S.D., and Russell, R.M. (1992) Sex Differences in Postabsorptive Plasma Vitamin A Transport, *Am. J. Clin. Nutr. 56:* 911–916.
48. Berr, F., and Kern, F. (1984) Plasma Clearance of Chylomicrons Labeled with Retinyl Palmitate in Healthy Human Subjects, *J. Lipid Res. 25:* 805–812.
49. Cardinault, N., Tyssandier, V., Grolier, P., Winklhofer-Roob, B.M., Ribalta, J., Bouteloup-Demange, C., Rock, E., and Borel, P. (2003) Comparison of the Postprandial Chylomicron Carotenoid Responses in Young and Older Subjects, *Eur. J. Nutr. 42:* 315–323.
50. Crompton, D.W., and Nesheim, M.C. (2002) Nutritional Impact of Intestinal Helminthiasis During the Human Life Cycle, *Annu. Rev. Nutr. 22:* 35–59.
51. Tripathy, K., Duque, E., Bolanos, O., Lotero, H., and Mayoral, L.G. (1972) Malabsorption Syndrome in Ascariasis, *Am. J. Clin. Nutr. 25:* 1276–1281.
52. Jalal, F., Nesheim, M.C., Agus, Z., Sanjur, D., and Habicht, J.P. (1998) Serum Retinol Concentrations in Children Are Affected by Food Sources of β-Carotene, Fat Intake, and Anthelmintic Drug Treatment, *Am. J. Clin. Nutr. 68:* 623–629.
53. Solomons, N.W. (1993) Pathways to the Impairment of Human Nutritional Status by Gastrointestinal Pathogens, *Parasitology 107* (*Suppl.*)*:* S19–S35.
54. Rao, C.N., and Rao, B.S.N. (1970) Absorption of Dietary Carotenes in Human Subjects, *Am. J. Clin. Nutr. 23:* 105–109.
55. Exton, M.S. (1997) Infection-Induced Anorexia: Active Host Defense Strategy, *Appetite 29:* 369–383.
56. Castetbon, K., Kadio, A., Bondurand, A., Boka, Y.A., Barouan, C., Coulibaly, Y., Anglaret, X., Msellati, P., Malvy, D., and Dabis, F. (1997) Nutritional Status and Dietary Intakes in Human Immunodeficiency Virus (HIV)-Infected Outpatients in Abidjan, Cote D'Ivoire, 1995, *Eur. J. Clin. Nutr. 51:* 81–86.
57. Mitra, A.K., Alvarez, J.O., Guay-Woodford, L., Fuchs, G.J., Wahed, M.A., and Stephensen, C.B. (1998) Urinary Retinol Excretion and Kidney Function in Children with Shigellosis, *Am. J. Clin. Nutr. 68:* 1095–1103.

58. Stephensen, C.B., Alvarez, J.O., Kohatsu, J., Hardmeier, R., Kennedy, J.I., Jr., and Gammon, R.B., Jr. (1994) Vitamin A Is Excreted in the Urine During Acute Infection, *Am. J. Clin. Nutr. 60:* 388–392.
59. Alvarez, J.O., Salazar, L.E., Kohatsu, J., Miranda, P., and Stephensen, C.B. (1995) Urinary Excretion of Retinol in Children with Acute Diarrhea, *Am. J. Clin. Nutr. 61:* 1273–1276.
60. Thurnham, D.I., and Singkamani, R. (1991) The Acute Phase Response and Vitamin A Status in Malaria, *Trans. R. Soc. Trop. Med. Hyg. 85:* 194–199.
61. Krinsky, N.I., Mayne, S.T., and Sies, H. (2004) *Carotenoids in Health and Disease*, Marcel Dekker, New York.
62. Blomhoff, R. (1994) *Vitamin A in Health and Disease*, Marcel Dekker, New York.
63. Underwood, B.A. (1994) Maternal Vitamin A Status and Its Importance in Infancy and Early Childhood, *Am. J. Clin. Nutr. 59:* 517S–522S.
64. Christian, P., West, K.P., Khatry, S.K., Kimbrough-Pradhan, E., LeClerq, S.C., Katz, J., Shrestha, S.R., Dali, S.M., and Sommer, A. (2000) Night Blindness During Pregnancy and Subsequent Mortality Among Women in Nepal: Effects of Vitamin A and β-Carotene Supplementation, *Am. J. Epidemiol. 152:* 542–547.
65. West, K.P. (2004) Vitamin A Deficiency As a Preventable Cause of Maternal Mortality in Undernourished Societies: Plausibility and Next Steps, *Int. J. Gynaecol. Obstet. 85:* S24–S27.
66. Suharno, D., West, C.E., Muhilal, Karyadi, D., and Hautvast, J.G.A.J. (1993) Supplementation with Vitamin A and Iron for Nutritional Anaemia in Pregnant Women in West Java, Indonesia, *Lancet 342:* 1325–1328.
67. Katz, J., West, K.P., Khatry, S.K., Pradhan, E.K., LeClerq, S.C., Christian, P., Wu, L.S.-F., Adhikari, R.K., Shrestha, S.R., and Sommer, A. (2000) Maternal Low-Dose Vitamin A or β-Carotene Supplementation Has No Effect on Fetal Loss and Early Infant Mortality: A Randomized Cluster Trial in Nepal, *Am. J. Clin. Nutr. 71:* 1570–1576.
68. Schmidt, M.K., Muslimatun, S., Schultink, W., West, C.E., and Hautvast, J.G.A.J. (2002) Randomised Double-Blind Trial of the Effect of Vitamin A Supplementation of Indonesian Pregnant Women on Morbidity and Growth of Their Infants During the First Year of Life, *Eur. J. Clin. Nutr. 56:* 338–346.
69. Stoltzfus, R.J., Hakimi, M., Miller, K.W., Rasmussen, K.M., Dawiesah, S., Habicht, J.P., and Dibley, M.J. (1993) High Dose Vitamin A Supplementation of Breast-Feeding Indonesian Mothers: Effects on the Vitamin A Status of Mother and Infant, *J. Nutr. 123:* 666–675.
70. Filteau, S.M., Rice, A.L., Ball, J.J., Chakraborty, J., Stoltzfus, R., de Francisco, A., and Willumsen, J.F. (1999) Breast Milk Immune Factors in Bangladeshi Women Supplemented Postpartum with Retinol or β-Carotene, *Am. J. Clin. Nutr. 69:* 953–958.
71. Rahmathullah, L., Tielsch, J.M., Thulasiraj, R.D., Katz, J., Coles, C., Devi, S., John, R., Prakash, K., Sadanand, A.V., Edwin, N., and Kamaraj, C. (2003) Impact of Supplementing Newborn Infants with Vitamin A on Early Infant Mortality: Community Based Randomised Trial in Southern India, *Br. Med. J. 327:* 254.
72. Humphrey, J.H., Agoestina, T., Wu, L., Usman, A., Nurachim, M., Subardja, D., Hidayat, S., Tielsch, J., West, K.P., and Sommer, A. (1996) Impact of Neonatal Vitamin A Supplementation on Infant Morbidity and Mortality, *J. Pediatr. 128:* 489–496.
73. West, K.P., Katz, J., Shrestha, S.R., LeClerq, S.C., Khatry, S.K., Pradhan, E.K., Adhikari, R., Wu, L.S., Pokhrel, R.P., and Sommer, A. (1995) Mortality of Infants < 6

Mo of Age Supplemented with Vitamin A: A Randomized, Double-Masked Trial in Nepal, *Am. J. Clin. Nutr. 62:* 143–148.

74. Martines, J. (1998) Randomised Trial to Assess Benefits and Safety of Vitamin A Supplementation Linked to Immunisation in Early Infancy, *Lancet 352:* 1257–1263.
75. Beaton, G.H., Martorell, R., Aronson, K.J., Edmonston, B., McCabe, G., Ross, A.C., and Harvey, B. (1993) Effectiveness of Vitamin A Supplementation in the Control of Young Child Morbidity and Mortality in Developing Countries, *ACC/SCN State-of-the-Art-series: Nutrition Policy Discussion Paper 13*, University of Toronto, Toronto.
76. Donnen, P., Dramaix, M., Brasseur, D., Bitwe, R., Vertongen, F., and Hennart, P. (1998) Randomized Placebo-Controlled Clinical Trial of the Effect of a Single High Dose or Daily Low Doses of Vitamin A on the Morbidity of Hospitalized, Malnourished Children, *Am. J. Clin. Nutr. 68:* 1254–1260.
77. Nacul, L.C., Kirkwood, B.R., Arthur, P., Morris, S.S., Magalhaes, M., and Fink, M.C.D.S. (1997) Randomised, Double Blind, Placebo Controlled Clinical Trial of Efficacy of Vitamin A Treatment in Non-Measles Childhood Pneumonia, *Br. Med. J. 315:* 505–510.
78. Stephensen, C.B., Franchi, L.M., Hernandez, H., Campos, M., Gilman, R.H., and Alvarez, J.O. (1998) Adverse Effects of High-Dose Vitamin A Supplements in Children Hospitalized with Pneumonia, *Pediatrics 101:* e3.
79. Sempertegui, F., Estrella, B., Camaniero, V., Betancourt, V., Izurieta, R., Ortiz, W., Fiallo, E., Troya, S., Rodriguez, A., and Griffiths, J.K. (1999) The Beneficial Effects of Weekly Low-Dose Vitamin A Supplementation on Acute Lower Respiratory Infections and Diarrhea in Ecuadorian Children, *Pediatrics 104:* e1.
80. Fawzi, W.W., Mbise, R., Spiegelman, D., Fataki, M., Hertzmark, E., and Ndossi, G. (2000) Vitamin A Supplements and Diarrheal and Respiratory Tract Infections Among Children in Dar Es Salaam, Tanzania, *J. Pediatr. 137:* 660–667.
81. Griffiths, J.K. (2000) The Vitamin A Paradox, *J. Pediatr. 137:* 604–607.
82. Ezzati, M., Lopez, A.D., Rodgers, A., Vander, H.S., and Murray, C.J. (2002) Selected Major Risk Factors and Global and Regional Burden of Disease, *Lancet 360:* 1347–1360.
83. Victora, C.G., Smith, P.G., Vaughan, J.P., Nobre, L.C., Lombardi, C., Teixeira, A.M., Fuchs, S.C., Moreira, L.B., Gigante, L.P., and Barros, F.C. (1989) Infant Feeding and Deaths Due to Diarrhea. A Case-Control Study, *Am. J. Epidemiol. 129:* 1032–1041.
84. Arifeen, S., Black, R.E., Antelman, G., Baqui, A., Caulfield, L., and Becker, S. (2001) Exclusive Breastfeeding Reduces Acute Respiratory Infection and Diarrhea Deaths Among Infants in Dhaka Slums, *Pediatrics 108:* E67.
85. WHO Collaborative Study Team on the Role of Breastfeeding in the Prevention of Infant Mortality (2000) Effect of Breastfeeding on Infant and Child Mortality Due to Infectious Diseases in Less Developed Countries: A Pooled Analysis, *Lancet 355:* 451–455.
86. Guerrant, R.L., Schorling, J.B., McAuliffe, J.F., and de Souza, M.A. (1992) Diarrhea As a Cause and an Effect of Malnutrition: Diarrhea Prevents Catch-Up Growth and Malnutrition Increases Diarrhea Frequency and Duration, *Am. J. Trop. Med. Hyg. 47:* 28–35.
87. Sommer, A., Katz, J., and Tarwotjo, I. (1984) Increased Risk of Respiratory Disease and Diarrhea in Children with Preexisting Mild Vitamin A Deficiency, *Am. J. Clin. Nutr. 40:* 1090–1095.
88. Jones, G., Steketee, R.W., Black, R.E., Bhutta, Z.A., and Morris, S.S. (2003) How Many Child Deaths Can We Prevent This Year? *Lancet 362:* 65–71.

89. Henning, B., Stewart, K., Zaman, K., Alam, A.N., Brown, K.H., and Black, R.E. (1992) Lack of Therapeutic Efficacy of Vitamin A for Non-Cholera, Watery Diarrhoea in Bangladeshi Children, *Eur. J. Clin. Nutr. 46:* 437–443.
90. Baqui, A.H., Black, R.E., Arifeen, S.E., Yunus, M., Chakraborty, J., Ahmed, S., and Vaughan, J.P. (2002) Effect of Zinc Supplementation Started During Diarrhoea on Morbidity and Mortality in Bangladeshi Children: Community Randomised Trial, *Br. Med. J. 325:* 1059.
91. Hersh, B.S., Tambini, G., Nogueira, A.C., Carrasco, P., and de Quadros, C.A. (2000) Review of Regional Measles Surveillance Data in the Americas, 1996–99, *Lancet 355:* 1943–1948.
92. Biellik, R., Madema, S., Taole, A., Kutsulukuta, A., Allies, E., Eggers, R., Ngcobo, N., Nxumalo, M., Shearley, A., Mabuzane, E., Kufa, E., and Okwo-Bele, J.M. (2002) First 5 Years of Measles Elimination in Southern Africa: 1996–2000, *Lancet 359:* 1564–1568.
93. Centers for Disease Control and Prevention (2004) Measles Mortality Reduction—West Africa, 1996–2002, *Morb. Mortal. Wkly. Rep. 53:* 28–30.
94. Hussey, G.D., and Clements, C.J. (1996) Clinical Problems in Measles Case Management, *Ann. Trop. Paediatr. 16:* 307–317.
95. Ellison, J.B. (1932) Intensive Vitamin Therapy in Measles, *Br. Med. J. 2:* 711.
96. Glasziou, P.P., and Mackerras, D.E. (1993) Vitamin A Supplementation in Infectious Diseases: A Meta-Analysis, *Br. Med. J. 306:* 366–370.
97. Duggan, M.B., Alwar, J., and Milner, R.D. (1986) The Nutritional Cost of Measles in Africa, *Arch. Dis. Child. 61:* 61–66.
98. Pepping, F., Hackenitz, E.A., West, C.E., Duggan, M.B., and Franken, S. (1988) Relationship Between Measles, Malnutrition, and Blindness, *Am. J. Clin. Nutr. 47:* 341–343.
99. Rice, A.L., Sacco, L., Hyder, A., and Black, R.E. (2000) Malnutrition As an Underlying Cause of Childhood Deaths Associated with Infectious Diseases in Developing Countries, *Bull. World Health Organ. 78:* 1207–1221.
100. Butler, J.C., Havens, P.L., Sowell, A.L., Huff, D.L., Peterson, D.E., Day, S.E., Chusid, M.J., Bennin, R.A., Circo, R., and Davis, J.P. (1993) Measles Severity and Serum Retinol (Vitamin A) Concentration Among Children in the United States, *Pediatrics 91:* 1176–1181.
101. Shankar, A.H. (2000) Nutritional Modulation of Malaria Morbidity and Mortality, *J. Infect. Dis. 182 (Suppl. 1):* S37–S53.
102. Verhoef, H., West, C.E., Veenemans, J., Beguin, Y., and Kok, F.J. (2002) Stunting May Determine the Severity of Malaria-Associated Anemia in African Children, *Pediatrics 110:* e48.
103. Shankar, A.H., Genton, B., Semba, R.D., Baisor, M., Paino, J., Tamja, S., Adiguma, T., Wu, L., Rare, L., and Tielsch, J.M. (1999) Effect of Vitamin A Supplementation on Morbidity Due to *Plasmodium falciparum* in Young Children in Papua New Guinea: A Randomised Trial, *Lancet 354:* 203–209.
104. Cooper, K.A., Adelekan, D.A., Esimai, A.O., Northrop-Clewes, C.A., and Thurnham, D.I. (2002) Lack of Influence of Red Palm Oil on Severity of Malaria Infection in Pre-School Nigerian Children, *Trans. R. Soc. Trop. Med. Hyg. 96:* 216–223.
105. UNAIDS (2003) *AIDS Epidemic Update: 2003*, pp. 1–13, UNAIDS, Geneva.
106. Coutsoudis, A., Pillay, K., Spooner, E., Kuhn, L., and Coovadia, H.M. (1999) Randomized Trial Testing the Effect of Vitamin A Supplementation on Pregnancy

Outcomes and Early Mother-to-Child HIV-1 Transmission in Durban, South Africa. South African Vitamin A Study Group, *AIDS 13:* 1517–1524.
107. Stephensen, C.B. (2003) Vitamin A, β-Carotene, and Mother-to-Child Transmission of HIV, *Nutr. Rev. 61:* 280–284.
108. Kennedy-Oji, C., Coutsoudis, A., Kuhn, L., Pillay, K., Mburu, A., Stein, Z., and Coovadia, H. (2001) Effects of Vitamin A Supplementation During Pregnancy and Early Lactation on Body Weight of South African HIV-Infected Women, *J. Health Popul. Nutr. 19:* 167–176.
109. Coutsoudis, A., Bobat, R.A., Coovadia, H.M., Kuhn, L., Tsai, W.Y., and Stein, Z.A. (1995) The Effects of Vitamin A Supplementation on the Morbidity of Children Born to HIV-Infected Women, *Am. J. Public Health 85:* 1076–1081.
110. Fawzi, W.W., Mbise, R.L., Hertzmark, E., Fataki, M.R., Herrera, M.G., Ndossi, G., and Spiegelman, D. (1999) A Randomized Trial of Vitamin A Supplements in Relation to Mortality Among Human Immunodeficiency Virus-Infected and Uninfected Children in Tanzania, *Pediatr. Infect. Dis. J. 18:* 127–133.
111. Fawzi, W.W., Msamanga, G.I., Wei, R., Spiegelman, D., Antelman, G., Villamor, E., Manji, K., and Hunter, D. (2003) Effect of Providing Vitamin Supplements to Human Immunodeficiency Virus-Infected, Lactating Mothers on the Child's Morbidity and CD^{4+} Cell Counts, *Clin. Infect. Dis. 36:* 1053–1062.
112. Fawzi, W.W., Msamanga, G.I., Spiegelman, D., Urassa, E.J., McGrath, N., Mwakagile, D., Antelman, G., Mbise, R., Herrera, G., Kapiga, S., Willett, W., and Hunter, D.J. (1998) Randomised Trial of Effects of Vitamin Supplements on Pregnancy Outcomes and T Cell Counts in HIV-1-Infected Women in Tanzania, *Lancet 351:* 1477–1482.
113. Jiamton, S., Pepin, J., Suttent, R., Filteau, S., Mahakkanukrauh, B., Hanshaoworakul, W., Chaisilwattana, P., Suthipinittharm, P., Shetty, P., and Jaffar, S. (2003) A Randomized Trial of the Impact of Multiple Micronutrient Supplementation on Mortality Among HIV-Infected Individuals Living in Bangkok, *AIDS 17:* 2461–2469.
114. Karyadi, E., Schultink, W., Nelwan, R.H., Gross, R., Amin, Z., Dolmans, W.M., van der Meer, J.W., Hautvast, J.GA.J., and West, C.E. (2000) Poor Micronutrient Status of Active Pulmonary Tuberculosis Patients in Indonesia, *J. Nutr. 130:* 2953–2958.
115. Hanekom, W.A., Potgieter, S., Hughes, E.J., Malan, H., Kessow, G., and Hussey, G.D. (1997) Vitamin A Status and Therapy in Childhood Pulmonary Tuberculosis, *J. Pediatr. 131:* 925–927.
116. Karyadi, E., West, C.E., Schultink, W., Nelwan, R.H., Gross, R., Amin, Z., Dolmans, W.M., Schlebusch, H., and van der Meer, J.W. (2002) A Double-Blind, Placebo-Controlled Study of Vitamin A and Zinc Supplementation in Persons with Tuberculosis in Indonesia: Effects on Clinical Response and Nutritional Status, *Am. J. Clin. Nutr. 75:* 720–727.
117. Williams, C., and Williams, C. (1868) On the Nature and Treatment of Pulmonary Consumption as Exemplified in Private Practice, *Lancet 92:* 3–4.
118. Williams, C., and Williams, C. (1868) On the Nature and Treatment of Pulmonary Consumption as Exemplified in Private Practice, *Lancet 91:* 369–370.
119. Biesalski, H.K. (1989) Comparative Assessment of the Toxicology of Vitamin A and Retinoids in Man, *Toxicology 57:* 117–161.
120. Underwood, B.A. (1989) Teratogenicity of Vitamin A, *Int. J. Vitam. Nutr. Res. 30:* 42–55.
121. Roos, N., Islam, M., and Thilsted, S.H. (2003) Small Fish Is an Important Dietary Source of Vitamin A and Calcium in Rural Bangladesh, *Int. J. Food Sci. Nutr. 54:* 329–339.

122. Arroyave, G., Mejia, L.A., and Aguilar, J.R. (1981) The Effect of Vitamin A Fortification of Sugar on the Serum Vitamin A Levels of Preschool Guatemalan Children: A Longitudinal Evaluation, *Am. J. Clin. Nutr. 34:* 41–49.
123. Institute of Medicine (2002) *Dietary Reference Intakes for Vitamin A, Vitamin K, Arsenic, Boron, Chromium, Copper, Iodine, Iron, Manganese, Molybdenum, Nickel, Silicon, Vanadium, and Zinc: a Report of the Panel on Dietary Antioxidants and Related Compounds, Subcommittees on Upper Reference Levels of Nutrients and Interpretation and Uses*, 2nd edn., p. 30, National Academy Press, Washington.
124. de Pee, S., West, C.E., Permaesih, D., Martuti, S., Muhilal, and Hautvast, J.G.A.J. (1998) Orange Fruit Is More Effective Than Are Dark-Green, Leafy Vegetables in Increasing Serum Concentrations of Retinol and β-Carotene in Schoolchildren in Indonesia, *Am. J. Clin. Nutr. 68:* 1058–1067.
125. Khan, N.C., West, C.E., de Pee, S., and Khôi, H.H. (1998) Comparison of Effectiveness of Carotenoids from Dark-Green Leafy Vegetables and Yellow and Orange Fruits in Improving Vitamin A Status of Breastfeeding Women in Vietnam, *Report of the XIIIth International Vitamin A Consultative Group meeting*, p. 92, ILSI, Washington.
126. Tang, G., Gu, X., Hu, S., Xu, Q., Qin, J., Dolnikowski, G.G., Fjeld, C.R., Gao, X., Russell, R.M., and Yin, S. (1999) Green and Yellow Vegetables Can Maintain Body Stores of Vitamin A in Chinese Children, *Am. J. Clin. Nutr. 70:* 1069–1076.
127. West, C.E., Eilander, A., and van Lieshout, M. (2002) Consequences of Revised Estimates of Carotenoid Bioefficacy for Dietary Control of Vitamin A Deficiency in Developing Countries, *J. Nutr. 132:* 2920S–2926S.

Chapter 18

Lycopene and Prostate Cancer

Ute C. Obermüller-Jevic[a] and Lester Packer[b]

[a]BASF Aktiengesellschaft, 67056 Ludwigshafen, Germany; [b]School of Pharmacy, Department of Molecular Pharmacology and Toxicology, University of Southern California, Los Angeles, CA 90089–9121

Introduction

Lycopene, a carotenoid consumed primarily from tomatoes, watermelons, and a few other plant-based foods, is one of the most promising chemopreventive agents found in the diet (1,2). Regular consumption of foods rich in lycopene is associated with a 30–40% lower risk of being diagnosed with prostate cancer (3–5). Lycopene is the major carotenoid detected in the human prostate gland, and the concentrations found in the human prostate are higher than in several other tissues (6). Although lycopene appears to have beneficial effects on prostate cancer, the molecular mechanisms involved remain unclear. Unlike β-carotene, lycopene is not a provitamin A carotenoid, but its acyclic hydrocarbon chain with 11 double bonds possesses powerful antioxidant activity as a singlet oxygen quencher (7) and lipid peroxyl radical scavenger (8). Moreover, lycopene inhibits the proliferation of cancer cells *via* modulation of cell signaling of growth factors and interference with cell cycle progression (9–14).

Prostate Cancer

Prostate cancer is the most common malignancy diagnosed in American men, and the second leading cause of cancer deaths (15). The occurrence of prostate cancer is related to a person's lifestyle and diet (16). Among carotenoids, only lycopene is inversely related to the risk of prostate cancer (3,17–19). In small clinical studies with cancer patients, it was shown that daily consumption of either a lycopene-rich meal or dietary supplements (isolated lycopene from tomato) before prostatectomy inhibited progression of the disease. In these studies, a dose of 30 mg lycopene/d taken for 3 wk before surgery lowered blood levels of prostate-specific antigen (PSA), a marker for tumor growth, and inhibited growth and invasiveness of the tumor. Lycopene intervention also diminished DNA damage in leukocytes and prostate tissue, and increased apoptotic cell death in the tumors of these patients (20,21).

In another clinical study in men with advanced, metastatic prostate cancer, who were treated with androgen ablation by castration (orchidectomy), lycopene supplements (4 mg/d) given in addition to therapy proved similarly effective. Patients administered lycopene in addition to orchidectomy for a time period of 2 y after surgery had significantly lower PSA levels after 6 mo, less progression of metastasis and a significant improvement in urinary flow than patients treated by orchidectomy alone. Although 35% of patients who participated in the study died within 2 y, only 13% were from the group receiving lycopene and orchidectomy compared with 22% with orchidectomy alone (22).

These clinical studies strongly suggest a therapeutic potential of lycopene in prostate cancer and demonstrate that both a diet rich in lycopene and supplemental lycopene could exert beneficial effects in patients. Lycopene seems to shrink the tumor burden as well as inhibit metastasis, improves clinical parameters, and may prolong life. Long-term studies are warranted to provide further evidence for the promising effects of lycopene in prostate cancer patients.

Several experimental studies support a beneficial role of lycopene in prostate cancer and help explain the potential mechanisms of lycopene action. Cell culture studies repeatedly showed that lycopene effectively inhibits proliferation of prostate cancer cells in different hormone-dependent and -independent cell lines (9,10).

Animal models further revealed potential therapeutic effects of lycopene in prostate cancer. The published data range from chemically or hormone-induced prostate cancer (23–25) to orthotopic tumor models in which prostate cancer cells are injected into the prostate to most adequately mimic development of tumors in their physiologic environment (26–28). Depending on the type of model used, the results from animal studies vary.

In the MatLyLu Dunning model of prostate cancer, lycopene (synthesized, nature-identical) was tested alone and in combination with vitamin E. In this rat model, prostate cancer cells from rats were injected into the ventral prostate. Feeding the rats lycopene (200 mg/kg diet), vitamin E (540 mg/kg diet) or a combination of both for 4 wk before and another 3 wk after injection of tumor cells caused increased necrosis in the tumor tissue. Microarray analysis revealed that lycopene given alone or in combination with vitamin E inhibited gene expression of 5-α-reductase 1, an enzyme that is involved in testosterone activation (yielding 5-α-dehydrotestosterone). Furthermore, lycopene inhibited expression of insulin-like growth factor-1 (IGF-1), which is involved in prostate carcinogenesis, and of interleukin-6, a growth factor for prostate tumors. Vitamin E, in contrast, did not modulate expression of 5-α-reductase 1 or IGF-1. Vitamin E downregulated aromatase expression, which may lead to inhibition of estrogen synthesis. It was postulated that both lycopene and vitamin E contributed individually to inhibiting prostate tumor growth on the level of cell signaling, steroid hormone metabolism, and activation of growth factors. The effects of lycopene and vitamin E were additive (26).

A similar animal study comparing synthesized, nature-identical lycopene, and vitamin E was reported recently (27,28). In a mouse xenograft model of prostate cancer with orthotopic tumors of PC-346C human cancer cells, lycopene [5 or 50 mg/kg body weight (BW)], vitamin E (5 or 50 mg/kg BW) or a combination of both (each 5 mg/kg BW) was tested. Tumor development was monitored *via* transrectal ultrasonography and related to the PSA levels in blood. Lycopene inhibited tumor growth, lowered PSA levels, and increased survival time of the mice. In the group administered the low dose of lycopene, the effects on tumor growth and survival time were significant. The inhibition was even stronger in the group administered lycopene plus vitamin E.

The effectiveness of lycopene from different sources was compared in a third animal study (25). In a model using *N*-methyl-*N*-nitrosourea (NMU) and testosterone to induce prostate tumors, rats were fed either tomato powder (13 mg lycopene/kg diet), synthesized and nature-identical lycopene (161 mg lycopene/kg diet), or placebo. Whole tomato powder was effective in this study in inhibiting tumor growth, whereas synthesized lycopene had no significant benefit. However, it is unfortunate that the doses used in this study were largely different; thus, the observed effects are not comparable. The study, therefore, does not show inefficacy of synthesized lycopene but rather suggests a dose-dependent effect of lycopene (the group receiving synthesized lycopene had a 10-fold higher dosage than the group fed tomato powder). Synthesized lycopene may have been given here in a supraoptimal dose (27).

From the published animal studies, the following can be concluded: (i) lycopene may inhibit the increase of PSA in blood and tumor development (as observed in humans); (ii) lycopene may exert its effects through inhibiting androgen signaling; and (iii) both natural-source and synthesized nature-identical lycopene seem to be effective, although further, more conclusive studies are warranted.

Other Diseases of the Prostate

In addition to prostate cancer, another common health issue is age-related enlargement of the prostate leading to a disease called benign prostate hyperplasia (BPH). BPH is more common with increasing age; it is diagnosed in 20% of men aged 40–50 y and in 80% of men aged 70–80 y (29). BPH leads to impaired urinary tract function and inflammatory processes and it is considered a risk factor for developing prostate cancer (30).

Epidemiologic studies revealed that dietary factors are likely to play a role in the development of BPH because the incidence of BPH in Asian men increases after they adopt a Western lifestyle. Whether lycopene is associated with a lower risk of BPH is unknown. It may be hypothesized that the beneficial effect of lycopene observed on the risk of prostate cancer may occur at early preneoplastic stages of disease development. No clinical or animal studies have been reported on lycopene in BPH, which could evaluate the role of lycopene in the primary prevention of prostate cancer. Only one trial in prostate cancer patients compared the effects of a lycopene-rich meal ingested daily for 3 wk before prostatectomy on apoptotic cell death in the tumor and the adjunct benign tissue. The lycopene-rich diet increased apoptotic cell death in both the tumor tissue and the benign tissue (31).

Hyperproliferation of prostate epithelial cells and inflammatory processes in the prostate as observed in BPH, and moreover in prostatitis, a chronic inflammatory disorder of the prostate, seem to be associated with the development of preneoplastic lesions. Until recently, it was largely unknown whether lycopene, which inhibits proliferation in cancer cells, also affected the growth of normal prostate cells. This may be important because hyperproliferation of epithelial cells in prostate tissue is a major reason for enlargement of the prostate in elderly men. Inhibition of epithelial cell proliferation is a common strategy to lower the risk of cancer (30,32).

We investigated whether lycopene (synthesized, nature-identical) modulated the growth of benign human prostate cells (33). PrEC cells, a nonimmortalized and primary cell strain of normal human prostate epithelial cells were used as a model. Lycopene significantly inhibited the growth of PrEC cells as determined by a $[^3H]$thymidine incorporation assay. Inhibition of cell proliferation by lycopene was dose dependent. A significant inhibition was observed at a dose ≥1 μmol/L lycopene ($P < 0.05$). The calculated 25 and 50% infective dose $(ID)_{25}$ and ID_{50} values for lycopene were 0.4 and 0.8 μmol/L, respectively, levels that are frequently found in humans consuming a diet rich in tomatoes and/or tomato products or supplementing their diets with carotenoids (21,34). This suggests that regular intake of lycopene may result in blood levels of lycopene likely to affect the growth of prostatic epithelial cells in men. A maximum inhibition of ~80% was achieved at a higher dose of 2 μmol/L lycopene. The observed inhibition of cell proliferation by lycopene of up to 82% in normal prostate epithelial cells was even higher than previous reports on prostate cancer cells. Pastori *et al.* (10) reported an ~20% inhibition of DU-145 prostate cancer cell proliferation using a dose of 5 μmol/L lycopene. Kotake-Nara *et al.* (9) reported an ~25% inhibition of three different prostate cancer cell lines (LNCaP, DU-145, PC-3) by 5 μmol/L lycopene. In our study, a 20% inhibition of PrEC cell proliferation was achieved at a dose of 0.3 μmol/L lycopene.

The observed growth inhibition of lycopene in benign prostate cells, if confirmed *in vivo*, could have an effect on reducing age-related enlargement of the prostate. Clinical trials are warranted in this regard. Also, it is of interest to investigate whether lycopene affects the growth of normal prostate in young men. During maturation of the prostate gland, enhanced cell proliferation is a physiologic process. No reports exist, however, on potential effects of consumption of lycopene-rich foods on prostate development and organ function.

Furthermore, it is not known whether the observed effects on normal prostate epithelial cells are specific for lycopene or may also be observed with other carotenoids. An *in vitro* study on prostate cancer cells revealed inhibition of cancer viability (as assessed by the MTT assay) by several carotenoids—to differing degrees, however (9). An even stronger inhibitory effect than that of lycopene on prostate cancer cells was caused by the 5,6-monoepoxy carotenoids, neoxanthin and fucoxanthin. Because lycopene is preferably accumulated in human prostate, it can be postulated that it is the major carotenoid in prostate health.

Concluding Remarks

There is clear evidence to support that lycopene is a promising nutrient in prostate health. Prostate cancer patients particularly may benefit from daily intake of high-dose lycopene (up to 30 mg/d). Because dietary sources are limited, intake of supplemental lycopene may be recommended. There is now experimental evidence that lycopene may play a role not only in cancer patients but also in age-related benign prostate disease. Several remaining key questions with regard to lycopene in prostate health are presented here.

1) Is Isolated Lycopene as Effective as Tomato Consumption?

In clinical trials with prostate cancer patients, tomato consumption and supplementation were effective in inhibiting disease progression as determined by PSA levels, tumor growth and invasiveness. Inhibition of cell proliferation by isolated lycopene in prostate cells was demonstrated repeatedly in cell culture studies including normal and prostate cancer cells. Animal studies show varying results, which may be due to the use of different models. It cannot be ruled out that tomatoes contain a number of other ingredients that exert additional benefits in prostate cancer. However, there is strong evidence showing that lycopene is the important substance from tomatoes, which is beneficial in prostate disease.

2) Is There a Difference Between Synthesized and Natural-Source Lycopene?

As outlined above, there is evidence from two animal studies that synthesized lycopene (chemically identical to the naturally occurring lycopene) inhibits tumor progression. Moreover, synthesized lycopene has antiproliferative effects in cultured prostate cancer cells and benign prostate cells. The findings suggest that synthesized lycopene is a promising supplement. Clinical studies are warranted to test the efficacy of such supplements for use in prostate cancer patients and in subjects with other age-related prostate diseases. The optimal dose range of lycopene in humans, independent of its source, and studies of the optimal method of its administration must be established.

References

1. Klein, E.A., and Thompson, I.M. (2004) Update on Chemoprevention of Prostate Cancer, *Curr. Opin. Urol. 14:* 143–149.

2. Kucuk, O. (2002) Chemoprevention of Prostate Cancer, *Cancer Metastasis Rev. 21:* 111–124.
3. Giovannucci, E. (2002) A Review of Epidemiologic Studies of Tomatoes, Lycopene, and Prostate Cancer, *Exp. Biol. Med. 227:* 852–859.
4. Lindsey, H. (2002) Consuming Tomato Products May Reduce Prostate-Cancer Risk, *Lancet Oncol. 3:* 198.
5. Wu, K., Erdman, J.W., Jr., Schwartz, S.J., Platz, E.A., Leitzmann, M., Clinton, S.K., DeGroff, V., Willett, W.C., and Giovannucci, E. (2004) Plasma and Dietary Carotenoids, and the Risk of Prostate Cancer: A Nested Case-Control Study, *Cancer Epidemiol. Biomark. Prev. 13:* 260–269.
6. Clinton, S., Emenhiser, C., Schwartz, S., Bostwick, D., Williams, A., Moore, B., and Erdman, J.J. (1996) *cis-trans* Lycopene Isomers, Carotenoids, and Retinol in the Human Prostate, *Cancer Epidemiol. Biomark. Prev. 5:* 823–833.
7. Heber, D., and Lu, Q. (2002) Overview of Mechanisms of Action of Lycopene, *Exp. Biol. Med. 227:* 920–923.
8. Tsuchiya, M., Scita, G., Thompson, D.F.T., and Packer, L. (1993) Retinoids and Carotenoids Are Peroxyl Radical Scavengers, in *Retinoids: Progress in Research and Clinical Applications*, Livrea, M.A., and Packer, L., eds., Marcel Dekker, New York, pp. 525–536.
9. Kotake-Nara, E., Kushiro, M., Zhang, H., Sugawara, T., Miyashita, K., and Nagao, A. (2001) Carotenoids Affect Proliferation of Human Prostate Cancer Cells, *J. Nutr. 131:* 3303–3306.
10. Pastori, M., Pfander, H., Boscoboinik, D., and Azzi, A. (1998) Lycopene in Association with Alpha-Tocopherol Inhibits at Physiological Concentrations Proliferation of Prostate Carcinoma Cells, *Biochem. Biophys. Res. Commun. 250:* 582–585.
11. Levy, J., Bosin, E., Feldman, B., Giat, Y., Miinster, A., Danilenko, M., and Sharoni, Y. (1995) Lycopene Is a More Potent Inhibitor of Human Cancer Cell Proliferation than Either Alpha-Carotene or Beta-Carotene, *Nutr. Cancer 24:* 257–266.
12. Ben-Dor, A., Nahum, A., Danilenko, M., Giat, Y., Stahl, W., Martin, H., Emmerich, T., Noy, N., Levy, J., and Sharoni, Y. (2001) Effects of Acyclo-Retinoic Acid and Lycopene on Activation of the Retinoic Acid Receptor and Proliferation of Mammary Cancer Cells, *Arch. Biochem. Biophys. 391:* 295–302.
13. Nahum, A., Hirsch, K., Danilenko, M., Watts, C., Prall, O., Levy, J., and Sharoni, Y. (2001) Lycopene Inhibition of Cell Cycle Progression in Breast and Endometrial Cancer Cells Is Associated with Reduction in Cyclin D Levels and Retention of p27(Kip1) in the Cyclin E-cdk2 Complexes, *Oncogene 20:* 3428–3436.
14. Karas, M., Amir, H., Fishman, D., Danilenko, M., Segal, S., Nahum, A., Koifmann, A., Giat, Y., Levy, J., and Sharoni, Y. (2000) Lycopene Interferes with Cell Cycle Progression and Insulin-Like Growth Factor I Signaling in Mammary Cancer Cells, *Nutr. Cancer 36:* 101–111.
15. American Cancer Society, Cancer Statistics 2004 [www.cancer.org].
16. Cohen, L. (2002) Nutrition and Prostate Cancer: A Review, *Ann. N.Y. Acad. Sci. 963:* 148–155.
17. Vogt, T., Mayne, S., Graubard, B., Swanson, C., Sowell, A., Schoenberg, J., Swanson, G., Greenberg, R., Hoover, R., Hayes, R., and Ziegler, R. (2002) Serum Lycopene, Other Serum Carotenoids, and Risk of Prostate Cancer in US Blacks and Whites, *Am. J. Epidemiol. 155:* 1023–1032.

18. Lu, Q., Hung, J., Heber, D., Go, V., Reuter, V., Cordon-Cardo, C., Scher, H., Marshall, J., and Zhang, Z. (2001) Inverse Associations Between Plasma Lycopene and Other Carotenoids and Prostate Cancer, *Cancer Epidemiol. Biomark. Prev. 10:* 749–756.
19. Gann, P., Giovannucci, E., Willett, W., Sacks, F., Hennekens, C., and Stampfer, M. (1999) Lower Prostate Cancer Risk in Men with Elevated Plasma Lycopene Levels: Results of a Prospective Analysis, *Cancer Res. 59:* 1225–1230.
20. Kucuk, O., Sarkar, F., Djuric, Z., Sakr, W., Pollak, M., Khachik, F., Banerjee, M., Bertram, J., and Wood, W.D., Jr. (2002) Effects of Lycopene Supplementation in Patients with Localized Prostate Cancer, *Exp. Biol. Med. 227:* 881–885.
21. Bowen, P., Chen, L., Stacewicz-Sapuntzakis, M., Duncan, C., Sharifi, R., Ghoch, L., Kim, H., Christov-Tzelkov, K., and van Breemen, R. (2002) Tomato Sauce Supplementation and Prostate Cancer: Lycopene Accumulation and Modulation of Biomarkers of Carcinogenesis, *Exp. Biol. Med. 227:* 886–893.
22. Ansari, M.S., and Gupta, N.P. (2003) A Comparison of Lycopene and Orchidectomy vs. Orchidectomy Alone in the Management of Advanced Prostate Cancer, *BJU Int. 92:* 375–378.
23. Imaida, K., Tamano, S., Kato, K., Ikeda, Y., Asamoto, M., Takahashi, S., Nir, Z., Murakoshi, M., Nishino, H., and Shirai, T. (2001) Lack of Chemopreventive Effects of Lycopene and Curcumin on Experimental Rat Prostate Carcinogenesis, *Carcinogenesis 22:* 467–472.
24. Guttenplan, J.B., Chen, M., Kosinska, W., Thompson, S., Zhao, Z., and Cohen, L.A. (2001) Effects of a Lycopene-Rich Diet on Spontaneous and Benzo[a]pyrene-Induced Mutagenesis in Prostate, Colon and Lungs of the lacZ Mouse, *Cancer Lett. 164:* 1–6.
25. Boileau, T.W., Liao, Z., Kim, S., Lemeshow, S., Erdman, J.W., Jr., and Clinton, S.K. (2003) Prostate Carcinogenesis in *N*-Methyl-*N*-nitrosourea (NMU)-Testosterone-Treated Rats Fed Tomato Powder, Lycopene, or Energy-Restricted Diets, *J. Natl. Cancer Inst. 95:* 1578–1586.
26. Siler, U., Barella, L., Spitzer, V., Schnorr, J., Lein, M., Goralczyk, R., and Wertz, K. (2004) Lycopene and Vitamin E Interfere with Autocrine/Paracrine Loops in the Dunning Prostate Cancer Model, *FASEB J. 18:* 1019–1021.
27. Limpens, J., van Weerden, W.M., Kraemer, K., Pallapier, D., Obermüller-Jevic, U.C., and Schroeder, F. (2004) Re: Prostate Carcinogenesis in *N*-Methyl-*N*-nitrosourea (NMU)-Testosterone-Treated Rats Fed Tomato Powder, Lycopene, or Energy-Restricted Diets [Comment], *J. Natl. Cancer Inst. 96:* 554.
28. Obermüller-Jevic, U., Olano-Martin, E., van Weerden, W., Kraemer, K., Cross, C.E., Schroeder, F.H., and Packer, L. (2004) Lycopene and Prostate Health, Oxygen Club of California World Congress, Santa Barbara, CA, March 10–13 (Abstr.).
29. Ziada, A., Rosenblum, M., and Crawford, E. (1999) Benign Prostatic Hyperplasia: An Overview, *Urology 53:* 1–6.
30. Heber, D. (2002) Prostate Enlargement: The Canary in the Coal Mine? *Am. J. Clin. Nutr. 75:* 605–606.
31. Kim, H.S., Bowen, P., Chen, L., Duncan, C., Ghosh, L., Sharifi, R., and Christov, K. (2003) Effects of Tomato Sauce Consumption on Apoptotic Cell Death in Prostate Benign Hyperplasia and Carcinoma, *Nutr. Cancer 47:* 40–47.
32. De Marzo, A., Putzi, M., and Nelson, W. (2001) New Concepts in the Pathology of Prostatic Epithelial Carcinogenesis, *Urology 57 (Suppl. 1):* 103–114.

33. Obermüller-Jevic, U., Olano-Martin, E., Crobacho, A.M., Eiserich, J.P., van der Vliet, A., Valacchi, G., Cross, C.E., and Packer, L. (2003) Lycopene Inhibits the Growth of Normal Human Prostate Epithelial Cells *In Vitro*, *J. Nutr. 133:* 3356–3360.
34. Olmedilla, B., Granado, F., Southon, S., Wright, A., Blanco, I., Gil-Martinez, E., Berg, H., Corridan, B., Roussel, A., Chopra, M., and Thurnham, D. (2001) Serum Concentrations of Carotenoids and Vitamins A, E, and C in Control Subjects from Five European Countries, *Br. J. Nutr. 85:* 27–238.

Chapter 19

Blood Response to β-Carotene Supplementation in Humans: An Evaluation Across Published Studies

Klaus Kraemer, Gerhard Krennrich, Ute Obermüller-Jevic, and Peter P. Hoppe

BASF Aktiengesellschaft, Ludwigshafen, Germany

Introduction

In human studies conducted with β-carotene (BC) supplements, the blood response varied widely, which was attributed to *inter alia*, dose, duration of supplementation, type of formulation, gender, smoking status, and differences in study design. Recently, negative results in smokers from the ATBC study (1) and beneficial effects on the progression of age-related macular degeneration from the AREDS study (2) emphasized the need to define a safe and effective BC dose. In the few studies investigating the relation between BC dose and steady-state blood concentration, the results were inconsistent. Although some studies found a significant dose-response relation (3,4), no relation was reported in others (5–7). The relationship between BC dose and blood response across studies is unknown. Moreover, the potential effect of different delivery forms of BC is not well known and there is little information on covariates affecting the response. For these reasons, we performed an evaluation of all published studies with the aim to determine the blood response to different delivery forms of isolated BC across studies and to identify covariates significantly affecting the blood BC response.

Methods

Bioavailability was defined as the steady-state concentration of BC in blood (plasma or serum) after ingestion over a period ≥26 d. Because the elimination half-life of supplemental BC is ~7 d (5,8), supplementation ≥26 d (*viz.*, ≥4 times the half-life) can be assumed to result in the steady-state concentration in blood.

A literature search in MEDLINE resulted in ~500 publications from 1982 to April 2004; these were screened for compatibility with the inclusion criteria: clinical trial with adult subjects in good health, supplementation with doses ≤30 mg/d, duration of supplementation for ≥26 d, supplement intake on a daily basis or on alternate days, and HPLC analysis of plasma or serum BC. Included were studies with isolated BC alone or in cosupplementation with one or more micronutrients, *viz.*, vitamin E, vitamin C, selenium, zinc, copper, and ginseng but excluding other added carotenoids. Moreover, studies were included if the source (synthetic or natural source extracted from algae or palm fruit) and the type of formulation (beadlets or oily dispersion) were specified in the publication or obtained from correspondence with authors. Also included were healthy control groups in intervention studies, from studies on patients diagnosed with disease conditions such as coronary heart disease, hypertension (9), chronic atrophic gastritis (10), previous colonic polyps (11–14), moderate cervical dysplasia (15), oral leukoplakia (16), asbestos-related lung fibrosis (17), and preneoplastic lesions (18). Excluded were studies on children, pregnant and lactating women, and studies in which results were shown as graphs only or in which blood BC was expressed relative to blood cholesterol only. Correction for blood lipid levels was not feasible because this information was lacking for the vast majority of publications. When the same data set was published in multiple publications, care was taken that only one data set was included. In studies with more than one dosage group, results from all dosage groups were entered. When blood was first sampled beyond 1 y, the first time point only was included. In studies with dosing on alternate days, the dose per day was calculated.

The following variables were recorded for each study: origin (natural-source, e.g., from palm oil or from *Dunaliella salina*, or synthetic), formulation (crystalline, dispersion in oil, or beadlet), dose (mg/d), dosing regimen (single- or multiple-unit), duration of intake (d), cosupplementation with other micronutrients, ingestion of supplement with or without food, intake of BC from diet, dietary recommendation (e.g., advice not to change the normal diets), compliance, number of subjects, age, gender, BMI, body weight, smoking status, alcohol consumption, prior BC depletion, mean serum or plasma BC at entry (baseline BC) and at ≥ 26 d (steady state), and mean blood BC increase (steady-state minus baseline). Because various variables were missing in individual studies, only dose, duration, mean baseline BC, mean steady-state BC, origin, formulation, and cosupplementation were used in the evaluation.

A total of 57 studies (1–7, 9–54) including two unpublished studies (Hoppe, and Dricker *et al.*) comprising 67 observations were evaluated. The higher number of observations than studies is due to studies with >1 dosage group. The number of subjects in the individual studies ranged from 3 to ~14,000. We refrained from weighting the individual studies according to the number of subjects to prevent the evaluation from being dominated by studies with a large sample size.

Results

Across studies, baseline BC concentration, postsupplementation BC concentration, and increase above baseline (mean and median) were 0.45 and 0.41, 3.09 and 2.71, and 2.64 and 2.23 μmol/L, respectively. Across studies, the duration of supplementation was 220 and 70 d (mean and median), respectively. Thus, it can be assumed that postsupplementation concentrations represented steady-state concentrations.

The mutual dependencies of the four continuous variables, dose, duration, baseline BC, and steady-state BC, were investigated by means of ordinary (Pearson) and rank (Spearman) correlation analysis. A significant relation between dose and steady-state BC was revealed ($r = 0.39$) as depicted in Figure 19.1. According to Figure 19.1, the values vary markedly for each dose. This is reflected by the low coefficient of determination ($R^2 = 0.15$) for the linear dose-response relation, indicating the existence of additional confounders. Note that the very high value of 9.25 μmol/L at 15 mg/d is from a study that used a multiple-unit dosing regimen (2 tablets given twice daily) (2).

Baseline BC and dose were significantly related ($r = 0.38$). Although this negative relation is certainly not of the cause and effect type because dose as a matter of timing cannot affect baseline BC, it may impose some bias on the dose effect because subjects with different baseline BC may respond differently to the same dose. The correlation coefficient between duration of supplementation and steady state BC ($r = 0.10$) was not significant, supporting the assumption that dosing for ≥ 26 d results in the steady-state concentration.

Frequency statistics of the categorical variables, origin, formulation, cosupplementation, regimen, gender, and smoking (Table 19.1), indicated a poor balance of

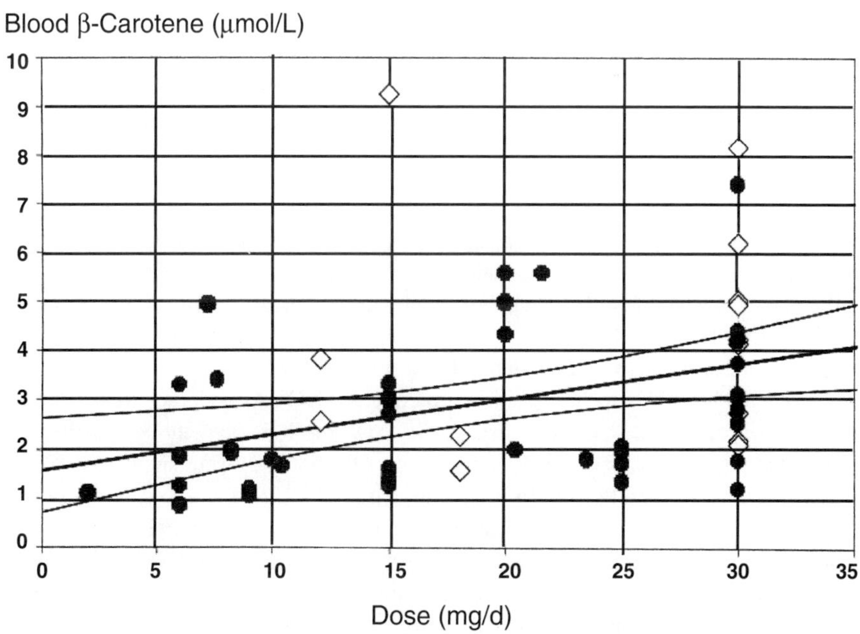

Fig. 19.1. Relation between β-carotene supplement dose and β-carotene steady-state concentration in blood serum or plasma (Scatter plot, n = 67 observations from 57 published studies; $y = 1.65 + 0.74x$). Points denote arithmetic means for treatment groups from studies with single-unit application (●) and multiple-unit application (◇).

TABLE 19.1
Frequency Statistics of the Categorical Variables: Origin, Formulation, Cosupplementation, Dosing Regimen, Gender, and Smoking

	Percentage
Origin	
Natural	16.7
Synthetic	83.3
Formulation	
Beadlet	65.0
Dispersion	35.0
Cosupplements	
No	65.2
Yes	34.8
Dosing regimen	
Multiple-unit	24.6
Single-unit	75.4
Gender	
Female	16.7
Male	45.4
Mixed	37.9
Smoking	
Mixed	29.8
Nonsmoking	59.7
Smoking	10.5

categorical variables impeding the chance to discriminate between different levels of the categorical variables. An exact Fisher test was used to test the categorical variables on pair-wise association. Smoking was found to be confounded and therefore was dropped from further analysis. Of the remaining variables, origin × formulation, origin × cosupplementation, cosupplementation × smoking, and regimen × gender were significantly associated ($P < 0.05$). The association of origin and formulation was particularly strong because the majority of beadlets are formulated from synthetic BC, whereas natural BC is used predominantly for making dispersions.

Potential variables to be used for subsequent ANOVA-modeling were selected using a nonparametric Wilcoxon test. In this way, the homogeneity of baseline BC was tested and gender ($P = 0.047$) and smoking ($P = 0.0042$) were significant. These factors were subsequently used for modeling baseline BC as a function of the potential covariates, i.e., baseline BC $= f$(gender, smoking). Using a Fisher test on pair-wise association, gender and smoking were not associated ($P_{\text{Fisher}} = 0.31$) and hence can be assumed to provide enough information on the interaction of gender and smoking. However, the interaction term was not significant ($P_{\text{gender*smoking}} = 0.09$) in the joint ANOVA model. The main-effect ANOVA model was well defined, explaining 37% of the variance of baseline BC. Gender and smoking had significant effects, with women having significantly higher plasma BC than men, and smokers having significantly lower plasma BC than nonsmokers (Table 19.2).

TABLE 19.2
Least-Square Means from the ANOVA Models for Baseline β-Carotene Concentration = *f*(gender, smoking) and for β-Carotene Increase = *f*(formulation, cosupplementation, dosing regimen)[a]

	LSM	SEM	LCL	UCL	n
	Baseline β-carotene				
Gender					
Male	0.34	0.03	0.27	0.40	25
Female	0.46	0.05	0.35	0.56	10
Mixed	0.46	0.03	0.38	0.53	21
Smoking					
Nonsmoking	0.50	0.03	0.44	0.55	33
Mixed	0.42	0.03	0.35	0.49	17
Smoking	0.34	0.06	0.21	0.46	6
	β-Carotene increase				
Formulation					
Dispersion	2.17	0.44	1.28	3.06	17
Beadlet	3.23	0.29	2.63	3.83	34
Cosupplements					
No	3.25	0.34	2.56	3.94	34
Yes	2.15	0.40	1.34	2.97	17
Dosing regimen					
Multiple-unit dose	3.48	0.46	2.53	4.42	14
Single unit dose	1.93	0.29	1.34	2.52	37

[a]LSM, least-square means; LCL, lower limit of confidence; SEM, standard error of mean; UCL, upper limit of confidence.

The pair-wise Wilcoxon test was used to analyze the increase in BC concentration (steady state–baseline) as a function of the covariates. The increase was significantly correlated with origin, formulation, cosupplementation, and dosing regimen. The increase of BC concentration was fitted by a multiple linear model using as covariates: formulation, cosupplementation, dosing regimen, and dose as well as all two-way interactions. Origin was not included in the model because of its strong association with formulation. Neither dose nor any interaction was found to be significant; therefore, dose and all higher-order terms were dropped from the model. The resulting ANCOVA model explained 30% variance of the increase ($DF_{error} = 47$). All main effects, namely, formulation ($P = 0.04$), cosupplementation ($P = 0.03$), and dosing regimen ($P < 0.01$), were significant. The least-square means are given in Table 19.2. The dosing regimen exerted by far the strongest effect, resulting in significantly higher plasma concentrations under a multiple-unit dosing regimen compared with single-unit dosing. Similarly, beadlets appeared to be more effective than the dispersion. Cosupplementation was associated with lower blood BC compared with exclusive BC treatment. Baseline BC was not significantly correlated with BC increase ($r = 0.07$, $P > 0.05$).

Discussion

A high variability in the response to BC supplements, within and between studies, was shown previously by others (10,28). Within the studies evaluated here, the mean response in individual studies ranged from 1.5 times baseline (19) to 18.5 times baseline (2). The wide variation is evident from Figure 19.1 showing, for example, that a steady-state concentration of ~1 µmol/L was elicited by a daily dose ranging from 2 to 30 mg. Moreover, a dose of 30 mg/d resulted in mean steady-state concentrations ranging from 1.2 to 8.2 µmol/L. Although baseline BC was not significantly related to the BC response in our analysis across studies (r = 0.07, $P > 0.05$), individual studies found that baseline BC was an important (20) or the most important predictor (55) of the response.

In the present analysis across studies, the formulation (oily dispersion or beadlet) was a significant predictor of plasma steady-state concentration. Similarly, various authors found that beadlet preparations appear to elicit a greater plasma response than BC in an oil matrix (3,5,13,56,57). This is likely due to the fact that in beadlet preparations, BC is more finely micronized than in oily dispersions, resulting in enhanced micellar solubility and thus, improved absorption (58). Compared with BC from vegetables, purified BC appears to have significantly higher bioavailability (59,60).

Across studies, dividing the daily dose over two or more meals (multiple-unit dosing regimen) was the most important predictor of the blood response. Similarly, a study where 51 mg/d BC was given once a day with a meal or in three divided doses with meals resulted in a blood response of 3 and 9 times the baseline concentration, respectively (61). We presume that, due to the hydrophobic nature of BC, multiple-unit application may result in a higher proportion of the total daily dose getting into micellar solution and thus enhancing absorption.

Across studies, cigarette smoking and gender were significant determinants of baseline BC, agreeing with earlier studies (55,62). We found that cosupplementation of BC with one or several micronutrients, *viz.*, vitamin E, vitamin C, selenium, zinc, copper, and ginseng, but excluding other carotenoids, attenuated the BC response. Although small, this effect merits further investigation.

Numerous anthropomorphic and dietary factors affect the bioavailability of BC as reviewed recently (59,60). However, many of these factors were not recorded in the studies analyzed, thus precluding entry in the statistical evaluation. For instance, intake of BC together with a main meal containing fat is crucial for absorption of carotenoids (59,61), yet no such advice was given in the majority of studies. Nonabsorbable fat replacers such as sucrose-polyester can decrease the plasma response markedly (63). Further unreported factors in many studies were compliance, age, BMI, body weight, the amount of dietary fat and of dietary BC, and alcohol consumption. Few study protocols included dietary recommendations (e.g., advice not to change the normal diets).

In a simple bivariate biometrical model, blood BC response across studies was significantly related to BC dose. However, upon applying a more sophisticated multivariate model, the dose effect vanished. This is in agreement with individual studies that found no significant dose effect (5–7) but in contrast to studies reporting a significant dose effect (3,4). This conflicting evidence indicates that the dose effect may be masked by covariates within a study and even more so, across studies. This points to the importance of controlling the multitude of potential covariates in BC bioavailability studies.

This evaluation had certain limitations. Apart from the lack of information on covariates as indicated above, no individual blood BC levels were obtainable from the included studies. Thus, it was not possible to use individual subjects as units of observation; therefore, the available data fell short of the preconditions for a *meta-analysis* as planned initially. Nevertheless, the comprehensive evaluation of a total of 57 studies provides useful new insight into the factors affecting the relation between BC supplementation and the blood response. Nevertheless, in spite of these limitations, our model explained 30% of the variance of the blood BC response. In a study with 582 subjects given 50 mg BC/d for 1 y (21), multivariate analysis accounted for relatively little of the variability of the plasma response ($R^2 = 0.14$).

Across studies, the dosing regimen was the most prominent factor affecting the response to BC. Thus, multiple-unit dosing was more effective than single-unit dosing, outweighing the effect of dose within the dosing range from 2 to 30 mg/d. This finding should be taken into consideration when planning BC availability trials or deriving maximum intake levels for BC supplementation.

Summary

In this evaluation across 57 human studies, cigarette smoking and gender exerted significant effects on baseline BC, with women revealing higher and smokers lower concentrations. BC increase was strongly affected by the dosing regimen with multiple-unit dosing being more effective than single-unit dosing. Beadlet formulation was more bioavailable than dispersion in oil. Cosupplementation with other micronutrients had a negative effect on the response to BC. Within the range of doses included in this evaluation (up to 30 mg/d) a significant effect of dose on blood BC increase was found in a simple bivariate model. However, the dose-effect disappeared after adjusting for the effects of other covariates in a joint ANCOVA model. The effectiveness of multiple-unit dosing should be considered in BC trial planning.

References

1. The Alpha-Tocopherol, Beta-Carotene Lung Cancer Prevention Study Group (1994) The Effect of Vitamin E and Beta Carotene on the Incidence of Lung Cancer and Other Cancers in Male Smokers, *N. Engl. J. Med. 330:* 1029–1035.
2. Age-Related Eye Disease Study Research Group (2001) A Randomized, Placebo-Controlled, Clinical Trial of High-Dose Supplementation with Vitamins C and E and Beta Carotene for Age-Related Cataract and Vision Loss, *Arch. Ophthalmol. 119:* 1439–1452.

3. Micozzi, M.S., Brown, E.D., Edwards, B.K., Bieri, J.G., Taylor, P.R., Khachik, F., Beecher, G.R., and Smith, J.C., Jr. (1992) Plasma Carotenoid Response to Chronic Intake of Selected Foods and Beta-Carotene Supplements in Men, *Am. J. Clin. Nutr. 55:* 1120–1125.
4. Ringer, T.V., DeLoof, M.J., Winterrowd, G.E., Francom, S.F., Gaylor, S.K., Ryan, J.A., Sanders, M.E., and Hughes, G.S. (1991) Beta-Carotene's Effects on Serum Lipoproteins and Immunologic Indices in Humans, *Am. J. Clin. Nutr. 53:* 688–694.
5. Thuermann, P.A., Steffen, J., Zwernemann, C., Aebischer, C.P., Cohn, W., Wendt, G., and Schalch, W. (2002) Plasma Concentration Response to Drinks Containing Beta-Carotene as Carrot Juice or Formulated as a Water Dispersible Powder, *Eur. J. Nutr. 41:* 228–235.
6. Abbey, M., Nestel, P.J., and Baghurst, P.A. (1993) Antioxidant Vitamins and Low-Density-Lipoprotein Oxidation, *Am. J. Clin. Nutr. 58:* 525–532.
7. Xu, M.J., Plezia, P.M., Alberts, D.S., Emerson, S.S., Peng, Y.M., Sayers, S.M., Liu, Y., Ritenbaugh, C., and Gensler, H.L. (1992) Reduction in Plasma or Skin Alpha-Tocopherol Concentration with Long-Term Oral Administration of Beta-Carotene in Humans and Mice, *J. Natl. Cancer Inst. 84:* 1559–1565.
8. Schwedhelm, E., Maas, R., Troost, R., and Boger, R.H. (2003) Clinical Pharmacokinetics of Antioxidants and Their Impact on Systemic Oxidative Stress, *Clin. Pharmacokinet. 42:* 437–459.
9. Galley, H.F., Thornton, J., Howdle, P.D., Walker, B.E., and Webster, N.R. (1997) Combination Oral Antioxidant Supplementation Reduces Blood Pressure, *Clin. Sci. 92:* 361–365.
10. Sasaki, S., Tsubono, Y., Okubo, S., Hayashi, M., Kakizoe, T., and Tsugane, S. (2000) Effects of 3 Months Oral Supplementation of β-Carotene and Vitamin C on Serum Concentrations of Carotenoids and Vitamins in Middle-Aged Subjects: A Pilot Study for a Randomized Controlled Trial to Prevent Gastric Cancer in High-Risk Japanese Population, *Jpn. J. Cancer Res. 91:* 464–470.
11. Frommel, T.O., Mobarhan, S., Doria, M., Halline, A.G., Luk, G.D., Bowen, P.E., Candel, A., and Liao, Y. (1995) Effect of Beta-Carotene Supplementation on Indices of Colonic Cell Proliferation, *J. Natl. Cancer Inst. 87:* 1781–1787.
12. Mobarhan, S., Shiau, A., Grande, A., Kolli, S., Stacewicz-Sapuntzakis, M., Oldham, T., Liao, Y., Bowen, P., Dyavanapalli, M., and Kazi, N. (1994) Beta-Carotene Supplementation Results in an Increased Serum and Colonic Mucosal Concentration of Beta-Carotene and a Decrease in Alpha-Tocopherol Concentration in Patients with Colonic Neoplasia, *Cancer Epidemiol. Biomark. Prev. 3:* 501–505.
13. Nierenberg, D.W., Stukel, T.A., Mott, L.A., and Greenberg, E.R. (1994) Steady-State Serum Concentration of Alpha Tocopherol Not Altered by Supplementation with Oral Beta Carotene. The Polyp Prevention Study Group, *J. Natl. Cancer Inst. 2:* 117–120.
14. Kazi, N., Radvany, R., Oldham, T., Keshavarzian, A., Frommel, T.O., Libertin, C., and Mobarhan, S. (1997) Immunomodulatory Effect of β-Carotene on T Lymphocyte Subsets in Patients with Resected Colonic Polyps and Cancer, *Nutr. Cancer 28:* 140–145.
15. Palan, P.R., Chang, C.J., Mikhail, M.S., Ho, G.Y., Basu, J., and Romney, S.L. (1998) Plasma Concentrations of Micronutrients During a Nine-Month Clinical Trial of β-Carotene in Women with Precursor Cervical Cancer Lesions, *Nutr. Cancer 30:* 46–52.
16. Kaugars, G.E., Silverman, S., Jr., Lovas, J.G., Brandt, R.B., Riley, W.T., Dao, Q., Singh, V.N., and Gallo, J. (1994) A Clinical Trial of Antioxidant Supplements in the Treatment of Oral Leukoplakia, *Oral Surg. Oral Med. Oral Pathol. 78:* 462–468.

17. Omenn, G.S., Goodman, G.E., Thornquist, M.D., Rosenstock, L., Barnhart, S., Gylys-Colwell, I., Metch, B., and Lund, B. (1993) The Carotene and Retinol Efficacy Trial (CARET) to Prevent Lung Cancer in High-Risk Populations: Pilot Study with Asbestos-Exposed Workers, *Cancer Epidemiol. Biomark. Prev. 2:* 381–387.
18. McLarty, J.W., Holiday, D.B., Girard, W.M., Yanagihara, R.H., Kummet, T.D., and Greenberg, S.D. (1995) Beta-Carotene, Vitamin A and Lung Cancer Chemoprevention: Results of an Intermediate Endpoint Study, *Am. J. Clin. Nutr. 62:* 1431S–1438S.
19. Malvy, D.J., Favier, A., Faure, H., Preziosi, P., Galan, P., Arnaud, J., Roussel, A.M., Briancon, S., and Hercberg, S. (2001) Effects of Two Years' Supplementation with Natural Antioxidants on Vitamin and Trace Element Status Biomarkers: Preliminary Data of the SU.VI.MAX Study, *Cancer Detect. Prev. 25:* 479–485.
20. Wright, A.J., Hughes, D.A., Bailey, A.L., and Southon, S.L. (1999) Beta-Carotene and Lycopene, but Not Lutein, Supplementation Changes the Plasma Fatty Acid Profile of Healthy Male Non-Smokers, *J. Lab. Clin. Med. 134:* 592–598.
21. Albanes, D., Virtamo, J., Rautalahti, M., Haukka, J., Palmgren, J., Gref, C.G., and Heinonen, O.P. (1992) Serum Beta-Carotene Before and After Beta-Carotene Supplementation, *Eur. J. Clin. Nutr. 46:* 15–24.
22. Allard, J.P., Royall, D., Kurian, R., Muggli, R., and Jeejeebhoy, K.N. (1994) Effects of Beta-Carotene Supplementation on Lipid Peroxidation in Humans, *Am. J. Clin. Nutr. 59:* 884–890.
23. Biesalski, H.K., Hemmes, C., Hopfenmuller, W., Schmid, C., and Gollnick, H.P. (1996) Effects of Controlled Exposure to Sunlight on Plasma and Skin Levels of β-Carotene, *Free Radic. Res. 24:* 215–242.
24. Blot, W.J., Li, J.-Y.X., Taylor, P.R., Guo, W., Dawsey, S., Wang, G.Q., Yang, C.S., Zheng, S.F., Gail, M., Li, G.Y., Yu, Y., Liu, B., Tangrea, J., Sun, Y., Liu, F., Fraumeni, J.F., Zhang, Y.-H., and Li, B. (1993) Nutrition Intervention Trials in Linxian, China: Supplementation with Specific Vitamin/Mineral Combinations, Cancer Incidence, and Disease-Specific Mortality in the General Population, *J. Nat. Cancer Inst. 85:* 1483–1492.
25. Calzada, C., Bruckdorfer, K.R., and Rice-Evans, C.A. (1997) The Influence of Antioxidant Nutrients on Platelet Functions in Healthy Volunteers, *Atherosclerosis 128:* 97–105.
26. Carroll, Y.L., Corridan, B.M., and Morissey, P.A. (2000) Lipoprotein Carotenoid Profiles and the Susceptibility of Low Density Lipoprotein to Oxidative Modification in Healthy Elderly Volunteers, *Eur. J. Clin. Nutr. 54:* 500–507.
27. Cheng, T., Zhu, Z., Masuda, S., and Morcos, N.C. (2001) Effects of Multinutrient Supplementation on Antioxidant Defence Systems in Healthy Human Beings, *J. Nutr. Biochem. 12:* 388–395.
28. Constantino, J.P., Kuller, L.H., Begg, L., Redmond, C.K., and Bates, M.W. (1988) Serum Level Changes After Administration of a Pharmacologic Dose of Beta-Carotene, *Am. J. Clin. Nutr. 48:*1277–1283.
29. Corridan, B.M., O'Donoghue, M.O., Hughes, D.A., and Morrissey, P.A. (2001) Low-Dose Supplementation with Lycopene or β-Carotene Does Not Enhance Cell-Mediated Immunity in Healthy Free-Living Elderly Humans, *Eur. J. Clin. Nutr. 55:* 627–635.
30. Faulks, R.M., Hart, D.J., Scott, J., and Southan, S.L. (1998) Changes in Plasma Carotenoid and Vitamin E Profile During Supplementation with Oil Palm Fruit Carotenoids, *J. Lab. Clin. Med. 132:* 507–511.

31. Fotouhi, N., Meydani, M., Santos, M.S., Meydani, S.N., Hennekens, C.H., and Gaziano, J.M. (1996) Carotenoid and Tocopherol Concentrations in Plasma, Peripheral Blood Mononuclear Cells, and Red Blood Cells After Long-Term Beta-Carotene Supplementation in Men, *Am. J. Clin. Nutr. 63:* 553–558.
32. Fuller, C.J., Faulkner, H., Bendich, A., Parker, R.S., and Roe, D.A. (1992) The Effect of β-Carotene on the Ultraviolet Light-Induced Suppression of Cellular Immune Function in Healthy Young Males, *Am. J. Clin. Nutr. 56:* 684–690.
33. Girodon, F., Blache, D., Monget, A.-L., Lombart, M., Brunet-Lecompte, P., Arnaud, J., Richard, M.J., and Galan, P. (1997) Effect of a Two-Year Supplementation with Low Doses of Antioxidant Vitamins and/or Minerals in Elderly Subjects on Levels of Nutrients and Antioxidant Defense Parameters, *J. Am. Coll. Nutr. 16:* 357–365.
34. Goodman, G.E., Omenn, G.S., Thornquist, M.D., Lund, B., Metch, B., and Gylys-Colwell, I. (1993) The Carotene and Retinol Efficacy Trial (CARET) to Prevent Lung Cancer in High-Risk Populations: Pilot Study with Cigarette Smokers, *Cancer Epidemiol. Biomarkers Prev. 2:* 389–396.
35. Gossage, C., Heyhim, M., Moser-Veillon, P.B., Douglas, L.W., and Kramer, T.R. (2000) Effect of β-Carotene Supplementation and Lactation of Carotenoid Metabolism and Mitogenic T Lymphocytes Proliferation, *Am. J. Clin. Nutr. 71:* 950–955.
36. Grievink, L., Jansen, S.M.A., van't Beer, P., and Brunekreef, B.L. (1998) Acute Effects of Ozone on Pulmonary Function of Cyclists Receiving Antioxidant Supplements, *Occup. Environ. Med. 55:* 13–17.
37. Herraiz, L.A., Hsieh, W.C., Parker, R.S., Swanson, J.E., Bendich, A., and Roe, D.A. (1998) Effect of UV Exposure and β-Carotene Supplementation on Delayed-Type Hypersensitivity Response in Healthy Older Men, *J. Am. Coll. Nutr. 17:* 617–624.
38. Hininger I.A, Meyer-Wenger, Moser, U., Wright, A., Southon, S., Thurnham, D., Chopra, M., van Den Berg, H., Olmedilla, B., Favier, A.E., and Roussel, A.M. (2001) No Significant Effects of Lutein, Lycopene or β-Carotene Supplementation on Biological Markers of Oxidative Stress and LDL Oxidizability in Healthy Adult Subjects, *J. Am. Coll. Nutr. 20:* 232–238.
39. Hughes, D.A., Wright, A.J.A., Finglas, P.M., Peerless, A.C., Bailey, A.L., Astley, S.B., Pinder, A.C., and Southon, S. (1997) The Effect of β-Carotene Supplementation on the Immune Function of Blood Monocytes from Healthy Male Nonsmokers, *J. Lab. Clin. Med. 129:* 309–317.
40. Kanter, M.M., Nolte, L.A., and Holloszy, J.O. (1993) Effects of an Antioxidant Vitamin Mixture on Lipid Peroxidation at Rest and Postexercise, *J. Appl. Physiol. 74:* 965–969.
41. Kardinaal, A.F., van't Beer, P., Brants, H.A., van den Berg, H., van Schoonhoven, J., and Hermus, R.J. (1995) Relations Between Antioxidant Vitamins in Adipose Tissue, Plasma, and Diet, *Am. J. Epidemiol. 5:* 440–450.
42. Kim, H.S., and Lee, B.M. (2001) Protective Effects of Antioxidant Supplementation on Plasma Lipid Peroxidation in Smokers, *J. Toxicol. Environ. Health 63:* 583–598.
43. Lee, B.M., Lee, S.K., and Kim, H.S.l. (1998) Inhibition of Oxidative DNA Damage, 8-OhdG, and Carbonyl Contents Treated with Antioxidants (Vitamin E, Vitamin C, β-Carotene and Red Ginseng), *Cancer Lett. 132:* 219–227.
44. Manetta, A., Schubbert, T., Chapman, J., Schell, M.J., Peng, Y.M., Liao, S.Y., and Meyskens, F.J., Jr. (1996) β-Carotene Treatment of Cervical Intraepithelial Neoplasia: A Phase II Study, *Cancer Epidemiol. Biomark. Prev. 5:* 929–932.

45. Meijer, E.P., Goris, A.H.C., Senden, J., van Dongen, J.L., Bast, A., and Westerterp, K.R. (2001) Antioxidant Supplementation and Exercise-Induced Oxidative Stress in the 60-Year-Old as Measured by Antipyrine Hydroxylates, *Br. J. Nutr. 86:* 569–575.
46. Meraji, S., Ziouzenkova, O., Resch, U., Khoschsorur, A., Tatzber, F., and Esterbauer, H. (1997) Enhanced Plasma Level of Lipid Peroxidation in Iranians Could Be Improved by Antioxidants Supplementation, *Eur. J. Clin. Nutr. 51:* 318–325.
47. Mobarhan, S., Bowen, P., Andersen, B., Evans, M., Stacewicz-Sapuntzakis, M., Sugerman, S., Simms, P., Lucchesi, D., and Friedman, H. (1990) Effects of Beta-Carotene Repletion on Beta-Carotene Absorption, Lipid Peroxidation, and Neutrophil Superoxide Formation in Young Men, *Nutr. Cancer 14:* 195–206.
48. Olmedilla, B., Granado, F., Southon, S., Wright, A.J., Blanco, I., Gil-Martinez, E., van den Berg, H., Thurnham, D., Corridan, B., Chopra, M., and Hininger, I. (2002) A European Multicentre, Placebo-Controlled Supplementation Study with α-Tocopherol, Carotene-Rich Palm Oil, Lutein or Lycopene: Analysis of Serum Responses, *Clin. Sci . 102:* 447–456.
49. Preziosi, P., Galan, P., Herbeth, B., Valeix, P., Roussel, A.M., Malvy, D., Paul-Dauphin, A., Arnaud, J., Richard, M.J., Briancon, S., Favier, A., and Hercberg, S. (1998) Effects of Supplementation with a Combination of Antioxidant Vitamins and Trace Elements, at Nutritional Doses, on Biochemical Indicators and Markers of the Antioxidant System in Adult Subjects, *J. Am. Coll. Nutr. 17:* 244–249.
50. Stahl, W., Heinrich, U., Jungmann, H., von Laar, J., Schietzel, M., Sies, H., and Tronnier, H. (1998) Increased Dermal Carotenoid Levels Assessed by Noninvasive Reflection Spectrophotometry Correlate with Serum Levels in Women Ingesting Betatene, *J. Nutr. 128:* 903–907.
51. Steinberg, F.M., and Chait, A. (1998) Antioxidant Vitamin Supplementation and Lipid Peroxidation in Smokers, *Am. J. Clin. Nutr. 68:* 319–327.
52. Sumida, S., Doi, T., Sakurai, M., Yoshioka, Y., and Okamura, K. (1997) Effect of a Single Bout of Exercise and β-Carotene Supplementation on the Urinary Excretion of 8-Hydroxy-deoxyguanosine in Humans, *Free Radic. Res. 27:* 607–618.
53. Tauler, P., Aguilo, A., Fuentespina, E., Tur, J.A., and Pons, A. (2002) Diet Supplementation with Vitamin E, Vitamin C and β-Carotene Cocktail Enhances Basal Neutrophil Antioxidant Enzymes in Athletes, *Pflugers Arch. 443:* 791–797.
54. Van Poppel, G., Kok, F.J., Duijzings, P., and de Vogel, N. (1992) No Influence of Beta-Carotene on Smoking-Induced DNA Damage as Reflected by Sister Chromatid Exchanges, *Int. J. Cancer 51:* 355–358.
55. Nierenberg, D.W., Stukel, T.A., Baron, J.A., Dain, B.J., and Greenberg, E.R. (Skin Cancer Prevention Study Group) (1991) Determinants of Increase in Plasma Concentration of β-Carotene After Chronic Oral Supplementation, *Am. J. Clin. Nutr. 53:* 1443–1449.
56. Toerroenen, R., Lehmusaho, M., and Häkinen, S. (1996) Serum β-Carotene Response to Supplementation with Raw Carrots, Carrot Juice or Purified β-Carotene in Healthy Non-Smoking Women, *Nutr. Res. 16:* 565–575.
57. Nierenberg, D.W., Dain, B.J., Mott, L.A., Baron, J.A., and Greenberg, E.R. (1997) Effects of 4 y of Oral Supplementation with β-Carotene on Serum Concentrations of Retinol, Tocopherol, and Five Carotenoids, *Am. J. Clin. Nutr. 66:* 315–319.
58. Horn, D., and Lüddecke, E. (1996) Preparation and Characterization of Nano-Sized Carotenoid Hydrosols, in *Fine Particles Science and Technology*, Pelizzetti. E., ed., Kluwer Academic Publisher, New York, pp. 761–775.

59. van het Hof, K., West, C.E., Westrate, J.A., and Hautvast, J.G.A.J. (2000) Dietary Factors That Affect the Bioavailability of Carotenoids, *J. Nutr. 130:* 503–506.
60. Castenmiller, J.J.M., and West, C.E. (1998) Bioavailability and Bioconversion of Carotenoids, *Annu. Rev. Nutr. 18:* 19–38.
61. Prince, M.R., and Frisoli, J.K. (1993) Beta-Carotene Accumulation in Serum and Skin, *Am. J. Clin. Nutr. 57:* 175–181.
62. Saintot, M., Astre, C., Scali, J., and Gerber, M. (1995) Within-Subjects Seasonal Variation and Determinants of Inter-Individual Variations of Plasma β-Carotene, *Intern. J. Vit. Nutr. Res. 65:* 169–174.
63. Schlagheck, T.G., Riccardi, K.A., Zorich, N.L., Torri, S.A., Dugan, L.D., and Peters, J.C. (1997) Olestra Dose Response on Fat Soluble and Water Soluble Nutrients in Humans, *J. Nutr. 127:* 1636S–1645S.

Chapter 20

New Horizons in Carotenoid Research

Helmut Sies and Wilhelm Stahl

Heinrich-Heine-University Düsseldorf, Institute of Biochemistry and Molecular Biology I, D-40001 Düsseldorf, Germany

Introduction

Epidemiologic studies revealed that the incidence of major types of cancer in the Western world, such as carcinoma of the lung, stomach, prostate, mouth, esophagus, colon or rectum, is inversely associated with the consumption of a diet rich in fruits and vegetables (1). The preventive effects of such a diet were also related to coronary heart disease, atherosclerosis, and stroke as well as cataract, age-related macular degeneration, obesity or diabetes. Among other micronutrients, a variety of carotenoids are major constituents of colored fruits and vegetables. However, disappointing results from intervention trials with β-carotene dampened the enthusiasm for the role of this compound in the etiology of degenerative diseases, and beneficial effects of carotenoids in general have been questioned (2). However, it should be recognized that as a result of this research, knowledge of carotenoids increased substantially and a more differentiated view of the implication of carotenoids in human health is warranted. Progress was made in understanding the mechanisms of prevention and toxicity, cellular signaling, regulation of gene expression or metabolism, and bioavailability and led to the development of new concepts in carotenoid research.

Lutein, (3*R*,3′*R*)-zeaxanthin, and (3*R*,3′*S*)-zeaxanthin (meso-zeaxanthin) are the predominant carotenoids of the macular pigment; they are apparently involved in the protection of this tissue against light-induced damage and probably prevent age-related macular degeneration (3) (see Chapter 7 in this book). The macular pattern of carotenoids and the spatial distribution within different areas were analyzed, with amounts of zeaxanthin increasing toward the center of the yellow spot. Applying noninvasive techniques, it was shown that the macular pigment density increases upon supplementation with lutein (4). There is evidence that lutein is partially converted to meso-zeaxanthin in this tissue. More sophisticated noninvasive techniques including Raman spectroscopy are under development and will be helpful to study the uptake, distribution, and metabolism of macular carotenoids (5) (see Chapter 6 in this book). The mechanism(s) underlying protection remain to be elucidated. Antioxidant properties, especially singlet oxygen quenching, as well as blue light filter effects were discussed (6,7). It is still not known why lutein and zeaxanthin are selectively enriched in the macula lutea, although β-carotene and lycopene are present in blood at higher levels and are efficient antioxidants and blue light filters as well. Specific physicochemical features directing the orientation of carotenoids in cellular membranes might be important (8). The mechanisms involved in the selective uptake and transport have not yet been identified. Characterization of carotenoid binding and metabolizing enzymes may provide motifs for selection. It should also be noted that it remains to be demonstrated whether the intake of dietary supplements of lutein or zeaxanthin lowers the incidence of age-related macular degeneration, which would be proof of their contribution to ocular health.

With respect to cancer prevention, it was postulated that supplementation with a single carotenoid at high doses is not sufficient to achieve effects. Assuming that antioxidant properties play a role, it was suggested that the entire antioxidant network must be strengthened. A combination of antioxidants as present in fruits and vegetables, rather than individual supplements, is more likely to be beneficial. Cooperative interactions of various dietary compounds, micronutrients, and macronutrients, may provide synergistic effects important for protection (9).

Dietary intake of tomatoes and tomato products containing lycopene as a major carotenoid was reported to be associated with a lower risk of prostate cancer, and a relationship between high plasma levels of lycopene and a decreased risk for prostate cancer was established (10) (see Chapter 18 in this book). In a meta-analysis including 11 case-control studies and 10 cohort studies or nested case-control studies, tomato products or lycopene and prostate cancer were evaluated (11). It was concluded that tomato products may play a role in the prevention of prostate cancer, although the effect was only moderate and was restricted to high amounts of tomato intake. Further research is warranted to verify the relation between tomato product consumption and prostate cancer and substantiate that lycopene is a biologically active ingredient involved in protection.

The adverse effects observed under long-term supplementation with β-carotene leading to an increased risk for lung cancer in heavy smokers deserve further attention. After animal studies, it was postulated that oxidation of β-carotene under conditions of smoking leads to the formation of unphysiologically high amounts of oxidation products interfering with retinoid-dependent signaling (12). The formation of oxidation products generated under conditions of smoking and β-carotene supplementation can be suppressed by the coapplication of vitamins E and C (13). Although disturbances of retinoic acid pathways likely play a role, the implication of retinoic acid receptors in β-carotene–induced lung cancer remains unclear. Prooxidant activities of carotenoids were demonstrated *in vitro* and were suggested to be involved in carcinogenic processes; however, no such activities have yet been proven *in vivo* (14). Most intervention studies analyzing biomarkers of oxidative damage provide evidence for antioxidant effects of carotenoids (see Chapter 13 in this book). In cell culture, β-carotene cleavage products increase levels of prooxidants (see Chapter 9 in this book). The development of biomarkers indicative of prooxidant activity of carotenoids and the relation of carotenoid exposure to biochemical and clinical markers of carcinogenesis will help reveal the contribution of carotenoids to cancer prevention.

There is increasing evidence that dietary antioxidants including carotenoids exhibit biological activities not directly related to their antioxidant properties (15). The parent compounds and/or their metabolites have an effect on cellular signaling pathways, influencing the expression of certain genes or may act as inhibitors of regulatory enzymes. Cooperative inhibitory effects on the growth of HL-60 leukemic cells were observed when lycopene was applied together with 1,25-dihydroxyvitamin D_3 (16). Cooperative activity of lycopene and tocopherol was found when the inhibition on the growth of human prostate carcinoma cell lines was investigated (17); lycopene alone was ineffective. Such cooperative effects on prostate cancer were also shown in animal models (see Chapter 18 of this book). Inhibition of cell proliferation by lycopene is associated with a delay in cell cycle progression related to diminished insulin-like growth factor (IGF)-I receptor signaling (18). Lycopene lowers cyclin D levels, leading to diminished phosphorylation of the retinoblastoma protein (19), likely associated with growth arrest. Carotenoids induce phase I and phase II metabolic enzymes, which play a role in the detoxification of carcinogens (15). There is evidence that lycopene stimulates the expression of phase II enzymes *via* pathways dependent on the transcription factor Nrf2 and antioxidant responsive elements (ARE) in the promoter region of the respective genes (see Chapter 16 of this book). Such effects on gene expression may be related to the protective or adverse properties of carotenoids and deserve further attention.

In the context of the cancer-preventive activities of carotenoids, other mechanisms with an effect on cell signaling were discussed (20). Particular attention has been given to the stimulatory effects exerted by carotenoids on gap junctional communication (GJC), a property which is not related to their antioxidant activity (21,22). GJC is implicated in the regulation of cell growth, differentiation, and apoptosis. Nontumorous cells are contact-inhibited and have functional GJC; most tumor cells, however, have dysfunctional homologous or heterologous GJC (23). Regulation of GJC is complex, and the mechanisms related to carotenoid activity are not fully understood. It remains to be investigated whether carotenoid metabolites are involved (see Chapters 5 and 11 of this book).

Epidemiologic and clinical data indicate that dietary carotenoids may protect against cardiovascular disease, and an inverse association between serum levels of β-carotene and other carotenoids and coronary heart disease was shown (24). Special attention was given to lycopene (see Chapter 12 of this book). However, clinical studies with carotenoid supplementation were equivocal, and in fact some major clinical trials with β-carotene supplementation showed either no or negative effects on cardiovascular disease and cancer. It was suggested that carotenoids are simply biomarkers for fruit and vegetable dietary intake, and that other plant-derived compounds may play a significant role. On the other hand, carotenoids prevent LDL oxidation, a key step in atherosclerosis.

As a light-exposed tissue, skin must be protected against photooxidation. β-Carotene is used for oral sun protection either in the prevention of sunburn or the treatment of erythropoietic protoporphyria. The effects of β-carotene were demonstrated in a series of studies, and it was shown that other carotenoids or mixtures of carotenoids are protective (25). When carotenoids were provided in sufficient amounts with the diet, an inhibition of UV-induced erythema was achieved. There are initiatives to develop functional food based on this property making use of these compounds as endogenous photoprotectors. It should be mentioned, however, that the extent of skin protection with ingested carotenoids is much lower than that obtained with a sunscreen. However, increasing the basal protection systemically contributes to the permanent defense against UV light–mediated skin damage (25).

Vitamin A deficiency is still a severe problem in developing countries, and even in the Western world, some segments of the population are at risk for an inadequate vitamin A supply. Depending on dietary habits, a significant part of the vitamin A supply is derived from provitamin A carotenoids ingested with fruits and vegetables. The key step for vitamin A formation is the oxidative cleavage of provitamin A carotenoids by carotene-15,15′-oxygenases. Carotenoid-cleaving enzymes were partially purified and kinetically characterized. Successful cloning and sequencing of enzymes with β,β-carotene 15,15′-oxygenase activity from *Drosophila*, chickens, mice, and humans were reported (26) (see Chapter 15 of this book) (27). Determination of the expression pattern in mice revealed that mRNA for the oxygenase was highly expressed in liver, testes, kidney, and small intestine. Additionally, a carotenoid-cleaving enzyme responsible for excentric cleavage was cloned and characterized in mammals including humans, mice, and ferrets (see Chapter 15 in this book). Further progress in this area is to be expected, stimulating research in carotenoid metabolism and contributing to the identification of carotenoid metabolites with biological activities.

Although carotenoid research related to human health has focused on β-carotene, other dietary carotenoids such as lutein, zeaxanthin, or lycopene were intensively investigated more recently. However, there are further carotenoids present in the human diet that deserve attention. Among them are carotenes such as phytoene or phytofluene and several oxo-carotenoids including capsorubin, capsanthin, or violerythrin. The latter are most likely metabolized or modified by hydrolysis. However, little is known about the fate and biological activity of these products. As becomes obvious from this brief overview, there is growing evidence that metabolites and chemical oxidation products of carotenoids have interesting properties. Thus, there is demand for the development of selective methods for chemical synthesis and sensitive analytical methods to follow the generation of these degradation products *in vivo* (see Chapter 4 of this book).

References

1. Riboli, E., and Norat, T. (2003) Epidemiologic Evidence of the Protective Effect of Fruit and Vegetables on Cancer Risk, *Am. J. Clin. Nutr. 78:* 559S–569S.
2. Albanes, D. (1999) β-Carotene and Lung Cancer: A Case Study, *Am. J. Clin. Nutr. 69 (Suppl.):* 1345S–1350S.

3. Landrum, J.T., and Bone, R.A. (2001) Lutein, Zeaxanthin, and the Macular Pigment, *Arch. Biochem. Biophys. 385:* 28–40.
4. Bone, R.A., Landrum, J.T., Guerra, L.H., and Ruiz, C.A. (2003) Lutein and Zeaxanthin Dietary Supplements Raise Macular Pigment Density and Serum Concentrations of These Carotenoids in Humans, *J. Nutr. 133:* 992–998.
5. Gellermann, W., and Bernstein, P.S. (2004) Noninvasive Detection of Macular Pigments in the Human Eye, *J. Biomed. Opt. 9:* 75–85.
6. Stahl, W., and Sies, H. (2003) Antioxidant Activity of Carotenoids, *Mol. Asp. Med. 24:* 345–351.
7. Junghans, A., Sies, H., and Stahl, W. (2001) Macular Pigments Lutein and Zeaxanthin as Blue Light Filters Studied in Liposomes, *Arch. Biochem. Biophys. 391:* 160–164.
8. Sujak, A., Gabrielska, J., Grudzinski, W., Borc, R., Mazurek, P., and Gruszecki, W.I. (1999) Lutein and Zeaxanthin as Protectors of Lipid Membranes Against Oxidative Damage: The Structural Aspects, *Arch. Biochem. Biophys. 371:* 301–307.
9. Stahl, W., Junghans, A., de Boer, B., Driomina, E., Briviba, K., and Sies, H. (1998) Carotenoid Mixtures Protect Multilamellar Liposomes Against Oxidative Damage: Synergistic Effects of Lycopene and Lutein, *FEBS Lett. 427:* 305–308.
10. Giovannucci, E. (2002) A Review of Epidemiologic Studies of Tomatoes, Lycopene, and Prostate Cancer, *Exp. Biol. Med. 227:* 852–859.
11. Etminan, M., Takkouche, B., and Caamano-Isorna, F. (2004) The Role of Tomato Products and Lycopene in the Prevention of Prostate Cancer: A Meta-Analysis of Observational Studies, *Cancer Epidemiol. Biomark. Prev. 13:* 340–345.
12. Wang, X.-D., Liu, C., Bronson, R.T., Smith, D.E., Krinsky, N.I., and Russell, R.M. (1999) Retinoid Signaling and Activator Protein-1 Expression in Ferrets Given β-Carotene Supplements and Exposed to Tobacco Smoke, *J. Natl. Cancer Inst. 91:* 60–66.
13. Liu, C., Russell, R.M., and Wang, X.D. (2004) α-Tocopherol and Ascorbic Acid Decrease the Production of β-Apo-Carotenals and Increase the Formation of Retinoids from β-Carotene in the Lung Tissues of Cigarette Smoke-Exposed Ferrets *In Vitro*, *J. Nutr. 134:* 426–430.
14. Young, A.J., and Lowe, G.M. (2001) Antioxidant and Prooxidant Properties of Carotenoids, *Arch. Biochem. Biophys. 385:* 20–27.
15. Stahl, W., Ale-Agha, N., and Polidori, M.C. (2002) Non-Antioxidant Properties of Carotenoids, *Biol. Chem. 383:* 553–558.
16. Amir, H., Karas, M., Giat, J., Danilenko, M., Levy, R., Yermiahu, T., Levy, J., and Sharoni, Y. (1999) Lycopene and 1,25-Dihydroxyvitamin D_3 Cooperate in the Inhibition of Cell Cycle Progression and Induction of Differentiation in HL-60 Leukemic Cells, *Nutr. Cancer 33:* 105–112.
17. Pastori, M., Pfander, H.P., Boscoboinik, D., and Azzi, A. (1998) Lycopene in Association with α-Tocopherol Inhibits at Physiological Concentrations Proliferation of Prostate Carcinoma Cells, *Biochem. Biophys. Res. Commun. 250:* 582–585.
18. Karas, M., Amir, H., Fishman, D., Danilenko, M., Segal, S., Nahum, A., Koifmann, A., Giat, Y., Levy, J., and Sharoni, Y. (2000) Lycopene Interferes with Cell Cycle Progression and Insulin-Like Growth Factor I Signaling in Mammary Cancer Cells, *Nutr. Cancer 36:* 101–111.
19. Nahum, A., Hirsch, K., Danilenko, M., Watts, C.K., Prall, O.W., Levy, J., and Sharoni, Y. (2001) Lycopene Inhibition of Cell Cycle Progression in Breast and Endometrial Cancer Cells Is Associated with Reduction in Cyclin D Levels and Retention of p27(Kip1) in the Cyclin E-cdk2 Complexes, *Oncogene 20:* 3428–3436.

20. Bertram, J.S. (1999) Carotenoids and Gene Regulation, *Nutr. Rev. 57:* 182–191.
21. Zhang, L.X., Cooney, R.V., and Bertram, J.S. (1992) Carotenoids Up-Regulate Connexin43 Gene Expression Independent of Pro-Vitamin A or Antioxidant Properties, *Cancer Res. 52:* 5707–5712.
22. Stahl, W., Nicolai, S., Briviba, K., Hanusch, M., Broszeit, G., Peters, M., Martin, H.D., and Sies, H. (1997) Biological Activities of Natural and Synthetic Carotenoids: Induction of Gap Junctional Communication and Singlet Oxygen Quenching, *Carcinogenesis 18:* 89–92.
23. Trosko, J.E., and Chang, C.C. (2000) Modulation of Cell-Cell Communication in the Cause and Chemoprevention/Chemotherapy of Cancer, *Biofactors 12:* 259–263.
24. Kritchevsky, S.B. (1999) β-Carotene, Carotenoids and the Prevention of Coronary Heart Disease, *J. Nutr. 129:* 5–8.
25. Sies, H., and Stahl, W. (2004) Nutritional Protection Against Skin Damage from Sunlight, *Annu. Rev. Nutr. 24:* 173–200.
26. von Lintig, J., and Vogt, K. (2004) Vitamin A Formation in Animals: Molecular Identification and Functional Characterization of Carotene Cleaving Enzymes, *J. Nutr. 134:* 251S–256S.
27. Wyss, A. (2004) Carotene Oxygenases: A New Family of Double Bond Cleavage Enzymes, *J. Nutr. 134:* 246S–250S.

Chapter 21

Carotenoids and Cardiovascular Disease

J. Michael Gaziano[a,b] and Howard D. Sessoa[a,b]

[a]Divisions of Aging and Preventive Medicine, Department of Medicine, Brigham & Women's Hospital and Harvard Medical School and [b]Massachusetts Veterans Epidemiology Research and Information Center, VA Boston Healthcare System, Boston, MA

Introduction

Nutritional epidemiologic research demonstrates that those who consume higher amounts of fruits and vegetables tend to have lower rates of heart and vascular diseases, including coronary heart disease and stroke (1–5). Short-term dietary intervention trials suggest that diets rich in fruit and vegetable intake lead to improvements in coronary risk factors and reduce cardiovascular mortality (6–8). For example, the DASH (Dietary Approaches to Stop Hypertension) diet, in part emphasizing increases in fruit and vegetable intake, resulted in improvements in blood pressure (6) and possibly HDL cholesterol (9). The mechanisms for these apparent protective effects are not clear, but attention has focused on the micronutrients with antioxidant properties that could be responsible for the associated lower rates of chronic diseases.

Carotenoids are abundant, plant-derived, fat-soluble pigments that can function as antioxidants (10). Carotenoids are stored in the liver or adipose tissue, and are lipid soluble by becoming incorporated into plasma lipoprotein particles during transport (11). For these reasons, carotenoids from plants may represent one plausible mechanism by which fruits and vegetables reduce the risk of chronic diseases such as cardiovascular disease (CVD).

In this chapter, we briefly review some of the proposed biological mechanisms by which carotenoids may be associated with heart and vascular diseases. We then provide a summary of the human observational epidemiology on both dietary and blood carotenoids, including α-carotene, β-carotene, lycopene, lutein/zeaxanthin, and β-cryptoxanthin, with the risk of CVD. Finally, we discuss the evidence from the primary and secondary prevention trials of β-carotene supplementation.

Biologic Mechanisms

Basic research supports the hypothesis that dietary antioxidants, such as carotenoids, might reduce the risk of atherosclerosis through inhibition of oxidative damage. Data from *in vitro* and *in vivo* studies suggest that oxidative damage to LDL promotes several steps in atherogenesis (12), including endothelial cell damage (13,14), foam cell accumulation (15–17) and growth (18,19), and synthesis of autoantibodies (20). Further, animal studies suggest that free radicals may directly damage arterial endothelium (21), promote thrombosis (22), and interfere with normal vasomotor regulation (23). As discussed elsewhere in this text, *in vitro* data demonstrated the possible role of these antioxidants in preventing or delaying various steps in atherogenesis by inhibiting oxidation of LDL or other free radical reactions.

Carotenoids may also reduce the risk of CVD by reducing inflammation. Some evidence suggests that carotenoids are consumed in various proinflammatory states such as smoking. Data from the cross-sectional Third National Health and Nutrition Examination Survey (NHANES III) showed that subjects in the upper 15% of C-reactive protein (CRP) distribution had significantly lower serum α-carotene, β-carotene, lutein/zeaxanthin, β-cryptoxanthin, and lycopene levels even among nonsmokers (24). In a study of elderly nuns, aged 77–99 y, the authors also found that elevated CRP levels were inversely associated with plasma α-carotene ($P = 0.02$), β-carotene ($P = 0.02$), lycopene ($P = 0.03$), and total carotenoid ($P = 0.01$) concentrations (25). These data suggest that carotenoids could be involved in protecting against various atherosclerotic processes.

Dietary Intake and Risk of Cardiovascular Diseases

Although the basic research supports a role for carotenoids in preventing or delaying atherogenesis, we must rely on human studies to determine the relevance of these findings. In this section, we review studies of dietary intake of carotenoids. In most cases, the intake of carotenoids is estimated using a food-frequency questionnaire. Based on the responses to this questionnaire, we can calculate approximate intake of total and individual carotenoids using micronutrient tables.

Total Dietary Carotenoid Intake

Several early studies explored the associations of dietary carotenoids with the risk of vascular disease using crude scores for intake of carotenoid-containing foods. Rimm *et al.* (26) examined data in a large cohort of men in the Health Professionals' Follow-Up Study, finding an inverse association between total dietary carotenoid intake and a lower risk of coronary artery disease. The multivariate relative risks (RR) for increasing quintiles of carotene intake were 1.00 (ref), 0.92, 0.91, 0.86, and 0.71 (P, trend = 0.03). Upon stratification by smoking status, the inverse association strengthened among former and current smokers.

In a cohort study of 5133 Finnish men and women aged 30–69 y, a similar inverse association was noted for dietary carotenoid intake, particularly in the highest tertile (27). A community-based study of 1299 elderly men and women in

Massachusetts had a follow-up of 4.75 y (28). Using a score based on the intake of carotene-containing fruits and vegetables, the multivariate RR of cardiovascular death for increasing quartiles of carotene-containing fruits and vegetables were 1.00 (ref), 0.77, 0.63, and 0.59 (*P*, trend = 0.014). In a cohort study of 747 Massachusetts residents ≥ 60 y old, the combined intake of the five primary dietary carotenoids was associated with a possible lower risk of coronary heart disease mortality after up to 12 y of follow-up (29). Subjects in the upper quintile of dietary carotenoid intake had a nonsignificant 36% reduction in coronary heart disease mortality compared with those in the lowest quintile of intake.

More recent studies calculated a total dietary carotenoid value, typically summed from intake of individual carotenoids including α-carotene, β-carotene, lycopene, lutein/zeaxanthin, and β-cryptoxanthin. Two separate reports presented data from the Iowa Women's Health Study of 34,000 postmenopausal women (30,31). Dietary carotenoid intake was not associated with the risk of death from coronary heart disease, with multivariate RR for increasing quintiles of 1.00 (ref), 1.26, 1.18, 1.04, and 1.03 (*P*, trend = 0.71) (30). For stroke death, there was a similar lack of association for dietary carotenoids in multivariate models (*P*, trend = 0.88) (31).

Individual Dietary Carotenoid Intake

Several large-scale observational studies reported results for individual carotenoids. β-Carotene and lycopene were studied widely, given the possible role in cancer prevention compared with other carotenoids, i.e., lutein/zeaxanthin, α-carotene, and β-cryptoxanthin. Studies often consider these three carotenoids as part of a larger study of dietary or plasma carotenoids in relation to heart and vascular diseases, but few have focused exclusively on these specific carotenoids.

A good example of this type of study is the Nurses' Health Study among 73,286 women with 12 y of follow-up (32). In overall multivariate models, potential inverse associations were noted for α-carotene and β-carotene, but not for lutein/zeaxanthin, lycopene, or β-cryptoxanthin. The RR of coronary artery disease for increasing quintiles of α-carotene were 1.00 (ref), 0.89, 0.69, 0.89, and 0.74 (*P*, trend = 0.05); for β-carotene, the RR were 1.00 (ref), 0.93, 0.91, 0.80, and 0.80 (*P*, trend = 0.04). Results stratified by baseline smoking status for each carotenoid did not appear to mediate the overall associations. In a parallel cohort of 43,738 male health professionals followed for ~8 y, during which 328 strokes occurred (210 ischemic, 70 hemorrhagic, 48 unclassified), there was generally no association between α-carotene, β-carotene, and lycopene and the risk of stroke (33). However, the pattern of RR for increasing quintiles of β-carotene intake suggested a possible nonlinear association, with RR of 1.00 (ref), 0.75, 0.83, 0.76, and 0.77 (*P*, trend > 0.2). Higher quintiles of dietary lutein were associated with lower risks of total stroke, with multivariate RR of 1.00 (ref), 0.89, 0.88, 0.87, and 0.70 (*P*, trend = 0.06). Results were similar for lutein and the risk of ischemic stroke.

We now detail some of the observational studies focusing on specific major carotenoids in the diet for their associations with heart and vascular diseases. In some of these studies, the dietary intake of the individual carotenoids was compared with blood levels of those same carotenoids, and correlations were high (34).

β-Carotene. For decades, β-carotene has been the most widely studied individual carotenoid for its possible association with heart and vascular disease risk. This was not without reason because β-carotene is one of the most abundant carotenoids and has provitamin A activity (35). Initial studies examining β-carotene, often in the context of fruit and vegetable intake, provided compelling evidence for a potential inverse association with the risk of heart and vascular diseases (36–39). Prospective epidemiologic data continue to emerge in support of β-carotene. For example, data in 4802 men and women free of baseline CVD from the Rotterdam Study indicated that those in increasing tertiles of dietary β-carotene intake had multivariate RR of 1.00 (ref), 0.72, and 0.55 (*P*, trend = 0.013) (40).

Lycopene. Although the majority of research to date regarding dietary, plasma, and adipose lycopene has focused on its potential role in the prevention of prostate cancer (41), burgeoning support has emerged for an additional role in the prevention of heart and vascular diseases. Because lycopene is found in high concentrations in a relatively small number of plant foods (tomato, watermelon, pink grapefruit, papaya, apricot), it may have greater appeal for targeted prevention efforts because tomato-based foods are the predominant food source. More than 80% of lycopene intake in the United States is from tomato products, including ketchup, tomato juice, and tomato sauces (42). Lycopene has significant antioxidant potential *in vitro* and has been hypothesized to play a prominent role in preventing CVD (43–45).

The association between dietary lycopene and the risk of CVD was investigated in the Women's Health Study, a cohort of more than 39,000 female health professionals with 7.2 y of follow-up (46). Women in increasing quintiles of lycopene had multivariate RR of CVD of 1.00 (ref), 1.11, 1.14, 1.15, and 0.90 (*P*, trend = 0.34). For the consumption of tomato-based products, women consuming 1.5 to <4, 4 to <7, 7 to <10, and ≥10 servings/wk had RR of CVD of 1.02, 1.04, 0.68, and 0.71 (*P*, trend = 0.029) compared with women consuming <1.5 servings/wk. Results from that study indicated that although dietary lycopene was not strongly associated with the risk of CVD, the possible inverse associations noted for higher levels of tomato-based products suggest that dietary lycopene may still play a role in heart and vascular disease prevention.

Small-scale human dietary intervention studies demonstrated the ability of various tomato products to increase plasma lycopene levels in middle-aged, healthy subjects (45,47,48). Some (45,48), but not all (47) of these studies with either lycopene-containing foods or lycopene supplementation demonstrated potential short-term improvements in LDL oxidation. However, lycopene supplementation over the course of several weeks did not reduce LDL cholesterol levels, despite improvements in LDL oxidation (45). Whether these apparent short-term benefits translate into long-term improvements in health, manifested by a reduction in the risk of heart and vascular diseases, remains unknown.

Lutein/zeaxanthin. Lutein/zeaxanthin is found in a wide variety of fruits and vegetables, including cooked spinach, lettuce, broccoli, peas, lima beans, orange juice, celery, string beans, and squash (49). Lutein, contained mainly in dark green vegetables (49,50), was found to protect against the development or progression of atherosclerosis. Studies showed that lutein is effective in reducing the effect of adhesion molecules along aortic endothelial cells, reflecting a possible role in the development of atherosclerosis (51). Data from the Health Professionals' Follow-Up Study, consisting of 43,738 male health professionals followed for ~8 y, reported that higher quintiles of dietary lutein were associated with lower risks of total stroke, with multivariate RR of 1.00 (ref), 0.89, 0.88, 0.87, and 0.70 (*P*, trend = 0.06) (33). The authors reported results for ischemic stroke that closely paralleled the overall stroke data.

α-Carotene and β-Cryptoxanthin. The food sources of α-carotene tend to closely parallel those of β-carotene. Carrots alone, whether raw or cooked, accounted for a very large part (55%) of the α-carotene intake in the Framingham Heart Study (52). In contrast, carrots were responsible for 37% of β-carotene intake. Based upon the foods and beverages listed in the Willett semiquantitative food-frequency questionnaire, there are only three major food contributors to dietary β-cryptoxanthin intake: orange juice (46.8%), oranges (32.4%), and peaches (14.2%) (52). This presents a unique advantage similar to that for lycopene intake, with a limited number of food sources allowing for an improved ability to isolate whether β-cryptoxanthin itself or some other dietary component in oranges and/or peaches may be important in the prevention of heart and vascular diseases. However, very limited data are available on either α-carotene or β-cryptoxanthin and their association with heart and vascular diseases.

Blood- and Tissue-Based Epidemiologic Studies of Carotenoids and Risk of Cardiovascular Diseases

A second epidemiologic approach with which to explore these associations is to use blood levels of carotenoids as the main exposure variable. Several studies examined plasma or serum levels of carotenoids and the subsequent risk of CVD. There are three main designs. In the cohort study, the carotenoids are measured for the entire cohort. Alternatively, in the nested case-control study, cases and matched controls are selected from a larger cohort and specimens are analyzed only for selected cases and controls. This is more efficient than measuring levels for the entire cohort and should yield the same result. Traditional case-control studies were also instrumental in building the literature on this hypothesis but tend to be more vulnerable to bias and error. In this section, we summarize results from some of the more recent studies that have examined plasma, serum, and tissue carotenoids and the risk of CVD.

A nested case-control study of 531 men diagnosed with myocardial infarction (MI) and paired controls did not find an inverse association between any of the five major carotenoids measured and MI risk (53). However, among current and former smokers, higher baseline levels of β-carotene tended to be associated with lower MI risk. The participants were followed for up to 13 y as part of the ongoing Physicians' Health Study (PHS).

The Rotterdam Study was a nested case-control study examining major serum carotenoids in 108 pairs of Dutch subjects with an average follow-up of 5.8 y for cases of aortic atherosclerosis (40). Although no significant linear trends were found for each carotenoid of interest, likely due in part to the small number of cases in each quartile, there was a possible association with serum lycopene and aortic atherosclerosis. Compared with the lowest quartile of serum lycopene, higher quartiles had multivariate RR of 0.89, 0.78, and 0.66 (*P*, trend = 0.28).

A case-control study of 104 cases of myocardial infarction and 106 unmatched controls revealed lower levels of lycopene and β-cryptoxanthin among cases, although the result was adjusted only for plasma cholesterol (54). A small cohort study of 638 elderly Dutch men and women (aged 65–85 y) found that β-cryptoxanthin and lutein were inversely associated with all-cause mortality (55). Ford and Giles (56) published data from a large cross-sectional analysis of serum carotenoids and the prevalence of angina pectoris using data from NHANES III. Among 11,327 middle-aged and older men and women, higher quartiles of α-carotene, β-carotene, and β-cryptoxanthin were all inversely associated with the prevalence of angina (all *P*, trend <0.05). D'Odorico *et al.* (57) examined plasma carotenoids and the risk of atherosclerosis in the carotid arteries in 392 middle-aged Italian men and women followed for ~5 y. Of the five major carotenoids, only a combined measure of α- and β-carotene was associated with a lower incidence of atherosclerotic lesions in the carotid arteries (*P* = 0.04).

α-Carotene is a hydrocarbon carotenoid, whereas β-cryptoxanthin is an oxygenated carotenoid, and baseline plasma levels of both were recorded as part of the Los Angeles Atherosclerosis Study (58). New data from this cohort suggest that the three oxygenated carotenoids measured (β-cyptoxanthin, zeaxanthin, and lutein) and one of the hydrocarbon carotenoids measured, α-carotene, may be protective against atherosclerosis. After adjusting for age, sex, smoking status, and potential confounders, including CRP, the change in intima-media thickness was significantly inversely related to all four carotenoids. For zeaxanthin and α-carotene, the association remained significant after including all four predictors in one regression model and controlling for additional atherosclerosis risk factors.

An association between lycopene specifically and the risk of heart and vascular disease emerged through several studies. Two reports from the EURAMIC study suggested that intimal wall thickness and the risk of myocardial infarction are reduced in persons with higher adipose tissue concentrations of lycopene (59,60). Another study of 520 middle-aged Finnish men and women as part of the Antioxidant Supplementation in Atherosclerosis Prevention study noted similar significant decreases in intimal wall thickness among men with higher plasma lycopene, but not among women (61). A recently published study among 1028 middle-aged Finnish men found that those in the lowest quartile of serum lycopene had a significantly greater mean intima-media thickness of the common carotid artery, with a trend across quintiles (62). A number of other studies reported possible inverse associations of higher serum lycopene levels and a reduced risk of MI (63,64), CVD (64–66), carotid atherosclerosis (67,68), and aortic atherosclerosis (40).

To date, however, data are particularly lacking in women with regard to plasma lycopene levels and the risk of heart and vascular disease. In a prospective, nested case-control study of 483 cases of CVD and 483 age- and smoking-matched controls free of CVD during an average of 4.8 y follow-up, plasma lycopene was measured (69). The multivariate RR of total CVD for women in increasing quartiles of plasma lycopene were 1.00 (ref), 0.94, 0.62, and 0.67 (*P*, trend = 0.05). This pattern in RR suggested a threshold effect in which women in the upper half of plasma lycopene had a significant 34% reduction in CVD risk. For CVD excluding angina, an L-shaped association was apparent because women in the upper three quartiles had a significant multivariate 50% risk reduction compared with those in the lowest quartile of plasma lycopene.

Two other key studies provided support for the role of lutein/zeaxanthin. First, 480 men and women 40–60 y old from the Los Angeles Atherosclerosis Study were followed for 18 mo for the progression of intima-media thickness of the common carotid arteries (70). Subjects in increasing quintiles of plasma lutein had a smaller progression of atherosclerosis (*P*, trend = 0.0007). A second observational study that examined carotid intima-media thickness included 231 case-control pairs from the Atherosclerosis Risk in Communities study cohort (67). Both lutein/zeaxanthin [odds ratio (OR) per SD increase: 0.76, 95% confidence interval (CI): 0.59–0.95] and β-cryptoxanthin (OR per SD increase: 0.75, 95% CI: 0.59–0.94) were significantly and inversely associated with carotid intima-media thickness.

Trials of β-Carotene Supplementation

Observational studies that suggested an inverse association of various carotenoids and CVD and cancer resulted in the initiation of several large clinical trials, such as the PHS (71), to examine the effect of β-carotene supplementation on chronic disease development. The early clinical trials were designed to study the promise of β-carotene supplementation for reducing the risk of cancer with CVD as a secondary outcome. More recent trials such as the ongoing primary [PHS II (72)] and secondary [Women's Antioxidant Cardiovascular Study (73)] prevention trials continue to test β-carotene supplementation and the risk of CVD as a primary outcome. We now focus our discussion on the completed primary and secondary prevention trials that have included β-carotene either as an individual study agent or as part of an antioxidant cocktail for the risk of CVD.

β-*Carotene as an Individual Supplement*

Results from large-scale randomized trials of β-carotene in the primary prevention of heart and vascular disease were largely disappointing, finding either no association or a weak positive association. The results from four of these trials suggested that β-carotene supplementation was not associated with the risk of heart and vascular diseases. The Skin Cancer Prevention Study randomized 1805 men and women with a history of skin cancer to 50 mg of β-carotene daily or placebo (74). After a median follow-up of 8.2 y, there was no association of β-carotene supplementation with cardiovascular mortality (RR, 1.15; 95% CI, 0.81–1.63). Next, the Alpha-Tocopherol, Beta-Carotene (ATBC) Cancer Prevention Study was a large-scale randomized trial of antioxidant vitamins in a well-nourished population. This 2 × 2 factorial trial tested the effect of synthetic vitamin E (50 mg/d) and β-carotene (20 mg/d) among 29,133 Finnish male smokers aged 50–69 y. There was a possible increase in ischemic heart disease mortality (RR, 1.12; 95% CI, 1.00–1.25) and no association with the risk of angina (RR, 1.06; 95% CI, 0.97–1.16) among those assigned to β-carotene (75).

Two other primary prevention trials examining β-carotene as an individual study agent revealed a null association with the risk of heart and vascular diseases. First, the PHS I was a randomized, double-blind, placebo-controlled trial of β-carotene (50 mg on alternate days) and low-dose aspirin among 22,071 U.S. male physicians aged 40–84 y (76). After 12 y, there were no differences in cardiovascular mortality (RR, 1.09; 95% CI, 0.93–1.27), MI (RR, 0.96; 95% CI, 0.84–1.09), stroke (RR, 0.96; 95% CI, 0.83–1.11), or a composite of the three endpoints (RR, 1.00; 95% CI, 0.91–1.09) associated with β-carotene assignment. A trial complementing the PHS I, the Women's Health Study (WHS), evaluated the effect of β-carotene (50 mg on alternate days), vitamin E, and low-dose aspirin on the development of CVD in 39,876 healthy female health professionals (77). Although the vitamin E and aspirin components of the trial are continuing, the β-carotene arm was terminated after just over a mean of 2 y follow-up, largely in response to the null findings on β-carotene reported in the PHS I. β-Carotene supplementation had no effect on cardiovascular mortality (RR, 1.17; 95% CI, 0.54–2.53), MI (RR, 1.08; 95% CI, 0.56–1.27), stroke (RR, 1.42; 95% CI, 0.96–2.10), or a composite of these endpoints (RR, 1.14; 95% CI 0.87–1.49). Smoking status did not modify the null association between β-carotene supplementation and the risk of heart and vascular diseases.

β-Carotene supplementation has been less well studied in secondary prevention than in primary prevention settings. Although no data are yet available from randomized trials specifically designed to answer whether β-carotene supplementation alone (as opposed to in combination with other antioxidants) is effective in secondary prevention of heart and vascular diseases, subgroup analyses in some of the aforementioned primary prevention trials allow an empirical examination of this issue.

In the ATBC trial, there were 1862 men who had a baseline history of MI and were randomized to β-carotene supplementation for 6 y. Among this subgroup of men, β-carotene was associated with a potential reduction in the risk of nonfatal myocardial infarction (RR, 0.67; 95% CI, 0.44–1.02), but a nonsignificant increased risk of fatal coronary heart disease (RR, 1.58; 95% CI, 1.05 to 2.40) (78). In PHS I, there were 333 men who reported a history of chronic stable angina or a coronary revascularization procedure before randomization. Among those in the β-carotene group, a reduction (RR = 0.46, 95% CI, 0.24–0.85) in the risk of major cardiovascular events was observed after 5 y, and a persistent although attenuated reduction was also found after 12 y (RR = 0.71, 95% CI, 0.47–1.07) (79).

β-Carotene as Part of a Combination Intervention

Other primary prevention trials have considered β-carotene as part of a nutrient combination. Although informative, it is difficult to extrapolate the findings for a mixture of several nutrients to the individual effects of β-carotene supplementation. There are three primary and three secondary prevention trials of heart and vascular disease that were completed and published. A fourth, the SUpplementation en VItamines et Mineraux AntioXydants (SU.VI.MAX) study (80) was recently completed, and publication of results is slated for later in 2004. [SU.VI.MAX evaluated the efficacy of a balanced combination of antioxidants (including 6 mg of β-carotene) and minerals in the primary prevention of cancer and CVD in 12,375 French men and women for 8 y.] According to information released before publication, there was no reduction in CVD risk among those taking the vitamin cocktail.

First, the Cancer Prevention Trial was a primary prevention trial conducted in a poorly nourished population in China that was at high risk of upper gastrointestinal cancers, presumably due to a low intake of micronutrients (81). Nearly 30,000 men and women were randomized to a cocktail of synthetic vitamin E (30 mg daily), β-carotene (15 mg daily), and selenium (50 mg daily). Participants taking this antioxidant cocktail had a RR of 0.90 (95% CI, 0.76–1.07) for cerebrovascular mortality and 0.91 (95% CI, 0.84–0.99) for total mortality compared with those administered placebo. Second, the Beta-Carotene and Retinol Efficacy Trial (CARET) evaluated a combined treatment of β-carotene (30 mg/d) and retinol (25,000 IU/d) among 18,314 men and women at elevated risk of lung cancer due to cigarette smoking and/or occupational exposure to asbestos (82). After 4 y, there was a borderline significant increased risk of cardiovascular death (RR, 1.26; 95% CI, 0.99–1.61) among individuals assigned to β-carotene and retinol. Finally, the Age-Related Eye Disease Study Group (AREDS) reported that an antioxidant cocktail of β-carotene, vitamin C, and vitamin E randomized to 4757 middle-aged and older subjects was not related to the risk of CVD (RR, 1.06; 95% CI, 0.84–1.33) (83).

The HDL-Atherosclerosis Treatment Study (HATS) (84), the Heart Protection Study (HPS) 85, and the Multivitamins and Probucol (MVP) study (86) are each secondary prevention trials investigating β-carotene supplementation as part of a mixture of antioxidants, with the results favoring no association between β-carotene and the risk of secondary events. In HATS, 160 men and women with coronary artery disease were randomly assigned in a 2 × 2 factorial design to simvastatin plus niacin, or to an antioxidant combination (800 IU of vitamin E, 1000 mg of vitamin C, 25 mg of natural β-carotene, and 100 μg selenium). After a mean follow-up of 14 mo, 21% of subjects taking the antioxidant combination developed CVD vs. 24% in the placebo group ($P > 0.05$) (84). In the HPS, a 2 × 2 factorial design tested a daily antioxidant cocktail (600 mg of synthetic vitamin E, 250 mg of vitamin C, and 20 mg of β-carotene) and simvastatin (40 mg) among 20,536 men and women with angina, stroke, claudication, or diabetes. Main-effects analyses showed neither a beneficial nor a deleterious effect of the antioxidant cocktail on cardiovascular outcomes over 5 y of follow-up (85). Finally, in the MVP study, a combination of vitamin C (1000 mg/d), vitamin E (1400 IU/d), and β-carotene (100 mg/d) had no effect on the rate and severity of restenosis (86).

Conclusions

Although >600 carotenoids have been identified (87), the majority of research in nutrition has focused on the five that are most abundant carotenoids from foods, i.e., α-carotene, β-carotene, lycopene, lutein/zeaxanthin, and β-cryptoxanthin. Basic research points to several mechanisms by which carotenoids could prevent or delay atherogenesis. Studies that examine fruit and vegetable intake have overwhelmingly supported a cardioprotective role of these foods. Published observational studies generally support an association between total carotenoids and some individual carotenoids, measured from dietary intake, supplement use, plasma, and serum, and lower risk of CVD.

β-Carotene has been the most widely studied carotenoid. However, recent observational research on the other major carotenoids, α-carotene, lycopene, lutein/zeaxanthin, and β-cryptoxanthin, has raised questions regarding their role in reducing the risk of CVD. Lycopene (tomatoes) and β-cryptoxanthin (oranges) have a limited number of food sources compared with their carotenoid counterparts. This makes lycopene and β-cryptoxanthin promising carotenoids whose intake can be increased with relatively simple dietary recommendations. Additional data from interventions of dietary approaches focused on increasing either total or specific carotenoid intake can provide critical information on the biologic mechanisms supporting the observational findings for carotenoids and a possible reduced risk of CVD.

β-Carotene is the only carotenoid that has been studied extensively in several large-scale primary and secondary prevention trials. These trials have generally not shown the anticipated reduced risks of CVD with β-carotene supplementation. There are several possible explanations for the discrepancy between the observational and trial evidence. First, CVD develops over decades and the trials may have been of insufficient duration to demonstrate this chronic effect. Second, the dose of β-carotene supplementation in these trials was arbitrary. Third, β-carotene may have to be consumed with other micronutrients to potentiate its efficacy. Finally, β-carotene may simply be a marker for another protective carotenoid or micronutrient from carotenoid-rich foods. More research is warranted to understand how β-carotene, other carotenoids, vitamins, and minerals all interact when consumed either as supplements or in the diet. Currently, however, based on a review of published studies, the U.S. Preventive Services Task Force does not recommend that people take β-carotene supplements to lower their risk of developing CVD or cancer (88).

References

1. Joshipura, K.J., Hu, F.B., Manson, J.E., Stampfer, M.J., Rimm, E.B., Speizer, F.E., Colditz, G., Ascherio, A., Rosner, B., Spiegelman, D., and Willett, W.C. (2001) The Effect of Fruit and Vegetable Intake on Risk for Coronary Heart Disease, *Ann. Intern. Med. 134:* 1106–1114.
2. Mozaffarian, D., Kumanyika, S.K., Lemaitre, R.N., Olson, J.L., Burke, G.L., and Siscovick, D.S. (2003) Cereal, Fruit, and Vegetable Fiber Intake and the Risk of Cardiovascular Disease in Elderly Individuals, *J. Am. Med. Assoc. 289:* 1659–1666.
3. Bazzano, L.A., He, J., Ogden, L.G., Loria, C.M., Vupputuri, S., Myers, L., and Whelton, P.K. (2002) Fruit and Vegetable Intake and Risk of Cardiovascular Disease in US Adults: The First National Health and Nutrition Examination Survey Epidemiologic Follow-up Study, *Am. J. Clin. Nutr. 76:* 93–99.
4. Liu, S., Manson, J.E., Lee, I.M., Cole, S.R., Hennekens, C.H., Willett, W.C., and Buring, J.E. (2000) Fruit and Vegetable Intake and Risk of Cardiovascular Disease: The Women's Health Study, *Am. J. Clin. Nutr. 72:* 922–928.
5. Joshipura, K.J., Ascherio, A., Manson, J.E., Stampfer, M.J., Rimm, E.B., Speizer, F.E., Hennekens, C.H., Spiegelman, D., and Willett, W.C. (1999) Fruit and Vegetable Intake in Relation to Risk of Ischemic Stroke, *J. Am. Med. Assoc. 282:* 1233–1239.
6. Sacks, F.M., Svetkey, L.P., Vollmer, W.M., Appel, L.J., Bray, G.A., Harsha, D., Obarzanek, E., Conlin, P.R., Miller, E.R., Simons-Morton, D.G., Karanja, N., and Lin, P.H. (2001) Effects on Blood Pressure of Reduced Dietary Sodium and the Dietary Approaches to Stop Hypertension (DASH) Diet. DASH-Sodium Collaborative Research Group, *N. Engl. J. Med. 344:* 3–10.
7. de Lorgeril, M., Salen, P., Martin, J.L., Monjaud, I., Delaye, J., and Mamelle, N. (1999) Mediterranean Diet, Traditional Risk Factors, and the Rate of Cardiovascular Complications After Myocardial Infarction: Final Report of the Lyon Diet Heart Study, *Circulation 99:* 779–785.
8. Trichopoulou, A., Costacou, T., Bamia, C., and Trichopoulos, D. (2003) Adherence to a Mediterranean Diet and Survival in a Greek Population, *N. Engl. J. Med. 348:* 2599–2608.

9. Obarzanek, E., Sacks, F.M., Vollmer, W.M., Bray, G.A., Miller, E.R., Lin, P.H., Karanja, N.M., Most-Windhauser, M.M., Moore, T.J., Swain, J.F., Bales, C.W., and Proschan, M.A. (2001) Effects on Blood Lipids of a Blood Pressure-Lowering Diet: The Dietary Approaches to Stop Hypertension (DASH) Trial, *Am. J. Clin. Nutr. 74:* 80–89.
10. Young, A.J., and Lowe, G.M. (2001) Antioxidant and Prooxidant Properties of Carotenoids, *Arch. Biochem. Biophys. 385:* 20–27.
11. Parker, R.S. (1989) Carotenoids in Human Blood and Tissues, *J. Nutr. 119:* 101–104.
12. Steinberg, D., Parthasarathy, S., Carew, T.E., Khoo, J.C., and Witztum, J.L. (1989) Beyond Cholesterol. Modifications of Low-Density Lipoprotein That Increase Its Atherogenicity, *N. Engl. J. Med. 320:* 915–924.
13. Hessler, J.R., Morel, D.W., Lewis, L.J., and Chisolm, G.M. (1983) Lipoprotein Oxidation and Lipoprotein-Induced Cytotoxicity, *Arteriosclerosis 3:* 215–222.
14. Yagi, K. (1984) Increased Serum Lipid Peroxides Initiate Atherogenesis, *Bioessays 1:* 58–60.
15. Quinn, M.T., Parthasarathy, S., and Steinberg, D. (1985) Endothelial Cell-Derived Chemotactic Activity for Mouse Peritoneal Macrophages and the Effects of Modified Forms of Low Density Lipoprotein, *Proc. Natl. Acad. Sci. USA 82:* 5949–5953.
16. Schaffner, T., Taylor, K., Bartucci, E.J., Fischer-Dzoga, K., Beeson, J.H., Glagov, S., and Wissler, R.W. (1980) Arterial Foam Cells with Distinctive Immunomorphologic and Histochemical Features of Macrophages, *Am. J. Pathol. 100:* 57–80.
17. Gerrity, R.G. (1981) The Role of the Monocyte in Atherogenesis: I. Transition of Blood-Borne Monocytes into Foam Cells in Fatty Lesions, *Am. J. Pathol. 103:* 181–190.
18. Fogelman, A.M., Shechter, I., Seager, J., Hokom, M., Child, J.S., and Edwards, P.A. (1980) Malondialdehyde Alteration of Low Density Lipoproteins Leads to Cholesteryl Ester Accumulation in Human Monocyte-Macrophages, *Proc. Natl. Acad. Sci. USA 77:* 2214–2218.
19. Goldstein, J.L., Ho, Y.K., Basu, S.K., and Brown, M.S. (1979) Binding Site on Macrophages That Mediates Uptake and Degradation of Acetylated Low Density Lipoprotein, Producing Massive Cholesterol Deposition, *Proc. Natl. Acad. Sci. USA 76:* 333–337.
20. Salonen, J.T., Yla-Herttuala, S., Yamamoto, R., Butler, S., Korpela, H., Salonen, R., Nyyssonen, K., Palinski, W., and Witztum, J.L. (1992) Autoantibody Against Oxidised LDL and Progression of Carotid Atherosclerosis, *Lancet 339:* 883–887.
21. Beckman, J.S., Beckman, T.W., Chen, J., Marshall, P.A., and Freeman, B.A. (1990) Apparent Hydroxyl Radical Production by Peroxynitrite: Implications for Endothelial Injury from Nitric Oxide and Superoxide, *Proc. Natl. Acad. Sci. USA 87:* 1620–1624.
22. Marcus, A.J., Silk, S.T., Safier, L.B., and Ullman, H.L. (1977) Superoxide Production and Reducing Activity in Human Platelets, *J. Clin. Investig. 59:* 149–158.
23. Saran, M., Michel, C., and Bors, W. (1990) Reaction of NO with O_2^-- Implications for the Action of Endothelium-Derived Relaxing Factor (EDRF), *Free Radic. Res. Commun. 10:* 221–226.
24. Kritchevsky, S.B., Bush, A.J., Pahor, M., and Gross, M.D. (2000) Serum Carotenoids and Markers of Inflammation in Nonsmokers, *Am. J. Epidemiol. 152:* 1065–1071.
25. Boosalis, M.G., Snowdon, D.A., Tully, C.L., and Gross, M.D. (1996) Acute Phase Response and Plasma Carotenoid Concentrations in Older Women: Findings from the Nun Study, *Nutrition 12:* 475–478.

26. Rimm, E.B., Stampfer, M.J., Ascherio, A., Giovannucci, E., Colditz, G.A., and Willett, W.C. (1993) Vitamin E Consumption and the Risk of Coronary Heart Disease in Men, *N. Engl. J. Med. 328:* 1450–1456.
27. Knekt, P., Reunanen, A., Jarvinen, R., Seppanen, R., Heliovaara, M., and Aromaa, A. (1994) Antioxidant Vitamin Intake and Coronary Mortality in a Longitudinal Population Study, *Am. J. Epidemiol. 139:* 1180–1189.
28. Gaziano, J.M., Manson, J.E., Branch, L.G., Colditz, G.A., Willett, W.C., and Buring, J.E. (1995) A Prospective Study of Consumption of Carotenoids in Fruits and Vegetables and Decreased Cardiovascular Mortality in the Elderly, *Ann. Epidemiol. 5:* 255–260.
29. Sahyoun, N.R., Jacques, P.F., and Russell, R.M. (1996) Carotenoids, Vitamins C and E, and Mortality in an Elderly Population, *Am. J. Epidemiol. 144:* 501–511.
30. Kushi, L.H., Folsom, A.R., Prineas, R.J., Mink, P.J., Wu, Y., and Bostick, R.M. (1996) Dietary Antioxidant Vitamins and Death from Coronary Heart Disease in Postmenopausal Women, *N. Engl. J. Med. 334:* 1156–1162.
31. Yochum, L.A., Folsom, A.R., and Kushi, L.H. (2000) Intake of Antioxidant Vitamins and Risk of Death from Stroke in Postmenopausal Women, *Am. J. Clin. Nutr. 72:* 476–483.
32. Osganian, S.K., Stampfer, M.J., Rimm, E., Spiegelman, D., Manson, J.E., and Willett, W.C. (2003) Dietary Carotenoids and Risk of Coronary Artery Disease in Women, *Am. J. Clin. Nutr. 77:* 1390–1399.
33. Ascherio, A., Rimm, E.B., Hernan, M.A., Giovannucci, E., Kawachi, I., Stampfer, M.J., and Willett, W.C. (1999) Relation of Consumption of Vitamin E, Vitamin C, and Carotenoids to Risk for Stroke Among Men in the United States, *Ann. Intern. Med. 130:* 963–970.
34. Michaud, D.S., Giovannucci, E.L., Ascherio, A., Rimm, E.B., Forman, M.R., Sampson, L., and Willett, W.C. (1998) Associations of Plasma Carotenoid Concentrations and Dietary Intake of Specific Carotenoids in Samples of Two Prospective Cohort Studies Using a New Carotenoid Database, *Cancer Epidemiol. Biomark. Prev. 7:* 283–290.
35. Pryor, W.A., Stahl, W., and Rock, C.L. (2000) Beta Carotene: From Biochemistry to Clinical Trials, *Nutr. Rev. 58:* 39–53.
36. Acheson, R.M., and Williams, D.R. (1983) Does Consumption of Fruit and Vegetables Protect Against Stroke? *Lancet 1:* 1191–1193.
37. Vollset, S.E., and Bjelke, E. (1983) Does Consumption of Fruit and Vegetables Protect Against Stroke? *Lancet 2:* 742.
38. Gey, K.F., Stahelin, H.B., and Eichholzer, M. (1993) Poor Plasma Status of Carotene and Vitamin C Is Associated with Higher Mortality from Ischemic Heart Disease and Stroke: Basel Prospective Study, *Clin. Investig. 71:* 3–6.
39. Armstrong, B.K., Mann, J.I., Adelstein, A.M., and Eskin, F. (1975) Commodity Consumption and Ischemic Heart Disease Mortality, with Special Reference to Dietary Practices, *J. Chronic Dis. 28:* 455–469.
40. Klipstein-Grobusch, K., Launer, L.J., Geleijnse, J.M., Boeing, H., Hofman. A., and Witteman, J.C. (2000) Serum Carotenoids and Atherosclerosis. The Rotterdam Study, *Atherosclerosis 148:* 49–56.
41. Giovannucci, E., Rimm, E.B., Liu, Y., Stampfer, M.J., and Willett, W.C. (2002) A Prospective Study of Tomato Products, Lycopene, and Prostate Cancer Risk, *J. Natl. Cancer Inst. 94:* 391–398.

42. Clinton, S.K. (1998) Lycopene: Chemistry, Biology, and Implications for Human Health and Disease, *Nutr. Rev. 56:* 35–51.
43. Arab, L., and Steck, S. (2000) Lycopene and Cardiovascular Disease, *Am. J. Clin. Nutr. 71:* 1691S–1695S.
44. Fuhrman, B., Elis, A., and Aviram, M. (1997) Hypocholesterolemic Effect of Lycopene and Beta-Carotene Is Related to Suppression of Cholesterol Synthesis and Augmentation of LDL Receptor Activity in Macrophages, *Biochem. Biophys. Res. Commun. 233:* 658–662.
45. Agarwal, S., and Rao, A.V. (1998) Tomato Lycopene and Low Density Lipoprotein Oxidation: A Human Dietary Intervention Study, *Lipids 33:* 981–984.
46. Sesso, H.D., Liu, S., Gaziano, J.M., and Buring, J.E. (2003) Dietary Lycopene, Tomato-Based Food Products and Cardiovascular Disease in Women, *J. Nutr. 133:* 2336–2341.
47. Dugas, T.R., Morel, D.W., and Harrison, E.H. (1999) Dietary Supplementation with Beta-Carotene, but Not with Lycopene, Inhibits Endothelial Cell-Mediated Oxidation of Low-Density Lipoprotein, *Free Radic. Biol. Med. 26:* 1238–1244.
48. Bub, A., Watzl, B., Abrahamse, L., Delincee, H., Adam, S., Wever, J., Muller, H., and Rechkemmer, G. (2000) Moderate Intervention with Carotenoid-Rich Vegetable Products Reduces Lipid Peroxidation in Men, *J. Nutr. 130:* 2200–2206.
49. Mangels, A.R., Holden, J.M., Beecher, G.R., Forman, M.R., and Lanza, E. (1993) Carotenoid Content of Fruits and Vegetables: An Evaluation of Analytic Data, *J. Am. Diet. Assoc. 93:* 284–296.
50. Mares-Perlman, J.A., Millen, A.E., Ficek, T.L., and Hankinson, S.E. (2002) The Body of Evidence to Support a Protective Role for Lutein and Zeaxanthin in Delaying Chronic Disease. Overview, *J. Nutr. 132:* 518S–524S.
51. Martin, K.R., Wu, D., and Meydani, M. (2000) The Effect of Carotenoids on the Expression of Cell Surface Adhesion Molecules and Binding of Monocytes to Human Aortic Endothelial Cells, *Atherosclerosis 150:* 265–274.
52. Tucker, K.L., Chen, H., Vogel, S., Wilson, P.W., Schaefer, E.J., and Lammi-Keefe, C.J. (1999) Carotenoid Intakes, Assessed by Dietary Questionnaire, Are Associated with Plasma Carotenoid Concentrations in an Elderly Population, *J. Nutr. 129:* 438–445.
53. Hak, A.E., Stampfer, M.J., Campos, H., Sesso, H.D., Gaziano, J.M., Willett, W., and Ma, J. (2003) Plasma Carotenoids and Tocopherols and Risk of Myocardial Infarction in a Low-Risk Population of US Male Physicians, *Circulation 108:* 802–807.
54. Ruiz Rejon, F., Martin-Pena, G., Granado, F., Ruiz-Galiana, J., Blanco, I., and Olmedilla, B. (2002) Plasma Status of Retinol, Alpha- and Gamma-Tocopherols, and Main Carotenoids to First Myocardial Infarction: Case Control and Follow-Up Study, *Nutrition 18:* 26–31.
55. De Waart, F.G., Schouten, E.G., Stalenhoef, A.F., and Kok, F.J. (2001) Serum Carotenoids, Alpha-Tocopherol and Mortality Risk in a Prospective Study Among Dutch Elderly, *Int. J. Epidemiol. 30:* 136–143.
56. Ford, E.S., and Giles, W.H. (2000) Serum Vitamins, Carotenoids, and Angina Pectoris: Findings from the National Health and Nutrition Examination Survey III, *Ann. Epidemiol. 10:* 106–116.
57. D'Odorico, A., Martines, D., Kiechl, S., Egger, G., Oberhollenzer, F., Bonvicini, P., Sturniolo, G.C., Naccarato, R., and Willett, J. (2000) High Plasma Levels of Alpha- and Beta-Carotene Are Associated with a Lower Risk of Atherosclerosis: Results from the Bruneck Study, *Atherosclerosis 153:* 231–239.

58. Dwyer, J.H., Paul-Labrador, M.J., Fan, J., Shircore, A.M., Merz, C.N., and Dwyer, K.M. (2004) Progression of Carotid Intima-Media Thickness and Plasma Antioxidants: The Los Angeles Atherosclerosis Study, *Arterioscler. Thromb. Vasc. Biol. 24:* 313–319.
59. Kohlmeier, L., Kark, J.D., Gomez-Gracia, E., Martin, B.C., Steck, S.E., Kardinaal, A.F., Ringstad, J., Thamm, M., Masaev, V., Riemersma, R., Martin-Moreno, J.M., Huttunen, J.K., and Kok, F.J. (1997) Lycopene and Myocardial Infarction Risk in the EURAMIC Study, *Am. J. Epidemiol. 146:* 618–626.
60. Gomez-Aracena, J., Sloots, J., and Garcia-Rodriguez, A. (1997) Antioxidants in Adipose Tissue and Myocardial Infarction in a Mediterranean Area, *Nutr. Metab. Cardiovasc. Dis. 7:* 376–382.
61. Rissanen, T., Voutilainen, S., Nyyssonen, K., Salonen, R., and Salonen, J.T. (2000) Low Plasma Lycopene Concentration Is Associated with Increased Intima-Media Thickness of the Carotid Artery Wall, *Arterioscler. Thromb. Vasc. Biol. 20:* 2677–2681.
62. Rissanen, T.H., Voutilainen, S., Nyyssonen, K., Salonen, R., Kaplan, G.A., and Salonen, J.T. (2003) Serum Lycopene Concentrations and Carotid Atherosclerosis: The Kuopio Ischaemic Heart Disease Risk Factor Study, *Am. J. Clin. Nutr. 77:* 133–138.
63. Street, D.A., Comstock, G.W., Salkeld, R.M., Schuep, W., and Klag, M.J. (1994) Serum Antioxidants and Myocardial Infarction. Are Low Levels of Carotenoids and Alpha-Tocopherol Risk Factors for Myocardial Infarction? *Circulation 90:* 1154–1161.
64. Rissanen, T.H., Voutilainen, S., Nyyssonen, K., Lakka, T.A., Sivenius, J., Salonen, R., Kaplan, G.A., and Salonen, J.T. (2001) Low Serum Lycopene Concentration Is Associated with an Excess Incidence of Acute Coronary Events and Stroke: The Kuopio Ischaemic Heart Disease Risk Factor Study, *Br. J. Nutr. 85:* 749–754.
65. Schmidt, R., Fazekas, F., Hayn, M., Schmidt, H., Kapeller, P., Roob, G., Offenbacher, H., Schumacher, M., Eber, B., Weinrauch, V., Kostner, G.M., and Esterbauer, H. (1997) Risk Factors for Microangiopathy-Related Cerebral Damage in the Austrian Stroke Prevention Study, *J. Neurol. Sci. 152:* 15–21.
66. Kristenson, M., Zieden, B., Kucinskiene, Z., Elinder, L.S., Bergdahl, B., Elwing, B., Abaravicius, A., Razinkoviene, L., Calkauskas, H., and Olsson, A.G. (1997) Antioxidant State and Mortality from Coronary Heart Disease in Lithuanian and Swedish Men: Concomitant Cross Sectional Study of Men Aged 50, *Br. Med. J. 314:* 629–633.
67. Iribarren, C., Folsom, A.R., Jacobs, D.R., Jr., Gross, M.D., Belcher, J.D., and Eckfeldt, J.H. (1997) Association of Serum Vitamin Levels, LDL Susceptibility to Oxidation, and Autoantibodies Against MDA-LDL With Carotid Atherosclerosis. A Case-Control Study. The ARIC Study Investigators. Atherosclerosis Risk in Communities, *Arterioscler. Thromb. Vasc. Biol. 17:* 1171–1177.
68. McQuillan, B.M., Hung, J., Beilby, J.P., Nidorf, M., and Thompson, P.L. (2001) Antioxidant Vitamins and the Risk of Carotid Atherosclerosis. The Perth Carotid Ultrasound Disease Assessment study (CUDAS), *J. Am. Coll. Cardiol. 38:* 1788–1794.
69. Sesso, H.D., Buring, J.E., Norkus, E.P., and Gaziano, J.M. (2004) Plasma Lycopene, Other Carotenoids, and Retinol and the Risk of Cardiovascular Disease in Women, *Am. J. Clin. Nutr. 79:* 47–53.
70. Dwyer, J.H., Navab, M., Dwyer, K.M., Hassan, K., Sun, P., Shircore, A., Hama-Levy, S., Hough, G., Wang, X., Drake, T., Merz, C.N., and Fogelman, A.M. (2001) Oxygenated Carotenoid Lutein and Progression of Early Atherosclerosis: The Los Angeles Atherosclerosis Study, *Circulation 103:* 2922–2927.

71. The Steering Committee of the Physicians' Health Study Research Group (1989) Final Report on the Aspirin Component of the Ongoing Physicians' Health Study, *N. Engl. J. Med. 321:* 129–135.
72. Christen, W.G., Gaziano, J.M., and Hennekens, C.H. (2000) Design of Physicians' Health Study II—A Randomized Trial of Beta-Carotene, Vitamins E and C, and Multivitamins, in Prevention of Cancer, Cardiovascular Disease, and Eye Disease, and Review of Results of Completed Trials, *Ann. Epidemiol. 10:* 125–134.
73. Manson, J.E., Gaziano, J.M., Spelsberg, A., Ridker, P.M., Cook, N.R., Buring, J.E., Willett, W.C., and Hennekens, C.H. (1995) A Secondary Prevention Trial of Antioxidant Vitamins and Cardiovascular Disease in Women. Rationale, Design, and Methods. The WACS Research Group, *Ann. Epidemiol. 5:* 261–269.
74. Greenberg, E.R., Baron, J.A., Karagas, M.R., Stukel, T.A., Nierenberg, D.W., Stevens, M.M., Mandel, J.S., and Haile, R.W. (1996) Mortality Associated with Low Plasma Concentration of Beta Carotene and the Effect of Oral Supplementation, *J. Am. Med. Assoc. 275:* 699–703.
75. Rapola, J.M., Virtamo, J., Haukka, J.K., Heinonen, O.P., Albanes, D., Taylor, P.R., and Huttunen, J.K. (1996) Effect of Vitamin E and Beta Carotene on the Incidence of Angina Pectoris. A Randomized, Double-Blind, Controlled Trial, *J. Am. Med. Assoc. 275:* 693–698.
76. Hennekens, C.H., Buring, J.E., Manson, J.E., Stampfer, M., Rosner, B., Cook, N.R., Belanger, C., LaMotte, F., Gaziano, J.M., Ridker, P.M., Willett, W., and Peto, R. (1996) Lack of Effect of Long-Term Supplementation with Beta Carotene on the Incidence of Malignant Neoplasms and Cardiovascular Disease, N. *Engl. J. Med. 334:* 1145–1149.
77. Lee, I.M., Cook, N.R., Manson, J.E., Buring, J.E., and Hennekens, C.H. (1999) Beta-Carotene Supplementation and Incidence of Cancer and Cardiovascular Disease: The Women-s Health Study, *J. Natl. Cancer Inst. 91:* 2102–2106.
78. Rapola, J.M., Virtamo, J., Ripatti, S., Huttunen, J.K., Albanes, D., Taylor, P.R., and Heinonen, O.P. (1997) Randomised Trial of Alpha-Tocopherol and Beta-Carotene Supplements on Incidence of Major Coronary Events in Men with Previous Myocardial Infraction, *Lancet 349:* 1715–1720.
79. Gaziano, J.M., Manson, J.E., Ridker, P.M., Buring, J.E., and Hennekens, C.H. (1990) Beta Carotene Therapy for Chronic Stable Angina, *Circulation 82 (4 Suppl. III):* 202 (Abstr.).
80. Hercberg, S., Preziosi, P., Briancon, S., Galan, P., Triol, I., Malvy, D., Roussel, A.M., and Favier, A. (1998) A Primary Prevention Trial Using Nutritional Doses of Antioxidant Vitamins and Minerals in Cardiovascular Diseases and Cancers in a General Population: The SU.VI.MAX Study—Design, Methods, and Participant Characteristics. SUpplementation en VItamines et Mineraux AntioXydants, *Control Clin. Trials 19:* 336–351.
81. Blot, W.J., Li, J.Y., Taylor, P.R., Guo, W., Dawsey, S., Wang, G.Q., Yang, C.S., Zheng, S.F., Gail, M., Li, G.Y., Yu, Y., Liu, B.-Q., Tangrea, J., Sun, Y.-H., Liu, F., Fraumeni, J.F., Zhang, Y.-H., and Li, B. (1993) Nutrition Intervention Trials in Linxian, China: Supplementation with Specific Vitamin/Mineral Combinations, Cancer Incidence, and Disease-Specific Mortality in the General Population, *J. Natl. Cancer. Inst. 85:* 1483–1492.
82. Omenn, G.S., Goodman, G.E., Thornquist, M.D., Balmes, J., Cullen, M.R., Glass, A., Keogh, J.P., Meyskens, F.L., Valanis, B., Williams, J.H., Barnhart, S., and Hammar, S. (1996) Effects of a Combination of Beta Carotene and Vitamin A on Lung Cancer and Cardiovascular Disease, *N. Engl. J. Med. 334:* 1150–1155.

83. Age-Related Eye Disease Study Research Group (2001) A Randomized, Placebo-Controlled, Clinical Trial of High-Dose Supplementation with Vitamins C and E and Beta Carotene for Age-Related Cataract and Vision Loss: AREDS Report No. 9, *Arch. Ophthalmol. 119:* 1439–1452.
84. Brown, B.G., Zhao, X.Q., Chait, A., Fisher, L.D., Cheung, M.C., Morse, J.S., Dowdy, A.A., Marino, E.K., Bolson, E.L., Alaupovic, P., Frohlich, J., and Albers, J.J. (2001) Simvastatin and Niacin, Antioxidant Vitamins, or the Combination for the Prevention of Coronary Disease, *N. Engl. J. Med. 345:* 1583–1592.
85. Heart Protection Study Collaborative Group (2002) MRC/BHF Heart Protection Study of Antioxidant Vitamin Supplementation in 20,536 High-Risk Individuals: A Randomised Placebo-Controlled Trial, *Lancet 360:* 23–33.
86. Tardif, J.C., Cote, G., Lesperance, J., Bourassa, M., Lambert, J., Doucet, S., Bilodeau, L., Nattel, S., and de Guise, P. (1997) Probucol and Multivitamins in the Prevention of Restenosis After Coronary Angioplasty. Multivitamins and Probucol Study Group, *N. Engl. J. Med. 337:* 365–372.
87. Holden, J.M., Eldridge, A.L., Beecher, G.R., Buzzard, I.M., Bhagwat, S., Davis, C.S., Douglass, L.W., Gebhardt, S., Haytowitz, D., and Schakel, S. (1999) Carotenoid Content of US Foods: An Update of the Database, *J. Food. Compos. Anal. 12:* 169–196.
88. U.S. Preventive Services Task Force (2003) Summaries for Patients Taking Vitamin Supplements to Prevent Cardiovascular Disease and Cancer: Recommendations from the U.S. Preventive Services Task Force, *Ann. Intern. Med. 139:* I–76.

Chapter 22

Safety of β-Carotene

Norman I. Krinsky

Department of Biochemistry, School of Medicine, and the Jean Mayer USDA Human Nutrition Research Center on Aging, Tufts University, Boston, MA 02111

Introduction

There have been several papers published over the years with titles similar to the one above (1–5). This manuscript represents one developed following a Round Table on the "Safety of β-Carotene" that was held on March 10, 2004 immediately before the Oxygen Club of California conference in Santa Barbara, CA. The participants at the Round Table included Carroll Cross, Marion Dietrich, Regina Goralczyk, John Hathcock, Klaus Kraemer, Norman Krinsky, Lester Packer, Helmut Sies, Olaf Sommerburg, Ute Obermüller-Jevic, Xiang-Dong Wang, and Hanspeter Witschi. In addition, the following participated by conference call: Michael Gaziano, John Erdman, Hans-Konrad Biesalski, Susan Mayne, and Andrew Palou. A wide range of topics was discussed, and these are summarized below. Included are the recommendations made at the conclusion of the Round Table.

Is β-Carotene from Natural Food Sources Safe?

There are no documented studies indicating that β-carotene from natural food sources, when ingested in moderation, causes harm to any population. There have been examples of ingestion of very large amounts of carotenoids from food sources, and the major side effect under these circumstances was the occasional appearance of carotenodermia, i.e., the accumulation of β-carotene in the skin with the effect of giving it a yellow or orange tint. Even the ingestion of 272 g/d cooked carrots, 180 g/d tomato juice, 300 g/d cooked broccoli, or 12 mg/d β-carotene for 6 wk did not result in the development of carotenodermia; however, a supplement of 30 mg/d for this period of time did produce carotenodermia in five of the participants when the plasma level of β-carotene exceeded 400 μg/dL (6). The condition appears to be harmless, and is rapidly reversed when the high consumption of carotenoids ceases.

We know that there are at-risk groups with respect to the ingestion of large supplements of β-carotene, which may result in an increased risk of lung cancer.

Because it would be unwise to have at-risk groups consume diets that are very high in fruits and vegetables rich in β-carotene, we can only use cohort studies to evaluate the influence of dietary carotenoids on diseases such as lung cancer. In a recent pooled analysis of seven cohort studies with 399,765 participants, Männistö *et al.* (7) reported that the intake of β-carotene was not associated with lung cancer risk. Other survey groups, such as the International Agency for Research on Cancer of the World Health Agency (8) also reported that "there is no evidence to suggest that β-carotene is toxic at the levels found in most diets."

Is β-Carotene from Supplements Safe?

The answer to this question depends on the population receiving such supplements. The evidence for nonsmokers is that supplemental β-carotene, as with that found in foods, is safe, with the exception of the occasional appearance of carotenodermia, depending on dose. However, the issue of the use of supplemental β-carotene in a normal, i.e., nonsmoking population was raised by a number of groups. The U.S. Preventive Services Task Force (9) recommended against the use of β-carotene supplements for the prevention of cancer or coronary heart disease. However, a report from the Women's Health Study (10) indicated that after a median period of 2.1 y, the use of a β-carotene supplement (50 mg every other day) resulted in no harm or benefit to a group of 39,876 women ≥45 y old, and the Physicians Health Study (11) reported that after 22,071 male physicians consumed 50 mg β-carotene every other day for 12 y, there was no early or late difference in the incidence of malignant neoplasms or cardiovascular disease, or in overall mortality. They also reported no difference among current or former smokers. Omenn (12) suggested that β-carotene supplement use be discouraged, due to the adverse effects in smokers and no evidence of benefit in nonsmokers. There is almost universal agreement with respect to the effects in smokers, but adverse effects in nonsmokers were not demonstrated.

Another example of the difference in response of smokers and nonsmokers to β-carotene supplementation was reported in a study of colorectal adenoma recurrence in 864 patients who had an adenoma removed and were polyp free at the start of the study. The subjects were administered 25 mg/d of β-carotene and/or vitamins C and E (1000 and 400 mg/d, respectively); in those who did not smoke cigarettes or drink alcohol, supplementation markedly decreased the risk of one or more recurrent adenomas. For those who both smoked cigarettes and drank >1 alcoholic drink/d, the risk of adenoma recurrence doubled (13).

There have been other suggestions of an adverse relation between alcohol consumption and β-carotene intake. An early analysis from the ATBC study reported that the excess lung cancer risk was evident primarily in the men consuming more alcohol (14), but a recent re-analysis of the ATBC study concluded that alcohol consumption was not a risk factor for lung cancer, and its effect was not significantly modified by β-carotene supplementation (15).

What Is the Contribution of β-Carotene to the Vitamin A Supply in Western Countries?

In the United States, β-carotene and the other provitamin A carotenoids account for 25–30% of the daily intake of vitamin A in the 19- to 30-y-old population group (16), and may account for up to 80–85% in less developed countries (5). This value is based on the new factor for converting β-carotene to vitamin A in humans. Older reports suggested that 45–50% of the dietary vitamin A supply came from provitamin A carotenoids (17).

How Do We Evaluate the β-Carotene Status of Individuals?

For many years, investigators depended on plasma/serum levels to serve as a marker of the overall status of β-carotene in the individual, even though these values reflected recent dietary intake (18,19). Some attempts were made to analyze tissue samples, but this technique has not become a routine sampling technique. Recently, noninvasive techniques have appeared to measure concentrations of total carotenoids in skin *in situ*, including both Resonance Raman spectroscopy [reviewed in (20)] and reflection photometry [reviewed in (21)]. Each technique has its own merits, but only reflection photometry was applied to evaluate the effectiveness of dietary supplementation of carotenoids on altering the minimal erythema dose (MED) of UV irradiation. Recently it was demonstrated that within 12 wk of supplementation with 24 mg/d β-carotene, erythema can be reduced significantly (22). These investigations not only permit an evaluation of the carotenoid status in skin, but also demonstrate a protective function for supplemental β-carotene with respect to sun-induced erythema.

Can a Safe Upper Level (UL) Be Established for β-Carotene?

The Panel on Antioxidants and Related Nutrients of the Food & Nutrition Board of the Institute of Medicine, National Academy of Sciences (USA) (23), as well as The Scientific Committee on Food (EU) (24) evaluated the evidence for a tolerable upper intake level (UL) for β-carotene. Because of a lack of sufficient data, both groups were unable to establish a UL for β-carotene.

The only attempt to determine a UL for β-carotene was presented by the Expert Group on Vitamins and Minerals (UK). They based their conclusions on the ATBC report in which an adverse effect (increased incidence of lung cancer) was reported in heavy smokers administered 20 mg/d β-carotene as a supplement (25). The lowest observed adverse effect level (LOAEL) (20 mg/d) was divided by an uncertainty factor of 3 to yield a UL of 7 mg/d supplemental β-carotene, which they applied to the general population, i.e., nonsmokers and those not exposed to asbestos (26).

However, in the ATBC studies of β-carotene supplementation, adverse effects were limited to long-term, heavy smokers, and were not observed either in nonsmokers or in men who smoked <20 cigarettes/d (27). On the basis of that report, it would appear that it is still difficult to establish a UL for nonsmokers due to a lack of data.

Which Population Groups Are Likely to Benefit from Supplemental β-Carotene?

Any compromise in the ability of the body to absorb lipids would result in a low plasma level of β-carotene. An example would be patients with cystic fibrosis, a genetic defect in lipid absorption that is characterized by extremely low levels of plasma β-carotene (28,29). Such patients are routinely supplemented with vitamins A, D, E, and K, and can achieve normal plasma levels of these nutrients with such supplements, but still demonstrate a decreased resistance to oxidative stress that can be reversed with β-carotene supplementation (30–32).

Other individuals exposed to oxidative stresses that decrease levels of β-carotene might benefit from supplemental β-carotene. However, the individuals most easily identified with low plasma β-carotene levels are smokers (33,34), and we know that they are susceptible to adverse effects with β-carotene supplementation. However, passive smokers also have low plasma β-carotene levels, and might benefit from supplementation (34,35).

What Additional Studies Should Be Carried Out?

1. The effects of inflammatory responses on oxidative stress and resultant alterations in β-carotene metabolism. Such effects will vary with the individual and cannot be predicted at this time.
2. What clinical and/or biological end-points can be used to study the effects of oxidative stress on carotenoid status?
3. What tissues should be studied to evaluate the actions of β-carotene, in addition to skin? The respiratory tract or tissues undergoing a localized, inflammatory reaction are possibilities.
4. Does the net amount of vitamin A in the body of individuals supplemented with β-carotene increase, even when there is little if any change in the level of circulating vitamin A?

Recommendations

- The attempts to develop a UL for β-carotene supplementation in the population that is not at high risk (nonsmokers and individuals not exposed to asbestos) remain controversial. As discussed above, two study groups concluded that insufficient data exist to establish a UL for supplemental β-carotene (23,24). Can new techniques be developed to establish a UL for the population that is not at high risk?

- There are large differences in the bioavailability of different types of β-carotene supplements, resulting in very different plasma responses. Should the plasma response (or a tissue such as skin) be used, rather than the intake, to help set limits on the amount of supplemental β-carotene added to the diet?

References

1. Scientific Committee on Food (2000) Opinion of the Scientific Committee on Food on the Safety of the Use of Beta-Carotene from All Dietary Sources, European Commission, Brussels.
2. Bendich, A. (1988) The Safety of β-Carotene, *Nutr. Cancer 11:* 207–214.
3. Diplock, A.T. (1997) The Safety of β-Carotene and the Antioxidant Vitamins C and E, in *Antioxidants and Disease Prevention*, Garewal, H.S., ed., CRC, Boca Raton, pp. 3–17.
4. Hathcock, J.N. (1997) Vitamins and Minerals: Efficacy and Safety, *Am. J. Clin. Nutr. 66:* 427–437.
5. Woutersen, R.A., Wolterbeek, A.P.M., Appel, M.J., van den Berg, H., Goldbohm, R.A., and Feron, V.J. (1999) Safety Evaluation of Synthetic β-Carotene, *Crit. Rev. Toxicol. 29:* 515–542.
6. Micozzi, M.S., Brown, E.D., Taylor, P.R., and Wolfe, E. (1988) Carotenodermia in Men with Elevated Carotenoid Intake from Foods and β-Carotene Supplements, *Am. J. Clin. Nutr. 48:* 1061–1064.
7. Männistö, S., Smith-Warner, S.A., Spiegelman, D., Albanes, D., Anderson, K., van den Brandt, P.A., Cerhan, J.R., Colditz, G., Feskanich, D., Freudenheim, J.L., Giovannucci, E., Goldbohm, R.A., Graham, S., Miller, A.B., Rohan, T.E., Virtamo, J., Willett, W.C., and Hunter, D.J. (2004) Dietary Carotenoids and Risk of Lung Cancer in a Pooled Analysis of Seven Cohort Studies, *Cancer Epidemiol. Biomark. Prev. 13:* 40–48.
8. International Agency for Research on Cancer (1998) IARC Handbooks of Cancer Prevention, Vol. 1: Carotenoids, International Agency for Research on Cancer, World Health Agency, Lyon, France.
9. US Preventive Services Task Force (2003) Routine Vitamin Supplementation to Prevent Cancer and Cardiovascular Disease: Recommendations and Rationale, *Ann. Intern. Med. 139:* 51–55.
10. Lee, I., Cook, N.R., Manson, J.E., Buring, J.E., and Hennekens, C.H. (1999) β-Carotene Supplementation and Incidence of Cancer and Cardiovascular Disease: The Women's Health Study, *J. Natl. Cancer Inst. 91:* 2102–2106.
11. Hennekens, C.H., Buring, J.E., Manson, J.E., Stampfer, M., Rosner, B., Cook, N.R., Belanger, C., LaMotte, F., Gaziano, J.M., Ridker, P.M., Willett, W., and Peto, R. (1996) Lack of Effect of Long-Term Supplementation with Beta Carotene on the Incidence of Malignant Neoplasms and Cardiovascular Disease, *N. Engl. J. Med. 334:* 1145–1149.
12. Omenn, G.S. (1998) Chemoprevention of Lung Cancer: The Rise and Demise of Beta-Carotene, *Annu. Rev. Public Health 19:* 73–99.
13. Baron, J.A., Cole, B.F., Mott, L., Haile, R., Grau, M., Church, T.R., Beck, G.J., and Greenberg, E.R. (2003) Neoplastic and Antineoplastic Effects of β-Carotene on Colorectal Adenoma Recurrence: Results of a Randomized Trial, *J. Natl. Cancer Inst. 95:* 717–722.
14. Albanes, D., Heinonen, O.P., Huttunen, J.K., Taylor, P.R., Virtamo, J., Edwards, B.K., Haapakoski, J., Rautalahti, M., Hartmen, A.M., Palmgren, J., and Greenwald, P. (1995)

Effects of α-Tocopherol and β-Carotene Supplements on Cancer Incidence in the Alpha-Tocopherol Beta-Carotene Study, *Am. J. Clin. Nutr. 62 (Suppl.):* 1427S–1430S.

15. Woodson, K., Albanes, D., Tangrea, J.A., Rautalahti, M., Virtamo, J., and Taylor, P.R. (1999) Association Between Alcohol and Lung Cancer in the Alpha-Tocopherol, Beta-Carotene Cancer Prevention Study in Finland, *Cancer Causes Control 10:* 219–226.
16. Food & Nutrition Board, Institute of Medicine (2002) Dietary Reference Intakes for Vitamin A, Vitamin K, Arsenic, Boron, Chromium, Copper, Iodine, Iron, Manganese, Molybdenum, Nickel, Silicon, Vanadium, and Zinc, National Academy Press, Washington, DC, pp. 1–486.
17. FAO/WHO Expert Consultation (1988) Requirements of Vitamin A, Iron, Folate and Vitamin B_{12}, Food and Nutrition Series No. 23, Rome.
18. Micozzi, M.S., Brown, E.D., Edwards, B.K., Bieri, J.G., Taylor, P.R., Khachik, F., Beecher, G.R., and Smith, J.C., Jr. (1992) Plasma Carotenoid Response in Men to Chronic Intake of Selected Foods and Beta-Carotene Supplements, *Am. J. Clin. Nutr. 55:* 1120–1125.
19. Yeum, K.-J., Booth, S., Sadowski, J., Lin, C., Tang, G., Krinsky, N.I., and Russell, R.M. (1996) Human Plasma Carotenoid Response to the Ingestion of Controlled Diets High in Fruits and Vegetables, *Am. J. Clin. Nutr. 64:* 594–602.
20. Bernstein, P.S., and Gellermann, W. (2004) Noninvasive Assessment of Carotenoids in the Human Eye and Skin, in *Carotenoids in Health and Disease*, Krinsky, N.I., Mayne, S.T., and Sies, H., eds., Marcel Dekker, New York, pp. 53–84.
21. Stahl, W., and Sies, H. (2004) Carotenoids in Systemic Protection Against Sunburn, in *Carotenoids in Health and Disease*, Krinsky, N.I., Mayne, S.T., and Sies, H., eds., Marcel Dekker, New York, pp. 491–502.
22. Heinrich, U., Gartner, C., Wiebusch, M., Eichler, O., Sies, H., Tronnier, H., and Stahl, W. (2003) Supplementation with β-Carotene or a Similar Amount of Mixed Carotenoids Protects Humans from UV-Induced Erythema, *J. Nutr. 133:* 98–101.
23. Food & Nutrition Board, Institute of Medicine (2000) Dietary Reference Intakes for the Antioxidant Nutrients: Vitamin C, Vitamin E, Selenium, and Carotenoids, National Academy Press, Washington, DC, pp. 1–506.
24. Scientific Committee on Food (2000) Opinion of the Scientific Committee on Food on the Tolerable Upper Intake Level of Beta-Carotene, European Commission, Brussels.
25. The Alpha-Tocopherol Beta-Carotene Study Group (1994) The Effect of Vitamin E and Beta Carotene on the Incidence of Lung Cancer and Other Cancers in Male Smokers, *N. Engl. J. Med. 330:* 1029–1035.
26. Expert Group on Vitamins and Minerals (2003) *Safe Upper Levels for Vitamins and Minerals*, HMSO, London.
27. Albanes, D., Heinonen, O.P., Taylor, P.R., Virtamo, J., Edwards, B.K., Rautalahti, M., Hartmen, A.M., Palmgren, J., Freedman, L.S., Haapakoski, J., Barrett, M.J., Pietinen, P., Malila, N., Tala, E., Liippo, K., Salomaa, E.-R., Tangrea, J.A., Teppo, L., Askin, F.B., Taskinen, E., Erozan, Y., Greenwald, P., and Huttunen, J.K. (1996) α-Tocopherol and β-Carotene Supplements and Lung Cancer Incidence in the Alpha-Tocopherol Beta-Carotene Prevention Study: Effects of Base-Line Characteristics and Study Compliance, *J. Natl. Cancer Inst. 88:* 1560–1570.
28. Homnick, D.N., Cox, J.H., DeLoof, M.J., and Ringer, T.V. (1991) Serum Carotenoid Levels in Normal Children and Children with Cystic Fibrosis, *Am. Rev. Respir. Dis. 143:* A295 (Abstr.).

29. Winklhofer-Roob, B.M., Shmerling, D.H., and Schimek, M.G. (1991) High Prevalence of Very Low Beta-Carotene in Patients with Cystic Fibrosis (CF), *Nutr. Rev. 49:* 271.
30. Winklhofer-Roob, B.M., Puhl, H., Khoschsorur, G., van't Hof, M.A., Esterbauer, H., and Shmerling, D.H. (1995) Enhanced Resistance to Oxidation of Low Density Lipoproteins and Decreased Lipid Peroxide Formation During β-Carotene Supplementation in Cystic Fibrosis, *Free Radic. Biol. Med. 18:* 849–859.
31. Lepage, G., Champagne, J., Ronco, N., Lamarre, A., Osberg, I., Sokol, R.J., and Roy, C.C. (1996) Supplementation with Carotenoids Corrects Increased Lipid Peroxidation in Children with Cystic Fibrosis, *Am. J. Clin. Nutr. 64:* 87–93.
32. Rust, P., Eichler, I., Renner, S., and Elmadfa, I. (1998) Effects of Long-Term Oral Beta-Carotene Supplementation on Lipid Peroxidation in Patients with Cystic Fibrosis, *Int. J. Vitam. Nutr. Res. 68:* 83–87.
33. Nierenberg, D.W., Stukel, T.A., Baron, J.A., Dain, B.J., and Greenberg, E.R. (1989) Determinants of Plasma Levels of Beta-Carotene and Retinol, *Am. J. Epidemiol. 130:* 511–521.
34. Dietrich, M., Block, G., Norkus, E.P., Hudes, M., Traber, M.G., Cross, C.E., and Packer, L. (2003) Smoking and Exposure to Environmental Tobacco Smoke Decrease Some Plasma Antioxidants and Increase γ-Tocopherol *In Vivo* After Adjustment for Dietary Antioxidant Intakes, *Am. J. Clin. Nutr. 77:* 160–166.
35. Alberg, A.J., Chen, J.C., Zhao, H., Hoffman, S.C., Comstock, G.W., and Helzlsouer, K.J. (2000) Household Exposure to Passive Cigarette Smoking and Serum Micronutrient Concentrations, *Am. J. Clin. Nutr. 72:* 1576–1582.

Index

Acyl coA:retinol acyltransferase, 9, 12
Adulthood, vitamin A requirements during, 34
Age-Related Eye Disease Study Group (AREDS), 329–330, 303
Age-related macular degeneration (AMD), 116–118
  and loss of macular pigment, 117
  possible mechanisms of, 117–118
  protective role of carotenoids, 65
Alcohol dehydrogenase (ADH), 9, 14
Aldehyde dehydrogenase (ALDH), 9
Alpha-Tocopherol, Beta-Carotene Cancer Prevention (ATBC) study, 130, 142, 149–150, 303, 329, 339, 340–341
Anchorage-independent cell growth, 192
Anhydroretinol, apoptosis induced by, 46
Antioxidant response element (ARE), 267
Antioxidants
  carotenoids as, 210–212
  synergy among, 211–212
Antioxidant Supplementation in Atherosclerosis Prevention (ASAP) study, 207, 208, 327
AP1, inhibited by carotenoids, 266
Apoptosis
  and gap-junction cellular communication, 77
  induced by anhydroretinol, 46, 55
  mechanisms of, 231–236
Astaxanthin, disodium disuccinate derivative of, 196–198
Atherosclerosis Risk in Communities (ARIC) study, 209–210, 327

Bacteriorhodopsin, photosynthesis *via*, 53
BAD
  in IGF-1 pathway, 169, 170
  phosphorylation of, enhanced by lycopene, 169, 173–175
Benign prostate hyperplasia (BPH), 297
Beta-Carotene and Retinol Efficacy Trial (CARET), 130, 142, 149–150, 329
Bioavailability, 196–198, 279, 280
Bioconversion, 279
Bioefficacy, 279, 280
Body temperature regulation, 30–31
Bone health and retinol intake, 35–36
Cancer prevention, 182–183
Cancer Prevention Trial, 329
Canthaxanthin
  cleavage products as gap-junction stimulators and receptor ligands, 80–81
  connexin expression upregulated by, 194–195
Carcinogenesis
  and gap-junctional intercellular communication, 185, 187–190, 262
  inhibited by expression of connexins, 191–193
  10T1/2 cell model of, 185
Cardax™, 196–198
Cardiovascular disease (CVD)
  dietary factors, 204–205
  epidemiological evidence of carotene's benefits, 206–210
  mechanisms of carotenoid action against, 210–214, 321–322
Carnosic acid, differentiation induced by,263
β-Carotene, 205. *See also* Carotenoids
  antioxidant efficacy of, 212
  AP1 activation inhibited by, 266
  cardiovascular health benefits of, epidemiological evidence for, 323–325, 325–327, 327–330
  cell-cycle progression delayed by, 262–263
  cell differentiation induced by, 263
  cleavage products are detrimental to cell function, 130–143, 175
  cleavage products formed by leukocytes, 158
  cleavage products formed *via* hypochlorous acid, 153–158
  cleavage products stimulate O_2^- release by leukocytes, 160–162
  detrimental for those at risk for lung disease, 130–131, 149–150, 175, 338–339
  gap-junction cell communication stimulated by, 78, 80
  genotoxicity of cleavage products, 138–140, 141–142
  mechanisms of action against cardiovascular disease, 210–214

supplementation with, evaluated, 303–309
supplementation with, risks and benefits of, 327–330, 338–341
Carotene-15,15′-monooxygenase (carotene cleavage enzyme), 7, 9
expression of genes for, 250–256
genes expressing, 245–250
Carotenodermia, 338
Carotenoids. *See also* Retinoids; Vitamin A
and the antioxidant response element (ARE), 267
bioavailability of, 196–198
as cancer-prevention agents, 183–184, 187–188
cardiovascular health benefits of, epidemiological evidence for, 206–210, 322–325, 325–327
cardiovascular health benefits of, mechanisms of, 321–322
chemical oxidation of, 65–66
connexin expression upregulated by, 193–196
dietary sources of, 2, 62–63, 205–206, 276–279
gap-junction cell communication modulated by, 78, 80–81
implicated in cellular pathways and processes, 261–264
in lens of eye, 118–119
of the macula, 116–116, 117
metabolites of, 63–65
optical properties of, 86–90
orphan receptors, interaction with, 268–269
oxidative stress reduced by, 210–212
peroxisomal proliferator-activated receptor, interaction with, 266–267
phase I detoxification enzymes induced by, 268
phase II antioxidant enzymes induced by, 267
in photosynthesis, 53
phytonutrients, synergy with, 264
resonance Raman scattering of, 86–90
retinoid receptors, interaction with, 265–266
in the skin, measurement of, 103–110
synergy among, 211–212
transcriptional activity of, 264–269
xenobiotic receptors, interaction with, 268–269
and the visual system, 3, 63, 65
CarRetPIE method, 279
Caspase activation, 234
induced by synthetic retinoids, 236
Catabolism of retinoids, 15–16
Cataracts, eye, 118–119
CD437 (synthetic retinoid), 229–230
caspase activation induced by, 236
mitochondrial permeability transition induced by, 233
structure of, 230
Childhood, vitamin A requirements during, 33–34
Chloroplasts, photosynthesis in, 53
Chylomicron, 7
Conjugation of retinoic acid, 16
Connexins, 77, 190–191
canthoxanthin upregulates expression of, 194–195
carcinogenesis inhibited by expression of, 191–193
lycopene upregulates expression of, 192–193, 194–195
Connexons, 76, 190–191
β-Cryptoxanthin, 205
stimulates gap-junction cell communication, 78, 80
CYP26 (retinoic acid hydroxylase), 11, 15–16

dAST (disodium disuccinate derivative of astaxanthin), 196–198
Dermal carotenoid levels, measuring, 103–110
Diarrhea, increased vitamin A requirement due to, 282
Dietary Approaches to Stop Hypertension (DASH) diet, 321
Drusen, formation of, 118

Epicatechin stimulates gap-junction activity, 82
European study on Antioxidants, Myocardial Infarction, and Breast Cancer (EURAMIC), 206–208, 326–327

Ferret model
of human carotenoid absorption, 196
of human lung response to smoke, 171
Fetal development, role of vitamin A in, 27–28
4HPR. *See* *N*-(4-Hydroxyphenyl)retinamide
Framingham Heart Study, 325

Gap-junctional communication, intercellular, 76–78

and carcinogenesis, 185, 187–190, 262
carotenoids as modulators of, 78, 80–81
lycopene enhaces, 213–214
polyphenols as modulators of, 82
thyroid hormone receptor ligands as modulators of, 82
vitamin D as modulators of, 81–92
Gap junction structure, 190–191
Genistein stimulates gap-function activity, 82

Haidinger's brushes, 116
HDL-Atherosclerosis Treatment Study (HATS), 330
Health Professionals' Follow-Up Study, 322, 325
Heart Protection Study (HPS), 330
HIV/AIDS, increased vitamin A requirement due to, 284–285
N-(4-Hydroxyphenyl)retinamide (4HPR, synthetic retinoid), 55, 229–230
caspase activation induced by, 236
mitochondrial permeability transition induced by, 233
structure of, 230
Hypochlorous acid, degradation of β-carotene by, 153–158

IGF-binding proteins
and BAD, 169, 170
role in cell regulation, 168–170
upregulated by lycopene, 169, 171–175, 262
Infancy, increased vitamin A requirement during, 281–282
Insulin-like growth factor (IGF), 262
interaction with lycopene, 169, 171–175
role in cell regulation, 168–170
Iowa Women's Health Study, 323

Kuopio Ischemic Heart Disease (KIHD) study, 207, 208

Lactation, increased vitamin A requirement due to, 281
Lecithin:retinol acyltransferase (LRAT), 7–12
Leukocytes
O_2^- release stimulated by β-carotene oxidation products, 160–162
oxidation of β-carotene by, 158–160
Linköping-Vilinus study, 207, 208
Lipofuscin, fluorescence of, 120
Liver, storage of retinol in, 10
Los Angeles Atherosclerosis Study, 326, 327
Lung development, neonatal, 31
Lung health
effect of b-carotene cleavage products on, 130–132, 149–162
effect of lycopene on, 170–171, 172
physiologic aspects of, 150–152
Lutein, 205
antioxidant efficacy of, 212
cell differentiation induced by, 263
dietary sources of, 205–206
gap-junction cell communication stimulated by, 78, 80
implicated in eye health, 118-119
oxidation of, in air, 67
photoprotective role of, 65
presence in retina, 115, 116, 117
structure of, 116
structure of metabolites of, 64
supplementation with, 122–125
Lycopene
anticancer effects, mechanisms of, 168–178
anti-inflammatory effect of, 213
antioxidant efficacy of, 212
AP1 activation inhibited by, 266
BAD phosphorylation downregulated by, 169, 173–175
bioavailability of, 205–206
cardiovascular health benefits of, 206–214
cell-cycle progression delayed by, 262–263
cell differentiation induced by, 263
cholestorol synthesis inhibited by, 213
connexin expression upregulated by, 192–193, 194–195
gap-junction cell communication stimulated by, 78, 80, 213–214
IGF-binding protein upregulated by, 169, 171–173, 262
and lung health, 170–171, 172, 175–177
and prostate health, 295–299
structure of metabolites of, 64
Macula of eye
carotenoids in, 115–116, 117
effect of lutein supplementation on, 122–125
pigment density in, *in vivo* measurement of, 119–122, 90–103
Measles, increased vitamin A requirement due to, 283–284
Metabolism
of carotenoids, 63–65
of retinoids, 6–9, 15–16

Mitochondria
- attacked by β-carotene cleavage products, 130–143
- and reactive oxygen species, production of, 55, 231
- and regulation of apoptosis, 229–238
- respiration of, inhibited, 135–138
- as target of potential anticancer agents, 237–238

Mitochondrial permeability transition (MPT), 232–233
- induced by synthetic retinoids, 233–235

Motion photometry, 120
Multivitamins and Probucol (MVP) study, 330

National Health and Nutrition Examination Survey (NHANES), 177, 322, 326
Neonatal development, 29–33
Nucleus of cell, attacked by β-carotene cleavage products, 130–143
Nurses' Health Study, 323

Orphan receptors, 265
- interaction with carotenoids, 268–269

Osteoporosis, risk of, 35–36
Oxidation
- of β-carotene, 149–162
- of carotenoids, by chemical means, 65–66
- of retinal, 14–15
- of retinoic acid, 15
- of retinol, 14–15

Oxidative stress
- definition of, 210
- diminished by carotenoids, 210–212
- effect of fruit- and vegetable-rich diet on, 218–227
- neutrophils as a source of, 152–153, 160–162
- promoted by β-carotene cleavage products, 130–132, 140–143

Peroxisome proliferator-activated receptor (PPAR), 6, 194, 254, 265
- interaction with carotenoids, 266–267

Phase I detoxification enzymes, induced by carotenoids, 268
Phase II antioxidant enzymes, induced by carotenoids, 267
Physicians' Health Study, 209, 326, 329, 339
Photosynthesis, carotenoids and retinoids in, 53
Phytonutrients, synergy with carotenoids, 264
Polyphenol, 82
Pregnancy
- increased vitamin A requirement during, 281
- role of vitamin A in, 26–28

Prostate health
- benign prostate hyperplasia (BPH), 297
- effect of lycopene on, 295–299
- prostatitis, 298

Prostate-specific antigen (PSA), 295, 297
Provitamin A carotenoids, 1, 276. *See also* β-Carotene; Vitamin A
- dietary sources of, 2
- intestinal processing of, 4, 7–9
- nutritional intake levels, optimum, 24–26, 36
- as source of vitamin A for the elderly, 36

Reactive oxygen species (ROS)
- production of, induced by synthetic retinoids, 55, 231–232
- zinc-finger structures as sensors of, 56

Reflection photometry, 340
Reflectometry, 120–122
- digital, 121–122

Reproduction, role of vitamin A in, 26–28
Resonance Raman scattering of carotenoids, 86–90
Resonance Raman spectroscopy, 120
- for evaluating β-carotene status, 340
- for measuring dermal carotenoid levels, 103–110
- for measuring macular pigment levels, 90–103

Retinal
- oxidation of, 14–15
- structure of, 2
- role in vision, 3, 12

Retinal dehydrogenase (RALDH), 14
all-*trans*-Retinoic acid
- catabolism of, 15–16
- ligand of RAR, 6, 229, 265
- structure of, 2, 230
- transcriptional activity of, 3–4

9-*cis*-Retinoic acid
- formation of, 2, 3
- ligand of RAR and RXR, 6, 229, 265
- structure of, 2, 230
- transcriptional activity of, 3–4

Retinoic acid formation, 2, 12–15
Retinoic acid receptor (RAR), 6, 7, 229
- canthaxanthin cleavage products as ligands of, 81
- interaction of carotenoids with, 265–266

Retinoid-binding proteins, 4–6
Retinoids. *See also* Carotenoids; Retinol; Vitamin A as cancer-prevention agents, 183–184, 187–188
cellular uptake of, 10–11
connexin expression upregulated by, 193–196
dietary sources of, 2
and fetal development, 27–28
formation of, 2, 3
forms of, in body, 2–4
hepatic storage of, 10–11
and neonatal health/development, 29–33
in photosynthesis, 53
all-*trans*-retinoic acid, 6
9-*cis*-retinoic acid, 6
role in reproduction, 26–28
structure of, 2
synthetic proapoptotic, 229–238
transcriptional activity of, 3–4, 6, 264–265
in visual system, 3, 53
Retinoid X receptors (RXR), 6, 7, 229
interaction of carotenoids with, 265–266
Retinol. *See also* Retinoids; Vitamin A
esterification of, 12
serine/threonine kinases as targets of, 47–50
signal-transducing (non-transcriptional) activity of, 42–43, 46–54
oxidation of, 12–14
and zinc-coordinated structures, 51, 53, 54
Retinol Activity Equivalent (RAE), 24
Retinyl ester hydrolase (REH), 9, 11
Retinyl ester (storage form of retinol), 2
hydrolysis to retinol, 6, 10–11
storage and transport of, 10–11
Rhodopsin, 3
ρ^0 (mitochondrial respiration–deficient) cancer cells, 232, 238
Rotterdam study (on carotenoids and atherosclerosis), 209, 326
Serine/threonine kinases
as retinol receptors, 47–50
zinc-finger structures as conformation-altering "linchpin" of, 52–53
SHetA2 (synthetic retinoid), 229–230
caspase activation induced by, 236
mitochondrial permeability transition induced by, 234
structure of, 230
Short-chain dehydrogenase/reductases (SDR), 9, 13–14
Smokers
β-carotene intervention detrimental to, 149–150, 175
effect of lycopene supplementation on, 175–177
Supplementation en Vitamines et Mineraux AntioXydants (SU.VI.MAX) study, 329
Supplementation, retinoid/carotenoid
with β-carotene, and cardiovascular health, 327–330
with β-carotene, comparative evaluation of, 303–309
with β-carotene is detrimental to those at risk for lung disease, 130–131, 149–150, 175, 338–339
with β-carotene, risks and benefits of, 338–341
with lutein, and eye health, 122–125
with lycopene, and lung health, 170–171, 172, 175–177
neonatal, 31
with vitamin A, conditions requiring, 281–285

10T1/2 cell model of carcinogenesis, 185
Thermogenesis, regulation of, 30
Thyroid hormone receptor (TR), 6, 265
ligands stimulate gap-junction activity, 82
Transmembrane potential, mitochondrial ($\Delta\Psi_m$), 233
Tuberculosis, 285
Tumor formation and downregulation of gap-junction communication, 77

UDP-glucuronosyl transferase, 11, 16

Vision
role of carotenoids in, 63, 65
role of retinal in, 3, 12, 53
Vitamin A. *See also* Carotenoids; Retinoids; Retinol
adulthood requirements, 34–36
alternative pathways of formation for, 249–250, 255–256
biosynthesis of, and carotene cleavage enzymes, 245–250
childhood requirements, 33
conditions leading to special requirements for, 281–285
content, of foods, 278–279
dietary sources of, 2, 276–279
efficacious supply of, 279–281

in fertility and reproduction, 26–28
and fetal development, 27–28
in the immune system, 43–46
intestinal processing of, 4, 7–9
and neonatal health/development, 29–33, 281–282
nutritional intake levels, optimum, 24–26, 36
in oxygen radical biology, 55–56
Vitamin A deficiency, 44, 275–276
Vitamin D receptor, 6, 265
ligands stimulate gap-junction activity, 81–82

Women's Health Study, 207, 208–209, 324, 339

Xanthophyll, 115
Xenobiotic receptors, interaction with carotenoids, 268–269

Zeaxanthin, 205
implicated in eye health, 118, 119
photoprotective role of, 65
presence in retina, 115, 116, 117
structure of, 116
structure of metabolites of, 64
Zinc-finger structures, 42, 51–53
as "linchpin" of kinase conformation, 52–53
as reactive oxygen sensors, 56